Springer Handbook of Enzymes Volume 26

Dietmar Schomburg and
Ida Schomburg (Eds.)

Springer Handbook of Enzymes

Volume 26

Class 1 · Oxidoreductases XI

EC 1.14.11–1.14.14

coedited by Antje Chang

Second Edition

Springer

Professor Dietmar Schomburg
e-mail: d.schomburg@uni-koeln.de

Dr. Ida Schomburg
e-mail: i.schomburg@uni-koeln.de

Dr. Antje Chang
e-mail: a.chang@uni-koeln.de

University to Cologne
Institute for Biochemistry
Zülpicher Strasse 47
50674 Cologne
Germany

Library of Congress Control Number: 2005928337

ISBN-10 3-540-26586-4 2nd Edition Springer Berlin Heidelberg New York

ISBN-13 978-3-540-26586-3 2nd Edition Springer Berlin Heidelberg New York

The first edition was published as Volume 10 (ISBN 3-540-59494-9) of the "Enzyme Handbook".

Springer is a part of Springer Science+Business Media
springeronline.com

Printed in Germany

Cover design: Erich Kirchner, Heidelberg
Typesetting: medionet AG, Berlin

Printed on acid-free paper 2/3141m-5 4 3 2 1 0

Preface

Today, as the full information about the genome is becoming available for a rapidly increasing number of organisms and transcriptome and proteome analyses are beginning to provide us with a much wider image of protein regulation and function, it is obvious that there are limitations to our ability to access functional data for the gene products – the proteins and, in particular, for enzymes. Those data are inherently very difficult to collect, interpret and standardize as they are widely distributed among journals from different fields and are often subject to experimental conditions. Nevertheless a systematic collection is essential for our interpretation of genome information and more so for applications of this knowledge in the fields of medicine, agriculture, etc. Progress on enzyme immobilisation, enzyme production, enzyme inhibition, coenzyme regeneration and enzyme engineering has opened up fascinating new fields for the potential application of enzymes in a wide range of different areas.

The development of the enzyme data information system BRENDA was started in 1987 at the German National Research Centre for Biotechnology in Braunschweig (GBF) and is now continuing at the University at Cologne, Institute of Biochemistry. The present book "Springer Handbook of Enzymes" represents the printed version of this data bank. The information system has been developed into a full metabolic database.

The enzymes in this Handbook are arranged according to the Enzyme Commission list of enzymes. Some 3,700 "different" enzymes are covered. Frequently enzymes with very different properties are included under the same EC-number. Although we intend to give a representative overview on the characteristics and variability of each enzyme, the Handbook is not a compendium. The reader will have to go to the primary literature for more detailed information. Naturally it is not possible to cover all the numerous literature references for each enzyme (for some enzymes up to 40,000) if the data representation is to be concise as is intended.

It should be mentioned here that the data have been extracted from the literature and critically evaluated by qualified scientists. On the other hand, the original authors' nomenclature for enzyme forms and subunits is retained. In order to keep the tables concise, redundant information is avoided as far as possible (e.g. if K_m values are measured in the presence of an obvious cosubstrate, only the name of the cosubstrate is given in parentheses as a commentary without reference to its specific role).

The authors are grateful to the following biologists and chemists for invaluable help in the compilation of data: Cornelia Munaretto and Dr. Antje Chang.

Cologne
September 2005

Dietmar Schomburg, Ida Schomburg

List of Abbreviations

A	adenine
Ac	acetyl
ADP	adenosine 5'-diphosphate
Ala	alanine
All	allose
Alt	altrose
AMP	adenosine 5'-monophosphate
Ara	arabinose
Arg	arginine
Asn	asparagine
Asp	aspartic acid
ATP	adenosine 5'-triphosphate
Bicine	N,N'-bis(2-hydroxyethyl)glycine
C	cytosine
cal	calorie
CDP	cytidine 5'-diphosphate
CDTA	trans-1,2-diaminocyclohexane-N,N,N,N-tetraacetic acid
CMP	cytidine 5'-monophosphate
CoA	coenzyme A
CTP	cytidine 5'-triphosphate
Cys	cysteine
d	deoxy-
D-	(and L-) prefixes indicating configuration
DFP	diisopropyl fluorophosphate
DNA	deoxyribonucleic acid
DPN	diphosphopyridinium nucleotide (now NAD^+)
DTNB	5,5'-dithiobis(2-nitrobenzoate)
DTT	dithiothreitol (i.e. Cleland's reagent)
EC	number of enzyme in Enzyme Commission's system
E. coli	Escherichia coli
EDTA	ethylene diaminetetraacetate
EGTA	ethylene glycol bis(-aminoethyl ether) tetraacetate
ER	endoplasmic reticulum
Et	ethyl
EXAFS	extended X-ray absorption fine structure
FAD	flavin-adenine dinucleotide
FMN	flavin mononucleotide (riboflavin 5'-monophosphate)
Fru	fructose
Fuc	fucose
G	guanine
Gal	galactose

GDP	guanosine 5'-diphosphate
Glc	glucose
GlcN	glucosamine
GlcNAc	N-acetylglucosamine
Gln	glutamine
Glu	glutamic acid
Gly	glycine
GMP	guanosine 5'-monophosphate
GSH	glutathione
GSSG	oxidized glutathione
GTP	guanosine 5'-triphosphate
Gul	gulose
h	hour
H4	tetrahydro
HEPES	4-(2-hydroxyethyl)-1-piperazineethane sulfonic acid
His	histidine
HPLC	high performance liquid chromatography
Hyl	hydroxylysine
Hyp	hydroxyproline
IAA	iodoacetamide
Ig	immunoglobulin
Ile	isoleucine
Ido	idose
IDP	inosine 5'-diphosphate
IMP	inosine 5'-monophosphate
ITP	inosine 5'-triphosphate
K_m	Michaelis constant
L-	(and D-) prefixes indicating configuration
Leu	leucine
Lys	lysine
Lyx	lyxose
M	mol/l
mM	millimol/l
m-	*meta-*
Man	mannose
MES	2-(N-morpholino)ethane sulfonate
Met	methionine
min	minute
MOPS	3-(N-morpholino)propane sulfonate
Mur	muramic acid
MW	molecular weight
NAD^+	nicotinamide-adenine dinucleotide
NADH	reduced NAD
$NADP^+$	NAD phosphate
NADPH	reduced NADP
NAD(P)H	indicates either NADH or NADPH
NBS	N-bromosuccinimide

NDP	nucleoside 5'-diphosphate
NEM	N-ethylmaleimide
Neu	neuraminic acid
NMN	nicotinamide mononucleotide
NMP	nucleoside 5'-monophosphate
NTP	nucleoside 5'-triphosphate
o-	*ortho-*
Orn	ornithine
p-	*para-*
PBS	phosphate-buffered saline
PCMB	*p*-chloromercuribenzoate
PEP	phosphoenolpyruvate
pH	$-\log10[H^+]$
Ph	phenyl
Phe	phenylalanine
PHMB	*p*-hydroxymercuribenzoate
PIXE	proton-induced X-ray emission
PMSF	phenylmethane-sulfonylfluoride
p-NPP	*p*-nitrophenyl phosphate
Pro	proline
Q_{10}	factor for the change in reaction rate for a 10 °C temperature increase
Rha	rhamnose
Rib	ribose
RNA	ribonucleic acid
mRNA	messenger RNA
rRNA	ribosomal RNA
tRNA	transfer RNA
Sar	N-methylglycine (sarcosine)
SDS-PAGE	sodium dodecyl sulfate polyacrylamide gel electrophoresis
Ser	serine
T	thymine
t_H	time for half-completion of reaction
Tal	talose
TDP	thymidine 5'-diphosphate
TEA	triethanolamine
Thr	threonine
TLCK	N^{α}-*p*-tosyl-L-lysine chloromethyl ketone
T_m	melting temperature
TMP	thymidine 5'-monophosphate
Tos-	tosyl-(*p*-toluenesulfonyl-)
TPN	triphosphopyridinium nucleotide (now $NADP^+$)
Tris	tris(hydroxymethyl)-aminomethane
Trp	tryptophan
TTP	thymidine 5'-triphosphate
Tyr	tyrosine
U	uridine
U/mg	μmol/(mg*min)

UDP	uridine 5'-diphosphate
UMP	uridine 5'-monophosphate
UTP	uridine 5'-triphosphate
Val	valine
Xaa	symbol for an amino acid of unknown constitution in peptide formula
XAS	X-ray absorption spectroscopy
Xyl	xylose

List of Deleted and Transferred Enzymes

Since its foundation in 1956 the Nomenclature Committee of the International Union of Biochemistry and Molecular Biology (NC-IUBMB) has continually revised and updated the list of enzymes. Entries for new enzymes have been added, others have been deleted completely, or transferred to another EC number in the original class or to different EC classes, catalyzing other types of chemical reactions. The old numbers have not been allotted to new enzymes; instead the place has been left vacant or cross-references given to the changes in nomenclature.

Deleted and Transferred Enzymes

For EC class 1.14.11–1.14.14 these changes are:

Recommended name	Old EC number	Alteration
5-hydroxymethyluracil, 2-oxoglutarate dioxygenase	1.14.11.5	deleted, included in EC 1.14.11.6
anthranilate 2,3-dioxygenase (deaminating)	1.14.12.2	transferred to EC 1.14.13.35
2-hydroxycyclohexanone 2-monooxygenase	1.14.12.6	transferred to EC 1.14.13.66
CMP-N-acetylneuraminate monooxygenase	1.14.13.45	transferred to EC 1.14.18.2
benzopyrene 3-monooxygenase	1.14.14.2	deleted, included in EC 1.14.14.1
choline monooxygenase	1.14.14.4	deleted, identical to EC 1.14.15.7

Index of Recommended Enzyme Names

Description of Data Fields

All information except the nomenclature of the enzymes (which is based on the recommendations of the Nomenclature Committee of IUBMB (International Union of Biochemistry and Molecular Biology) and IUPAC (International Union of Pure and Applied Chemistry) is extracted from original literature (or reviews for very well characterized enzymes). The quality and reliability of the data depends on the method of determination, and for older literature on the techniques available at that time. This is especially true for the fields Molecular Weight and Subunits.

The general structure of the fields is: **Information – Organism – Commentary – Literature**

The information can be found in the form of numerical values (temperature, pH, K_m etc.) or as text (cofactors, inhibitors etc.).

Sometimes data are classified as Additional Information. Here you may find data that cannot be recalculated to the units required for a field or also general information being valid for all values. For example, for Inhibitors, Additional Information may contain a list of compounds that are not inhibitory.

The detailed structure and contents of each field is described below. If one of these fields is missing for a particular enzyme, this means that for this field, no data are available.

1 Nomenclature

EC number

The number is as given by the IUBMB, classes of enzymes and subclasses defined according to the reaction catalyzed.

Systematic name

This is the name as given by the IUBMB/IUPAC Nomenclature Committee

Recommended name

This is the name as given by the IUBMB/IUPAC Nomenclature Committee

Synonyms

Synonyms which are found in other databases or in the literature, abbreviations, names of commercially available products. If identical names are frequently used for different enzymes, these will be mentioned here, cross references are given. If another EC number has been included in this entry, it is mentioned here.

CAS registry number

The majority of enzymes have a single chemical abstract (CAS) number. Some have no number at all, some have two or more numbers. Sometimes two enzymes share a common number. When this occurs, it is mentioned in the commentary.

2 Source Organism

For listing organisms their systematic name is preferred. If these are not mentioned in the literature, the names from the respective literature are used. For example if an enzyme from yeast is described without being specified further, yeast will be the entry. This field defines the code numbers for the organisms in which the enzyme with the respective EC number is found. These code numbers (form <_>) are displayed together with each entry in all fields of Brenda where organism-specific information is given.

3 Reaction and Specificity

Catalyzed reaction

The reaction as defined by the IUBMB. The commentary gives information on the mechanism, the stereochemistry, or on thermodynamic data of the reaction.

Reaction type

According to the enzyme class a type can be attributed. These can be oxidation, reduction, elimination, addition, or a name (e.g. Knorr reaction)

Natural substrates and products

These are substrates and products which are metabolized in vivo. A natural substrate is only given if it is mentioned in the literature. The commentary gives information on the pathways for which this enzyme is important. If the enzyme is induced by a specific compound or growth conditions, this will be included in the commentary. In *Additional information* you will find comments on the metabolic role, sometimes only assumptions can be found in the references or the natural substrates are unknown.

In the listings, each natural substrate (indicated by a bold **S**) is followed by its respective product (indicated by a bold **P**). Products are given with organisms and references included only if the respective authors were able to demonstrate the formation of the specific product. If only the disappearance of the substrate was observed, the product is included without organisms of references. In cases with unclear product formation only a ? as a dummy is given.

Substrates and products

All natural or synthetic substrates are listed (not in stoichiometric quantities). The commentary gives information on the reversibility of the reaction,

on isomers accepted as substrates and it compares the efficiency of substrates. If a specific substrate is accepted by only one of several isozymes, this will be stated here.
The field *Additional Information* summarizes compounds that are not accepted as substrates or general comments which are valid for all substrates. In the listings, each substrate (indicated by a bold **S**) is followed by its respective product (indicated by a bold **P**). Products are given with organisms and references included if the respective authors demonstrated the formation of the specific product. If only the disappearance of the substrate was observed, the product will be included without organisms or references. In cases with unclear product formation only a ? as a dummy is given.

Inhibitors

Compounds found to be inhibitory are listed. The commentary may explain experimental conditions, the concentration yielding a specific degree of inhibition or the inhibition constant. If a substance is activating at a specific concentration but inhibiting at a higher or lower value, the commentary will explain this.

Cofactors, prosthetic groups

This field contains cofactors which participate in the reaction but are not bound to the enzyme, and prosthetic groups being tightly bound. The commentary explains the function or, if known, the stereochemistry, or whether the cofactor can be replaced by a similar compound with higher or lower efficiency.

Activating Compounds

This field lists compounds with a positive effect on the activity. The enzyme may be inactive in the absence of certain compounds or may require activating molecules like sulfhydryl compounds, chelating agents, or lipids. If a substance is activating at a specific concentration but inhibiting at a higher or lower value, the commentary will explain this.

Metals, ions

This field lists all metals or ions that have activating effects. The commentary explains the role each of the cited metal has, being either bound e.g. as Fe-S centers or being required in solution. If an ion plays a dual role, activating at a certain concentration but inhibiting at a higher or lower concentration, this will be given in the commentary.

Turnover number (min^{-1})

The k_{cat} is given in the unit min^{-1}. The commentary lists the names of the substrates, sometimes with information on the reaction conditions or the type of reaction if the enzyme is capable of catalyzing different reactions with a single substrate. For cases where it is impossible to give the turnover number in the defined unit (e.g., substrates without a defined molecular weight, or an undefined amount of protein) this is summarized in Additional Information.

Specific activity (U/mg)

The unit is micromol/minute/milligram of protein. The commentary may contain information on specific assay conditions or if another than the natural substrate was used in the assay. Entries in Additional Information are included if the units of the activity are missing in the literature or are not calculable to the obligatory unit. Information on literature with a detailed description of the assay method may also be found.

K_m-Value (mM)

The unit is mM. Each value is connected to a substrate name. The commentary gives, if available, information on specific reaction condition, isozymes or presence of activators. The references for values which cannot be expressed in mM (e.g. for macromolecular, not precisely defined substrates) are given in Additional Information. In this field we also cite literature with detailed kinetic analyses.

K_i-Value (mM)

The unit of the inhibition constant is mM. Each value is connected to an inhibitor name. The commentary gives, if available, the type of inhibition (e.g. competitive, non-competitive) and the reaction conditions (pH-value and the temperature). Values which cannot be expressed in the requested unit and references for detailed inhibition studies are summerized under Additional information.

pH-Optimum

The value is given to one decimal place. The commentary may contain information on specific assay conditions, such as temperature, presence of activators or if this optimum is valid for only one of several isozymes. If the enzyme has a second optimum, this will be mentioned here.

pH-Range

Mostly given as a range e.g. 4.0–7.0 with an added commentary explaining the activity in this range. Sometimes, not a range but a single value indicating the upper or lower limit of enzyme activity is given. In this case, the commentary is obligatory.

Temperature optimum (°C)

Sometimes, if no temperature optimum is found in the literature, the temperature of the assay is given instead. This is always mentioned in the commentary.

Temperature range (°C)

This is the range over which the enzyme is active. The commentary may give the percentage of activity at the outer limits. Also commentaries on specific assay conditions, additives etc.

4 Enzyme Structure

Molecular weight

This field gives the molecular weight of the holoenzyme. For monomeric enzymes it is identical to the value given for subunits. As the accuracy depends on the method of determination this is given in the commentary if provided in the literature. Some enzymes are only active as multienzyme complexes for which the names and/or EC numbers of all participating enzymes are given in the commentary.

Subunits

The tertiary structure of the active species is described. The enzyme can be active as a monomer a dimer, trimer and so on. The stoichiometry of subunit composition is given. Some enzymes can be active in more than one state of complexation with differing effectivities. The analytical method is included.

Posttranslational modifications

The main entries in this field may be proteolytic modification, or side-chain modification, or no modification. The commentary will give details of the modifications e.g.:

- proteolytic modification <1> (<1>, propeptide Name) [1];
- side-chain modification <2> (<2>, N-glycosylated, 12% mannose) [2];
- no modification [3]

5 Isolation / Preparation / Mutation / Application

Source / tissue

For multicellular organisms, the tissue used for isolation of the enzyme or the tissue in which the enzyme is present is given. Cell-lines may also be a source of enzymes.

Localization

The subcellular localization is described. Typical entries are: cytoplasm, nucleus, extracellular, membrane.

Purification

The field consists of an organism and a reference. Only references with a detailed description of the purification procedure are cited.

Renaturation

Commentary on denaturant or renaturation procedure.

Crystallization

The literature is cited which describes the procedure of crystallization, or the X-ray structure.

Cloning

Lists of organisms and references, sometimes a commentary about expression or gene structure.

Engineering

The properties of modified proteins are described.

Application

Actual or possible applications in the fields of pharmacology, medicine, synthesis, analysis, agriculture, nutrition are described.

6 Stability

pH-Stability

This field can either give a range in which the enzyme is stable or a single value. In the latter case the commentary is obligatory and explains the conditions and stability at this value.

Temperature stability

This field can either give a range in which the enzyme is stable or a single value. In the latter case the commentary is obligatory and explains the conditions and stability at this value.

Oxidation stability

Stability in the presence of oxidizing agents, e.g. O_2, H_2O_2, especially important for enzymes which are only active under anaerobic conditions.

Organic solvent stability

The stability in the presence of organic solvents is described.

General stability information

This field summarizes general information on stability, e.g., increased stability of immobilized enzymes, stabilization by SH-reagents, detergents, glycerol or albumins etc.

Storage stability

Storage conditions and reported stability or loss of activity during storage.

References

Authors, Title, Journal, Volume, Pages, Year.

γ-Butyrobetaine dioxygenase 1.14.11.1

1 Nomenclature

EC number
1.14.11.1

Systematic name
4-trimethylammoniobutanoate,2-oxoglutarate:oxygen oxidoreductase (3-hydroxylating)

Recommended name
γ-butyrobetaine dioxygenase

Synonyms
α-butyrobetaine hydroxylase
butyrobetaine hydroxylase
γ-butyrobetaine hydroxylase
oxygenase, γ-butyrobetaine di-

CAS registry number
9045-31-2

2 Source Organism

<1> *Rattus norvegicus* (Sprague-Dawley rats [17]) [1, 3, 8, 9, 15, 17-21]
<2> *Pseudomonas sp.* (<2> AK1 [2, 5, 10, 11, 13, 16]) [2, 5, 6, 10-13, 16]
<3> *Bos taurus* (<3> calf [4]) [4, 13]
<4> *Homo sapiens* (<4> three isoforms [7]) [5, 7, 16]
<5> *Cavia porcellus* [14]

3 Reaction and Specificity

Catalyzed reaction
4-trimethylammoniobutanoate + 2-oxoglutarate + O_2 = 3-hydroxy-4-trimethylammoniobutanoate + succinate + CO_2 (<2,3> srereochemistry of reaction, pro-R hydrogen atom abstraction [13])

Reaction type
oxidation
oxidative decarboxylation
redox reaction
reduction

Natural substrates and products

S 4-trimethylammoniobutanoate + 2-oxoglutarate + O_2 <3> (<3> γ-butyrobetaine, terminal reaction in the pathway of L-carnitine biosynthesis [4]) (Reversibility: ? <3> [4]) [4]

P 3-hydroxy-4-trimethylammoniobutanoate + succinate + CO_2

Substrates and products

S 3-trimethylaminopropionic acid + 2-oxoglutarate + O_2 <1> (<1> 20% of the hydroxylation rate of γ-butyrobetaine [15]) (Reversibility: ? <1> [15]) [15]

P ?

S 4-dimethylaminobutyric acid + 2-oxoglutarate + O_2 <1> (<1> poor substrate [15]) (Reversibility: ? <1> [15]) [15]

P 4-dimethylamino-3-hydroxybutyric acid + succinate + CO_2 <1> [15]

S 4-trimethylammoniobutanoate + 2-oxoglutarate + O_2 <1-5> (<1> i.e. γ-butyrobetaine [1]) (Reversibility: ? <1-5> [1-15, 17]) [1-15]

P 3-hydroxy-4-trimethylammoniobutanoate + succinate + CO_2 <1-5> (<1> i.e. L-carnitine [1]) [1-15, 17]

S 5,5-dimethylhexanoate + 2-oxoglutarate + O_2 <2> (<2> poor substrate, 20% of the decarboxylation events lead to hydroxylation [10]) (Reversibility: ? <2> [10]) [10]

P ?

S 6-N-trimethyl-L-lysine + 2-oxoglutarate + O_2 <5> (Reversibility: ? <5> [14]) [14]

P ?

S Additional information <2, 4> (<2,4> L-carnithine is an uncoupler [5]; <2> overview on properties of substrate analogues, binding sites [11]) [5, 11]

P ?

Inhibitors

2-oxoglutarate <1> (<1> at concentrations above 1 mM [19]) [19]
3,4-dihydroxybenzoate <2> [11]
3-(2,2,2-trimethylhydrazinium)propionate <1> (<1> complete inhibition at 0.05 mM [19]) [19]
3-(2,2-dimethylcyclopropyl)propionic acid <2> (<2> mechanism-based inhibitor [10]) [10]
3-bromo-2-oxoglutarate <1> (<1> noncompetitive inhibition, 2-oxoglutarate as variable substrate [9]) [9]
3-glutathione-2-oxoglutarate <1> (<1> noncompetitive to 2-oxoglutarate [9]) [9]
3-trimethylaminopropyl-1-sulfonate <1> [1]
Ba^{2+} <3> [4]
Ca^{2+} <3> [4]
Cd^{2+} <3> [4]
Co^{2+} <3> [4]
Cu^{2+} <3> [4]
D-carnitine <2> (<2> uncoupling agent [5]) [5]

DL-carnitine <1, 4> (<4> uncouples the decarboxylation from the hydroxylation, mammalian enzyme [5]) [5, 7, 9]
FMN <1> (<1> in high concentrations [15]) [15]
H_2O_2 <1, 2> [6, 15]
Hg^{2+} <3> [4]
Mg^{2+} <3> [4]
Mn^{2+} <3> [4]
N-ethylmaleimide <1> (<1> less effective than *p*-chloromercuriphenylsulfonate [1]) [1]
Ni^{2+} <3> [4]
O_2 <2> (<2> irreversible [6]) [6]
Pb^{2+} <3> [4]
Sn^{2+} <3> [4]
Zn^{2+} <3> [4]
α,α'-bipyridyl <3> (<3> no activity at 2 mM [4]) [4]
arsenite <1> [1]
ascorbate <1, 2> (<2> irreversible inactivation during preincubation [6]) [6, 15]
cyclopropyl-substituted γ-butyrobetaines <2> [12]
dioxane <3> [4]
γ-butyrobetaine <1> (<1> at concentrations above 0.2 mM [19]) [19]
iodoacetate <1> (<1> less effective than *p*-chloromercuriphenylsulfonate [1]) [1]
iodosobenzoate <1> (<1> less effective than *p*-chloromercuriphenylsulfonate [1]) [1]
p-chloromercuribenzoate <1> (<1> inactivates enzyme completely at 0.1 mM [1]) [1]
p-chloromercuriphenylsulfonate <1> (<1> inactivates enzyme completely at 0.1 mM [1]) [1]
phosphate <1> [9]
pyridine-2,4-dicarboxylate <2> [11]
quinacrine <1> [15]
structure analogues of γ-butyrobetaine and 2-oxoglutarate <1, 2> [1, 11]
succinic semialdehyde <1> [1]
Additional information <2> (<2> inhibition by cyclopropyl-substituted γ-butyrobetains [12]) [12]

Cofactors/prosthetic groups

2,6-dichlorophenolindophenol <1> [1]
ascorbate <1-5> (<1> activation, keeps Fe^{2+} in the reduced state [1]) [1, 4-6, 8, 9, 13-15, 19]
Additional information <1> (<1> KCl, nicotinamide, $MgCl_2$ or catalase not required for full activity [3]) [3]

Activating compounds

2-amino-6,7-dimethyl-4-hydroxy-5, 6,7,8-tetrahydropteridine <1> [1]
Cs^+ <1> (<1> increase of activity [9]) [9]

GSH/GSH-peroxidase <1> (<1> increase of activity, more efficient in assay and during preincubation than catalase, protects the enzyme from increasing phosphate concentrations [8, 9]) [8, 9]
K^+ <1> (<1> efficient coupling of decarboxylation and hydroxylation, stimulation [9]) [9]
L-histidine <2> [6]
NADPH <1> (<1> NADPH-regenerating system, increase of activity, no absolute requirement [15]) [15]
NH_4^+ <1> (<1>, increase of activity [9]) [9]
Rb^+ <1> (<1> increase of activity [9]) [9]
catalase <1-3, 5> (<1,2> increase of activity [6,19]; <1> protection from inactivation [8]) [1, 4, 6, 8, 9, 13-15, 19]
isoascorbate <1> (<1> increase of activity [1]) [1]
microsomal preparation <1> (<1> increase of activity [15]) [15]
nicotinamide <1> (<1> increase of activity [15]) [15]
Additional information <2> (<2> various proteins such as lactoperoxidase, horseradish peroxidase, hemoglobin, myoglobin, cytochrome c, bovine serum albumin activate [6]) [6]

Metals, ions

Fe^{2+} <1-4> (<1-4> requirement [1-9,15,19]) [1-9, 15, 19]

Turnover number (min^{-1})

2040 <2> (γ-butyrobetaine) [2]

Specific activity (U/mg)

0.0014 <1> [15]
0.003 <1> [1]
0.0274 <1> [8]
0.053 <3> [4]
0.162 <4> (<4> human kidney [5]) [5]
1.02-1.74 <4> (<4> three isozymes of mammalian enzyme [7]) [7]
21.6 <2> [2]
25 <2> [11]

K_m-Value (mM)

0.012 <3> (Fe^{2+}) [4]
0.018 <2> (2-oxoglutarate) [11]
0.029 <1> (γ-butyrobetaine) [9]
0.052 <1> (L-carnitine) [9]
0.08 <1> (γ-butyrobetaine) [19]
0.1-0.13 <4> (2-oxoglutarate, <4> γ-butyrobetaine, values about the same for all three isozymes [7]) [7]
0.125 <1> (2-oxo-glutarate) [19]
0.25 <4> (2-oxoglutarate) [5]
0.47 <1> (D-carnitine) [9]
0.5 <1> (2-oxoglutarate) [1]
0.51 <3> (γ-butyrobetaine) [4]

0.82 <3> (2-oxoglutarate) [4]
2.4 <2> (γ-butyrobetaine) [10]
Additional information <1, 3> [1, 4]

K_i-Value (mM)
0.0002 <2> (pyridine-2,4-dicarboxylate) [11]
0.0006 <2> (3,4-dihydroxybenzoate) [11]

pH-Optimum
6.7 <1> [1]

pH-Range
6-7.5 <1> (<1> half-maximal activity at pH 6.0 and 7.5 [1]) [1]
6-8.4 <1> (<1> half-maximal activity at pH 6.0 and 8.4, partially purified preparation [15]) [15]

Temperature optimum (°C)
37 <1-4> (<1-4> assay at [1-13,15]) [1-13, 15]

4 Enzyme Structure

Molecular weight
64000 <4> (<4> gel filtration, 3 isozymes, detection by chromatofocusing and anion exchange chromatography [7]) [7]
68000 <1> (<1> gel filtration [19]) [19]
80000 <3> (<3> gel filtration [4]) [4]
86000 <2> (<2> gel filtration [2]) [2]
86000-95000 <2> (<2> gel filtration, sedimentation equilibrium centrifugation [2]) [2]

Subunits
? <1, 4> (<4> x * 42000, amino acid composition [16]; <1> x * 44000, SDS-PAGE [18]) [16, 18]
dimer <1-4> (<2> 1 * 39000 + 1 * 37000, SDS-PAGE, amino acid analysis, N-terminal amino acid sequence [2]; <3> 2 * 46000, SDS-PAGE [4]; <3> 2 * 42000, SDS-PAGE [4]; <4> SDS-PAGE, same value for all three isozymes, dimeric combinations of two subunits [7]; <2> 2 * 43000, amino acid analysis, a chain of 383 amino acids and a truncated chain of 382 amino acids [16]; <1> 2 * 43000, SDS-PAGE [19]) [2, 4, 7, 16, 19]

5 Isolation/Preparation/Mutation/Application

Source/tissue
blood <1> [8]
heart <5> [14]
kidney <4> [5, 7]
liver <1, 3-5> [1, 3, 4, 7-9, 13-15, 17-21]

skeletal muscle <1, 5> [8, 14]
Additional information <1> (<1> no activity found: muscle, kidney [15]) [15]

Localization
cytoplasm <1-4> [1-11, 15]
peroxisome <1> [17]

Purification
<1> (partial [15]) [1, 8, 15, 18, 19]
<2> [2, 11]
<3> [4]
<4> [7]

Cloning
<1> [21]

Application
medicine <1> (<1> effects of hyperthyroidism and hypothyroidism on enzyme activity [20]) [20]

6 Stability

Temperature stability
50 <4> (<4> 12 min half-life [5]) [5]
60 <1> (<1> complete inactivation after 15 min [15]) [15]

Oxidation stability
<1, 2>, O_2, irreversible inactivation during preincubation [6, 15]

Organic solvent stability
acetone <3> (<3> powder, retains full activity of fresh liver homogenate [4]) [4]

General stability information
<2>, DTT, prevents inactivation by O_2, ascorbate or H_2O_2 [6]
<2>, His, stabilizes [6]
<3>, dialysis, 4°C, 50% loss of activity overnight [4]
<3>, ultra-filtration, 4°C, 50% loss of activity overnight [4]

Storage stability
<1>, 4°C, 50 mM sodium phosphate pH 7.4, 20 mM KCl, 10 mM DTT, 200g/l glycerol are best conservation conditions [18]
<1>, addition of EDTA causes total loss of activity [18]
<2>, -20°C to 4°C, 50% loss of activity in a week [2]
<2>, -60° to -170°C, stable for 6 months in potassium phosphate buffer, pH 6.5 [2]
<3>, -90°C, stable for a year or more as acetone powder, stable for 3 months as purified and lyophilized enzyme [4]

References

[1] Lindstedt, G.; Lindstedt, S.: Cofactor requirements of γ-butyrobetaine hydroxylase from rat liver. J. Biol. Chem., **245**, 4178-4186 (1970)
[2] Lindstedt, G.; Lindstedt, S.; Nordin, I.: Purification and properties of γ-butyrobetaine hydroxylase from Pseudomonas sp AK 1. Biochemistry, **16**, 2181-2188 (1977)
[3] Carter, A.L.; Stratman, F.W.: A rapid and sensitive assay of γ-butyrobetaine hydroxylase. FEBS Lett., **111**, 112-114 (1980)
[4] Kondo, A.; Blanchard, J.S.; Englard, S.: Purification and properties of calf liver γ-butyrobetaine hydroxylase. Arch. Biochem. Biophys., **212**, 338-346 (1981)
[5] Holme, E.; Lindstedt, S.; Nordin, I.: Uncoupling in the γ-butyrobetaine hydroxylase reaction by D- and L-carnitine. Biochem. Biophys. Res. Commun., **107**, 518-524 (1982)
[6] Blanchard, J.S.; Englard, S.; Kondo, A.: γ-Butyrobetaine hydroxylase: A unique protctive effect of catalase. Arch. Biochem. Biophys., **219**, 327-334 (1982)
[7] Lindstedt, S.; Nordin, I.: Multiple forms of γ-butyrobetaine hydroxylase (EC 1.14.11.1). Biochem. J., **223**, 119-127 (1984)
[8] Punekar, N.S.; Wehbie, R.S.; Lardy, H.A.: γ-Butyro*β*ine hydroxylase and the protective role of glutathione peroxidase. J. Biol. Chem., **262**, 6720-6724 (1987)
[9] Wehbie, R.S.; Punekar, N.S.; Lardy, H.A.: Rat liver γ-butyrobetaine hydroxylase catalyzed reaction: influence of potassium, substrates, and substrate analogues on hydroxylation and decarboxylation. Biochemistry, **27**, 2222-2228 (1988)
[10] Ziering, D.L.; Pascal, R.A.: Mechanism-based inhibition of bacterial γ-butyrobetaine hydroxylase. J. Am. Chem. Soc., **112**, 834-838 (1990)
[11] Ng, S.F.; Hanauske-Abel, H.M.; Englard, S.: Cosubstrate binding site of Pseudomonas sp. AK1 γ-butyrobetaine hydroxylase. Interactions with structural analogs of α-ketoglutarate. J. Biol. Chem., **266**, 1526-1533 (1991)
[12] Petter, R.C.; Banerjee, S.; Englard, S.: Inhibition of γ-butyrobetaine hydroxylase by cyclopropyl-substituted γ-butyrobetaines. J. Org. Chem., **55**, 3088-3097 (1990)
[13] Englard, S.; Blanchard, J.S.; Midelfort, C.F.: γ-Butyro*β*ine hydroxylase: stereochemical course of the hydroxylation reaction. Biochemistry, **24**, 1110-1116 (1985)
[14] Dunn, W.A.; Rettura, G.; Seifter, E.; Englard, S.: Carnitine biosynthesis from γ-butyrobetaine and from exogenous protein-bound 6-N-trimethyl-L-lysine by the perfused guinea pig liver. Effect of ascorbate deficiency on the in situ activity of γ-butyrobetaine hydroxylase. J. Biol. Chem., **259**, 10764-10770 (1984)
[15] Lindstedt, G.: Hydroxylation of γ-butyrobetaine to carnitine in rat liver. Biochemistry, **6**, 1271-1282 (1967)

[16] Ruetschi, U.; Nordin, I.; Odelhog, B.; Jornvall, H.; Lindstedt, S.: γ-Butyrobetaine hydroxylase. Structural characterization of the Pseudomonas enzyme. Eur. J. Biochem., **213**, 1075-1080 (1993)
[17] Paul, H.S.; Sekas, G.; Adibi, S.A.: Carnitine biosynthesis in hepatic peroxisomes. Demonstration of γ-butyrobetaine hydroxylase activity. Eur. J. Biochem., **203**, 599-605 (1992)
[18] Vaz, F.M.; van Gool, S.; Ofman, R.; Ijlst, L.; Wanders, R.J.: Carnitine biosynthesis. Purification of γ-butyrobetaine hydroxylase from rat liver. Adv. Exp. Med. Biol., **466**, 117-124 (1999)
[19] Galland, S.; Le Borgne, F.; Guyonnet, D.; Clouet, P.; Demarquoy, J.: Purification and characterization of the rat liver γ-butyrobetaine hydroxylase. Mol. Cell. Biochem., **178**, 163-168 (1998)
[20] Galland, S.; Georges, B.; Le Borgne, F.; Conductier, G.; Dias, J.V.; Demarquoy, J.: Thyroid hormone controls carnitine status through modifications of g-butyrobetaine hydroxylase activity and gene expression. Cell. Mol. Life Sci., **59**, 540-545 (2002)
[21] Galland, S.; Le Borgne, F.; Bouchard, F.; Georges, B.; Clouet, P.; Grand-Jean, F.; Demarquoy, J.: Molecular cloning and characterization of the cDNA encoding the rat liver γ-butyrobetaine hydroxylase. Biochim. Biophys. Acta, **1441**, 85-92 (1999)

Procollagen-proline dioxygenase 1.14.11.2

1 Nomenclature

EC number

1.14.11.2

Systematic name

procollagen-L-proline,2-oxoglutarate:oxygen oxidoreductase (4-hydroxylating)

Recommended name

procollagen-proline dioxygenase

Synonyms

collagen proline hydroxylase
hydroxylase, collagen proline
peptidyl proline hydroxylase
procollagen prolyl 4-hydroxylase
proline hydroxylase
proline protocollagen hydroxylase
proline, 2-oxoglutarate dioxygenase
proline,2-oxoglutarate 4-dioxygenase
prolyl 4-hydroxylase
prolyl hydroxylase
prolyl-glycyl-peptide, 2-oxoglutarate:oxygen oxidoreductase, 4-hydroxylating
prolylprotocollagen dioxygenase
prolylprotocollagen hydroxylase
protocollagen hydroxylase
protocollagen proline 4-hydroxylase
protocollagen proline dioxygenase
protocollagen proline hydroxylase
protocollagen prolyl hydroxylase

CAS registry number

9028-06-2

2 Source Organism

<1> *Gallus gallus* [1-9, 11-20, 22, 23, 25, 27-30, 32, 33, 35, 37, 38, 47-51, 53, 58, 60, 65, 70]
<2> *Ascaris lumbricoides* [52]
<3> *Homo sapiens* [1-3, 6-9, 25, 34, 56, 57, 61, 62, 64-66, 68, 70]

<4> *Mytilus edulis* (marine mussel [10]) [10]
<5> *Rattus norvegicus* (Sprague-Dawley strain [23, 24, 70]; female, oestradiol-treated [70]) [1, 21, 23-25, 31, 46, 48, 70]
<6> *Panagrellus silusiae* [26]
<7> *Mus musculus* [3, 9, 36, 62, 66]
<8> *Phaseolus vulgaris* (french bean [39, 67]) [39, 67]
<9> *Chlamydomonas reinhardtii* [40, 41]
<10> *Vinca rosea* [42, 43]
<11> *Lolium multiflorum* [42, 54]
<12> *Volvox carteri* (green algae [3, 44]) [3, 44]
<13> *Enteromorpha intestinalis* [44]
<14> *Helianthus tuberosum* [44]
<15> *Persea americana* [44]
<16> *Daucus carota* [42, 45]
<17> *Lumbricus terrestris* (earthworm [46, 55]) [46, 55]
<18> *Cavia porcellus* [48]
<19> *Caenorhabditis elegans* [59]
<20> *Streptomyces griseoviridus* [63]
<21> *Tropaeolum majus* [67]
<22> *Drosophila melanogaster* [68]
<23> *Paramecium bursaria Chlorella virus 1* (virus-1, eukaryotic algal virus [69]) [69]
<24> *Onchocerca volvulus* [71]
<25> *Brugia malayi* [71]
<26> *Oryctolagus cuniculus* [72]
<27> *Arabidopsis thaliana* [73]
<28> *Onchocerca volvulus* [71]
<29> *Arabidopsis thaliana* [73]
<30> *Arabidopsis thaliana* [73]
<31> *Arabidopsis thaliana* [73]
<32> *Arabidopsis thaliana* [73]
<33> *Arabidopsis thaliana* [73]

3 Reaction and Specificity

Catalyzed reaction

procollagen L-proline + 2-oxoglutarate + O_2 = procollagen trans-4-hydroxy-L-proline + succinate + CO_2 (<1, 3> mechanism [1, 20]; <1> mechanism, role of cosubstrates [28]; <1> mechanism, kinetic analysis of the reaction sequence [29])

Reaction type

decarboxylation
hydroxylation
oxidation
redox reaction
reduction

Natural substrates and products

S procollagen L-proline + 2-oxoglutarate + O_2 <1, 3> (<1, 3> key enzyme in collagen biosynthesis, catalyzes the conversion of selected prolyl residues to trans-hydroxyproline in nascent or completed pro-α chains of procollagen [2]; <1, 3> the biological substrate for the enzyme is a proline residue in an appropriate sequence of a growing or newly synthesized polypeptide chain [3]) (Reversibility: ? <1, 3> [2, 3]) [2, 3]
P procollagen trans-4-hydroxy-L-proline + succinate + CO_2 <1, 3> [2, 3]
S Additional information <26> (<26> molecular mechanism of intracellular degradation of type I collagen in normal corneal endothelial cells, role of the enzyme and protein-disulfide isomerase, which is the β subunit of the enzyme, during procollagen I biosynthesis [72]) [72]
P ?

Substrates and products

S $(Ala\text{-}Pro\text{-}Gly)_5$ + 2-oxoglutarate + O_2 <27, 29-33> (<27, 29-33> recombinant enzyme [73]) (Reversibility: ? <27, 29-33> [73]) [73]
P 4-hydroxyproline containing peptide + succinate + CO_2 <27, 29-33> [73]
S $(Ala\text{-}Thr\text{-}Pro\text{-}Pro\text{-}Pro\text{-}Val)_3$ + 2-oxoglutarate + O_2 <27, 29-33> (Reversibility: ? <27, 29-33> [73]) [73]
P 4-hydroxyproline containing peptide + succinate + CO_2 <27, 29-33> [73]
S $(Gly\text{-}Ala\text{-}Pro)_n$ + 2-oxoglutarate + O_2 <5, 17> (<17> less than 10% relative activity with respect to $(Gly\text{-}Pro\text{-}Ala)_n$ [46]; <5> active substrate [46]) (Reversibility: ? <5, 17> [46]) [46]
P 4-hydroxyproline containing peptide + succinate + CO_2 <5, 17> [46]
S $(Gly\text{-}Pro\text{-}4Hyp)_5$ + 2-oxoglutarate + O_2 <27, 29-33> (Reversibility: ? <27, 29-33> [73]) [73]
P 4-hydroxyproline containing peptide + succinate + CO_2 <27, 29-33> [73]
S $(Gly\text{-}Pro\text{-}Ala)_n$ + 2-oxoglutarate + O_2 <5, 17> (<17> the best substrate [46]; <5> poor substrate [46]) (Reversibility: ? <5, 17> [46]) [46]
P 4-hydroxyproline containing peptide + succinate + CO_2 <5, 17> [46]
S $(L\text{-prolylglycyl-}L\text{-prolyl})_n$ + 2-oxoglutarate + O_2 <1> (<1> molecular weight of the peptide substrate: about 4000 [38]) (Reversibility: ? <1> [38]) [38]
P 4-hydroxyproline containing peptide <1> [38]
S $(Pro\text{-}Ala\text{-}Gly)_5$ + 2-oxoglutarate + O_2 <27, 29-33> (Reversibility: ? <27, 29-33> [73]) [73]
P 4-hydroxyproline containing peptide + succinate + CO_2 <27, 29-33> [73]
S $(Pro\text{-}Ala\text{-}Pro\text{-}Lys)_n$ + 2-oxoglutarate + O_2 <23, 27, 29-33> (<23> n: 1-10, recombinant enzyme, $(Pro\text{-}Ala\text{-}Pro\text{-}Lys)_{10}$ is a particularly good substrate [69]; <27,29-33> n: 3,10 [73]) (Reversibility: ? <23, 27, 29-33> [69, 73]) [69, 73]
P 4-hydroxyproline containing peptide + succinate + CO_2 <23, 27, 29-33> [69, 73]
S $(Pro\text{-}Glu\text{-}Pro\text{-}Pro\text{-}Ala)_5$ + 2-oxoglutarate + O_2 <23, 27, 29-33> (<23> recombinant enzyme [69]) (Reversibility: ? <23, 27, 29-33> [69, 73]) [69, 73]

P 4-hydroxyproline containing peptide + succinate + CO_2 <23, 27, 29-33> [69, 73]

S (Pro-Gly-Pro)$_n$ + 2-oxoglutarate + O_2 <5, 17> (<5, 17> active substrate [46]) (Reversibility: ? <5, 17> [46]) [46]

P 4-hydroxyproline containing peptide + succinate + CO_2 <5, 17> [46]

S (Pro-Pro-Gly)$_n$ + 2-oxoglutarate + O_2 <1, 3-5, 7, 9, 10, 19, 22, 23, 27, 29-33> (<1, 3-5, 7, 19, 22, 23, 27, 29-33> n: 1,5,10 [3, 4, 7, 8, 10, 12, 15-19, 25, 27-29, 33-35, 37, 40-43, 59-62, 65, 68, 69, 73]; <1> structural requirement for proline hydroxylation in collagen and related peptides: supersecondary structure, consisting of PP-II followed by a β-turn, as the conformational determinant for proline hydroxylation in nascent collagen and related substrates of the enzyme, in elastin peptides, the β-structure appears to substitute for the PP-II structure [12]; <1> n: 20 [37]; <9> a low hydroxylation rate is found with denatured (Pro-Pro-Gly)$_{10}$ [40]; <10> n: 5,10, poor substrates [43]; <23> recombinant enzyme [69]; <27, 29-33> the enzyme acts in (Pro-Pro-Gly)$_{10}$ preferentially on prolines in Y positions in the X-Y-Gly triplets [73]) (Reversibility: ? <1, 3-5, 7, 9, 10, 19, 22, 23, 27, 29-33> [3, 4, 7, 8, 10, 15-19, 25, 27-29, 33-35, 37, 40-43, 59-62, 65, 68, 69, 73]) [3, 4, 7, 8, 10, 12, 15-19, 25, 27-29, 33-35, 37, 40-43, 59-62, 65, 68, 69, 73]

P (Pro-4-hydroxy-Pro-Gly)$_n$ + succinate + CO_2 <1, 3-5, 7, 9, 10, 19, 22, 23, 27, 29-33> (<1, 3-5, 7, 19, 22, 23, 27, 29-33> n: 1,5,10 [3, 4, 7, 8, 10, 12, 15-19, 25, 27-29, 33-35, 37, 40-43, 59-62, 65, 68, 69, 73]; <1> n: 20 [37]) [3, 4, 7, 8, 10, 12, 15-19, 25, 27-29, 33-35, 37, 40-43, 59-62, 65, 68, 69, 73]

S (Ser-Pro-Lys-Pro-Pro)$_5$ + 2-oxoglutarate + O_2 <23, 27, 29-33> (<23> recombinant enzyme, (Ser-Pro-Lys-Pro)$_5$ is a particularly good substrate [69]) (Reversibility: ? <23, 27, 29-33> [69, 73]) [69, 73]

P 4-hydroxyproline containing peptide + succinate + CO_2 <23, 27, 29-33> [69, 73]

S 2-oxoglutarate + O_2 + ascorbate <1, 9> (<1> uncoupled oxidative decarboxylation [3,15,28,40]) (Reversibility: ? <1, 9> [3, 15, 28, 40]) [3, 15, 28, 40]

P succinate + dehydroascorbate + CO_2 + H_2O <1, 9> [3, 15, 28, 40]

S Arg-Gly-(Pro-Pro-Gly)$_5$ + 2-oxoglutarate + O_2 <1> (Reversibility: ? <1> [37]) [37]

P 4-hydroxyproline containing peptide + succinate + CO_2 <1> [37]

S Asp-Ala-Leu-Thr-Leu-Leu-Ala-Pro-Ala-Ala-Gly-Asp-Thr-Ile-Ile-Ser-Leu-Phe-Gly + 2-oxoglutarate + O_2 <27, 29-33> (<27, 29-33> peptide representing hypoxia-inducible transcription factor α sequences [73]) (Reversibility: ? <27, 29-33> [73]) [73]

P 4-hydroxyproline containing peptide + succinate + CO_2 <27, 29-33> [73]

S Asp-Leu-Asp-Leu-Glu-Met-Leu-Ala-Pro-Tyr-Ile-Pro-Met-Asp-Asp-Asp-Phe-Gln-Leu + 2-oxoglutarate + O_2 <27, 29-33> (<27, 29-33> peptide representing hypoxia-inducible transcription factor α sequences [73]) (Reversibility: ? <27, 29-33> [73]) [73]

P 4-hydroxyproline containing peptide + succinate + CO_2 <27, 29-33> [73]

S Glu-Gly-(Pro-Pro-Gly)$_5$ + 2-oxoglutarate + O_2 <1> (Reversibility: ? <1> [37]) [37]

P 4-hydroxyproline containing peptide + succinate + CO_2 <1> [37]

S Gly-Pro$_8$ + 2-oxoglutarate + O_2 <10> (<10> 96% relative activity with respect to poly(L-proline) of MW 2000 [42, 43]) (Reversibility: ? <10> [42, 43]) [42, 43]

P 4-hydroxyproline containing peptide <10> [42, 43]

S Gly-Val-Pro-Gly-Val + 2-oxoglutarate + O_2 <3> (Reversibility: ? <3> [65]) [65]

P Gly-Val-4-hydroxyproline-Gly-Val + succinate + CO_2 <3> [65]

S L-proline + 2-oxoglutarate + O_2 <20> (<20> 2-oxoglutarate is essential for hydroxylation [63]) (Reversibility: ? <20> [63]) [63]

P 4-hydroxyproline + succinate + CO_2 <20> [63]

S Lys-Pro-Ala + 2-oxoglutarate + O_2 <23> (<23> recombinant enzyme [69]) (Reversibility: ? <23> [69]) [69]

P Lys-4-hydroxyproline-Ala + succinate + CO_2 <23> [69]

S Ser-Pro-Pro-Pro-Pro-Val-Ser-Pro-Pro-Pro-Val-Ser-Pro-Pro-Pro-Pro-Val + 2-oxoglutarate + O_2 <27, 29-33> (Reversibility: ? <27, 29-33> [73]) [73]

P 4-hydroxyproline containing peptide + succinate + CO_2 <27, 29-33> [73]

S Ser-Pro-Pro-Pro-Val-Tyr-Lys-Ser-Pro-Pro-Pro-Pro-Val-Lys-His-Tyr-Ser-Pro-Pro-Pro-Val + 2-oxoglutarate + O_2 <27, 29-33> (Reversibility: ? <27, 29-33> [73]) [73]

P 4-hydroxyproline containing peptide + succinate + CO_2 <27, 29-33> [73]

S bradykinin + 2-oxoglutarate + O_2 <9> (Reversibility: ? <9> [41]) [41]

P 4-hydroxyproline containing bradykinin <9> [41]

S collagen + 2-oxoglutarate + O_2 <6> (Reversibility: ? <6> [26]) [26]

P 4-hydroxyproline containing collagen + succinate + CO_2 <6> [26]

S lysine hydroxylated protocollagen + 2-oxoglutarate + O_2 <1> (Reversibility: ? <1> [51]) [51]

P lysine 4-hydroxyproline containing protocollagen + succinate + CO_2 <1> [51]

S octa-L-proline + 2-oxoglutarate + O_2 <10> (<10> 81.8% relative activity with respect to poly(L-proline) of MW 2000 [42,43]) (Reversibility: ? <10> [42, 43]) [42, 43]

P 4-hydroxyproline containing peptide <10> [42, 43]

S poly(L-Pro) + 2-oxoglutarate + O_2 <3, 8-15, 23, 27, 29-33> (<1, 4> not [10, 38]; <9> MW 7000 [41]; <10> MW 2000, 6000 and 12000 [42, 43]; <19> do not serve as substrate, assayed with recombinant enzyme [59]; <3> MW: 7000 and 44000 [65]; <23, 27, 29-33> recombinant enzyme [69, 73]) (Reversibility: ? <3, 8-15, 23, 27, 29-33> [39, 40-44, 65, 69, 73]) [39, 40-44, 65, 69, 73]

P poly(4-hydroxyproline) + succinate + CO_2 <3, 8-15, 23, 27, 29-33> [39, 40-44, 65, 69, 73]

S procollagen L-proline + 2-oxoglutarate + O_2 <1, 3, 5> (Reversibility: ? <1, 3, 5> [4, 20, 21, 23, 34]) [4, 20, 21, 23, 34]

P procollagen trans-4-hydroxy-L-proline + succinate + CO_2 <1, 3, 5> [4, 20, 21, 23, 34]

S proline containing peptide + 2-oxoglutarate + O_2 <1, 3, 5> (<1, 3> no hydroxylation of free proline, minimum sequence required X-Pro-Gly, best substrates are those where Pro precedes Gly, which can be substituted by Ala or β-alanine. The amino acid preceding Pro can be Pro, Ala, Leu, Arg, Val, Glu, but not Gly or Ser. Additionally the sequence, the conformation and the peptide chain length influence the rate of hydroxylation [1]; <1, 3> the presence of 2-oxoglutarate is an absolute and highly specific requirement, the formation of 4-hydroxyproline is accompanied by a stoichiometric decarboxylation of 2-oxoglutarate, the oxygen of the hydroxyl group is derived from molecular oxygen, the other atom of the O_2 molecule being incorporated into the succinate, the activated form of oxygen is probably superoxide [1,9]; <1> chelation to the enzyme-bound metal ion is important for proper binding of the substrate 2-oxoglutarate [5]; <1> synthetic peptides [12]) (Reversibility: ? <1, 3, 5> [1, 5, 12, 25, 65]) [1, 5, 9, 12, 25]

P 4-hydroxyproline containing peptide + succinate + CO_2 <1, 3, 5> [1, 5, 12, 25, 65]

S protocollagen + 2-oxoglutarate + O_2 <1, 3, 5, 6, 10, 11, 16, 24, 28> (<1> 2-oxoadipate can replace 2-oxoglutarate as cosubstrate [16]) (Reversibility: ? <1, 3, 5, 6, 10, 11, 16, 24, 28> [2, 11, 16, 17, 24, 26, 32, 42, 47, 51, 58, 60, 65, 71]) [2, 11, 16, 17, 24, 26, 32, 42, 47, 51, 58, 65, 71]

P 4-hydroxyproline containing protocollagen + succinate + CO_2 <1, 3, 5, 6, 10, 11, 16, 24, 28> [2, 11, 16, 17, 24, 26, 32, 42, 47, 51, 58, 60, 65, 71]

S tert-butyloxycarbonyl-Gly-Val-Pro-Gly-Val-OH + 2-oxoglutarate + O_2 <1> (Reversibility: ? <1> [12]) [12]

P 4-hydroxyproline containing peptide + succinate + CO_2 <1> [12]

S tert-butyloxycarbonyl-Pro-Pro-Ala-Pro-OH + 2-oxoglutarate + O_2 <1> (Reversibility: ? <1> [12]) [12]

P 4-hydroxyproline containing peptide + succinate + CO_2 <1> [12]

S tert-butyloxycarbonyl-Pro-Pro-Gln-Pro-OCH_3 + 2-oxoglutarate + O_2 <1> (Reversibility: ? <1> [12]) [12]

P 4-hydroxyproline containing peptide + succinate + CO_2 <1> [12]

S tert-butyloxycarbonyl-Pro-Pro-Gly-Pro + 2-oxoglutarate + O_2 <1> (Reversibility: ? <1> [12]) [12]

P 4-hydroxyproline containing peptide + succinate + CO_2 <1> [12]

S tert-butyloxycarbonyl-Pro-Pro-Gly-Pro-$NHCH_3$ + 2-oxoglutarate + O_2 <1> (Reversibility: ? <1> [12]) [12]

P 4-hydroxyproline containing peptide + succinate + CO_2 <1> [12]

S tert-butyloxycarbonyl-Pro-Pro-Gly-Pro-Pro-OH + 2-oxoglutarate + O_2 <1> (Reversibility: ? <1> [12]) [12]

P 4-hydroxyproline containing peptide + succinate + CO_2 <1> [12]

S tert-butyloxycarbonyl-Pro_8 + 2-oxoglutarate + O_2 <10> (<10> 91.4% relative activity with respect to poly(L-proline) of MW 2000 [42, 43]) (Reversibility: ? <10> [42, 43]) [42, 43]

P 4-hydroxyproline containing peptide <10> [42, 43]

S tert-butyloxycarbonyl-Val-Pro-Gly-Val-OH + 2-oxoglutarate + O_2 <1> (Reversibility: ? <1> [12]) [12]

P 4-hydroxyproline containing peptide + succinate + CO_2 <1> [12]

S Additional information <1, 4, 5, 6, 9-12, 16, 17, 20, 27, 29-33> (<4> no activity with poly-L-proline or the unhydroxylated decapeptide analog (Ala-Lys-Pro-Ser-Tyr-Pro-Pro-Thr-Tyr-Lys) of the polyphenolic protein [10]; <1> in the absence of a peptidylproline substrate, the oxidative decarboxylation of 2-oxoglutarate by the enzyme is stoichiometrically coupled to the oxidation of ascorbate [15]; <1> characterization of the 2-oxoglutarate binding site of the enzyme, 2-oxosuccinate, 2-oxovalerate, 2-oxobutyrate, 3-oxoglutarate or pyridine 2,5-dicarboxylate do not replace 2-oxoglutarate as cosubstrates [16]; <1, 5, 6> thermal denaturing of the triple-helical conformation of the substrate before hydroxylation [16-19, 24, 26-28, 33, 37, 60]; <6> effect of substrate on activity [26]; <1> the enzyme is specific for proline in the second position after glycine, the position in which the hydroxyproline in collagen is found. No reaction with: free proline, glycyl-L-prolyl-L-proline or poly-L-proline II with a molecular weight of about 15000 [38]; <9> protocollagen and triple-helical (Pro-Pro-Gly)$_{10}$ do not serve as substrate. 2-oxoadipate, 2-oxosuccinate or 2-oxovalerate do not replace 2-oxoglutarate [40]; <10, 11, 16> no reaction with: free proline, prolyl peptides whose residues number is four or less [42, 43]; <12> proline or the dipeptides Ser-Pro or Ala-Pro do not serve as substrates [44]; <17> the enzyme clearly prefers X position proline to Y position proline in the sequences of (Gly-Pro-Ala)$_n$ versus (Gly-Ala-Pro)$_n$ [46]; <1> the presence of lysyl hydroxylase in the reaction mixture has no effect on the activity of the enzyme [51]; <20> 2-oxoglutarate cannot be replaced by oxalylglycine, 2-oxopentanoate, 2-oxoadipate, pyruvate or 2-oxomalonate [63]; <27, 29-33> the enzyme preferentially hydroxylates proline residues preceding glycines in (X-Y-Gly)$_n$ peptides. Peptides representing the N- and C-terminal hydroxylation sites present in hypoxia-inducible transcription factor α also serve as substrates [73]) [10, 15-19, 24, 26-28, 33, 37, 38, 40, 42-44, 46, 51, 60, 63, 73]

P ?

Inhibitors

(+)-mandelate <1> [17]

(-)-mandelate <1> [17]

(Gly-Pro-Gly)$_n$ <1, 3> (<1,3> competitive inhibitors with respect to the polypeptide substrate [1]) [1]

(Pro-Ala-Gly)$_n$ <1, 3> (<1,3> competitive inhibitors with respect to the polypeptide substrate [1]) [1]

(Pro-Pro-Gly)$_5$ <1> (<1> at concentrations higher than 0.56 mM, substrate inhibition observed [19]) [19]

1,10-phenanthroline <5, 6, 25> (<5> inhibits at a concentration higher than the Fe^{2+} concentration in the reaction mixture [21]; <5> 91% inhibition at 0.02 mM [48]) [21, 26, 48, 71]

1,2,3-trihydroxybenzene <1> [17]

1,2-dihydroxybenzene <1> (<1> competitive with respect to 2-oxoglutarate and ascorbate, noncompetitive with respect to Fe^{2+} [17]) [17]

1,3-dihydroxybenzene <1> (<1> competitive with respect to 2-oxoglutarate and ascorbate, noncompetitive with respect to Fe^{2+} [17]) [17]
1,4-dihydrophenanthroline-4-one-3-carboxylic acid <1, 5> (<1> purified enzyme, competitive inhibitor of 2-oxoglutarate [70]; <5> potent competitive inhibitor of collagen hydroxylation in the oestradiol-stimulated uterus in vivo [70]) [70]
1,4-dihydroxybenzene <1> [17]
2,2'-dipyridyl <1, 3-6, 8> (<5> inhibits at a concentration higher than the Fe^{2+} concentration in the reaction mixture [21]; <5> 87% inhibition at 0.02 mM [48]; <8> 50% inhibition at 0.06 mM [67]) [2, 10, 21, 26, 39, 48, 67]
2,3-dihydroxybenzoate <1, 3> (<1> competitive with respect to 2-oxoglutarate and ascorbate, noncompetitive with respect to Fe^{2+} [17]) [9, 17]
2,4-dihydroxybenzoate <1> (<1> competitive with respect to 2-oxoglutarate and ascorbate, noncompetitive with respect to Fe^{2+} [17]) [17]
2,5-dihydroxybenzoate <1> (<1> competitive with respect to 2-oxoglutarate and ascorbate, noncompetitive with respect to Fe^{2+} [17]) [17]
2,6-dihydroxybenzoate <1> (<1> competitive with respect to 2-oxoglutarate and ascorbate, noncompetitive with respect to Fe^{2+} [17]) [17]
2,7,8-trihydroxyanthraquinone <1> (<1> 50% inhibition at 0.047 mM, competitive inhibitor with respect to 2-oxoglutarate, non-competitive with regard to ascorbate, uncompetitive with regard to protocollagen. The inhibition is greatly enhanced in the absence of Fe^{2+}, structural requirements for inhibition [11]) [11]
2-hydroxybenzoate <1> (<1> competitive with respect to 2-oxoglutarate and ascorbate, noncompetitive with respect to Fe^{2+} [17]) [17]
2-oxoadipate <9> [40]
2-oxoadipinate <1> (<1> competitive inhibitor with respect to 2-oxoglutarate, noncompetitive with respect to Fe^{2+} [16]) [16]
2-oxobutyrate <1> (<1> competitive inhibitor with respect to 2-oxoglutarate, noncompetitive with respect to Fe^{2+} [16]) [16]
2-oxoglutarate <20> (<20> at concentrations higher than 0.5 mM, decreases activity [63]) [63]
2-oxosuccinate <1, 9> (<1> competitive inhibitor with respect to 2-oxoglutarate, noncompetitive with respect to Fe^{2+} [16]; <1,9> i.e. oxaloacetate [16,28,40]) [16, 28, 40]
2-oxovalerate <1> (<1> competitive inhibitor with respect to 2-oxoglutarate, noncompetitive with respect to Fe^{2+} [16]) [16]
3,4,5-trihydroxybenzoate <1> [17]
3,4-dihydroxybenzoate <1, 3, 9, 20> (<1> competitive with respect to 2-oxoglutarate and ascorbate, noncompetitive with respect to Fe^{2+} [17]) [3, 8, 9, 17, 18, 40, 63]
3,4-dihydroxycinnamate <1> (<1> competitive with respect to 2-oxoglutarate and ascorbate [17]) [17]
3,4-dihydroxymandelate <1> (<1> competitive with respect to 2-oxoglutarate and ascorbate and noncompetitive with respect to Fe^{2+} [17]) [17]

3,4-dihydroxyphenylacetate <1, 3, 12-15> (<1> competitive with respect to 2-oxoglutarate and ascorbate, noncompetitive with respect to Fe^{2+} [17,44]) [9, 17, 44]

3,4-dihydroxyphenylpropionate <1, 12-15> (<1> competitive with respect to 2-oxoglutarate and ascorbate [17, 44]) [17, 44]

3,5-dihydroxybenzoate <1> (<1> competitive with respect to 2-oxoglutarate and ascorbate, noncompetitive with respect to Fe^{2+} [17]) [17]

3-hydroxybenzoate <1> [17]

3-hydroxybutyrate <1> (<1> less than 10% inhibition [28]) [28]

3-oxoglutarate <1> (<1> competitive inhibitor with respect to 2-oxoglutarate, noncompetitive with respect to Fe^{2+} [16]) [16]

4-hydroxybenzoate <1, 3> (<1> competitive with respect to 2-oxoglutarate, and noncompetitive with respect to ascorbate [17]) [8, 9, 17]

5-azidopyridine-2-carboxylic acid <1> (<1> incorporated in the α subunit of the enzyme, complete inactivation of the enzyme by incorporation of 2 mol of photoaffinity label per mol of tetramer [5]) [5, 50]

8-(N-butyl-N-ethylcarbamoyl)-1,4-dihydrophenanthrolin-4-one-3-carboxylic acid <1, 5> (<1> purified enzyme, competitive inhibitor of 2-oxoglutarate [70]; <5> potent competitive inhibitor of collagen hydroxylation in the oestradiol-stimulated uterus in vivo [70]) [70]

8-hydroxyquinoline <5> (<5> inhibits at a concentration higher than the Fe^{2+} concentration in the reaction mixture [21]) [21]

ADP-ribose <1> (<1> weak inhibitor [13]) [13]

CO_2 <1> (<1> 5% inhibition at 3.6 mM, 35% inhibition at 7.2 mM and 75% inhibition at 12 mM [29]) [29]

Co^{2+} <20> (<20> complete inhibition [63]) [63]

Cu^{2+} <1, 3, 20> (<20> causes a 20-30% fall in activity [63]) [9, 32, 63]

EDTA <1, 4, 5, 8> (<1> 98-100% inhibition at 0.01 mM [19];<5> 64% inhibition at 0.02 mM [48]) [10, 19, 39, 48]

EDTA <5, 8> (<5> inhibits at a concentration higher than the Fe^{2+} concentration in the reaction mixture [21]; <8> 50% inhibition at 0.15 mM [67]) [21, 39, 67]

EGTA <8> [39]

Fe^{2+} <1, 20> (<1> inhibitory at high concentrations [28]; <20> inhibitory at a concentration higher than 0.5 mM [63]) [28, 63]

H_2O_2 <1> (<1> 94% inhibition at 1 M, dissociation of the enzyme, 12% of the enzyme remains in the tetrameric form [47]) [47]

Hg^{2+} <20> (<20> causes a 20-30% fall in activity [63]) [63]

L-galactono γ-lactone <1> [17]

Mn^{2+} <20> (<20> causes a 20-30% fall in activity [63]) [63]

N,N'-ethylamide of pyridine 2,4-dicarboxylate <1, 3> [3]

N-(4-azido-2-nitrophenyl)-glycyl-(Pro-Pro-Gly)$_5$ <1> (<1> loss of enzyme activity with (Pro-Pro-Gly)$_5$ as a substrate upon photoaffinity labeling [4]) [4, 49]

N-hydroxyethylenediaminetriacetic acid <5> (<5> inhibits at a concentration higher than the Fe^{2+} concentration in the reaction mixture [21]) [21]

NaCl <9, 10> (<9> at 0.5 mM [41]; <10> more than 0.3 M [43]) [41, 43]

Pd^{2+} <1> (<1> strong irreversible inhibition, competitive with respect to Fe^{2+} [32]) [32]

Zn^{2+} <1, 3, 4, 9, 19, 20> (<1> competitive inhibition with respect to Fe^{2+}, non-competitive with respect to the polypeptide substrate and 2-oxoglutarate [28]; <19> recombinant enzyme, competitive inhibition with respect to Fe^{2+} and 2-oxoglutarate [59]; <20> complete inhibition [63]) [1, 3, 9, 10, 28, 40, 59, 63]

adipinate <1> (<1> competitive inhibitor with respect to 2-oxoglutarate, noncompetitive with respect to Fe^{2+} [16]) [16]

antimycin A <5> (<5> 18% inhibition at 0.02 mM [48]) [48]

ascorbate <1, 20> (<1> at high concentrations [28]; <20> the inhibition may result from competition for binding at the 2-oxoacid binding site between 2-oxoglutarate and L-ascorbate [63]) [28, 63]

benzene 1,2-dicarboxylate <1> (<1> competitive inhibitor with respect to 2-oxoglutarate, noncompetitive with respect to Fe^{2+} [16]) [16]

benzene 1,3-dicarboxylate <1, 9> (<1> competitive inhibitor with respect to 2-oxoglutarate, noncompetitive with respect to Fe^{2+} [16]) [16, 40]

benzene 1,4-dicarboxylate <1> (<1> competitive inhibitor with respect to 2-oxoglutarate, noncompetitive with respect to Fe^{2+} [16]; <1> competitive inhibition with respect to ascorbate [17]) [16, 17]

benzoate <1> (<1> competitive inhibitor with respect to 2-oxoglutarate, noncompetitive with respect to Fe^{2+} [16]) [16]

benzyloxycarbonyl-Phe-Opr-Gly-benzyl ester <1, 3> (<1, 3> 50% inactivation in 1 h at 0.0008 mM, the most effective inhibitor within oxaproline peptides [6]) [6]

benzyloxycarbonyl-Phe-oxaproline-Gly-benzyl ester <1, 3> [3]

β-lactam antibiotics <1, 3> [3]

bradykinin analogs <1, 3> (<1, 3> especially those in which the proline in the -X-Pro-Gly- triplet is replaced by certain proline analogues, the addition of a glutamyl residue to the N-terminal end of 3,4-dehydroprolyl- or trans-4-hydroxyprolyl-bradykinin considerably increases their effectiveness [1]) [1]

catechol analogues <1, 3> (<1, 3> inhibitor of the reaction due in part to the chelation of Fe^{2+} [1]) [1]

citrate <1> (<1> competitive inhibition with respect to 2-oxoglutarate [28]) [28]

collagen <1> (<1> product inhibitor, noncompetitive with respect to all substrates of the reaction [29]) [29]

compound ZM 226681 <8> (<8> oxaloglycine derivative, inhibitor [67]) [67]

concanavalin A <1> [30]

concavalin A <1> (<1> partially inhibits, when the enzyme is assayed in the absence of bovine serum albumin [30]) [30]

coumalic acid <1, 3, 25> (<1, 3> competitive inhibitor, potential syncatalytic inhibitor, time-dependent inactivation, increasing concentrations of Fe^{2+} enhance the inactivation [8]; <1, 3> i.e. 2-oxo-1,2H-pyran-5-carboxylic acid [3, 8]) [3, 8, 71]

cupferron <5> (<5> 28% inhibition at 0.02 mM [48]) [48]

daunorubicin <1, 3, 25> (<1,3> irreversible inhibitor, 50% inhibition after 1 h at 0.06 mM, effect dependent on the presence of iron ions [7]) [3, 7, 71]
diethyl dicarbonate <20> (<20> 98% inhibition at 1 mM [63]) [63]
diethyldithiocarbamate <5> (<5> 10% inhibition at 0.02 mM [48]) [48]
diethylenetriaminepentaacetic acid <5> (<5> inhibits at a concentration higher than the Fe^{2+} concentration in the reaction mixture [21]) [21]
dilantin <1, 3> (<1,3> inhibitor of the reaction due in part to the chelation of Fe^{2+} [1]) [1]
dimethyloxalylalanine <1> (<1> 50% inhibition in chicken embryo calvaria at 1 mM [60]) [60]
dimethyloxalylglycine <1> (<1> 50% inhibition in chicken embryo calvaria at 0.002 mM, inhibitor of hydroxyproline synthesis in embryonic chicken lung [60]) [60]
dithiothreitol <1, 5> (<5> powerful inhibitor at 1 mM [24]; <1> 95-100% inhibition at 0.45 mM [27, 47]) [24, 27, 47]
doxorubicin <1, 3, 25> (<1,3> irreversible inhibitor, 50% inhibition after 1 h at 0.06 mM, effect dependent on the presence of iron ions [7]; <8> not inhibitory at concentration up to 0.5 mM [67]) [3, 7, 71]
epinephrine <1> (<1> competitive inhibition with respect to Fe^{2+} [28]) [28]
ethylpyridine-2,4-dicarboxylate <1, 3> [3]
fumarate <1> (<1> competitive inhibition with respect to 2-oxoglutarate [28]) [28]
gelatin <1, 5, 18> (<1, 5, 18> commercial, inhibitor [48]) [48]
glutamyl-3,4-dehydroprolyl-bradykinin <1, 3> [9]
glutarate <9> [40]
glutarate <1, 9> (<1> competitive inhibitor with respect to 2-oxoglutarate, noncompetitive with respect to Fe^{2+} [16]) [16, 40]
hydralazine <1, 3> (<1,3> inhibitor of the reaction due in part to the chelation of Fe^{2+} [1]) [1]
hydroxybenzene <1> [9]
isocitrate <1> (<1> competitive inhibition with respect to 2-oxoglutarate [28]) [28]
ketomalonate <1, 5, 18> (<1, 5, 18> 100% inhibition at 1 mM, acts by chelating ferrous ion rather than by competing with α-ketoglutarate [48]) [48]
lactate <1> (<1> less than 10% inhibition [28]) [28]
levulinate <1> (<1> competitive inhibitor with respect to 2-oxoglutarate, noncompetitive with respect to Fe^{2+} [16]) [16]
malate <1> (<1> competitive inhibition with respect to 2-oxoglutarate [28]) [28]
malonate <1, 5, 18> (<1> competitive inhibitor with respect to 2-oxoglutarate, noncompetitive with respect to Fe^{2+} [16]; <1, 5, 18> 19% inhibition at 1 mM [48]) [16, 48]
nitroblue tetrazolium <1, 3> (<1, 3> is capable of scavenging superoxide, competitive inhibitor with respect to O_2 [1]; <1> competitive inhibition with respect to O_2 [28]) [1, 28]
oxalate <1, 5, 18> (<1, 5, 18> 34% inhibition at 1 mM [48]) [48]

oxaloacetate <1> (<1> competitive inhibition with respect to 2-oxoglutarate [28]) [28]

oxalyl-β-alanine <1> (<1> competitive inhibition with respect to 2-oxoglutarate [60]) [60]

oxalylalanine <1> (<1> inhibits purified enzyme, competitive inhibition with respect to 2-oxoglutarate, 50% inhibition of microsomal enzyme at 0.123 mM [60]) [60]

oxalylcystine <1> (<1> competitive inhibition with respect to 2-oxoglutarate [60]) [60]

oxalylglycine <1, 20> (<1> inhibits purified enzyme, competitive inhibition with respect to 2-oxoglutarate, 50% inhibition of microsomal enzyme at 0.023 mM [60]) [60, 63]

oxalylproline <1> [60]

oxalylsarcosine <1> (<1> noncompetitive inhibition with respect to 2-oxoglutarate [60]) [60]

oxalylvaline <1> [60]

phenanthrolines <1, 3, 5> (<1> potent competitive inhibitors inhibitory of purified enzyme [70]; <1> potent competitive inhibitors of collagen hydroxylation in embryonic tendon cells in vitro [70]; <3> potent competitive inhibitors of collagen hydroxylation in foreskin fibroblasts in vitro [70]) [70]

phenylacetate <1> [17]

phosphoribosyl adenosine monophosphate <1> (<1> 46% inhibition at 25 nM, 87% inhibition at 50 nM [13]) [13]

poly(ADP-ribose) <1> (<1> near complete inhibition at 6 nM, the effect is noncompetitive with respect to the binding of the cofactors ascorbate and α-ketoglutarate or of the substrate [13]) [13]

poly(L-proline) <1, 3, 4, 6, 7, 22> (<1,3> competitive inhibitors with respect to the polypeptide substrate, the inhibition increases with chain length [1,3]; <4> competitive inhibition [10]; <1> competitive inhibitor with respect to the polypeptide substrate and uncompetitive with respect to Fe^{2+} and 2-oxoglutarate [29]; <19> not inhibitory [59]; <3> inhibitory, type I enzyme tetramer, MW: 7000 and 44000 [62]; <7> inhibitor, recombinant type II enzyme tetramer, MW: 7000 and 44000 [62]; <3> type I und II enzyme [65]; <3,22> MW: 7000 and 44000 [68]) [1, 3, 9, 10, 26, 28, 29, 62, 65, 68]

poly-L-hydroxyproline <10> (<10> MW: 30000, inhibitory [43]) [43]

potassium phosphate <10> [42, 43]

propyl gallate <1, 5, 18> (<1,5,18> 100% inhibition at 2 mM [48]) [48]

pyridine 2,3-dicarboxylate <1> (<1> competitive inhibitor with respect to 2-oxoglutarate, noncompetitive with respect to Fe^{2+} [16]) [16]

pyridine 2,4-dicarboxylate <1, 3, 7-9, 12, 19, 20> (<1> competitive inhibitor with respect to 2-oxoglutarate, noncompetitive with respect to Fe^{2+} [16]; <1> uncompetitive inhibition with respect to ascorbate [17]; <19> recombinant enzyme, competitive inhibitor with respect to Fe^{2+} and 2-oxoglutarate [59]; <3> inhibits wild-type enzyme and enzyme tetramer containing the histidine 501 to serine mutant α subunit [61]; <3> inhibitory, type I enzyme tetramer [62]; <7> inhibitory, recombinant type II enzyme tetramer [62]; <8> effective inhibitor [67]; <25> not inhibitory [71]) [3, 8, 9, 15-17, 40, 44, 59-63, 67]

pyridine 2,5-dicarboxylate <1, 3, 9, 12, 20> (<1> competitive inhibitor with respect to 2-oxoglutarate, noncompetitive with respect to Fe^{2+} [16]; <1> uncompetitive inhibition with respect to ascorbate [17]) [3, 8, 9, 16, 17, 40, 44, 63]
pyridine 2,6-dicarboxylate <1> (<1> competitive with respect to Fe^{2+} and noncompetitive with respect to 2-oxoglutarate [16]) [16]
pyridine 2-carboxylate <1, 9> (<1> competitive inhibitor with respect to 2-oxoglutarate, noncompetitive with respect to Fe^{2+} [16]; <1> uncompetitive inhibition with respect to ascorbate [17]) [5, 9, 16, 17, 40]
pyridine 3,4-dicarboxylate <1> (<1> competitive inhibitor with respect to 2-oxoglutarate, noncompetitive with respect to Fe^{2+} [16]) [16]
pyridine 3,5-dicarboxylate <1> (<1> competitive inhibitor with respect to 2-oxoglutarate, noncompetitive with respect to Fe^{2+} [16]) [16]
pyridine 3-carboxylate <1> (<1> competitive inhibitor with respect to 2-oxoglutarate, noncompetitive with respect to Fe^{2+} [16]) [16]
pyridine 4-carboxylate <1> (<1> competitive inhibitor with respect to 2-oxoglutarate, noncompetitive with respect to Fe^{2+} [16]) [16]
pyruvate <1> (<1> competitive inhibition with respect to 2-oxoglutarate [28]) [28]
pyruvate <1, 9> (<1> competitive inhibitor with respect to 2-oxoglutarate, noncompetitive with respect to Fe^{2+} [16]) [16, 40]
ribosyl-ribosyl-adenine <1> (<1> 43% inhibition at 25 nM, 85% inhibition at 50 nM [13]) [13]
ribosyl-ribosyl-hypoxanthine <1> (<1> 40% inhibition at 25 nM, 86% inhibition at 50 nM [13]) [13]
salicylyl hydroxamate <8> [39]
sodium acetate <10> [42, 43]
sodium pyrocatechol disulfonate <1> (<1> 98-100% inhibition at 0.1 mM [19]) [19]
succinate <1, 5, 9, 10, 18> (<1> competitive inhibitor with respect to 2-oxoglutarate, noncompetitive with respect to Fe^{2+} [16,29]; <10> 11.5% inhibition at 0.5 mM and 38.3% inhibition at 3 mM [43]; <1,5,18> 51% inhibition at 1 mM [48]) [16, 29, 40, 43, 48]
tetracyclin <1, 3> (<1,3> inhibitor of the reaction due in part to the chelation of Fe^{2+} [1]) [1]
trifluorothienylbutanedione <5> (<5> 48% inhibition at 0.05 mM [48]) [48]
Additional information <1, 3> (<1,3> synthetic peptides containing the unphysiologic amino acid 5-oxaproline in the sequence R_1-Xaa-oxaproline-Gly-OR_2 are specific syncatalytic inactivators, noncompetitive inhibition with respect to peptide substrate and ascorbate, compounds with aromatic substituents R1 and R2 are particularly effective when compared with those with an aliphatic group, inactivation is only observed in the presence of Fe^{2+} and 2-oxoglutarate [6]) [6]
Additional information <1, 3, 4, 10> (<4,10> high concentrations of salts are inhibitory [10,43]; <1> the synthesis and degradation of ADP-ribose moieties may possibly regulate prolyl hydroxylase activity in vivo [13]; <1> inhibition of the purified enzyme and of collagen hydroxylation in embryonic tendon

fibroblast by novel phenanthrolines, structure-activity relationships [70]; <3> inhibition of the secretion of procollagen in foreskin fibroblasts by novel phenanthrolines, structure-activity relationships [70]) [10, 13, 43, 70]

Cofactors/prosthetic groups

2-oxoglutarate <1, 3-6, 9, 10, 17, 18, 24, 28> (<10> K_m: 0.09 mM [42,43]; <1, 5, 18> no compound can replace α-ketoglutarate [48]) [2, 10, 13, 21, 26, 30, 35, 37, 38, 41-43, 46, 48, 71]
5,6-isopropylideneascorbate <1, 9> (<1, 9> can replace ascorbate [17, 40]) [17, 40]
D-isoascorbate <1, 9, 5, 18> (<1, 9, 5, 18> can replace ascorbate [17, 40, 48]) [17, 40, 48]
ascorbate <1, 3-7, 9-19, 22-24, 27-33> (<1, 3> probably required to prevent oxidation of the enzyme-bound Fe^{2+} or free enzyme between catalytic cycles, replaceability by certain reduced pteridines and thiols. K_m with biological substrate: 0.1 mM, K_m with synthetic substrate: 0.3 mM [1]; <1, 3, 5, 18> requirement [1-3, 28, 48]; <1> pure enzyme, high specificity, dithiothreitol and L-cysteine are the only compounds that give more than 10% of the activity found with the optimal ascorbate concentration, some reduced pteridines give values ranging from 3 to 9% [1]; <1,3> highly specific for [9]; <1,3> oxygen acceptor in the decarboxylation of 2-oxoglutarate without subsequent hydroxylation of peptide substrate [3,9]; <1> completely dependent, required for activity [19]; <1> K_m: 0.2 mM [28]; <9> K_m: 0.240 mM [40]; <12> K_m: 0.14 mM [44]; <1,5,18> K_m: 0.2-1 mM, dependent on the concentrations of α-ketoglutarate and Fe^{2+}, reduced pteridines can partially replace ascorbate [48]; <19> recombinant enzyme, K_m: 0.3 mM [59]; <3> K_m for wild-type enzyme tetramer: 0.33 mM, enzyme tetramer containing the histidine 501 to serine mutant α subunit, K_m: 0.4 mM [61]; <3> type I enzyme tetramer: K_m 0.33 mM [62]; <7> recombinant type II enzyme tetramer: K_m 0.34 mM [62]; <3> pure type II enzyme: K_m 0.34 mM, pure type I enzyme: K_m 0.33 mM [65]; <22> K_m: 0.3 mM [68]; <3> type I enzyme: K_m 0.32 mM, type II enzyme: K_m 0.34 mM [68]; <23,27,29-33> recombinant enzyme: K_m 0.3 mM [69,73]) [1-4, 6-13, 15-19, 20, 25-30, 33, 35, 37, 38, 40, 42-46, 48, 59-62, 65, 68, 69, 71, 73]
Additional information <1, 10> (<1> analysis of the activity in the absence and presence of ascorbate [18]; <1> the ascorbate requirement cannot be replaced by tetrahydrofolic acid, dithiothreitol, $NADH_2$ or dithionite to any significant extent [28]; <10> α-ketoglutarate cannot be replaced by pyruvate and oxaloacetate [43]) [18, 28, 43]

Activating compounds

1,10-phenanthroline <5> (<5> stimulates if it reaches an equimolar concentration with Fe^{2+} [21]) [21]
8-hydroxyquinoline <5> (<5> stimulates if it reaches an equimolar concentration with Fe^{2+} [21]) [21]
D-isoascorbate <9> (<9> can partially replace ascorbate [40]) [40]
EDTA <5> (<5> 1.4fold stimulation at 0.2 mM, equimolar with Fe^{2+} [21]) [21]

N-hydroxyethylenediaminetriacetic acid <5> (<5> 5fold stimulation at 0.2 mM, equimolar with Fe^{2+} [21]) [21]
Triton X-100 <8> (<8> activation at 0.1% v/v [39]) [39]
α,α'-dipyridyl <5> (<5> stimulates if it reaches an equimolar concentration with Fe^{2+} [21]) [21]
bleomycin <1, 3> (<1,3> activation [2]) [2]
bovine serum albumin <1, 3, 4, 5, 9, 10, 17, 24, 28> (<1, 3, 4, 5, 9, 10, 17, 24, 28> activation [1, 2, 4, 6-8, 10-12, 15-19, 20, 27-30, 33, 35, 37, 40-43, 46, 60, 71]) [1, 2, 4, 6-8, 10-12, 15-19, 20, 27-30, 33, 35, 37, 40-43, 46, 60, 71]
catalase <1, 3, 4-6, 9, 20, 24, 28> (<1, 3, 4-6, 9, 20, 24, 28> activation [1, 2, 4, 6-8, 10-12, 15-17, 19-21, 26-30, 33, 35, 37, 40, 60, 63, 71]) [1, 2, 4, 6-8, 10-12, 15-17, 19-21, 26-30, 33, 35, 37, 40, 60, 63, 71]
chelating agents <1, 3> (<1,3> activation [2]) [2]
cysteine <1, 3> (<1,3> activation [9]) [9]
diethylenetriaminepentaacetic acid <5> (<5> 5fold stimulation at 0.2 mM, equimolar with Fe^{2+} [21]) [21]
dithiothreitol <1, 3, 4-6, 9, 10, 17, 24, 28> (<1, 3, 4-6, 9, 10, 17, 24, 28> activation [1, 2, 4, 6-12, 15-17, 19, 20, 26-30, 33, 35, 37, 40, 42, 43, 46, 60, 71]) [1, 2, 4, 6-12, 15-17, 19, 20, 26-30, 40, 33, 35, 37, 40, 42, 43, 46, 60, 71]
glycoletherdiamine tetraacetic acid <5> (<5> 1.4fold stimulation at 0.2 mM, equimolar with Fe^{2+}, and 2.8fold stimulation at 1 mM [21]) [21]
lactate <1> (<1> stimulates, the activation of the enzyme in lactate-treated fibroblasts is correlated with a reduction in the total ADP-ribosylation [13]) [13]
nitrilotriacetic acid <5> (<5> 6fold stimulation at 0.2 mM, equimolar with Fe^{2+}, and 7.6fold stimulation at 1 mM [21]) [21]
nucleoside triphosphates <1, 3, 5> (<1,3,5> stimulate [2,21]) [2, 21]
pyridine 2,6-dicarboxylate <20> (<20> activator [63]) [63]
thymol <1, 3> (<1,3> activation [2]) [2]
Additional information <1> (protein disulfide-isomerase is able to activate the purified enzyme, the activation is probably due to the repairing of disulfide exchanges ocurring in the prolyl 4-hydroxylase structure during purification and storage. Protein disulfide-isomerase is also able to reactivate the enzyme inactivated by mild H_2O_2 treatment [47]) [47]

Metals, ions

Fe^{2+} <1, 3-20, 22-24, 27-33> (<1> very specific requirement, Fe^{2+} is apparently not firmly bound, the enzyme may bind 4 mol of Fe^{2+} at its maximum activity, there is a positive co-operativity in this binding, binding may occur to one or more SH-groups [1]; <3> purified enzyme does not require exogenous Fe^{2+} to obtain full enzymic activity, the iron in the holomeric placental enzyme appears to be more tightly bound than in the chick embryo enzyme [2]; <1> bound: 2 mol/mol [3,49]; <1> 1 mol/mol, firmly bound, required for catalytic activity, the iron is not part of a 2Fe-2S or a 4Fe-4S cluster [19]; <1> absolute requirement, K_m: 0.005 mM [28]; <9> K_m: 0.01-0.03 mM [40]; <1> K_m with biological substrate: 0.002 mM, K_m with synthetic substrate: 0.004 mM [1]; <9> K_m: 0.028 mM [40]; <12> K_m: 0.02 mM [44];

<1,5,18> no compound can replace Fe^{2+}, K_m: 0.03 mM [48]; <19> recombinant enzyme: K_m 0.005 mM [59]; <3> enzyme tetramer containing the histidine 501 to serine mutant α subunit: K_m 0.005 mM [61]; <3> type I enzyme tetramer: K_m 0.004 mM [62]; <7> recombinant type II enzyme tetramer: K_m 0.004 mM [62]; <20> the maximal rate of hydroxylation is attained at a ferrous ion concentration of approx. 0.03 mM and proceeds at approx. 8% of the maximal rate in the absence of exogenous iron, further suggesting the presence of residual enzyme-bound iron [63]; <3> pure type II enzyme: K_m 0.006 mM, pure type I enzyme: K_m 0.004 mM [65]; <22> K_m: 0.003 mM [68]; <3> type I enzyme: K_m 0.003 mM, type II enzyme: K_m 0.004 mM [68]; <23> recombinant enzyme: K_m 0.0004 mM [69]; <27, 29-33> K_m: 0.016 mM [73]) [1-3, 6, 7, 9-12, 15, 16, 18-21, 25-30, 33, 35, 37-46, 49, 48, 59-63, 65, 68, 69, 71, 73]
Mg^{2+} <1> (<1> 3% relative activity with respect to Fe^{2+} [48]) [48]
Mn^{2+} <1> (<1> 10% relative activity with respect to Fe^{2+} [48]; <20> no activation, causes a 20-30% fall in activity [63]) [48]

Turnover number (min^{-1})

4.2 <1> (tert-butyloxycarbonyl-Pro-Pro-Gln-Pro-OCH_3) [12]
6.6 <1> (tert-butyloxycarbonyl-Pro-Pro-Ala-Pro-OH) [12]
12 <1> (tert-butyloxycarbonyl-Pro-Pro-Gly-Pro) [12]
22.8 <1> (tert-butyloxycarbonyl-Pro-Pro-Gly-Pro-$NHCH_3$) [12]
47.4 <1> (tert-butyloxycarbonyl-Pro-Pro-Gly-Pro-Pro-OH) [12]
52.8 <1> (tert-butyloxycarbonyl-Val-Pro-Gly-Val-OH) [12]
139.8 <1> (tert-butyloxycarbonyl-Gly-Val-Pro-Gly-Val-OH) [12]
240 <1> (protocollagen) [58]
288 <1> (procollagen L-proline) [20]
Additional information <1> [19]

Specific activity (U/mg)

0.000066 <9> (<9> purified enzyme, 6 M-urea-elution and dialysis [41]) [41]
0.000786 <9> (<9> purified enzyme, poly-L-proline-eluted [41]) [41]
0.00162 <4> (<4> after affinity chromatography, 6 M-urea-eluted [10]) [10]
0.052 <4> (<4> after affinity chromatography, poly-L-proline-eluted [10]) [10]
0.0714 <8> (<8> homogenate [39]) [39]
0.13 <1> (<1> reaction mixture lacking catalase [19]) [19]
0.138 <8> (<8> 15000 g pellet [39]) [39]
0.162 <8> (<8> microsomal fraction, rough membrane [39]) [39]
0.2 <1> (<1> reaction mixture lacking serum albumin [19]) [19]
0.27 <8> (<8> 1000 g pellet [39]) [39]
0.6 <1> (<1> reaction mixture lacking dithiothreitol [19]) [19]
0.63 <8> (<8> microsomal fraction [39]) [39]
0.66-1.53 <1> (<1> purified enzyme [33]) [33]
0.75 <1> (<1> reaction mixture lacking Fe^{2+} [19]) [19]
1.02-1.38 <3> (<3> purified enzyme [34]) [34]
1.295 <5> (<5> purified enzyme [31]) [31]
1.3 <3, 5> (<3,5> purified enzyme [25]) [25]

1.5 <1> (<1> purified enzyme [25]) [25]
1.668 <8> (<8> microsomal fraction, smooth membrane [39]) [39]
1.8-2.6 <1> [19]
2.328 <8> (<8> purified enzyme, 6 M-urea eluted, then dialysed [39]) [39]
2.4 <1> (<1> ascorbate added before the addition of 2-oxoglutarate [18]) [18]
77.8 <3> [34]
87.53 <8> (<8> purified enzyme, poly-L-proline-eluted [39]) [39]
Additional information <1, 3, 5, 6, 8, 12-15, 17, 22> (<1,3,5> assay method [19,25]; <1> effect of preincubation with 2-oxoglutarate on the rate and extent of oxygen uptake by the ascorbate-independent enzyme activity [18]; <1> enzyme activity after dialysis against iron chelators [19]; <5> specific activity of the purified β-subunit [25]; <1> specific activity of the enzyme in the presence and absence of dithiothreitol [27]; <1> activity before and after treatment with protein disulfide-isomerase [47]; <3> enzyme activity of Triton X-100 extracts from cells expressing various mutant α subunits together with the wild-type protein disulfide-isomerase/β subunit [61]; <3> activities of recombinant type II and I enzymes in Triton X-100 extracts from insect cells [65]; <22> enzyme activity of mutants [68]) [18, 19, 25, 26, 27, 33, 38, 39, 44, 46, 47, 58, 61, 65, 68]

K$_m$-Value (mM)

0.00000001 <1, 3, 5> (protocollagen) [25]
0.0000016 <1> (procollagen L-proline) [20]
0.000002 <1> (protocollagen, <1> random-coil from of substrate [58]) [58]
0.0000024 <1> (protocollagen) [51]
0.0000025 <1> (lysine hydroxylated protocollagen) [51]
0.0002 <27, 29-33> (poly(L-proline), <27> MW: 10000-20000 [73]) [73]
0.0002 <3> (protocollagen, <3> type I enzyme [65]) [65]
0.0002 <1, 3> (protocollagen type I) [3]
0.0011 <3> (protocollagen, <3> type II enzyme [65]) [65]
0.002 <27, 29-33> ((Pro-Ala-Pro-Lys)10) [73]
0.002 <9> (poly(L-Pro), <9> MW 7000 [41]) [41]
0.002 <27, 29-33> (poly(L-proline), <27> MW: 5000-10000 [73]) [73]
0.002 <27, 29-33> (poly(L-proline), <27,29-33> MW: 5000-10000 [73]) [73]
0.004 <27, 29-33> ((Pro-Glu-Pro-Pro-Ala)5) [73]
0.005 <1> (2-oxoglutarate, <1> biological substrate [1]) [1]
0.005 <8> (poly(L-Pro), <8> MW 30000 [39]) [39]
0.005 <6> (protocollagen) [26]
0.007 <9> (poly(L-Pro), <9> MW 31000 [40]) [40]
0.008 <1> (2-oxoglutarate) [19]
0.01 <27, 29-33> (Ser-Pro-Pro-Pro-Pro-Val-Ser-Pro-Pro-Pro-Val-Ser-Pro-Pro-Pro-Pro-Val) [73]
0.01 <12> (poly(L-proline)) [44]
0.01-0.03 <1, 3, 5> ((Pro-Pro-Gly)$_{10}$) [25]
0.011 <9> (poly(L-Pro), <9> MW 19000 [40]) [40]
0.012 <7> (2-oxoglutarate, <7> recombinant type II enzyme tetramer [62]) [62]

0.015 <19> ((Pro-Pro-Gly)$_{10}$, <19> recombinant enzyme [59]) [59]
0.018 <3> ((Pro-Pro-Gly)$_{10}$, <3> wild-type enzyme tetramer and enzyme tetramer containing the histidine 501 to serine mutant α subunit [61]; <3> type I enzyme tetramer [62]; <3> type I enzyme [65]) [61, 62, 65]
0.02 <27, 29-33> ((Ala-Thr-Pro-Pro-Pro-Val)$_3$) [73]
0.02 <23> ((Pro-Ala-Pro-Lys)$_{10}$, <23> recombinant enzyme [69]) [69]
0.02 <27, 29-33> ((Ser-Pro-Lys-Pro-Pro)$_5$) [73]
0.02 <23> ((Ser-Pro-Lys-Pro-Pro)$_5$, <23> recombinant enzyme [69]) [69]
0.02 <1> (2-oxoglutarate, <1> partial reaction or complete reaction with in the presence of (Pro-Pro-Gly)$_n$ as a peptide substrate [15]) [15]
0.02 <19> (2-oxoglutarate, <19> recombinant enzyme [59]) [59]
0.02 <23> (2-oxoglutarate, <23> recombinant enzyme [69]) [69]
0.021 <3> ((Pro-Pro-Gly)$_{10}$, <3> type I enzyme [68]) [68]
0.022 <1> (2-oxoglutarate, <1> synthetic substrate [1]) [1]
0.022 <3> (2-oxoglutarate, <3> type I and type II enzymes [68]) [68]
0.022 <1, 3> (2-oxoglutarate, <3> type I enzyme tetramer [62]; <3> type I and type II enzyme [65]) [8, 62, 65]
0.023 <9> (poly(L-Pro), <9> MW 7000 [40]) [40]
0.03 <9> (2-oxoglutarate) [40]
0.03 <1, 5, 18> (O_2) [48]
0.032 <20> (2-oxoglutarate) [63]
0.04 <27, 29-33> (Ser-Pro-Pro-Pro-Val-Tyr-Lys-Ser-Pro-Pro-Pro-Pro-Val-Lys-His-Tyr-Ser-Pro-Pro-Pro-Val) [73]
0.043 <1> (O_2, <1> synthetic substrate [1]) [1]
0.045 <7> ((Pro-Pro-Gly)$_{10}$, <7> recombinant type II enzyme tetramer [62]) [62]
0.05 <23> ((Pro-Ala-Pro-Lys)$_5$, <23> recombinant enzyme [69]) [69]
0.05 <1, 3> ((Pro-Pro-Gly)$_{10}$) [3]
0.05 <3> (2-oxoglutarate, <3> wild-type enzyme tetramer [61]) [61]
0.06 <27, 29-33> ((Pro-Pro-Gly)$_{10}$) [73]
0.06 <10> (O_2) [42, 43]
0.06 <10> (O_2) [42, 43]
0.084 <22> (2-oxoglutarate, <22> wild-type enzyme [68]) [68]
0.088 <3> ((Pro-Pro-Gly)$_{10}$, <3> type I enzyme [68]) [68]
0.09 <27, 29-33> ((Pro-Ala-Pro-Lys)$_3$) [73]
0.095 <3> ((Pro-Pro-Gly)$_{10}$, <3> type II enzyme [65]) [65]
0.1 <27, 29-33> ((Ala-Pro-Gly)$_5$) [73]
0.1 <23> (poly(L-proline), <23> MW: 40000, recombinant enzyme [69]) [69]
0.1-0.5 <1, 3, 5> ((Pro-Pro-Gly)$_5$) [25]
0.1-0.5 <6> (carboxymethylated collagen) [26]
0.106 <22> (2-oxoglutarate, <22> R490H mutant [68]) [68]
0.106 <22> (2-oxoglutarate, <22> R490S mutant [68]) [68]
0.12 <27, 29-33> ((Pro-Pro-Gly)$_5$) [73]
0.13 <27, 29-33> (2-oxoglutarate) [73]
0.14 <1> (2-oxoadipinate, <1> as cosubstrate [16]) [16]
0.15 <3> (2-oxoglutarate, <3> enzyme tetramer containing the histidine 501 to serine mutant α subunit [61]) [61]

0.17 <4> ((Pro-Pro-Gly)$_{10}$) [10]
0.17 <12> (2-oxoglutarate) [44]
0.19 <1> (O_2, <1> complete reaction with (Pro-Pro-Gly)$_5$ as a peptide substrate [15]) [15]
0.23 <10> (poly(L-proline)) [42, 43]
0.26 <22> ((Pro-Pro-Gly)$_{10}$) [68]
0.28 <27, 29-33> ((Pro-Ala-Gly)$_5$) [73]
0.31 <23> ((Pro-Ala-Pro-Lys)3, <23> recombinant enzyme [69]) [69]
0.4-0.5 <1> ((Pro-Pro-Gly)$_5$) [19]
0.445 <20> (L-proline) [63]
0.5 <23> (poly(L-proline), <23> MW: 13000, recombinant enzyme [69]) [69]
0.95 <23> ((Pro-Ala-Pro-Lys)$_2$, <23> recombinant enzyme [69]) [69]
1 <23> ((Pro-Glu-Pro-Pro-Ala)$_5$, <23> recombinant enzyme [69]) [69]
1.5 <1> (O_2, <1> partial reaction [15]) [15]
1.8 <1, 3> ((Pro-Pro-Gly)$_5$, <1> O_2 with cosubstrate polyprolin [15]) [3, 15]
2.9 <23> ((Pro-Pro-Gly)$_{10}$, <23> recombinant enzyme [69]) [69]
4.8 <23> (Pro-Ala-Pro-Lys, <23> recombinant enzyme [69]) [69]
7.8 <3> (Gly-Val-Pro-Gly-Val, <3> type I enzyme [65]) [65]
8.6 <23> (Lys-Pro-Ala, <23> recombinant enzyme [69]) [69]
10.7 <1> (tert-butyloxycarbonyl-Val-Pro-Gly-Val-OH) [12]
20 <1, 3> (Pro-Pro-Gly) [3]
22 <3> (Gly-Val-Pro-Gly-Val, <3> type II enzyme [65]) [65]
22.2 <1> (tert-butyloxycarbonyl-Gly-Val-Pro-Gly-Val-OH) [12]
24 <1> (tert-butyloxycarbonyl-Pro-Pro-Ala-Pro-OH) [12]
32.2 <1> (tert-butyloxycarbonyl-Pro-Pro-Gln-Pro-OCH_3) [12]
37.2 <1> (tert-butyloxycarbonyl-Pro-Pro-Gly-$NHCH_3$) [12]
37.4 <1> (tert-butyloxycarbonyl-Pro-Pro-Gly-Pro-Pro-OH) [12]
38.9 <1> (tert-butyloxycarbonyl-Pro-Pro-Gly-Pro-OH) [12]
Additional information <1, 3, 6, 27, 29-33> (<1,3> effect of amino acids in other parts of the peptide chain, effect of peptide chain length and asymmetry in the hydroxylation of (Pro-Pro-Gly)$_n$ [1]; <1> effect of photoaffinity labeling with N-(4-azido-2-nitrophenyl)glycyl-(Pro-Pro-Gly)$_5$, substrates: (Pro-Pro-Gly)$_5$ or procollagen [4]; <1> K_m decreases markedly with chain-length, it remains the same for the triple-helical conformation and for the random-coil form, and decreases by the presence of a basic amino acid in the NH_2^-terminal end [37]; <27,29-33> K_m values of enzyme for peptides representing transcription factor HIF α sequences and containing only one proline residue [73]) [1, 4, 6, 26, 29, 35, 37, 73]

K_i-Value (mM)

0.0000015 <1> (poly(ADP-ribose)) [13]
0.0000162 <1> (phosphoribosyl adenosine monophosphate) [13]
0.0000162 <1> (ribosyl-ribosyl-adenine) [13]
0.0000166 <1> (8-(N-butyl-N-ethylcarbamoyl)-1,4-dihydrophenanthrolin-4-one-3-carboxylic acid, <1> purified enzyme [70]) [70]
0.00002 <1,3> (poly(L-proline), <1,3> molecular weight of 15000 [3]) [3]

0.00002 <3> (poly(L-proline), <3> MW: 44000 Da, human type I enzyme [68]) [68]
0.00002 <3> (poly(L-proline), <3> MW: 44000 Da, type I enzyme [65]) [65]
0.00002 <3> (poly(L-proline), <3> MW: 44000 Da, type I enzyme tetramer [62]) [62]
0.0004 <20> (oxalylglycine) [63]
0.0005 <3> (poly(L-proline), <3> MW: 7000 Da, type I enzyme [65]) [65]
0.0005 <3> (poly(L-proline), <3> MW: 7000 Da, type I enzyme tetramer [62]) [62]
0.0006 <1,3> (Zn^{2+}) [3,28]
0.0006 <3> (poly(L-proline), <3> MW: 7000 Da, human type I enzyme [68]) [68]
0.00065 <1> (1,4-dihydrophenanthroline-4-one-3-carboxylic acid, <1> purified enzyme [70]) [70]
0.0008 <1,3> (pyridine-2,5-dicarboxylate) [3,8,16]
0.001 <7> (pyridine 2,4-dicarboxylate, <3> recombinant type II enzyme tetramer [62]) [62]
0.0019-0.007 <1> (oxalylglycine) [60]
0.002 <1> (pyridine 2,4-dicarboxylate) [15]
0.002 <3> (pyridine 2,4-dicarboxylate, <3> type I enzyme tetramer [62]) [62]
0.002 <3> (pyridine 2,4-dicarboxylate, <3> wild-type enzyme tetramer [61]) [61]
0.002 <1,3> (pyridine-2,4-dicarboxylate) [3,8,16]
0.0029 <22> (poly(L-proline), <22> MW: 44000 Da [68]) [68]
0.005 <1> (3,4-dihydroxybenzoate) [17]
0.005 <1,3> (3,4-dihydroxybenzoate) [8]
0.005 <1> (3,4-dihydroxybenzoate, <1> competitive inhibitor with respect to ascorbate [3]) [3]
0.005 <20> (pyridine 2,4-dicarboxylate) [63]
0.005 <3> (pyridine 2,4-dicarboxylate, <3> enzyme tetramer containing the histidine 501 to serine mutant α subunit [61]) [61]
0.008 <9,12> (pyridine 2,4-dicarboxylate) [40,44]
0.009 <1> (3,4-dihydroxymandelate) [17]
0.009 <1> (3,4-dihydroxyphenylacetate) [17]
0.013 <1> (3,4-dihydroxyphenylpropionate) [17]
0.018 <1> (3,4-dihydroxycinnamate) [17]
0.018 <22> (poly(L-proline), <22> MW: 7000 Da [68]) [68]
0.02 <1> (Fe^{2+}) [32]
0.02 <3> (poly(L-proline), <3> MW: 44000 Da, type II enzyme [65]) [65]
0.02 <9,12> (pyridine 2,5-dicarboxylate) [40,44]
0.022 <3> (poly(L-proline), <3> MW: 44000 Da, human type II enzyme [68]) [68]
0.024 <12> (3,4-dihydroxyphenylpropionate) [44]
0.025 <1> (1,2-dihydroxybenzene) [17]
0.025 <1> (pyridine 2-carboxylate) [16]
0.03 <1> (1,2,3-trihydroxybenzene) [17]
0.03 <1> (3,4,5-trihydroxybenzoate) [17]

0.03 <7> (poly(L-proline), <3> MW: 44000 Da, recombinant type II enzyme tetramer [62]) [62]
0.032 <20> (3,4-dihydroxybenzoate) [63]
0.04 <12> (3,4-dihydroxyphenylacetate) [44]
0.04 <1,3> (doxorubicin, <1,3> competitive inhibitor with respect to 2-oxoglutarate, in concentrations not exceeding 0.02 mM, competitive inhibition is also found with respect to ascorbate [7]) [7]
0.04 <1> (nitroblue tetrazolium) [28]
0.04 <1> (oxalylalanine) [60]
0.0403 <1> (2,7,8-trihydroxyanthraquinone) [11]
0.049 <20> (pyridine 2,5-dicarboxylate) [63]
0.05 <9> (pyridine 2-carboxylate) [40]
0.07 <1> (2,3-dihydroxybenzoate) [17]
0.07 <9> (3,4-dihydroxybenzoate) [40]
0.09 <1> (5-azidopyridine-2-carboxylic acid) [5]
0.095 <3> (poly(L-proline), <3> MW: 7000 Da, type II enzyme [65,68]) [65,68]
0.1 <1> (2-oxosuccinate) [16]
0.1 <1> (epinephrine) [28]
0.2 <9> (2-oxoadipinate) [40]
0.21 <1> (pyridine 2,3-dicarboxylate) [16]
0.3 <7> (poly(L-proline), <3> MW: 7000 Da, recombinant type II enzyme tetramer [62]) [62]
0.4 <9> (2-oxosuccinate) [40]
0.4 <1> (succinate) [16]
0.4-0.5 <1> (citrate, <1> calculated from the slopes or intercepts [28]) [28]
0.5 <1> (pyridine 3-carboxylate) [16]
0.5-0.6 <1> (fumarate, <1> calculated from the slopes or intercepts [28]) [28]
0.53 <1> (2,4-dihydroxybenzoate) [17]
0.6 <1> (Fe^{2+}) [28]
0.6 <1> (oxaloacetate) [28]
0.8 <1> (3-oxoglutarate) [16]
0.9 <1> (2-oxoadipinate) [16]
0.9 <1> (pyridine 3,5-dicarboxylate) [16]
0.9 <1> (pyridine 4-carboxylate) [16]
1 <1> (oxalylsarcosine) [60]
1.1 <1> (2,6-dihydroxybenzoate) [17]
1.2-1.3 <1> (malate, <1> calculated from the slopes or intercepts [28]) [28]
1.25 <1> (oxalylcystine) [60]
1.4 <1> (2,5-dihydroxybenzoate) [17]
1.4 <1> (oxalyl-β-alanine) [60]
1.4-1.6 <1> (succinate, <1> calculated from the slopes or intercepts [29]) [29]
1.5 <9> (glutarate) [40]
1.5-2.1 <1> (isocitrate, <1> calculated from the slopes or intercepts [28]) [28]
1.6 <9> (succinate) [40]
2.2 <1> (pyridine 3,4-dicarboxylate) [16]
2.3 <1> (2-hydroxybenzoate) [17]

2.3-2.5 <1> (poly(L-proline), <1> calculated from the slopes or intercepts [29]) [29]
3 <12> (poly(L-hydroxyproline)) [44]
3.2 <1> (4-hydroxybenzoate) [17]
3.2 <1,3> (4-hydroxybenzoate) [8]
3.3 <1> (benzene 1,4-dicarboxylate) [16]
4.2 <1> (glutarate) [16]
5.8 <1> (benzene 1,3-dicarboxylate) [16]
5.9 <1> (malonate) [16]
7 <1> (ascorbate) [28]
7.6 <1> (2-oxobutyrate) [16]
9 <1> (3,5-dihydroxybenzoate) [17]
9.4 <1> (2-oxovalerate) [16]
9.9 <1> (benzoate) [16]
10 <1> (benzene 1,2-dicarboxylate, <1> value above [16]) [16]
10 <1> (oxalylproline, <1> value above [60]) [60]
10 <1> (oxalylvaline, <1> value above [60]) [60]
10.2 <1> (1,3-dihydroxybenzene) [17]
11 <9> (benzene 1,3-dicarboxylate) [40]
14 <1> (1,4-dihydroxybenzene) [17]
15 <1> (adipinate, <1> value above [16]) [16]
15 <1> (levulinate, <1> value above [16]) [16]
15 <1> (pyruvate) [16]
15-18 <1> (pyruvate, <1> calculated from the slopes or intercepts [28]) [28]
17 <1,3> (coumalic acid, <1,3> S-parabolic competitive inhibition with respect to 2-oxoglutarate [8]) [8]
20 <1> ((+)-mandelate, <1> value above [17]) [17]
20 <1> ((-)-mandelate, <1> value above [17]) [17]
20 <1> (3-hydroxybenzoate, <1> value above [17]) [17]
20 <1> (L-galactono γ-lactone, <1> value above [17]) [17]
20 <1> (phenylacetate, <1> value above [17]) [17]
20 <9> (pyruvate) [40]
40 <1> (collagen, <1> from basement membranes, calculated from the slopes or intercepts [29]) [29]
170-180 <1> (collagen, <1> from skin, calculated from the slopes or intercepts [29]) [29]

pH-Optimum

6.6-6.8 <12> [44]
6.8 <10> [42, 43]
6.8-6.9 <9> [40]
7 <9, 23> [41, 69]
7.2 <6> [26]
7.4 <1, 3, 5> (<1> assay at [12]) [12, 19, 20, 25]
7.5 <11> [42]
7.6 <1> (<1> assay at [11]) [11]
7.7 <1> (<1> assay at [15,18,20]) [15, 18, 20]

7.8 <1> (<1> assay at [16,17]) [16, 17]
7.8-8.3 <4> [10]

Temperature optimum (°C)

20 <4> [10]
26 <20> (<20> assay at [63]) [63]
30 <1, 9, 5, 18, 27, 29-33> (<27,29-33> assay at [73]) [40, 41, 48, 73]
32 <1> (<1> assay at [20]) [20]
37 <1, 3, 5, 6, 24, 28> (<1,3,5,24,28> assay at [4, 6-8, 11, 12, 15-17, 19, 21, 23, 24, 27, 29, 33, 35, 37, 38, 60, 71]) [4, 6-8, 11, 12, 15-17, 19, 21, 23, 24, 26, 27, 29, 30, 33, 35, 37, 38, 60, 71]
40 <12> [44]

4 Enzyme Structure

Molecular weight

27200 <23> (<23> recombinant enzyme, calculated with the N-terminal histidine tag and the thrombin cleavage site [69]) [69]
30000 <23> (<23> recombinant enzyme, gel filtration [69]) [69]
37700 <20> (<20> gel filtration [63]) [63]
40000 <9, 12, 13> (<9,12,13> gel filtration [40,44]) [40, 44]
85000-105000 <7> (<7> inactive precursor of the enzyme, gel filtration [36]) [36]
120000 <19> (<19> recombinant enzyme, gel filtration [59]) [59]
200000 <1> (<1> gel filtration [38]) [38]
230000-240000 <1, 3, 4> (<1> sedimentation equilibrium centrifugation [1,25,33,35]; <4> gel filtration [10]; <3,5> gel filtration [1,25,34]) [1, 10, 25, 33-35]
250000 <9> (<9> native gel electrophoresis [41]) [41]
260000-300000 <7> (<7> gel filtration [36]) [36]
285000 <6> (<6> gel filtration [26]) [26]
300000 <8> (<8> gel filtration [39]) [39]
350000 <1> (<1> gel filtration [33,35]) [33, 35]
400000-600000 <1, 3> (<1,3> value depending on salt concentration of buffer [2]) [2]

Subunits

? <8> (<8> x * 65000, SDS-PAGE [39]) [39]
dimer <10, 19> (<19> recombinant enzyme, α β, ratio: 1 to 1 [59]) [42, 59]
monomer <12, 20, 23, 27, 29-33> (<12> the algal enzyme is clearly structurally related to the α subunit of the vertebrate enzyme [3]; <12> 1 * 65000, SDS-PAGE [44]; <20> 1 * 35000, SDS-PAGE [63]; <23> recombinant enzyme, SDS-PAGE [69]) [3, 44, 63, 69, 73]
tetramer <1, 3-7, 9, 22, 26> (<1,3,5> α_2 β_2, α: 64000, β: 60000, ratio 1 to 1, SDS-PAGE [1,2,9,22,25,31,33,35]; <1,3> the tetramer appears to contain one active site per pair of dissimilar subunits, the 2-oxoglutarate and peptide binding sites of the enzyme are located on the α subunit, whereas the ascor-

bate binding site may be built up of both α and β subunits [3]; <1,3,7> there are two forms of the α subunit [3,9,14]; <3> characterization of the β subunit, the β subunit is a multifunctional polypeptide, having disulfide isomerase activity [3]; <3> α: 64000, β: 62000, ratio 1 to 2 [2]; <9> 2 * 60000 + 2 * 65000 [41]; <6> 4 * 67000, SDS-PAGE [26]; <1> fibroblasts, SDS-PAGE, due to a larger carbohydrate content the α-subunit also exists in larger forms [9]; <3> liver, SDS-PAGE, due to a larger carbohydrate content the α-subunit also exists in larger forms [9]; <7> newborn, SDS-PAGE, due to a larger carbohydrate content the α-subunit also exists in larger forms [9]; <1,3> the 2-oxoglutarate-binding site of the enzyme is located within the α-subunit [8]; <4> α_2 β_2, α: 60000, β: 57000, ratio 1.3 to 4, SDS-PAGE [10]; <1> 2 * 60000 + 2 * 64000, SDS-PAGE [11]; <1> 2 * 58000 + 2 * 62000, SDS-PAGE [19]; <3> 2 * 61000 + 2 * 640000 [34]; <8> doublet with MW of 65000, SDS-PAGE [39]; <9> 2 * 60000 + 2 * 65000, SDS-PAGE [41]; <3> type II tetramer represents about 30% of the total enzyme [65]; <1> type II tetramer represents about 5-15% in various tissues [65]; <22> the α subunit forms enzyme α_2 β_2 tetramers with the Drosophila and human protein-disulfide isomerase polypeptides, nondenaturing-PAGE and Coomassie Blue-staining [68]; <26> α_2 β_2, α: 65000, β: 60000, SDS-PAGE [72]) [1-3, 8-11, 14, 19, 22, 25, 26, 31, 33-35, 39, 41, 65, 68, 72]

Additional information <1, 3, 22> (<1,3> structure of the tetramer, physicochemical properties of the subunits [1,2]; <1> about 65% of the enzyme is present in the form of active enzyme tetramers, and about 35% in a form corresponding in molecular weight to the enzyme monomers when studied by gel filtration. The monomer-size protein in the cell represents, at least in part, precursors of the enzyme tetramers, and it can be associated to active tetramers after its ribosomal biosynthesis [27]; <1> the enzyme is dissociated into both monomers and dimers by either dithiothreitol or mercaptoethanol, indicating that the structural integrity of the enzyme is maintained in part by either intrachain or interchain disulfide bonds [35]; <22> amino acid sequence of the α subunit and its comparison with those of the human α-I and α-II subunits and the Caenorhabditis elegans α subunit [68]) [1, 2, 27, 35, 68]

Posttranslational modification

glycoprotein <1, 3> (<1,3> carbohydrate composition [2,9]; <1> 35.4 residues mannose and 3.9 residues N-acetyl-glucosamine per 240 kDa [2]; <3> 29.8 residues mannose and 3.8 residues N-acetyl-glucosamine per 240 kDa [2]; <1> the two forms of the α-subunit differ in carbohydrate content, the larger α subunit, α', contains two N-linked high mannose oligosaccharides, each containing eight mannose units, the smaller subunit, α, contains a single seven-mannose N-linked oligosaccharide [14]; <1> subunit α: contains 16 residues of mannose, 1 residue of galactose and at least 2 residues of N-acetylglucosamine, subunit β: contains 2 residues of mannose and 3 residues of galactose [22,25]) [1, 2, 9, 14, 22, 25, 29]

5 Isolation/Preparation/Mutation/Application

Source/tissue

aorta <1> (<1> embryo [23]) [23]
bone <1, 7> (<1> embryo [23,60]; <7> represents about 45% of total enzyme activity [66]) [23, 60, 66]
brain <3, 7> (<7> expression of α-I and α-II subunits mRNAs [62]; <3> expression of enzyme α-I subunit mRNA [65]) [62, 65]
calvarium <1> (<1> embryo [60,65]) [60, 65]
cartilage <1, 7> (<1> embryo [1,23]; <7> represents 80% of total enzyme activity, the type II enzyme represents the main enzyme form [66]) [1, 23, 66]
cell suspension culture <8, 10> [39, 43]
chondrocyte <7> (<7> represents at least about 70% of the total enzyme activity, the type II enzyme represents the main enzyme form [66]) [66]
corneal endothelium <26> [72]
cranium <1> (<1> embryo [23]) [23]
cuticle <17> [46]
egg <2> (<2> developing [52]) [52]
embryo <1> (<1> tendon cell, cartilage, skin, spleen [1]; <1> leg [11,70]) [1, 2, 6-9, 11-20, 22, 23, 25, 27-30, 32, 33, 35, 38, 53, 47, 48, 51, 58, 60, 65, 70]
epithelium <17> (<17> subcuticular [55]) [55]
femur <1> (<1> embryo, high enzyme activity [23]) [23]
fetus <3, 5> (<3> skin [1]; <3> skin, mixture of fetal tissues [34]; <3> foot [66]) [1, 21, 34, 66]
fibroblast <3, 7> (<3> adult and fetal skin, the type I enzyme is the main enzyme form [66]; <3> foreskin [70]) [36, 65, 66, 70]
foot <3> (<3> fetus [66]) [66]
foot muscle <4> [10]
heart <1, 3, 7> (<1> embryo [23]; <7> expression of enzyme α-I and α-II subunits mRNAs [62]; <3> expression of enzyme α-I and α-II subunit mRNAs [65]; <7> the type I enzyme is the main enzyme form [66]) [23, 62, 65, 66]
kidney <1, 3, 5, 7> (<3> high expression of enzyme α-I subunit mRNA and low expression of enzyme α-II subunit mRNA [65]; <7> the type I enzyme is the main enzyme form [66]) [23, 24, 51, 65, 66]
larva <22> [68]
leg <1> (<1> embryo [11]) [11]
lens <1> (<1> embryo, lowest enzyme activity [23]) [23]
liver <1, 3, 5, 7> (<5> lowest enzyme activity [23]; <3> high expression of enzyme α-I subunit mRNA, low expression of enzyme α-II subunit mRNA [65]; <7> the type I enzyme is the main enzyme form [66]) [1, 9, 23, 31, 65, 66]
lung <1, 3, 5, 7> (<5> highest enzyme activity [23]; <1> embryo [60]; <7> expression of α-I and α-II subunits mRNAs [62]; <3> expression of enzyme α-I and α-II subunit mRNAs [65]) [23, 60, 62, 65]

pancreas <3> (<3> expression of enzyme α-I and α-II subunits mRNAs [65]) [65]
placenta <3> (<3> expression of enzyme α-I and α-II subunits mRNAs [65]) [2, 6, 7, 9, 65]
seed <21> [67]
skeletal muscle <3, 7> (<3> high expression of enzyme α-I subunit mRNA and low expression of enzyme α-II subunit mRNA [65]; <7> the type I enzyme is the main enzyme form [66]) [65, 66]
skin <1, 3, 5> (<1> embryo [1,23]; <3> fetus, adult [34]; <5> highest enzyme activity [23]; <5> new born [25,46]; <3,5> fetus [34,48]) [1, 25, 34, 23, 46, 48]
spleen <1> (<1> embryo [1]) [1]
sternum <1> (<1> embryo, high enzyme activity [23]) [23, 51, 65]
tendon <1> (<1> embryo, highest activity [23]; <1> embryo [1,27,51,70]) [1, 14, 23, 27, 51, 65, 70]
tibia <1> (<1> embryo, high enzyme activity [23]; <1> embryo [32]) [23, 32]
uterus <5> [70]
vascular system <3> (<3> type II enzyme represents the main or only enzyme form [66]) [66]
whole body <5> (<5> newborn [31]) [31]
Additional information <5> (<5> the enzyme activity decreases with age in all the tissues studied [23]) [23]

Localization

Golgi apparatus <8, 11, 21> (<11> 70% of the enzyme activity [42]) [42, 54, 67]
endoplasmic reticulum <1, 3, 5, 8, 11, 21, 26> (<1,3> luminal site [2]; <1,3,5> cisternae [25]; <11> 30% of the enzyme activity [42]; <26> co-localization with procollagen I: perinuclear site [72]) [2, 14, 25, 39, 42, 67, 72]
microsome <1, 10> (<10> 40.7% of total activity [43]; <1> embryonic bone [60]) [43, 47, 60]
microsome <9> (<9> loosely associated to membrane [41]) [41]
mitochondrion <10> (<10> 47.9% of total activity [43]) [43]
soluble <10> (<10> 11.4% of total activity [43]) [43]
Additional information <1> (<1> subcellular distribution, differential centrifugation [38]) [38]

Purification

<1> (using fractionating precipitation, ion-exchange chromatography and gel filtration [1,25]; using affinity chromatography on a column containing a polypeptide substrate of the enzyme linked to agarose, elution of the enzyme with a second peptide substrate and separation of the enzyme from this peptide by gel filtration [1, 25, 35]; using affinity chromatography on a column containing poly-(L-proline) linked to agarose, elution with the same polypeptide of a lower molecular weight, and gel filtration [1,25,33]; isolation of subunits [9]; isolation of subunits by ion exchange chromatography on DEAE-cellulose in 8 M urea [22]; using chromatography on DEAE-cellulose column, affinity column, anion-exchange chromatography and a second chromatography on DEAE-cellulose column [2]; using ammonium sulfate fractionation,

affinity chromatography on poly(L-proline) coupled to Sepharose 4B, DEAE-cellulose chromatography, and gel filtration [6-8]; using affinity chromatography, ion-exchange chromatography and gel filtration [9,12]; using ammonium sulfate precipitation, calcium phosphate batch fractionations and anion exchange chromatography on DEAE-agarose followed by gel filtration on agarose [13]; using affinity chromatography and column chromatography on DEAE-cellulose [14]; using ammonium sulfate fractionation, fractionation with calcium phosphate gel, fractionation with alumina gel, chromatography on DEAE-Sephadex and polyacrylamide gel filtration [38]; using ammonium sulfate fractionation, affinity chromatography on poly-L-proline coupled to Sepharose 4B and DEAE-cellulose chromatography [60]) [1, 2, 6-9, 12-14, 19, 22, 25, 33, 35, 38, 53, 60]
<3> (using affinity chromatography on a column containing poly-(L-proline) linked to agarose, elution with the same polypeptide of a lower molecular weight, and gel filtration [1,25,34]; using chromatography on DEAE-cellulose column, affinity column, anion-exchange chromatography and a second chromatography on DEAE-cellulose column [2]; using ammonium sulfate fractionation, affinity chromatography on poly(L-proline) coupled to Sepharose 4B, DEAE-cellulose chromatography, and gel filtration [6-8]; using affinity chromatography, ion-exchange chromatography and gel filtration [9]; of the recombinant enzyme, using an affinity purification procedure based on the use of a histidine tag in the N terminus of the protein disulfide-isomerase/β polypeptide [65]) [1, 2, 6-9, 25, 34, 65]
<4> (using column chromatography on DEAE-cellulose, gel filtration and affinity chromatography [10]) [10]
<5> (using fractionating precipitations, ion-exchange chromatographies and gel filtrations [1,25]; using affinity chromatography on a column containing a polypeptide substrate of the enzyme linked to agarose, elution of the enzyme with a second peptide substrate and separation of the enzyme from this peptide by gel filtration [1,25]; using affinity chromatography on a column containing poly-(L-proline) linked to agarose, elution with the same polypeptide of a lower molecular weight, and gel filtration [1,25]; using ammonium sulfate fractionation and affinity chromatography [21]; separation of 4-hydroxylase activity from 3-hydroxylase activity using affinity chromatography and gel filtration [24]; of the β-subunit using ammonium sulfate fractionation, chromatography on preimmune IgG column and immunoadsorbent column, and gel filtration [25]; using ammonium sulfate precipitation, affinity chromatography on poly(L-proline) column and gel filtration [31]) [1, 21, 24, 25, 31]
<6> (partial, using ammonium sulfate precipitation and calcium phosphate gel ion exchange [26]) [26]
<7> (separation of the enzyme from the inactive precursor of the enzyme, using ammonium sulfate precipitation and column chromatography on DEAE-Sephadex A-50 and Agarose A-1.5m columns [36]) [36]
<8> (using ion-exchange chromatography on DEAE-cellulose and affinity chromatography on a column of poly-L-proline-Sepharose 4B [39]) [39]

<9> (partial, using ion-exchange chromatography on DEAE-cellulose and gel filtration [40]; partial, using ion-exchange chromatography on DEAE-cellulose, affinity chromatography with poly-L-proline coupled to tresyl-activated Sepharose-4B and gel filtration [41]) [40, 41]
<10> (partial, using ammonium sulfate precipitation, column chromatography on DEAE-Sephadex A-50 and affinity chromatography on agarose-linked poly(L-proline) column [42]; using ammonium sulfate precipitation and column chromatography on DEAE-Sephadex A-50 [43]) [42, 43]
<12> (using ion-exchange chromatography on DEAE-cellulose, and affinity chromatography on poly(L-hydroxyproline), 3,4-dihydroxyphenylacetate or 3,4-dihydroxyphenylpropionate linked to Sepharose [44]) [44]
<17> (using ammonium sulfate fractionation, column chromatography on DEAE-cellulose and calcium phosphate gel fractionation [46]) [46, 55]
<19> (of the recombinant enzyme, using anion exchange chromatography on a DEAE-cellulose column and two gel filtration steps [59]) [59]
<20> (using chromatography on DEAE-Sepharose Fast flow column, ammonium sulfate precipitation, chromatography on phenyl-Superose column, second ammonium sulfate precipitation and column chromatography on Superdex G75 HR [63]) [63]
<23> (using Ni^{2+}-chelate affinity chromatography [69]) [69]

Renaturation

<1> (protein disulfide-isomerase may be an enzyme that is able to retain the prolyl 4-hydroxylase structure in its native conformation, but it seems to lack the ability to associate the enzyme monomers into the active tetrameric form [47]) [47]

Cloning

<3> (preparation of cDNA clones for the two subunits, expression in Escherichia coli [9]; α-subunit [56]; β-subunit [57]; expression of histidine and cysteine mutant α subunits together with the wild-type β subunit in insect cells by means of baculovirus vectors [61]; infection of cells with recombinant viruses encoding type XII minicollagen together with various combinations of recombinant viruses encoding the wild-type or mutant enzyme, analysis of the contributions of the various cellular post-translational modifications to the assembly of trimeric type XII minicollagen, study of the conditions required for the production of disulfide-bonded type XII minicollagen in insect cells using the baculovirus expression system [64]; cloning of the α-II subunit, coexpression with the β subunit in insect cells [65]) [9, 56, 57, 61, 64, 65]
<7> (cloning of the α-I subunit and of a second mouse α-subunit isoform, termed the α-II subunit, expression of the α-II subunit together with human protein disulfide isomerase/β subunit in insect cells by baculovirus vectors [62]) [62]
<19> (cloning of the α subunit of the enzyme, expression of the α subunit together with human protein disulfide isomerase/β subunit in insect cells by baculovirus vectors [59]) [59]
<22> (cloning of the α subunit of the enzyme, coexpression in insect cells with the Drosophila protein-disulfide isomerase polypeptide produces an ac-

tive enzyme tetramer, coexpression in insect cells with human protein-disulfide isomerase polypeptide produces also small amounts of a hybrid tetramer [68]) [68]
<23> (expression in Escherichia coli, the genome of virus-1 encodes a 242-amino acid polypeptide that shows a distinct amino acid sequence similarity to the C-terminal half of the catalytic α subunit of animal enzymes, the genome contains many open reading frames for proteins with proline-rich repeats [69]) [69]
<24, 28> (identification of three cDNAs encoding putative α subunits of the enzyme, co-expression of one of these with protein-disulfide isomerase in a catalytically active form in an insect system, genomic structure [71]) [71]
<27, 29-33> (the genome encodes six enzyme-like polypeptides, that show an identity of 21-27% to the catalitically important C-terminal regions of the human enzyme α-I and α-II subunits. Cloning of one of the genes, recombinant enzyme expressed in insect cells and in Escherichia coli [73]) [73]

Engineering

C150S <3> (<3> the mutation has no major effect on tetramer assembly, but the amount of tetramer is slightly reduced, being about 80% of that of the wild-type enzyme [61]) [61]
C486S <3> (<3> the mutation totally prevents tetramer assembly [61]) [61]
C511S <3> (<3> the mutation totally prevents tetramer assembly [61]) [61]
H141S <3> (<3> the mutation has no effect on enzyme activity and does not inhibit tetramer assembly [61]) [61]
H165S <3> (<3> the mutation produces a reduction of about 60% in enzyme activity per unit extractable cell protein relative to that obtained with the wild-type α subunit, the amount of tetramer is reduced by about 20-25%, the K_m values for Fe^{2+}, 2-oxoglutarate, ascorbate and the peptide substrate with the mutant are identical to those with the wild-type enzyme [61]) [61]
H180A <27, 29-33> (<27,29-33> the mutation completely inactivates the enzyme [73]) [73]
H180Q <27, 29-33> (<27,29-33> the mutation completely inactivates the enzyme [73]) [73]
H221S <3> (<3> the mutation produces a reduction of about 30% in enzyme activity per unit extractable cell protein relative to that obtained with the wild-type α subunit, the amount of tetramer is reduced by about 20-25%, the K_m values for Fe^{2+}, 2-oxoglutarate, ascorbate and the peptide substrate with the mutant are identical to those with the wild-type enzyme [61]) [61]
H260A <27, 29-33> (<27,29-33> the mutation completely inactivates the enzyme [73]) [73]
H260Q <27, 29-33> (<27,29-33> the mutation completely inactivates the enzyme [73]) [73]
H296S <3> (<3> the mutation has no effect on enzyme activity and does not inhibit tetramer assembly [61]) [61]
H324S <3> (<3> the mutation totally prevents tetramer assembly [61]) [61]

H412S <3> (<3> the mutation causes a complete inactivation of the enzyme with no effect on tetramer assembly or binding of the tetramer to poly(L-proline), role in the binding of Fe^{2+} to a catalytic site [61]) [61]
H483S <3> (<3> the mutation causes a complete inactivation of the enzyme with no effect on tetramer assembly or binding of the tetramer to poly(L-proline), role in the binding of Fe^{2+} to a catalytic site [61]) [61]
H501S <3> (<3> the mutation reduces the enzyme activity to about 4% with no effect on tetramer assembly or binding of the tetramer to poly(L-proline), role in the binding of Fe^{2+} to a catalytic site, the K_m values for Fe^{2+}, ascorbate and the peptide substrate with the mutant are identical to those with the wild-type enzyme, but the K_m for 2-oxoglutarate is about 2.5fold higher. The main difference is that the V_{max} determined from kinetic plots is consistently less than about 5% of that of the wild-type enzyme [61]) [61]
H63S <3> (<3> the mutation has no effect on enzyme activity and does not inhibit tetramer assembly [61]) [61]
K270A <27, 29-33> (<27,29-33> the mutation completely inactivates the enzyme [73]) [73]
K270R <27, 29-33> (<27,29-33> the mutation completely inactivates the enzyme [73]) [73]
N182A <27, 29-33> (<27,29-33> the mutation completely inactivates the enzyme [73]) [73]
N182Q <27, 29-33> (<27,29-33> the mutation completely inactivates the enzyme [73]) [73]
N96Q/N242Q <3> (<3> the amount of enzyme activity observed with the double mutant α subunit is identical to that of the wild-type enzyme, the size of the double mutant α subunit is distinctly smaller than that of either the diglycosylated or monoglycosylated α subunit present in the wild-type enzyme, the difference being consistent with loss of all the carbohydrate [61]) [61]
R278A <27, 29-33> (<27,29-33> the mutation completely inactivates the enzyme [73]) [73]
R278H <27, 29-33> (<27,29-33> the mutation reduces the enzyme activity to approximately 26% [73]) [73]
R490H <22> (<22> the mutation reduces the percentage of uncoupled decarboxylation [68]) [68]
R490S <22> (<22> the mutation increases the K_m for 2-oxoglutarate, reduces the reaction velocity and increases the percentage of uncoupled decarboxylation [68]) [68]
S272A <27, 29-33> (<27,29-33> the mutation reduces the enzyme activity by 83% [73]) [73]

6 Stability

pH-Stability

7.5 <1, 3> (<1,3> increased spontaneous inactivation at pH values above in the absence of an inactivating compound [8]) [8]

Temperature stability

30 <4> (<4> denaturation above [10]) [10]

General stability information

<1, 3>, unstable in absence of 2-oxoglutarate and either Fe^{2+} or ascorbate [8]
<1, 3, 5>, labile in tissue extracts [25]
<1, 3, 5>, stabilization by NaCl, detergents [25]

Storage stability

<1>, -20°C, in microsomes, less than 10% inactivation, 10 months [47]
<6>, -30°C, half-life of ammonium sulfate precipitated enzyme: 7.3 days, stability of enzyme purified with calcium phosphate gel: 21 days, 20-30% loss of activity [26]
<7>, inactive precursor of the enzyme, 0°C, 50% loss of activity, 1 week [36]
<7>, inactive precursor of the enzyme, liquid N_2, at least 1 month [36]
<10>, -80°C, dialyzed against 25% glycerol, 1 month [43]
<10>, 4°C, unstable [43]
<1, 10>, freeze-thawing, unstable [43, 58]

References

[1] Kivirikko, K.I.; Myllylä, R.: The hydroxylation of prolyl and lysyl residues. Enzymol. Post-transl. Modif. Proteins (Freedman, R.B., Hawkins, H.C., eds.) Academic Press, New York, **1**, 5304 (1980)
[2] Guzman, N.A.; Ascari, W.Q.; Cutroneo, K.R.; Desnick, R.J.: Comparison between avian and human prolyl 4-hydroxylases: studies on the holomeric enzymes and their constituent subunits. J. Cell. Biochem., **48**, 172-189 (1992)
[3] Kivirikko, K.I.; Myllylä, R.; Pihlajaniemi, T.: Protein hydroxylation: prolyl 4-hydroxylase, an enzyme with four cosubstrates and a multifunctional subunit. FASEB J., **3**, 1609-1617 (1989)
[4] De Waal, A.; de Jong, L.: Processive action of the two peptide binding sites of prolyl 4-hydroxylase in the hydroxylation of procollagen. Biochemistry, **27**, 150-155 (1988)
[5] De Waal, A.; Hartog, A.F.; de Jong, L.: Localisation and quantification of the 2-oxoglutarate binding sites of prolyl 4-hydroxylase by photoaffinity labeling with tritiated 5-azidopyridine-2-carboxylic acid. Biochim. Biophys. Acta, **953**, 20-25 (1988)
[6] Gunzler, V.; Brocks, D.; Henke, S.; Myllylä, R.; Geiger, R.; Kivirikko, K.I.: Syncatalytic inactivation of prolyl 4-hydroxylase by synthetic peptides containing the unphysiologic amino acid 5-oxaproline. J. Biol. Chem., **263**, 19498-19504 (1988)
[7] Gunzler, V.; Hanauske-Abel, H.M.; Myllylä, R.; Kaska, D.D.; Hanauske, A.; Kivirikko, K.I.: Syncatalytic inactivation of prolyl 4-hydroxylase by anthracyclines. Biochem. J., **251**, 365-372 (1988)
[8] Gunzler, V.; Hanauske-Abel, H.M.; Myllylä, R.; Mohr, J.; Kivirikko, K.I.: Time-dependent inactivation of chick-embryo prolyl 4-hydroxylase by cou-

malic acid. Evidence for a syncatalytic mechanism. Biochem. J., **242**, 163-169 (1987)
[9] Kivirikko, K.I.; Myllylä, R.: Recent developments in posttranslational modification: intracellular processing. Methods Enzymol., **144**, 96-114 (1987)
[10] Marumo, K.; Waite, J.H.: Prolyl 4-hydroxylase in the foot of the marine mussel Mytilus edulis L.: Purification and characterization. J. Exp. Zool., **244**, 365-374 (1987)
[11] Cunliffe, C.J.; Franklin, T.J.: Inhibition of prolyl 4-hydroxylase by hydroxyanthraquinones. Biochem. J., **239**, 311-315 (1986)
[12] Atreya, P.L.; Ananthanarayanan, V.S.: Interaction of prolyl 4-hydroxylase with synthetic peptide substrates. A conformational model for collagen proline hydroxylation. J. Biol. Chem., **266**, 2852-2858 (1991)
[13] Hussain, M.Z.; Ghani, Q.P.; Hunt, T.K.: Inhibition of prolyl hydroxylase by poly(ADP-ribose) and phosphoribosyl-AMP. Possible role of ADP-ribosylation in intracellular prolyl hydroxylase regulation. J. Biol. Chem., **264**, 7850-7855 (1989)
[14] Kedersha, N.L.; Tkacz, J.S.; Berg, R.A.: Characterization of the oligosaccharides of prolyl hydroxylase, a microsomal glycoprotein. Biochemistry, **24**, 5952-5960 (1985)
[15] De Jong, L.; Kemp, A.: Stoicheiometry and kinetics of the prolyl 4-hydroxylase partial reaction. Biochim. Biophys. Acta, **787**, 105-111 (1984)
[16] Majamaa, K.; Hanauske-Abel, H.M.; Gunzler, V.; Kivirikko, K.I.: The 2-oxoglutarate binding site of prolyl 4-hydroxylase. Identification of distinct subsites and evidence for 2-oxoglutarate decarboxylation in a ligand reaction at the enzyme-bound ferrous ion. Eur. J. Biochem., **138**, 239-245 (1984)
[17] Majamaa, K.; Gunzler, V.; Hanauske-Abel, H.M.; Myllylä, R.; Kivirikko, K.I.: Partial identity of the 2-oxoglutarate and ascorbate binding sites of prolyl 4-hydroxylase. J. Biol. Chem., **261**, 7819-7823 (1986)
[18] De Jong, L.; Albracht, S.P.J.; Kemp.; A.: Prolyl 4-hydroxylase activity in relation to the oxidation state of enzyme-bound iron. The role of ascorbate in peptidyl proline hydroxylation. Biochim. Biophys. Acta, **704**, 326-332 (1982)
[19] Nietfeld, J.J.; Kemp, A.: Properties of prolyl 4-hydroxylase containing firmly-bound iron. Biochim. Biophys. Acta, **613**, 349-358 (1980)
[20] De Jong, L.; van der Kraan, I.; de Waal, A.: The kinetics of the hydroxylation of procollagen by prolyl 4-hydroxylase. Proposal for a processive mechanism of binding of the dimeric hydroxylating enzyme in relation to the high k_{cat}/K_m ratio and a conformational requirement for hydroxylation of -X-Pro-Gly- sequences. Biochim. Biophys. Acta, **1079**, 103-111 (1991)
[21] Takeda, K.; Katoh, F.; Kawai, S.; Konno, K.: Stimulation of prolyl hydroxylase activity by chelating agents. Arch. Biochem. Biophys., **197**, 273-276 (1979)
[22] Berg, R.A.; Kedersha, N.L.; Guzman, N.A.: Purification and partial characterization of the two nonidentical subunits of prolyl hydroxylase. J. Biol. Chem., **254**, 3111-3118 (1979)
[23] Tryggvason, K.; Majamaa, K.; Kivirikko, K.I.: Prolyl 3-hydroxylase and 4-hydroxylase activities in certain rat and chick-embryo tissues and age-related changes in their activities in the rat. Biochem. J., **178**, 127-131 (1979)

[24] Tryggvason, K.; Risteli, J.; Kivirikko, K.I.: Separation of prolyl 3-hydroxylase and 4-hydroxylase activities and the 4-hydroxyproline requirement for synthesis of 3-hydroxyproline. Biochem. Biophys. Res. Commun., **76**, 275-281 (1977)
[25] Kivirikko, K.I.; Myllylä, R.: Posttranslational enzymes in the biosynthesis of collagen: intracellular enzymes. Methods Enzymol., **82**, 245-304 (1982)
[26] Leushner, J.R.A.; Pasternak, J.: Partial purification and characterization of prolyl hydroxylase from the free-living nematode Panagrellus silusiae. Can. J. Zool., **56**, 159-165 (1978)
[27] Tuderman, L.; Oikarinen, A.; Kivirikko, K.I.: Tetramers and monomers of prolyl hydroxylase in isolated chick-embryo tendon cells. The association of inactive monomers to active tetramers and a preliminary characterization of the intracellular monomer-size protein. Eur. J. Biochem., **78**, 547-556 (1977)
[28] Tuderman, L.; Myllylä, R.; Kivirikko, K.I.: Mechanism of the prolyl hydroxylase reaction. 1. Role of co-substrates. Eur. J. Biochem., **80**, 341-348 (1977)
[29] Myllylä, R.; Tuderman, L.; Kivirikko, K.I.: Mechanism of the prolyl hydroxylase reaction. 2. Kinetic analysis of the reaction sequence. Eur. J. Biochem., **80**, 349-357 (1977)
[30] Guzman, N.A.; Berg. R.A.; Prockop, D.J.: Concanavalin A binds of purified prolyl hydroxylase and partially inhibits its enzymic activity. Biochem. Biophys. Res. Commun., **73**, 279-285 (1976)
[31] Risteli, J.; Tuderman, L.; Kivirikko, K.I.: Intracellular enzymes of collagen biosynthesis in rat liver as a function of age and in hepatic injury induced by dimethylnitrosamine. Purification of rat prolyl hydroxylase and comparison of changes in prolyl hydroxylase activity with changes in immunoreactive prolyl hydroxylase. Biochem. J., **158**, 369-376 (1976)
[32] Rapaka, R.S.; Sorensen, K.R.; Lee, S.D.; Bhatnagar, R.S.: Inhibition of hydroxyproline synthesis by palladium ions. Biochim. Biophys. Acta, **429**, 63-71 (1976)
[33] Tuderman, L.; Kuutti, E.R.; Kivirikko, K.I.: An affinity-column procedure using poly(L-proline) for the purification of prolyl hydroxylase. Purification of the enzyme from chick embryos. Eur. J. Biochem., **52**, 9-16 (1975)
[34] Kuutti, L.R.; Tuderman, L.; Kivirikko, K.I.: Human prolyl hydroxylase. Purification, partial characterization and preparation of antiserum to the enzyme. Eur. J. Biochem., **57**, 181-188 (1975)
[35] Berg, R.A.; Prockop, D.J.: Affinity column purification of protocollagen proline hydroxylase from chick embryos and further characterization of the enzyme. J. Biol. Chem., **248**, 1175-1182 (1973)
[36] McGee, J. O'D.; Udenfriend, S.: Partial purification and characterization of peptidyl proline hydroxylase precursor from mouse fibroblasts. Arch. Biochem. Biophys., **152**, 216-221 (1972)
[37] Kivirikko, K.I.; Kishida, Y.; Sakakibara, S.; Prockop, D.J.: Hydroxylation of $(X\text{-}Pro\text{-}Gly)_n$ by protocollagen proline hydroxylase. Effect of chain length, helical conformation and amino acid sequence in the substrate. Biochim. Biophys. Acta, **271**, 347-356 (1972)

[38] Kivirikko, K.I.; Prockop, D.J.: Purification and partial characterization of the enzyme for the hydroxylation of proline in protocollagen. Arch. Biochem. Biophys., **118**, 611-618 (1967)

[39] Bolwell, G.P.; Robbins, M.P.; Dixon, R.A.: Elicitor-induced prolyl hydroxylase from French bean (Phaseolus vulgaris). Localization, purification and properties. Biochem. J., **229**, 693-699 (1985)

[40] Kaska, D.D.; Gunzler, V.; Kivirikko, K.I.; Myllylä, R.: Characterization of a low-relative-molecular-mass prolyl 4-hydroxylase from the green alga Chlamydomonas reinhardii. Biochem. J., **241**, 483-490 (1987)

[41] Blankenstein, P.; Lang, W.C.; Robinson, D.G.: Isolation and characterization of prolyl hydroxylase from Chlamydomonas reinhardii. Planta, **169**, 238-244 (1986)

[42] Chrispeels, M.J.: Prolyl hydroxylase in plants. Methods Enzymol., **107**, 361-369 (1984)

[43] Tanaka, M.; Shibata, H.; Uchida, T.: A new prolyl hydroxylase acting on poly-L-proline, from suspension cultured cells of Vinca rosea. Biochim. Biophys. Acta, **616**, 188-198 (1980)

[44] Kaska, D.D.; Myllylä, R.; Gunzler, V.; Gibori, A.; Kivirikko, K.I.: Prolyl 4-hydroxylase from Volvox carteri. A low-Mr enzyme antigenically related to the α subunit of the vertebrate enzyme. Biochem. J., **256**, 257-263 (1988)

[45] Kuttan, R.; Radhakrishnan, A.N.: Biochemistry of the hydroxyprolines. Adv. Enzymol. Relat. Areas Mol. Biol., **37**, 273-347 (1973)

[46] Adams, E.; Lamon, M.: Prolyl hydroxylase of earthworms. Substrate specificity of an enzyme from the subcuticular epithelium. J. Biol. Chem., **252**, 7591-7597 (1977)

[47] Koivu, J.; Myllylä, R.: Protein disulfide-isomerase retains procollagen prolyl 4-hydroxylase structure in its native conformation. Biochemistry, **25**, 5982-5986 (1986)

[48] Hutton, J.J.; Tappel, A.L.; Udenfried, S.: Cofactor and substrate requirements of collagen proline hydroxylase. Arch. Biochem. Biophys., **118**, 231-240 (1967)

[49] De Waal, A.; de Jong, L.; Hartog, A.F.; Kemp, A.: Photoaffinity labeling of peptide binding sites of prolyl 4-hydroxylase with N-(4-azido-2-nitrophenyl)glycyl-(Pro-Pro-Gly)$_5$. Biochemistry, **24**, 6493-6499 (1985)

[50] De Waal, A.; Hartog, A.F.; de Jong, L.: Photoaffinity labeling of the 2-oxoglutarate binding site of prolyl 4-hydroxylase with 5-azidopyridine-2-carboxylic acid. Biochim. Biophys. Acta, **912**, 151-155 (1987)

[51] Turpeenniemi-Hujanen, T.; Myllylä, R.: Concomitant hydroxylation of proline and lysine residues in collagen using purified enzymes in vitro. Biochim. Biophys. Acta, **800**, 59-65 (1984)

[52] Cain, G.D.; Fairbairn, D.: Protocollagen proline hydroxylase and collagen synthesis in developing eggs of Ascaris lumbricoides. Comp. Biochem. Physiol. B, **40**, 165-179 (1971)

[53] Guzman, N.A.; Oronsky, A.L.; Suarez, G.; Meyerson, L.R.; Cutroneo, K.R.; Olsen, B.R.; Prockop, D.J.: Prolyl 4-hydroxylase from human placenta. Simultaneous isolation of immunoglobulin G which binds to Ascaris cuticle collagen. Collagen Relat. Res., **2**, 365-380 (1982)

[54] Cohen, P.B.; Schibeci, A.; Fincher, G.B.: Biosynthesis of arabinogalactan-protein in Lolium multiflorum (ryegrass) endosperm cells. III. Subcellular distribution of prolyl hydroxylase. Plant Physiol., **72**, 754-758 (1983)
[55] Rao, N.V.; Adams, E.: Fed. Proc., **33**, 1535 (1974)
[56] Helaakoski, T.; Vuori, K.; Myllyla, R.; Kivirikko, K.I.; Pihlajaniemi, T.: Molecular cloning of the a-subunit of human prolyl 4-hydroxylase: the complete cDNA-derived amino acid sequence and evidence for alternative splicing of RNA transcripts. Proc. Natl. Acad. Sci. USA, **86**, 4392-4396 (1989)
[57] Pihlajaniemi, T.; Helaakoski, T.; Tasanen, K.; Myllylä, R.; Huhtala, M.L.; Koivu, J.; Kivirikko, K.I.: Molecular cloning of the β-subunit of human prolyl 4-hydroxylase. This subunit and protein disulfide isomerase are products of the same gene. EMBO J., **6**, 643-649 (1978)
[58] Berg, R.A.; Prockop, D.J.: Purification of (^{14}C)-protocollagen and its hydroxylation by prolyl-hydroxylase. Biochemistry, **12**, 3395-3401 (1973)
[59] Veijola, J.; Koivunen, P.; Annunen, P.; Pihlajaniemi, T.; Kivirikko, K.I.: Cloning, baculovirus expression, and characterization of the a subunit of prolyl 4-hydroxylase from the nematode Caenorhabditis elegans. This a subunit forms an active ab dimer with the human protein disulfide isomerase/b subunit. J. Biol. Chem., **269**, 26746-26753 (1994)
[60] Baader, E.; Tschank, G.; Baringhaus, K.H.; Burghard, H.; Guenzler, V.: Inhibition of prolyl 4-hydroxylase by oxalyl amino acid derivatives in vitro, in isolated microsomes and in embryonic chicken tissues. Biochem. J., **300**, 525-530 (1994)
[61] Lamberg, A.; Pihlajaniemi, T.; Kivirikko, K.I.: Site-directed mutagenesis of the a subunit of human prolyl 4-hydroxylase. Identification of three histidine residues critical for catalytic activity. J. Biol. Chem., **270**, 9926-9931 (1995)
[62] Helaakoski, T.; Annunen, P.; Vuori, K.; MacNeil, I.A.; Pihlajaniemi, T.; Kivirikko, K.I.: Cloning, baculovirus expression, and characterization of a second mouse prolyl 4-hydroxylase a-subunit isoform: formation of an a2b2 tetramer with the protein disulfide-isomerase/b subunit. Proc. Natl. Acad. Sci. USA, **92**, 4427-4431 (1995)
[63] Lawrence, C.C.; Sobey, W.J.; Field, R.A.; Baldwin, J.E.; Schofield, C.J.: Purification and initial characterization of proline 4-hydroxylase from Streptomyces griseoviridis P8648: a 2-oxoacid, ferrous-dependent dioxygenase involved in etamycin biosynthesis. Biochem. J., **313**, 185-191 (1996)
[64] Mazzorana, M.; Snellman, A.; Kivirikko, K.I.; van der Rest, M.; Pihlajaniemi, T.: Involvement of prolyl 4-hydroxylase in the assembly of trimeric minicollagen XII. Study in baculovirus expression system. J. Biol. Chem., **271**, 29003-29008 (1996)
[65] Annunen, P.; Helaakoski, T.; Myllyharju, J.; Veijola, J.; Pihlajaniemi, T.; Kivirikko, K.I.: Cloning of the human prolyl 4-hydroxylase a subunit isoform a(II) and characterization of the type II enzyme tetramer. The a(I) and a(II) subunits do not form a mixed a(I)a(II)b2 tetramer. J. Biol. Chem., **272**, 17342-17348 (1997)
[66] Annunen, P.; Autio-Harmainen, H.; Kivirikko, K.I.: The novel type II prolyl 4-hydroxylase is the main enzyme form in chondrocytes and capillary en-

dothelial cells, whereas the type I enzyme predominates in most cells. J. Biol. Chem., **273**, 5989-5992 (1998)
[67] Wojtaszek, P.; Smith, C.G.; Bolwell, G.P.: Ultrastructural localization and further biochemical characterization of prolyl 4-hydroxylase from Phaseolus vulgaris: comparative analysis. Int. J. Biochem. Cell Biol., **31**, 463-477 (1999)
[68] Annunen, P.; Koivunen, P.; Kivirikko, K.I.: Cloning of the a subunit of prolyl 4-hydroxylase from Drosophila and expression and characterization of the corresponding enzyme tetramer with some unique properties. J. Biol. Chem., **274**, 6790-6796 (1999)
[69] Eriksson, M.; Myllyharju, J.; Tu, H.; Hellman, M.; Kivirikko, K.I.: Evidence for 4-hydroxyproline in viral proteins. Characterization of a viral prolyl 4-hydroxylase and its peptide substrates. J. Biol. Chem., **274**, 22131-22134 (1999)
[70] Franklin, T.J.; Morris, W.P.; Edwards, P.N.; Large, M.S.; Stephenson, R.: Inhibition of prolyl 4-hydroxylase in vitro and in vivo by members of a novel series of phenanthrolinones. Biochem. J., **353**, 333-338 (2001)
[71] Merriweather, A.; Guenzler, V.; Brenner, M.; Unnasch, T.R.: Characterization and expression of enzymatically active recombinant filarial prolyl 4-hydroxylase. Mol. Biochem. Parasitol., **116**, 185-197 (2001)
[72] Ko, M.K.; Kay, E.P.: Subcellular localization of procollagen I and prolyl 4-hydroxylase in corneal endothelial cells. Exp. Cell Res., **264**, 363-371 (2001)
[73] Hieta, R.; Myllyharju, J.: Cloning and characterization of a low molecular weight prolyl 4-hydroxylase from Arabidopsis thaliana. Effective hydroxylation of proline-rich, collagen-like, and hypoxia-inducible transcription factor a-like peptides. J. Biol. Chem., **277**, 23965-23971 (2002)

Pyrimidine-deoxynucleoside 2'-dioxygenase 1.14.11.3

1 Nomenclature

EC number
1.14.11.3

Systematic name
2'-deoxyuridine,2-oxoglutarate:oxygen oxidoreductase (2'-hydroxylating)

Recommended name
pyrimidine-deoxynucleoside 2'-dioxygenase

Synonyms
deoxyuridine 2'-dioxygenase
deoxyuridine 2'-hydroxylase
pyrimidine deoxyribonucleoside 2'-hydroxylase
thymidine 2'-dioxygenase
thymidine 2'-hydroxylase
thymidine 2-oxoglutarate dioxygenase
thymidine dioxygenase
Additional information (cf. EC 1.14.11.10)

CAS registry number
9076-89-5

2 Source Organism

<1> *Rhodotorula glutinis* (strain ATCC 2527 [1]) [1, 2, 4]
<2> *Neurospora crassa* (strain STA 4 [3,5]; strain 1A [6,7,8]; cell-free preparation [8]) [3, 5, 6, 7, 8]

3 Reaction and Specificity

Catalyzed reaction
2'-deoxyuridine + 2-oxoglutarate + O_2 = uridine + succinate + CO_2

Reaction type
oxidation
oxidative decarboxylation
redox reaction
reduction

Natural substrates and products

S thymidine + 2-oxoglutarate + O_2 <2> (<2> conversion to the ribonucleotide prior to its utilization for DNA synthesis [7]) (Reversibility: ? <2> [7]) [7]

P 5-methyluridine + succinate + CO_2 <2> [7]

Substrates and products

S 5-bromodeoxyuridine + 2-oxoglutarate + O_2 <2> (Reversibility: ? <2> [5]) [5]

P 5-bromouridine + succinate + CO_2

S 5-hydroxymethyldeoxyuridine + 2-oxoglutarate + O_2 <2> (Reversibility: ? <2> [5]) [5]

P 5-hydroxymethyluridine + succinate + CO_2

S 6-azathymidine + 2-oxoglutarate + O_2 <2> (Reversibility: ? <2> [5]) [5]

P 5-mehyluridine + succinate + CO_2

S thymidylate + 2-oxoglutarate + O_2 <2> (Reversibility: ? <2> [5]) [5]

P 5-methyluridine 5'-phosphate + succinate + CO_2

S Additional information <2> (<2> only pyrimidine deoxyribonucleosides with oxygen at carbon 2 and 4 are active [5]) [5]

P ?

Inhibitors

1,10-phenanthroline <2> (at 0.02 mM complete inhibition [5]) [5]
2-ketoadipate <2> (<2> 38% inhibition at 2.5 mM [5]) [5]
2-ketopimelate <2> (<2> 26% inhibition at 2.5 mM [5]) [5]
3-ketoadipate <2> (<2> 50% inhibition at 2.5 mM [5]) [5]
3-oxoglutarate <2> (<2> 27% inhibition at 2.5 mM [5]) [5]
Co^{2+} <2> (<2> 95% inhibition at 0.25 mM [5]) [5]
Cu^{2+} <2> (<2> 95% inhibition at 0.25 mM [5]) [5]
Mn^{2+} <2> (<2> 60% inhibition at 0.25 mM [5]) [5]
N_2 <2> (<2> complete inhibition by exclusion of oxygen [7]) [7]
Ni^{2+} <2> (<2> 95% inhibition at 0.25 mM [5]) [5]
Zn^{2+} <2> (<2> 95% inhibition at 0.25 mM [5]) [5]
diglycolate <2> (<2> 18% inhibition at 2.5 mM [5]) [5]
glutamate <2> (<2> 20% inhibition at 2.5 mM [5]) [5]
iminodiacetate <2> (<2> 28% inhibition at 2.5 mM [5]) [5]
thiodiglycolate <2> (<2> 24% inhibition at 2.5 mM [5]) [5]

Activating compounds

ascorbate <1, 2> (<1> no enzyme activity detected when omitted from standard incubation mixture [4]; <2> 3fold activation at 5 mM [5]) [4, 5, 6, 7, 8]
catalase <1, 2> (<1> 30% activation at 0.33 mg per ml catalase [4]; <2> 1.5fold activation at 2 mg per ml [5]) [4, 5, 6]

Metals, ions

Fe^{2+} <1, 2> [4, 5, 6, 7, 8]

Specific activity (U/mg)

0.053 <2> (<2> with thymidine as substrate [6]) [6]
0.35 <1> [1]
0.97 <1> [2]

K_m-Value (mM)

0.07 <2> (thymidylate) [5]
0.09 <2> (thymidine) [5]
0.15 <2> (5-bromodeoxyuridine) [5]
0.19 <2> (deoxyuridine) [5]
0.29 <2> (5-hydroxymethyldeoxyuridine) [5]
0.29 <2> (6-azathymidine) [5]
0.3 <2> (2-oxoglutarate) [5]
0.45 <2> (Fe^{2+}, <2> at pH 7.2 in 50 mM phosphate buffer [5]) [5]

pH-Optimum

6.5 <2> [5]

4 Enzyme Structure

Molecular weight

47000 <2> (<2> gel filtration [3,5]) [3, 5]
64600 <1> (<1> gel filtration [2]) [2]

Subunits

monomer <2> (<2> 1 * 46000, SDS-PAGE [3]) [3]

5 Isolation/Preparation/Mutation/Application

Purification

<1> (partial purification [1]) [1, 2]
<2> [3]

References

[1] Stubbe, J.: Identification of two α-ketoglutarate-dependent dioxygenases in extracts of Rhodotorula glutinis catalyzing deoxyuridine hydroxylation. J. Biol. Chem., **260**, 9972-9975 (1985)
[2] Warn-Cramer, B.J.; Macrander, L.A.; Abbott, M.T.: Markedly different ascorbate dependencies of the sequential α-ketoglutarate dioxygenase reactions catalyzed by an essentially homogeneous thymine 7-hydroxylase from Rhodotorula glutinis. J. Biol. Chem., **258**, 10551-10557 (1983)
[3] Holme, E.: Purification of thymidine 2'-hydroxylase from Neurospora crassa. Acta Chem. Scand., **37**, 743-745 (1983)

[4] Wondrack, L.M.; Hsu, C.A.; Abbott, M.T.: Thymine 7-hydroxylase and pyrimidine desoxyribonucleoside 2'-hydroxylase activities in Rhodotorula glutinis. J. Biol. Chem., **253**, 6511-6515 (1978)
[5] Bankel, L.; Lindstedt, G.; Lindstedt, S.: Thymidine 2-hydroxylation in Neurospora crassa. J. Biol. Chem., **247**, 6128-6134 (1972)
[6] Liu, C.K.; Shaffer, P.M.; Slaughter, R.S.; McCroskey, R.P.; Abbott, M.T.: Stoichiometry of the pyrimidine deoxyribonucleoside 2-hydroxylase reaction and of the conversions of 5-hydroxymethyluracil to 5-formyluracil and of the latter to uracil-5-carboxylic acid. Biochemistry, **11**, 2172-2176 (1972)
[7] Shaffer, P.M.; McCroskey, R.P.; Abbott, M.T.: Substrate specificity of the hydroxylase reaction in which thymidine is converted to thymine ribonucleoside. Biochim. Biophys. Acta, **258**, 387-394 (1972)
[8] Shaffer, P.M.; McCroskey, R.P.; Palmatier, R.D.; Midgett, R.J.; Abbott, M.T.: The cell-free conversion of a deoxyribonucleoside to a ribonucleoside without detachment of the deoxyribose. Biochem. Biophys. Res. Commun., **33**, 806-811 (1968)

Procollagen-lysine 5-dioxygenase 1.14.11.4

1 Nomenclature

EC number
1.14.11.4

Systematic name
procollagen-L-lysine,2-oxoglutarate:oxygen oxidoreductase (5-hydroxylating)

Recommended name
procollagen-lysine 5-dioxygenase

Synonyms
collagen lysine hydroxylase
lysine hydroxylase
lysine, 2-oxoglutarate 5-dioxygenase
lysine-2-oxoglutarate dioxygenase
lysyl hydroxylase <3> (<3> lysyl hydroxylase 1 and 2 [17]; <3> lysyl hydroxylase 3 [18]) [17, 18]
lysylprotocollagen dioxygenase
oxygenase, protocollagen lysine, di-
peptidyl-lysine, 2-oxoglutarate: oxygen oxidoreductase
peptidyllysine, 2-oxoglutarate:oxygen 5-oxidoreductase
protocollagen lysine hydroxylase
protocollagen lysyl hydroxylase

CAS registry number
9059-25-0

2 Source Organism

<1> *Gallus gallus* [1-7, 9-15, 22]
<2> *Sus scrofa* [8]
<3> *Homo sapiens* (HT 1080 sarcoma cells [3]; three isoenzymes [19]) [2, 3, 16-19, 22]
<4> *Caenorhabditis elegans* [20]
<5> *Rattus norvegius* [21]
<6> *Rattus norvegius* [22]

3 Reaction and Specificity

Catalyzed reaction

procollagen L-lysine + 2-oxoglutarate + O_2 = procollagen 5-hydroxy-L-lysine + succinate + CO_2 (<1> mechanism [1,6])

Reaction type

decarboxylation
hydroxylation
redox reaction

Natural substrates and products

S collagen + 2-oxoglutarate + O_2 <1, 3, 5> (<1, 3, 5> influence the integrity and stability of collagen [9, 17, 19, 21]; <3, 5> enzyme required during collagen biosynthesis [16, 17, 18, 19, 21]) (Reversibility: ? <1, 3, 5> [1, 9, 16-19, 21]) [1, 9, 16-19, 21]

P 5-hydroxylysyl-collagen + succinate + CO_2 <1, 3, 5> [1, 9, 16-19, 21]

Substrates and products

S (Ile-Lys-Gly)$_3$ + 2-oxoglutarate + O_2 <3> (Reversibility: ? <3> [18, 19]) [18, 19]

P (Ile-5-hydroxylyseine-Gly)$_3$ + succinate + CO_2 <3> [18, 19]

S (Pro-Pro-Gly)$_4$-Ala-Arg-Gly-Met-Lys-Gly-His-Arg-Gly-(Pro-Pro-Gly)$_4$ + 2-oxoglutarate + O_2 <1> (Reversibility: ? <1> [11]) [11]

P (Pro-Pro-Gly)$_4$-Ala-Arg-Gly-Met-5-hydroxylysine-Gly-His-Arg-Gly-(Pro-Pro-Gly)$_4$ + succinate + CO_2 <1> [11]

S 2-oxoglutarate + O_2 + ascorbate <1> (<1> uncoupled decarboxylation in absence of peptide substrate [5]) (Reversibility: ? <1> [5]) [5]

P succinate + CO_2 + dehydroascorbate + H_2O <1> [5]

S Ala-Arg-Gly-Ile-Lys-Gly-Ile-Arg-Gly-Phe-Ser-Gly + 2-oxoglutarate + O_2 <1, 3> (Reversibility: ? <1, 3> [11, 18]) [11, 18]

P Ala-Arg-Gly-Ile-5-hydroxylysine-Gly-Ile-Arg-Gly-Phe-Ser-Gly + succinate + CO_2 <1, 3> [11, 18]

S Ala-Arg-Gly-Met-Lys-Gly-His-Arg-Gly-(Pro-Pro-Gly)$_4$ + 2-oxoglutarate + O_2 <1> (Reversibility: ? <1> [11]) [11]

P Ala-Arg-Gly-Met-5-hydroxylysine-Gly-His-Arg-Gly-(Pro-Pro-Gly)$_4$ + succinate + CO_2 <1> [11]

S peptidyl-L-lysine + 2-oxoglutarate + O_2 <1-6> (<1, 2> protocollagen [7, 8, 13, 15]; <1, 3> minimum sequence required: Xaa-Lys-Gly [1, 2]; <1> in e.g. lysinevasopressin, lysine-rich histone [1]; <1> helical regions of collagen [4]; <1, 3> synthetic peptides [11, 13, 17-19]) (Reversibility: ? <1-6> [1-4, 5, 7, 8, 11-13, 17-22]) [1-4, 5, 7, 8, 11-13, , 15, 17-22]

P peptidyl-5-hydroxy-L-lysine + succinate + CO_2 <1-6> (<1-6> hydroxy-L-lysine [7, 8, 9, 10-13, 17, 18, 20-22]) [5, 7, 8, 9, 10-13, 17, 18, 20-22]

Inhibitors

CO_2 <1> [6, 10]
Ca^{2+} <1> [10]

Cd^{2+} <1> [10]
Cu^{2+} <1> [10]
DL-serine 2-[(2,3,4-trihydroxyphenyl)methyl]hydrazide <1> (<1> most potent [9]) [9]
N-ethylmaleimide <2> [8]
Zn^{2+} <1> [6, 10]
adrenochrome <1> (<1> slight [9]) [9]
catechol <1> [9]
dehydroascorbate <1> [6]
dopamine <1> [9]
ephedrine <1> (<1> slight [9]) [9]
epinephrine <1> [6, 9]
homogentisic acid <1> [6]
hydroxylysine-rich peptides <1> [6]
iodoacetamide <2> [8]
malaoxon <5> (<5> mechanism of inhibition [21]) [21]
malathion <5> (<5> mechanism of inhibition [21]) [21]
nitroblue tetrazolium <1> [6]
norepinephrine <1> [9]
p-chloromercuribenzoate <2> [8]
p-mercuribenzoate <1> [13]
phenylalanine <1> (<1> slight [9]) [9]
phenylephedrine <1> (<1> slight [9]) [9]
pyridine 2,4-dicarboxylate <1> [14]
pyridine 2,5-dicarboxylate <1> [14]
pyridine 2-carboxylate <1> [14]
pyrogallol <1> (<1> most potent [9]) [9]
succinate <1> [6]
tyrosine <1> (<1> slight [9]) [9]

Cofactors/prosthetic groups

ascorbate <1, 3> (<1, 3> required [3, 5, 13, 15, 17]) [1, 2, 3, 5, 13, 15, 17]

Activating compounds

2-amino-4-hydroxy-6,7-dimethyltetrahydropteridine <1> (<1> can replace ascorbate) [5]
2-mercaptoethanol <1> (<1> can replace ascorbate [5]) [5]
L-cysteine <1> (<1> can replace ascorbate [5]) [5]
Triton X-100 <1> (<1> activation [10]) [10]
bovine serum albumin <1, 2> (<1,2> activation [8,12]) [8, 10, 12, 13]
catalase <1, 2> (<2> activation [8]) [8, 13]
dithiothreitol <1, 2> (<1> can replace ascorbate [5]; <1> no indication [12]) [5, 8, 12, 13]
lysolecithin <1> (<1> activation [10]) [10]

Metals, ions

Fe^{2+} <1, 3> (<1, 3> required, K_m: 0.001-0.005 mM [3, 5, 7, 13, 15, 17, 19]) [1-3, 5, 7, 9, 13, 15, 17, 19]

Turnover number (min^{-1})

156-252 <1> (lysine, <1> in synthetic peptides [7]) [7]
Additional information <1, 3> [2]

Specific activity (U/mg)

Additional information <1-3, 6> (<1,3> assay methods [2]; <1,6> specific activity of the recombinant enzyme three-fold higher than the specific activity of native enzyme [22]) [2, 7, 8, 10, 13, 22]

K_m-Value (mM)

0.04-0.05 <1> (O_2) [5]
0.05 <1> (2-oxoglutarate) [5]
0.05 <1> (ascorbate) [13]
0.09-0.12 <1> (2-oxoglutarate) [5]
0.1 <3> (2-oxoglutarate, <3> lysyl hydroxylases 1 and 3 [18]; <3> lysyl hydroxylase 3 [19]) [18, 19]
0.1 <6> (protocollagen) [22]
0.106 <6> (2-oxoglutarate) [22]
0.12 <3> (2-oxoglutarate, <3> lysyl hydroxylase 1 [19]) [19]
0.122 <6> (ascorbate) [22]
0.19-0.22 <1> (ascorbate) [5]
0.2 <1> ((Pro-Pro-Gly)$_4$-Ala-Arg-Gly-Met-Lys-Gly-His-Arg-Gly-(Pro-Pro-Gly)$_4$) [11]
0.2 <1> (Ala-Arg-Gly-Met-Lys-Gly-His-Arg-Gly-(Pro-Pro-Gly)$_4$) [11]
0.3 <3> (ascorbate, <3> lysyl hydroxylase 3 [18]) [18]
0.35 <3> (ascorbate, <3> lysyl hydroxylase 1 [18]; <3> lysyl hydroxylases 1 and 3 [19]) [18, 19]
0.4 <3> ((Ile-Lys-Gly)$_3$, <3> lysyl hydroxylase 1 [19]) [19]
0.4 <1> (Ala-Arg-Gly-Ile-Lys-Gly-Ile-Arg-Gly-Phe-Ser-Gly) [11]
0.4-0.5 <1, 3> (protocollagen) [2, 5]
0.43 <3> ((IIle-Lys-Gly)$_3$, <3> lysyl hydroxylase 3 [19]) [19]
0.5 <3> (Ala-Arg-Gly-Ile-Lys-Gly-Ile-Arg-Gly-Phe-Ser-Gly, <3> lysyl hydroxylase 1 [18]) [18]
0.6 <3> (Ala-Arg-Gly-Ile-Lys-Gly-Ile-Arg-Gly-Phe-Ser-Gly, <3> lysyl hydroxylase 3 [18]) [18]
0.7 <3> ((Ile-Lys-Gly)$_3$, <3> lysyl hydroxylase 1 [18]) [18]
0.8 <3> ((Ile-Lys-Gly)$_3$, <3> lysyl hydroxylase 3 [18]) [18]
Additional information <1, 3> (<1,3> comparison of K_m of 2-oxoglutarate, ascorbate, Fe^{2+} and type-1,2 and -4 protocollagen substrate with type-1,2 and -4 enzyme [3]) [3]

K_i-Value (mM)

0.015 <1> (catechol) [9]
0.047 <5> (malaoxon) [21]
0.059 <5> (malathion) [21]

pH-Optimum

7.4 <1, 3> [2, 13]
8 <2> [8]
8-8.4 <1> [12]

Temperature optimum (°C)

30 <2> [8]
Additional information <1> [12]

4 Enzyme Structure

Molecular weight

150000 <3> (<3> gel filtration [16]) [16]
180000 <3> (<3> gel filtration, isoforms 1-3 [19]) [19]
200000 <1> (<1> gel filtration [7,13]) [1, 2, 7, 13]
220000 <1> (<1> gel filtration, peak 1 [10]) [10]
550000 <1> (<1> gel filtration, peak 2 [10,13]) [10, 13]

Subunits

? <2, 5, 6> (<2> x * 70000 + x * 115000, SDS-PAGE [8]; <5,6> x * 85000, SDS-PAGE and Western blotting [21,22]) [8, 21, 22]
dimer <1> (<1> 2 * 83000-85000, SDS-PAGE [7]) [1, 2, 7]
homodimer <3> (<3> 2 * 88000 or 97000 SDS-PAGE, recombinant enzymes: isoform 2, possibly two forms are due to variation in the glycosylation of enzymes [17]; <3> 2 * 82380, calculation from amino acid sequence, recombinant enzyme: isoform 3 [18]; <3> 2 * 80000-85000, SDS-PAGE, isoforms 1-3 [19]) [17, 18, 19]

Posttranslational modification

glycoprotein <1, 3> [1, 2, 17]
side-chain modification <1, 3-5> (<1, 3, 5> hydroxylation of lysyl residue in -X-Lys-Gly [9, 17, 18, 19, 21]) [9, 17, 18, 19, 20, 21]

5 Isolation/Preparation/Mutation/Application

Source/tissue

embryo <1, 3> (<1> tendons and sterna [3]; <1> homogenates and cartilage [10]) [1-7, 9-16]
lung <6> [22]
placenta <3> [2]
skin <2, 3> (<2> fetus [8]; <3> fibroblasts [16]) [8, 16]
Additional information <1, 3> (<1,3> collagen type-specific isoenzymes [3]; <3> existence of isoforms [17-19]) [3, 17-19]

Localization

endoplasmic reticulum <1, 3> (<3> lumen [16]) [2, 7, 16]
membrane <3> (<3> enzyme binds to membrane via weak electrostatic interactions [16]) [16]

Purification

<2> [8]
<3> (three recombinant isoenzymes [19]) [19]
<1, 6> (two alternative procedures [7]) [2, 4, 7, 10-13, 15, 22]

Cloning

<3> (expression in various human tissues [17,18]; expression in insect cells using a baculovirus vector [17,18,19]) [17, 18, 19]
<4> (expression in Escherichia coli, isoform 3 gene product possesses the collagen glycosyltransferase activity, but not isoform 1 and 2 [20]) [20]
<5> (expression in insect cells using a baculovirus vector [21]) [21]
<6> (expression in insect cells using a baculovirus vector [22]) [22]

Engineering

C144I <4> (<4> isoform 3, reduces glycosyltransferase activity [20]) [20]
L208I <4> (<4> isoform 3, reduces glycosyltransferase activity [20]) [20]

Application

medicine <3, 6> (<3,6> deficiency in enzyme activity causes the Ehler-Danlos syndrome type 6 [18, 21, 22]) [18, 21, 22]

6 Stability

General stability information

<1>, enzyme formes aggregates in low ionic strength buffer [13]
<1>, stabilization by glycine [12]
<1, 3>, inactivation by freezing/thawing [2]
<1, 3>, labile in tissue extracts [2]
<1, 3>, loss of activity during concentration [2, 7]
<1, 3>, stabilization by detergents, NaCl [2]

Storage stability

<1>, -20°C, enzyme purified by collagen-agarose column-chromatography stable, enzyme from Bio-gel column unstable [7]
<1>, 0-4°C [13]
<1, 3>, -20°C [2, 4]

References

[1] Kivirikko, K.I.; Myllylä, R.: The hydroxylation of prolyl and lysyl residues. Enzymol. Post-transl. Modif. Proteins (Freedman, R.B., Hawkins, H.C., eds.) Academic Press, New York, **1**, 53-104 (1980)

[2] Kivirikko, K.I.; Myllylä, R.: Posttranslational enzymes in the biosynthesis of collagen: intracellular enzymes. Methods Enzymol., **82**, 245-304 (1982)
[3] Puistola, U.: Catalytic properties of lysyl hydroxylase from cells synthesizing genetically different collagen types. Biochem. J., **201**, 215-219 (1982)
[4] Royce, P.M.; Barnes, M.J.: Failure of highly purified lysyl hydroxylase to hydroxylate lysyl residues in the non-helical regions of collagen. Biochem. J., **230**, 475-480 (1985)
[5] Puistola, U.; Turpeenniemi-Hujanen, T.M.; Myllylä, R.; Kivirikko, K.I.: Studies on the lysyl hydroxylase reaction. I. Initial velocity kinetics and related aspects. Biochim. Biophys. Acta, **611**, 40-50 (1980)
[6] Puistola, U.; Turpeenniemi-Hujanen, T.M.; Myllylä, R.; Kivirikko, K.I.: Studies on the lysyl hydroxylase reaction. II. Inhibition kinetics and the reaction mechanism. Biochim. Biophys. Acta, **611**, 51-60 (1980)
[7] Turpeenniemi-Hujanen, T.M.; Puistola, U.; Kivirikko, K.I.: Isolation of lysyl hydroxylase, an enzyme of collagen synthesis, from chick embryos as a homogeneous protein. Biochem. J., **189**, 247-253 (1980)
[8] Miller, R.L.; Varner, H.H.: Purification and enzymatic properties of lysyl hydroxylase from fetal procine skin. Biochemistry, **18**, 5928-5932 (1979)
[9] Murray, J.C.; Cassell, R.H.; Pinnell, S.R.: Inhibition of lysyl hydroxylase by catechol analogs. Biochim. Biophys. Acta, **481**, 63-70 (1977)
[10] Ryhänen, L.: Lysyl hydroxylase. Further purification and characterization of the enzyme from chick embryos and chick embryo cartilage. Biochim. Biophys. Acta, **438**, 71-89 (1976)
[11] Kivirikko, K.I.; Shudo, K.; Sakakibara, S.; Prockop, D.J.: Studies on protocollagen lysine hydroxylase. Hydroxylation of synthetic peptides and the stoichiometric decarboxylation of α-ketoglutarate. Biochemistry, **11**, 122-129 (1972)
[12] Popenoe, E.A.; Aronson, R.B.: Partial purification and properties of collagen lysine hydroxylase from chick embryos. Biochim. Biophys. Acta, **258**, 380-386 (1972)
[13] Kivirikko, K.I.; Prockop, D.J.: Partial purification and characterization of protocollagen lysine hydroxylase from chick embryos. Biochim. Biophys. Acta, **258**, 366-379 (1972)
[14] Majamaa, K.; Turpeenniemi-Hujanen, T.M.; Latipää, P.; Gunzler, V.; Hanauske-Abel, H.M.; Hassinen, I.E.; Kivirikko, K.I.: Differences between collagen hydroxylases and 2-oxoglutarate dehydrogenase in their inhibition by structural analogues of 2-oxoglutarate. Biochem. J., **229**, 127-133 (1985)
[15] Hausmann E.: Cofactor requirements for the enzymatic hydroxylation of lysine in a polypeptide precursor of collagen. Biochim. Biophys. Acta, **133**, 591-593 (1967)
[16] Kellokumpu, S.; Sormunen, R.; Heikkinen, J.; Myllyla, R.: Lysyl hydroxylase, a collagen processing enzyme, exemplifies a novel class of luminally-oriented peripheral membrane proteins in the endoplasmic reticulum. J. Biol. Chem., **269**, 30524-30529 (1994)
[17] Valtavaara, M.; Papponen, H.; Pirttilae, A.M.; Hiltunen, K.; Helander, H.; Myllylae, R.: Cloning and characterization of a novel human lysyl hydroxy-

lase isoform highly expressed in pancreas and muscle. J. Biol. Chem., **272**, 6831-6834 (1997)
[18] Passoja, K.; Rautavuoma, K.; Ala-Kokko, L.; Kosonen, T.; Kivirikko, K.I.: Cloning and characterization of a third human lysyl hydroxylase isoform. Proc. Natl. Acad. Sci. USA, **95**, 10482-10486 (1998)
[19] Rautavuoma, K.; Takaluoma, K.; Passoja, K.; Pirskanen, A.; Kvist, A.P.; Kivirikko, K.I.; Myllyharju, J.: Characterization of three fragments that constitute the monomers of the human lysyl hydroxylase isoenzymes 1-3: The 30-kDa N-terminal fragment is not required for lysyl hydroxylase activity. J. Biol. Chem., **277**, 23084-23091 (2002)
[20] Wang, C.; Risteli, M.; Heikkinen, J.; Hussa, A.K.; Uitto, L.; Myllyla, R.: Identification of amino acids important for the catalytic activity of the collagen glucosyltransferase associated with the multifunctional lysyl hydroxylase 3 (LH3). J. Biol. Chem., **277**, 18568-18573 (2002)
[21] Samimi, A.; Last, J.A.: Mechanism of inhibition of lysyl hydroxylase activity by the organophosphates malathion and malaoxon. Toxicol. Appl. Pharmacol., **176**, 181-186 (2001)
[22] Armstrong, L.C.; Last, J.A.: Rat lysyl hydroxylase: molecular cloning, mRNA distribution and expression in a baculovirus system. Biochim. Biophys. Acta, **1264**, 93-102 (1995)

5-Hydroxymethyluracil,2-oxoglutarate dioxygenase — 1.14.11.5

1 Nomenclature

EC number
1.14.11.5 (deleted, included in EC 1.14.11.6)

Recommended name
5-hydroxymethyluracil,2-oxoglutarate dioxygenase

Thymine dioxygenase 1.14.11.6

1 Nomenclature

EC number
1.14.11.6

Systematic name
thymine,2-oxoglutarate:oxygen oxidoreductase (7-hydroxylating)

Recommended name
thymine dioxygenase

Synonyms
5-hydroxy-methyluracil dioxygenase
5-hydroxy-methyluracil oxygenase
thymine 7-hydroxylase
thymine dioxygenase

CAS registry number
37256-67-0

2 Source Organism

<1> *Rhodotorula glutinis* (grown with thymine as nitrogen source [5]) [1, 2, 3, 4, 5, 8, 9]
<2> *Neurospora crassa* (STA 4 [6, 10, 11, 13]; uc-1 [7]; strain 1A [12, 14, 15]) [6, 7, 10, 11, 12, 13, 14, 15]

3 Reaction and Specificity

Catalyzed reaction
thymine + 2-oxoglutarate + O_2 = 5-hydroxymethyluracil + succinate + CO_2

Reaction type
hydroxylation
oxidation
redox reaction
reduction

Natural substrates and products
S 5-formyluracil + 2-oxoglutarate + O_2 <1, 2> (Reversibility: ? <1, 2> [1, 2, 3, 5, 6, 9, 10, 11]) [1, 2, 3, 5, 6, 9, 10, 11]
P 5-carboxyuracil + succinate + CO_2 <1> [1, 2, 3, 5, 11]

- **S** 5-hydroxymethyluracil + 2-oxoglutarate + O_2 <1, 2> (Reversibility: ? <1, 2> [1, 2, 3, 5, 6, 9, 10, 11]) [1, 2, 3, 5, 6, 9, 10, 11]
- **P** 5-formyluracil + succinate + CO_2 <1, 2> (<1,2> 5-formyluracil reacts subsequently to 5-carboxyuracil [1, 2, 3, 6]) [1, 2, 3, 5, 6, 9, 11]
- **S** thymine + 2-oxoglutarate + O_2 <1, 2> (<1> stereospecific removal of the pro-S hydrogen [3]) (Reversibility: ? <1, 2> [1, 2, 3, 5, 6, 7, 8, 9, 10, 11, 14, 15]) [1, 2, 3, 5, 6, 7, 8, 9, 10, 11, 14, 15]
- **P** 5-hydroxymethyluracil + succinate + CO_2 <1, 2> (<1, 2> 5-hydroxymethyluracil reacts subsequently to 5-formyluracil [1, 2, 3, 5, 6, 9]) [1, 2, 3, 5, 6, 7, 8, 9, 10, 11, 15]

Substrates and products

- **S** 1-ethylthymine + 2-oxoglutarate + O_2 <2> (Reversibility: ? <2> [10]) [10]
- **P** ? + succinate + CO_2 <2> [10]
- **S** 1-ethyluracil + 2-oxoglutarate + O_2 <2> (Reversibility: ? <2> [10]) [10]
- **P** ? + succinate + CO_2 <2> [10]
- **S** 1-methylthymine + 2-oxoglutarate + O_2 <1, 2> (Reversibility: ? <1, 2> [3, 6, 10]) [3, 6, 10]
- **P** 1-methyl-5-(hydroxymethyl)uracil + thymine + formaldehyde + succinate + CO_2 <1> [3]
- **S** 1-methyluracil + 2-oxoglutarate + O_2 <2> (Reversibility: ? <2> [10]) [10]
- **P** ? + succinate + CO_2 <2> [10]
- **S** 5,6-dihydrothymine + 2-oxoglutarate + O_2 <1> (Reversibility: ? <1> [3]) [3]
- **P** 5,6-dihydro-5-(hydroxymethyl)uracil + succinate + CO_2 <1> [3]
- **S** 5-(2-hydroxyethyl)uracil + 2-oxoglutarate + O_2 <1> (Reversibility: ? <1> [3, 4]) [3, 4]
- **P** 5-(1,2-dihydroxyethyl)uracil + succinate + CO_2 <1> [3, 4]
- **S** 5-(methylthio)uracil + 2-oxoglutarate + O_2 <1> (Reversibility: ? <1> [3]) [3]
- **P** 5-(methylsulfonyl)uracil + 5-(methylsulfinyl)uracil + succinate + CO_2 <1> [3]
- **S** 5-aminouracil + 2-oxoglutarate + O_2 <2> (Reversibility: ? <2> [7, 10]) [7, 10]
- **P** ? + succinate + CO_2 <2> [7, 10]
- **S** 5-bromo-1-methyluracil + 2-oxoglutarate + O_2 <2> (Reversibility: ? <2> [10]) [10]
- **P** ? + succinate + CO_2 <2> [10]
- **S** 5-ethynyluracil + 2-oxoglutarate + O_2 + H_2O <1> (Reversibility: ? <1> [1, 2, 4]) [1, 2, 4]
- **P** 5-(carboxymethyl)uracil + succinate + CO_2 <1> [1, 2, 4]
- **S** 5-fluorouracil + 2-oxoglutarate + O_2 <1, 2> (Reversibility: ? <1, 2> [3, 6]) [3, 6]
- **P** ? + succinate + CO_2 <1, 2> [3, 6]
- **S** 5-formyluracil + 2-oxoglutarate + O_2 <1, 2> (<1,2> 5-formyluracil reacts subsequently to 5-carboxyuracil [1,2,3,6]) (Reversibility: ? <1, 2> [1, 2, 3, 5, 6, 9, 10, 11]) [1, 2, 3, 5, 6, 9, 10, 11]

P 5-carboxyuracil + succinate + CO_2 <1> [1, 2, 3, 5, 11]
S 5-hydroxymethyluracil + 2-oxoglutarate + O_2 <1, 2> (Reversibility: ? <1, 2> [1, 2, 3, 5, 6, 9, 10, 11]) [1, 2, 3, 5, 6, 9, 10, 11]
P 5-formyluracil + succinate + CO_2 <1, 2> [1, 2, 3, 5, 6, 9, 11]
S 5-methylcytosine + 2-oxoglutarate + O_2 <1> (Reversibility: ? <1> [9]) [9]
P ?
S 5-vinyluracil + 2-oxoglutarate + O_2 <1> (Reversibility: ? <1> [3, 4]) [3, 4]
P 5-(1,2-dihydroxyethyl)uracil + succinate + CO_2 <1> [3, 4]
S 6-azathymine + 2-oxoglutarate + O_2 <2> (Reversibility: ? <2> [10]) [10]
P ? + succinate + CO_2 <2> [10]
S thymine + 2-oxoglutarate + O_2 <1, 2> (<1> stereospecific removal of the pro-S hydrogen [3]) (Reversibility: ? <1, 2> [1, 2, 3, 5, 6, 7, 8, 9, 10, 11, 14, 15]) [1, 2, 3, 5, 6, 7, 8, 9, 10, 11, 14, 15]
P 5-hydroxymethyluracil + succinate + CO_2 <1, 2> (<1,2> 5-hydroxymethyluracil reacts subsequently to 5-formyluracil [1, 2, 3, 5, 6, 9]) [1, 2, 3, 5, 6, 7, 8, 9, 10, 11, 15]
S uracil + 2-oxoglutarate + O_2 <2> (Reversibility: ? <2> [7]) [7]
P ? + succinate + CO_2 <2> [7]

Inhibitors

1-ethylthymine <2> (<2> 70% inhibition at 5 mM with thymine as substrate [10]) [10]
1-ethyluracil <2> (<2> 60% inhibition at 5 mM with thymine as substrate [10]) [10]
1-methylthymine <2> (<2> 73% inhibition at 5 mM with thymine as substrate [10]) [10]
1-methyluracil <2> (<2> 80% inhibition at 5 mM with thymine as substrate [10]) [10]
2,4-dimethoxy-5-methylpyrimidine <2> (<2> 68% inhibition at 5 mM with thymine as substrate [10]) [10]
5,6-dihydrothymine <2> (<2> 73% inhibition at 5 mM with thymine as substrate [10]) [10]
5-acetyluracil <1> (<1> poor competitive inhibitor [3]) [3]
5-aminouracil <2> (<2> 100% inhibition at 5 mM with thymine as substrate [10]) [10]
5-bromouracil <2> (<2> 95% inhibition at 5 mM with thymine as substrate [10]) [10]
5-carboxyuracil <2> [11]
5-ethynyluracil <1> (<1> binds covalently to a phenylalanine residue of the enzyme [1]; <1> mechanism-based inactivator [2]) [1, 2]
5-fluorouracil <2> (<2> linear non-competitive inhibitor [6]) [6]
5-formyluracil <1, 2> (<1,2> competitive inhibitor with respect to thymine [3, 12]; <2> 79% inhibition at 2.5 mM with thymine as substrate [10]; <2> competitive inhibitor with respect to 5-hydroxymethyluracil [12]) [3, 10, 12]
5-hydroxymethyluracil <2> (<2> competitive inhibitor against thymine, non-competitive against oxygen [11, 12]) [11, 12]

5-hydroxyuracil <2> (<2> 92% inhibition at 5 mM with thymine as substrate [10]) [10]
5-mercaptouracil <2> (<2> 60% inhibition at 5 mM with thymine as substrate [10]) [10]
5-methylaminouracil <2> (<2> 100% inhibition at 5 mM with thymine as substrate [10]) [10]
5-nitrouracil <2> (<2> 70% inhibition at 5 mM with thymine as substrate [10]) [10]
6-azathymine <2> (<2> 94% inhibition at 5 mM with thymine as substrate [10]) [10]
6-methylthymine <2> (<2> 41% inhibition at 5 mM with thymine as substrate [10]) [10]
6-methyluracil <2> (<2> 18% inhibition at 5 mM with thymine as substrate [10]) [10]
uracil <2> (<2> non-competitive inhibitor [7]; <2> inhibitor of the three reactions, 50% inhibition at 1 mM [12]) [7, 12]

Activating compounds

(β,γ-methylene)adenosine 5'-triphosphate <1> (<1> at 1 mM 2.5fold stimulation of the thymine dependent reaction [8]) [8]
ADP <1> (<1> at 1 mM 2.5fold stimulation of the thymine dependent reaction [8]) [8]
ATP <1> (<1> at 5 mM 4fold stimulation of the thymine and 5-formyluracil-dependent reactions, only 20% stimulation of the 5-hydroxymethyluracil dependent reaction [5]; <1> at 1 mM 3.7fold stimulation of the thymine dependent reaction [8]) [5, 8]
GTP <1> (<1> at 1 mM 2fold stimulation of the thymine dependent reaction [8]) [8]
UTP <1> (<1> at 1 mM 2fold stimulation of the thymine dependent reaction [8]) [8]
adenyl-5'-ylimidodiphosphate <1> (<1> at 1 mM 2fold stimulation of the thymine dependent reaction [8]) [8]
ascorbate <1, 2> (<1> at 3 mM 5fold stimulation of the thymine and the 5-formyluracil-dependent reactions, only 15% stimulation of the 5-hydroxymethyluracil dependent reaction [5]; <1> at 1 mM 10fold stimulation of the thymine dependent reaction [8]) [5, 6, 7, 8, 9]
catalase <2> [7, 9]

Metals, ions

Fe^{2+} <1, 2> (<1> non-heme [2, 3]) [1, 2, 3, 5, 6, 7, 8, 9, 12, 13, 14]

Turnover number (min^{-1})

1000 <1> (thymine) [2]

Specific activity (U/mg)

0.042 <2> [14]
16.1 <1> [5]
30 <1> [2]

K_m-Value (mM)

0.0011 <1> (Fe^{2+}) [3]
0.024 <1> (2-oxoglutarate) [3]
0.036 <1> (O_2) [3]
0.058 <1> (thymine) [3]
0.087 <1> (5,6-dihydroxythymine) [3]
0.1 <2> (thymine) [7]
0.118 <1> (5-hydroxymethyluracil) [3]
0.124 <1> (5-(2-hydroxyethyl)uracil) [3]
0.2 <2> (thymine) [10]
0.2 <2> (uracil) [7]
0.22 <2> (5-formyluracil) [10]
0.25 <2> (6-azathymine) [10]
0.322 <1> (5-fluorouracil) [3]
0.45 <2> (1-methylthymine) [10]
0.5 <2> (1-methyluracil) [10]
0.6 <2> (1-ethylthymine) [10]
1 <2> (1-ethyluracil) [10]
1 <2> (5-hydroxymethyluracil) [10]
1.3 <1> (1-methylthymine) [3]
2.34 <1> (5-formyluracil) [3]

K_i-Value (mM)

0.022 <1> (5-ethynyluracil) [2]
0.2 <2> (uracil) [7]
0.83 <1> (5-acetyluracil) [3]
1.39 <1> (5-formyluracil, <1> with respect to thymine [3]) [3]

4 Enzyme Structure

Molecular weight

40300 <1> (<1> disc gel electrophoresis under denaturing conditions [5]) [5]

Subunits

monomer <1> (<1> 1 * 42700, gel filtration [5]) [5]

5 Isolation/Preparation/Mutation/Application

Source/tissue

mycelium <2> [7, 12, 13, 15]

Purification

<1> [3, 5, 8]
<2> (partial [6]) [6, 12, 14]

6 Stability

Temperature stability

60 <2> (<2> 10% activity after 2 min [12]) [12]
98 <1> (<1> 50% activity after 5 min of ascorbate or ATP-dependent stimulation of thymine dependent reaction [8]; <1> 20% activity after 15 min of ascorbate or ATP-dependent stimulation of thymine dependent reaction [8]) [8]
100 <2> (<2> rapid inactivation [15]) [15]

General stability information

<2>, stabilized by glycine and dithiothreitol [13]

Storage stability

<1>, -70°C, ammonium sulfate precipitate, several weeks, no activity loss [9]
<1>, 0°C, ammonium sulfate precipitate, 5 h, quite stable [9]
<1>, 20 mM potassium phosphate pH 7.5, 100 mM glycine and 0.1 mM EDTA, stable [2]
<2>, 4°C, after dialysis against distilled water, one day or more [15]
<2>, freezing at -79°C, storage at -39°C, 50 mM Tris-HCl buffer, pH 8.0, 1 mM glutathione, 1 mM ascorbate, 10 mM KCl, 8 days, 96% activity [14]

References

[1] Lai, M.T.; Wu, W.; Stubbe, J.: Characterization of a novel, stable norcaradiene adduct resulting from the inactivation of thymine hydroxylase by 5-ethynyluracil. J. Am. Chem. Soc., **117**, 5023-5030 (1995)
[2] Thornburg, L.D.; Stubbe, J.: Mechanism-based inactivation of thymine hydroxylase, an α-ketoglutarate-dependent dioxygenase, by 5-ethynyluracil. Biochemistry, **32**, 14034-14042 (1993)
[3] Thornburg, L.D.; Lai, M.T.; Wishnok, J.S.; Stubbe, J.: A non-heme iron protein with heme tendencies: An investigation of the substrate specificity of thymine hydroxylase. Biochemistry, **32**, 14023-14033 (1993)
[4] Thornburg, L.D.; Stubbe, J.: Mechanism-based inhibition of thymine hydroxylase. J. Am. Chem. Soc., **111**, 7632-7633 (1989)
[5] Warn-Cramer, B.J.; Macrander, L.A.; Abbott, M.T.: Markedly different ascorbate dependencies of the sequential α-ketoglutarate dioxygenase reactions catalyzed by an essentially homogeneous thymine 7-hydroxylase from Rhodotorula glutinis. J. Biol. Chem., **258**, 10551-10557 (1983)
[6] Holme, E.; Lindstedt, S.: Studies on the partial reaction of thymine 7-hydroxylase in the presence of 5-fluorouracil. Biochim. Biophys. Acta, **704**, 278-283 (1982)
[7] Hsu, C.A.; Saewert, M.D.; Polsinelli Jr.; L.F.; Abbott, M.T.: Uracils uncoupling of the decarboxylation of α-ketoglutarate in the thymine 7-hydroxylase reaction of Neurospora crassa. J. Biol. Chem., **256**, 6098-6101 (1981)

[8] Wondrack, L.M.; Warn, B.J.; Saewert, M.D.; Abbott, M.T.: Substitution of nucleoside triphosphates for ascorbate in the thymine 7-hydroxylase reaction of Rhodotorula glutinis. J. Biol. Chem., **254**, 26-29 (1979)
[9] Wondrack, L.M.; Hsu, C.A.; Abbott, M.T.: Thymine 7-hydroxylase and pyrimidine desoxyribonucleoside 2'-hydroxylase activities in Rhodotorula glutinis. J. Biol. Chem., **253**, 6511-6515 (1978)
[10] Bankel, L.; Lindstedt, G.; Lindstedt, S.: Thymine 7-hydroxylase from Neurospora crassa. Substrate specificity studies. Biochim. Biophys. Acta, **481**, 431-437 (1977)
[11] Holme, E.: A kinetic study of thymine 7-hydroxylase from Neurospora crassa. Biochemistry, **14**, 4999-5003 (1975)
[12] Liu, C.K.; Hsu, C.A.; Abbott, M.T.: Catalysis of three sequential dioxygenase reactions by thymine 7-hydroxylase. Arch. Biochem. Biophys., **159**, 180-187 (1973)
[13] Bankel, L.; Holme, E.; Lindstedt, G.; Lindstedt, S.: Oxygenases involved in thymine and thymidine metabolism in Neurospora crassa. FEBS Lett., **21**, 135-138 (1972)
[14] McCroskey, R.P.; Griswold, W.R.; Sokoloff, R.L.; Sevier, E.D.; Lin, S.; Liu, C.K.; Shaffer, P.M.; Palmatier, R.D.; Parker, T.S.; Abbott, M.T.: Studies pertaining to the purification and properties of thymine 7-hydroxylase. Biochim. Biophys. Acta, **277**, 264-277 (1971)
[15] Abbott, M.T.; Kadner, R.J.; Fink, R.M.: Conversion of thymine to 5-hydroxymethyluracil in a cell-free system. J. Biol. Chem., **239**, 156-159 (1964)

Procollagen-proline 3-dioxygenase 1.14.11.7

1 Nomenclature

EC number

1.14.11.7

Systematic name

procollagen-L-proline,2-oxoglutarate:oxygen oxidoreductase (3-hydroxylating)

Recommended name

procollagen-proline 3-dioxygenase

Synonyms

oxygenase, protocollagen proline 3-di-
proline,2-oxoglutarate 3-dioxygenase
prolyl 3-hydroxylase
prolyl-4-hydroxyprolyl-glycyl-peptide, 2-oxoglutarate: oxygen oxidoreductase, 3-hydroxylating
protocollagen proline 3-hydroxylase

CAS registry number

63551-75-7

2 Source Organism

<1> *Gallus gallus* [1, 2]
<2> *Rattus norvegicus* [3, 4]

3 Reaction and Specificity

Catalyzed reaction

procollagen L-proline + 2-oxoglutarate + O_2 = procollagen trans-3-hydroxy-L-proline + succinate + CO_2

Reaction type

decarboxylation
hydroxylation
redox reaction

Natural substrates and products

S procollagen + 2-oxoglutarate + O_2 <1, 2> (<1, 2>, the enzyme catalyzes the synthesis of 3-hydroxyproline in collagen by the hydroxylation of prolyl residues [1, 3]) (Reversibility: ? <1, 2> [1, 3]) [1, 3]

P procollagen trans-3-hydroxy-L-proline + succinate + CO_2

Substrates and products

S procollagen L-proline + 2-oxoglutarate + O_2 <1, 2> (<1>, proline-labelled polypeptide substrate [2]; <2>, 2,3-T-L-proline-labeled polypeptide substrate [3]; <2>, chicken embryo tendon protocollagen and procollagen or cartilage protocollagen. The formation of 3-hydroxyproline is affected by chain length and the conformation of the substrate, in that longer polypeptide chains proved better substrates, while the native triple-helical conformation of protocollagen or procollagen completely prevents the reaction [4]) (Reversibility: ? <1, 2> [2, 3, 4]) [2, 3, 4]

P procollagen trans-3-hydroxy-L-proline + succinate + CO_2

S protocollagen containing 4-hydroxyproline + 2-oxoglutarate + O_2 <1, 2> (Reversibility: ? <1, 2> [1, 4]) [1, 4]

P ?

Inhibitors

2-oxoadipate <1> [2]
2-oxobutyrate <1> [2]
2-oxpentanoate <1> [2]
3-oxoglutarate <1> [2]
adipate <1> [2]
benzene-1,2-dicarboxylate <1> [2]
benzene-1,3-dicarboxylate <1> [2]
benzene-1,4-dicarboxylate <1> [2]
benzoate <1> [2]
concanavalin A <1> [1]
glutarate <1> [2]
laevulinate <1> [2]
malonate <1> [2]
oxaloacetate <1> [2]
pyridine-2,3-dicarboxylate <1> [2]
pyridine-2,4-dicarboxylate <1> [2]
pyridine-2,5-dicarboxylate <1> [2]
pyridine-2-carboxylate <1> [2]
pyridine-3,4-dicarboxylate <1> [2]
pyridine-3,5-dicarboxylate <1> [2]
pyridine-3-carboxylate <1> [2]
pyridine-4-carboxylate <1> [2]
pyruvate <1> [2]
succinate <1> [2]

Cofactors/prosthetic groups

ascorbate <1, 2> (<1, 2>, required [1, 3, 4]; <1>, K_m: 0.12 mM [1]) [1, 3, 4]

Metals, ions

Fe^{2+} <1, 2> (<1, 2>, required [1, 3, 4]; <1>, K_m: 0.002 mM [1]) [1, 3, 4]

Specific activity (U/mg)

Additional information <1, 2> (<2>, rapid assay method [3]) [1, 3]

K_m-Value (mM)

0.000034 <2> (procollagen) [3]
0.003 <1> (2-oxoglutarate) [1]
0.03 <1> (O_2) [1]

K_i-Value (mM)

0.003 <1> (pyridine-2,4-dicarboxylate, <1>, competitive inhibition with respect to 2-oxoglutarate and noncompetitive with respect to Fe^{2+} and the peptide substrate [2]) [2]
0.015 <1> (pyridine-2,5-dicarboxylate, <1>, competitive inhibition with respect to 2-oxoglutarate and noncompetitive with respect to Fe^{2+} and the peptide substrate [2]) [2]
0.2 <1> (pyridine-2-carboxylate, <1>, competitive inhibition with respect to 2-oxoglutarate and noncompetitive with respect to Fe^{2+} and the peptide substrate [2]) [2]
0.3 <1> (pyridine-3-carboxylate, <1>, competitive inhibition with respect to 2-oxoglutarate and noncompetitive with respect to Fe^{2+} and the peptide substrate [2]) [2]
0.5 <1> (benzene-1,3-dicarboxylate, <1>, competitive inhibition with respect to 2-oxoglutarate and noncompetitive with respect to Fe^{2+} and the peptide substrate [2]) [2]
0.5 <1> (benzene-1,4-dicarboxylate, <1>, competitive inhibition with respect to 2-oxoglutarate and noncompetitive with respect to Fe^{2+} and the peptide substrate [2]) [2]
0.5 <1> (oxaloacetate, <1>, competitive inhibition with respect to 2-oxoglutarate and noncompetitive with respect to Fe^{2+} and the peptide substrate [2]) [2]
0.5 <1> (pyridine-3,5-dicarboxylate, <1>, competitive inhibition with respect to 2-oxoglutarate and noncompetitive with respect to Fe^{2+} and the peptide substrate [2]) [2]
0.5 <1> (pyridine-3-carboxylate, <1>, competitive inhibition with respect to 2-oxoglutarate and noncompetitive with respect to Fe^{2+} and the peptide substrate [2]) [2]
0.7 <1> (pyridine-2,3-dicarboxylate, <1>, competitive inhibition with respect to 2-oxoglutarate and noncompetitive with respect to Fe^{2+} and the peptide substrate [2]) [2]
0.8 <1> (succinate, <1>, competitive inhibition with respect to 2-oxoglutarate and noncompetitive with respect to Fe^{2+} and the peptide substrate [2]) [2]
1 <1> (2-oxoadipate, <1>, competitive inhibition with respect to 2-oxoglutarate and noncompetitive with respect to Fe^{2+} and the peptide substrate [2]) [2]

1.3 <1> (benzene-1,2-dicarboxylate, <1>, competitive inhibition with respect to 2-oxoglutarate and noncompetitive with respect to Fe^{2+} and the peptide substrate [2]) [2]
2 <1> (pyridine-3,4-dicarboxylate, <1>, competitive inhibition with respect to 2-oxoglutarate and noncompetitive with respect to Fe^{2+} and the peptide substrate [2]) [2]
2.8 <1> (3-oxoglutarate, <1>, competitive inhibition with respect to 2-oxoglutarate and noncompetitive with respect to Fe^{2+} and the peptide substrate [2]) [2]
3.1 <1> (benzoate, <1>, competitive inhibition with respect to 2-oxoglutarate and noncompetitive with respect to Fe^{2+} and the peptide substrate [2]) [2]
3.6 <1> (glutarate, <1>, competitive inhibition with respect to 2-oxoglutarate and noncompetitive with respect to Fe^{2+} and the peptide substrate [2]) [2]
4.2 <1> (pyruvate, <1>, competitive inhibition with respect to 2-oxoglutarate and noncompetitive with respect to Fe^{2+} and the peptide substrate [2]) [2]
6 <1> (adipate) [2]
7.4 <1> (malonate, <1>, competitive inhibition with respect to 2-oxoglutarate and noncompetitive with respect to Fe^{2+} and the peptide substrate [2]) [2]
8 <1> (laevulinate, <1>, competitive inhibition with respect to 2-oxoglutarate and noncompetitive with respect to Fe^{2+} and the peptide substrate [2]) [2]
9.9 <1> (2-oxobutyrate, <1>, competitive inhibition with respect to 2-oxoglutarate and noncompetitive with respect to Fe^{2+} and the peptide substrate [2]) [2]
15 <1> (2-oxopentanoate, <1>, above, competitive inhibition with respect to 2-oxoglutarate and noncompetitive with respect to Fe^{2+} and the peptide substrate [2]) [2]

Temperature optimum (°C)
20 <2> [4]

4 Enzyme Structure

Molecular weight
160000 <1> (<1>, gel filtration [1]) [1]

Posttranslational modification
glycoprotein <1> [1]

5 Isolation/Preparation/Mutation/Application

Source/tissue
embryo <1> [1, 2]
kidney <2> (<2>, cortex [3, 4]) [3, 4]

Purification
<1> (partial [1]) [1]
<2> [3, 4]

References

[1] Tryggvason, K.; Majamaa, K.; Risteli, J.; Kivirikko, K.I.: Partial purification and characterization of chick-embryo prolyl 3-hydroxylase. Biochem. J., **183**, 303-307 (1979)
[2] Majamaa, K.; Turpeenniemi-Hujanen, T.M.; Latipää, P.; Gunzler, V.; Hanauske-Abel, H.M.; Hassinen, I.E.; Kivirikko, K.I.: Differences between collagen hydroxylases and 2-oxoglutarate dehydrogenase in their inhibition by structural analogues of 2-oxoglutarate. Biochem. J., **229**, 127-133 (1985)
[3] Risteli, J.; Tryggvason, K.; Kivirikko, K.I.: A rapid assay for prolyl 3-hydroxylase activity. Anal. Biochem., **84**, 423-431 (1978)
[4] Risteli, J.; Tryggvason, K.; Kivirikko, K.I.: Prolyl 3-hydroxylase: partial characterization of the enzyme from rat kidney cortex. Eur. J. Biochem., **73**, 485-492 (1977)

Trimethyllysine dioxygenase 1.14.11.8

1 Nomenclature

EC number
1.14.11.8

Systematic name
N^6,N^6,N^6-trimethyl-L-lysine,2-oxoglutarate:oxygen oxidoreductase (3-hydroxylating)

Recommended name
trimethyllysine dioxygenase

Synonyms
ε-trimethyllysine 2-oxoglutarate dioxygenase
oxygenase, trimethyllysine di-
TML dioxygenase
TML hydroxylase
TML-α-ketoglutarate dioxygenase
TMLD
trimethyllysine α-ketoglutarate dioxygenase

CAS registry number
84012-77-1

2 Source Organism

<1> *Rattus norvegicus* [1, 3]
<2> *Bos taurus* [2]
<3> *Neurospora crassa* [4]

3 Reaction and Specificity

Catalyzed reaction
N^6,N^6,N^6-trimethyl-L-lysine + 2-oxoglutarate + O_2 = 3-hydroxy-N^6,N^6,N^6-trimethyl-L-lysine + succinate + CO_2

Reaction type
redox reaction

Natural substrates and products

S N^6,N^6,N^6-trimethyl-L-lysine + 2-oxoglutarate + O_2 <1> (carnitine biosynthesis) [1]

Substrates and products

S N^6,N^6,N^6-trimethyl-L-lysine + 2-oxoglutarate + O_2 <1-3> (N^6,N^6, N^6-trimethyl-L-lysine is ε-N-trimethyl-L-lysine) [1-4]

P 3-hydroxy-N^6,N^6,N^6-trimethyl-L-lysine + succinate + CO_2 <1> [1]

Inhibitors

Mg^{2+} <1> [3]
Mn^{2+} <1> [3]
Zn^{2+} <1, 2> [2, 3]
α-N-acetyltrimethyllysine <2> [2]
citrate <2> [2]
hydroxytrimethyllysine <2> (4 stereoisomers) [2]
isocitrate <2> [2]
lactate <2> [2]
malonate <2> [2]
p-chloromercuribenzoate <2> [2]
succinate <2> [2]

Cofactors/prosthetic groups

ascorbate <1-3> (required, <1,2> [1-3]; stimulation, <3> [4]) [1-4]

Activating compounds

Dithiothreitol <2> (sparing as well as augmenting effect) [2]

Metals, ions

Ca^{2+} <1> (stimulation, <1> [3]; no stimulation, <2> [2]) [3]
Fe^{2+} <1-3> (required, <1,2> [1, 2]; stimulation, <3> [4]; preincubation with Fe^{2+} in absence of other cofactors inactivates (ascorbate or dithiothreitol prevents inactivation), <2> [2]) [1, 2, 4]
$FeSO_4$ <1> (required) [3]

Specific activity (U/mg)

Additional information <2> [2]

K_m-Value (mM)

0.13 <2> (N^6,N^6,N^6-trimethyl-L-lysine) [2]
0.22 <2> (α-ketoglutarate) [2]
1.6 <1> (trimethyllysine) [3]

pH-Optimum

6.8 <2> (assay at) [2]

Temperature optimum (°C)

37 <2> (assay at) [2]

4 Enzyme Structure

Molecular weight
180000 <2> (<2> gel filtration) [2]

Subunits
? <2> (<2> x * 52000 + x * 65000, bovine, SDS-PAGE) [2]

5 Isolation/Preparation/Mutation/Application

Source/tissue
kidney <1, 2> (cortex (67%), medulla (33%)) [2, 3]
liver <1, 2> [1, 2]

Localization
cytosol <3> [4]
mitochondrion <1, 2> [1, 2, 3]
Additional information <1> (no activity found in microsomal or soluble fractions of liver) [1]

Purification
<2> (partial) [2]

6 Stability

Temperature stability
0 <2> (half-life: 5 days) [2]

Storage stability
<2>, 0°C, half-life: 5 days [2]

References

[1] Hulse, J.D.; Ellis, S.R.; Henderson, L. M.: Carnitine biosynthesis. β-Hydroxylation of trimethyllysine by an α-ketoglutarate-dependent mitochondrial dioxygenase. J. Biol. Chem., **253**, 1654-1659 (1978)
[2] Henderson, L.M.; Nelson, P.J.; Henderson, L.: Mammalian enzymes of trimethyllysine conversion to trimethylaminobutyrate. Fed. Proc., **41**, 2843-2847 (1982)
[3] Sachan, D.S.; Hoppel, C.L.: Carnitine biosynthesis. Hydroxylation of N^6-trimethyl-lysine to 3-hydroxy-N^6-trimethyl-lysine. Biochem. J., **188**, 529-534 (1980)
[4] Sachan, D.S.; Broquist, H.P.: Synthesis of carnitine from ε-N-trimethyllysine in post mitochondrial fractions of Neurospora crassa. Biochem. Biophys. Res. Commun., **96**, 870-875 (1980)

Naringenin 3-dioxygenase 1.14.11.9

1 Nomenclature

EC number
1.14.11.9

Systematic name
naringenin,2-oxoglutarate:oxygen oxidoreductase (3-hydroxylating)

Recommended name
naringenin 3-dioxygenase

Synonyms
(2S)-flavanone 3-hydroxylase
F3H
FHTPH
FS I
flavanone 3-hydroxylase
flavanone 3β-hydroxylase
flavanone synthase I
oxygenase, flavanone 3-di-

CAS registry number
75991-43-4

2 Source Organism

<1> *Dahlia variabilis* (enzyme is present in cyanic strain, absent in acyanic strain [8]) [8]
<2> *Streptocarpus hybrida* (enzyme is present in cyanic strain, absent in acyanic strain) [8]
<3> *Verbena hybrida* (enzyme is present in cyanic strain, absent in acyanic strain) [8]
<4> *Zinnia elegans* (enzyme is present in cyanic strain, absent in acyanic strain) [8]
<5> *Lycopersicon esculentum* [11]
<6> *Petroselinum hortense* [1]
<7> *Petunia hybrida* [2, 7, 12, 13, 14, 15]
<8> *Tulipa sp.* (c.v. Apeldoorn) [3]
<9> *Sinapis alba* [4]
<10> *Antirrhinum majus* [5, 14]
<11> *Matthiola incana* [6, 9, 14]

<12> *Medicago sativa* [10]
<13> *Dianthus caryophyllus* [14]
<14> *Callistephus chinensis* [14]
<15> *Hordeum vulgare* [14]
<16> *Arabidopsis thaliana* [16]

3 Reaction and Specificity

Catalyzed reaction

naringenin + 2-oxoglutarate + O_2 = dihydrokaempferol + succinate + CO_2 (<6>, enzyme catalyzes double-bond formation by direct abstraction of vicinal hydrogen atoms at C-2 and C-3 of the substrate [1])

Reaction type

oxidation
redox reaction
reduction

Natural substrates and products

S naringenin + 2-oxoglutarate + O_2 <1-4, 7, 9, 11> (<11>, enzyme is involved in anthocyanin biosynthesis [9]; <9>, induction of the enzyme by light [4]; <1-4, 5, 11>, enzyme is involved in anthocyanin pathway [8, 9, 11]; <7>, enzyme is involved in the biosynthesis of flavonoids, catechins, and anthocyanidins [12,13]; <16>, the F^3H gene is coordinately expressed with chalcone synthase and chalcone isomerase in seedlings. The F3H gene may represent a pivotal point in the regulation of flavonoid biosynthesis [16]) (Reversibility: ? <1-4, 7, 9, 11> [4, 8, 9, 11, 12, 13, 16]) [4, 8, 9, 11, 12, 13, 16]

P (2R,3R)-dihydrokaempferol + succinate + CO_2

Substrates and products

S (2S)-naringenin + 2-oxoglutarate + O_2 <6-8, 11> (<7>, His220, His278 and Asp222 are part of the 2-oxoglutarate binding site [12]) (Reversibility: ? <6-8, 11> [1, 3, 6, 7, 12]) [1, 3, 6, 7, 12]

P (2R,3R)-dihydrokaempferol + succinate + CO_2 <7> [7]

S eriodictyol + 2-oxoglutarate + O_2 <6-8, 11> (<7>, (2S)-eriodictyol [7]; <8>, 50% of the activity with naringenin [3]) (Reversibility: ? <6-8, 11> [1, 3, 6, 7]) [1, 3, 6, 7]

P (2R,3R)-dihydroquercitin + succinate + CO_2 <7> [7]

S naringenin + 2-oxoadipate + O_2 <7> (Reversibility: ? <7> [12]) [12]

P dihydrokaempferol + pentanedioate + CO_2 <7> [12]

S Additional information <6, 7> (<6>, no activity with (+)-dihydrokaempferol and 2R-naringenin [1]; <7>, no activity with (2R)-naringenin and 5,7,3',4',5'-pentahydroxyflavanone [7]) [1, 7]

P ?

Inhibitors

(+)-dihydrokaempferol <7> (<7>, product inhibition [7]) [7]
3-bromo-2-oxoglutarate <7> (<7>, 1 mM 92% inhibition [7]) [7]
Cu^{2+} <6, 7> (<6>, in presence of 0.01 mM Fe^{2+} [1]) [1, 7]
EDTA <1-4, 7, 8, 11> (<8>, 2 mM, complete inhibition [3]; <11>, 1 mM, 75% inhibition [6]; <7>, 1 mM, 21% inhibition [7]) [3, 6-8]
Fe^{3+} <7> (<7>, 1 mM, 56% inhibition [7]) [7]
KCN <1-4, 7, 8, 11> (<8>, 5 mM, 72% inhibition [3]; <11>, 5 mM, complete inhibition [6]; <7>, 1 mM, 11% inhibition [7]) [3, 6-8]
Zn^{2+} <6> (<6>, in presence of 0.01 mM Fe^{2+} [1]) [1]
diethyldicarbonate <7, 8, 11> (<8, 11>, slight [3, 6]; <8>, 2 mM, complete inhibition [3]; <8>, 0.5 mM, 10% inhibition [3]; <11>, 0.5 mM, 15% inhibition [6]; <7>, ascorbate protects against inactivation [12]) [3, 6, 12, 14]
diethyldithiocarbamate <1-4, 7, 11> (<11>, 2 mM, complete inhibition [6]; <7>, 2 mM, 81% inhibition [7]) [6-8]
p-chloromercuribenzoate <7, 11> (<8>, no inhibition [3]; <11>, 0.1 mM, 5% inhibition [6]; <7>, 1 mM, 91% inhibition [7]) [6, 7]
pyridine-2,4-dicarboxylate <6, 7> (<6>, most potent competitive inhibitor [1]; <7>, 0.002 mM, 50% inhibition [7]) [1, 7]
pyridine-2,5-dicarboxylate <7> (<7>, 0.01 mM, 40% inhibition [7]) [7]

Activating compounds

ascorbate <6, 7, 11> (<6, 7, 11> required [1, 6, 7]) [1, 6, 7]
catalase <7> (<7>, stimulates [7]) [7]

Metals, ions

Co^{2+} <7> (<7>, can partially replace Fe^{2+} [7]) [7]
Fe^{2+} <6, 7, 8, 11> (<6, 7, 11>, required [1, 3, 6, 7, 12]) [1, 3, 6, 7, 12]
iron <7> (<7>, non-heme iron protein. His220, His278 and Asp222 constitute three of the possible ligands for iron binding in the active site of the enzyme [12]) [12]

Specific activity (U/mg)

Additional information <6, 7, 11> [1, 2, 6, 7, 12]

K_m-Value (mM)

0.0019 <7> (2-oxoglutarate, <7>, wild-type enzyme [13]) [13]
0.005 <6> ((2S)-naringenin) [1]
0.0056 <7> ((2S)-naringenin) [7]
0.008 <6> ((2S)-eriodictyol) [1]
0.012 <7> ((2S)-eriodictyol) [7]
0.016 <6> (2-oxoglutarate, <6>, reaction with naringenin [1]) [1]
0.02 <7> (2-oxoglutarate, <7>, reaction with naringenin [7]) [7]
0.0256 <7> (2-oxoglutarate, <7>, mutant enzyme S290T [13]) [13]
0.08 <7> (2-oxoglutarate, <7>, mutant enzyme S290A [13]) [13]
0.1896 <7> (2-oxoglutarate, <7>, mutant enzyme S290V [13]) [13]
1.4 <7> (2-oxoadipate) [12]
Additional information <7> (<7>, varying K_m-values, depending on the purification procedure [7]; <7>, K_m-values of mutant enzymes [12]) [7, 12]

K_i-Value (mM)

0.0012 <7> (pyridine 2,4-dicarboxylate) [7]
0.0018 <6> (2,4-pyridinedicarboxylate) [1]
0.04 <7> (pyridine 2,5-dicarboxylate) [7]

pH-Optimum

6 <7> (<7>, wild-type enzyme, and second lower optimum at pH 8.0 [12]) [12]
7.5 <8> [3]
8.5 <7> [7]
8.5-8.6 <6> [1]
Additional information <7> (<7>, pH-optima of mutant enzymes [12]) [12]

4 Enzyme Structure

Molecular weight

39200 <7> (<7>, sedimentation equilibrium analysis [15]) [15]
48000 <6> (<6>, gel filtration [1]) [1]
74000 <7> (<7>, gel filtration [7]) [7]

Subunits

? <7, 11, 15> (<7>, x * 35000 + x * 37000, two-dimensional SDS-PAGE [7]; <11>, x * 40000, calculation from nucleotide sequence [14]; <15>, x * 43600, calculation from nucleotide sequence [14]) [7, 14]
dimer <6> (<6>, 2 * 24000-25000, SDS-PAGE [1]) [1]
Additional information <7> (<7>, enzyme exists as functional monomeric and oligomeric forms. The monomeric polypeptide comprises the catalytically active flavanone 3β-hydroxylase [15]) [15]

5 Isolation/Preparation/Mutation/Application

Source/tissue

anther <8> (<8>, tapetum-bound [3]) [3]
bud <11> [9]
flower <1-4, 10, 11, 12> [5, 6, 8, 9, 10]
flower bud <11, 13, 14> [14]
hypocotyl <5> [11]
nodule <11> [9]
root <11> [9]
seedling <9, 16> (<9>, enzyme shows a drastic increase in activity after illumination and reaches a maximum at the time of rapid anthocyanin accumulation [4]; <16>, etiolated [16]) [4, 16]

Localization

soluble <1-4, 7, 10, 11> [5, 6, 8, 12, 13]

Purification

<6> [1]
<7> (wild-type and mutant enzymes [13]; recombinant enzyme [14]) [2, 7, 12, 13, 14]

Cloning

<7> (wild-type and mutant enzymes, expression in Escherichia coli [12,13]) [12, 13]
<11> [14]
<13> (expression in a reticulocyte system [14]) [14]
<14> (expression in a reticulocyte system [14]) [14]
<16> [16]

Engineering

H220Q <7> (<7>, catalytic activity is reduced to about 0.15% of that of the wild-type enzyme. Slightly increased K_m-value with respect to iron binding, as compared to the wild-type enzyme [12]) [12]
H278Q <7> (<7>, mutant enzyme has no detectable enzyme activity [12]) [12]
N222N <7> (<7>, catalytic activity is reduced to about 0.15% of that of the wild-type enzyme. Slightly increased K_m-value with respect to iron binding, as compared to the wild-type enzyme [12]) [12]
R288K <7> (<7>, decrease in catalytic activity and a 5fold increase in K_m-value for 2-oxoglutarate [12]) [12]
R288Q <7> (<7>, decrease in catalytic activity and a 160fold increase in K_m-value for 2-oxoglutarate [12]) [12]
S290A <7> (<7>, activity is reduced to 8% of that of the wild-type enzyme [13]) [13]
S290T <7> (<7>, activity is reduced to 20% of that of the wild-type enzyme [13]) [13]
S290V <7> (<7>, activity is reduced to 1% of that of the wild-type enzyme [13]) [13]

6 Stability

Oxidation stability

<7>, partially stabilized under anaerobic conditions in presence of ascorbate [7]

Storage stability

<7>, -70°C, in presence of 20 mM ascorbate, stable for more than 6 months [7]

References

[1] Britsch, L.: Purification and characterization of flavone synthase I, a 2-oxoglutarate-dependent desaturase. Arch. Biochem. Biophys., **282**, 152-160 (1990)

[2] Britsch, L.: Purification of flavanone 3 β-hydroxylase from Petunia hybrida: antibody preparation and characterization of a chemogenetically defined mutant. Arch. Biochem. Biophys., **276**, 348-354 (1990)

[3] Beerhues, L.; Forkmann, G.; Schoepker, H.; Stotz, G.; Wiermann, R.: Flavanone 3-hydroxylase and dihydroflavonol oxygenase activities in anthers of Tulipa. The significance of the tapetum fraction in flavonoid metabolism. J. Plant Physiol., **133**, 743-746 (1989)

[4] Takeda, K.; Fischer, D.; Grisebach, H.: Anthocyanin composition of Sinapis alba, light induction of enzymes and biosynthesis. Phytochemistry, **27**, 1351-1353 (1988)

[5] Forkmann, G.; Stotz, G.: Genetic control of flavanone 3-hydroxylase activity and flavonoid 3'-hydroxylase activity in Antirrhinum majus (Snapdragon). Z. Naturforsch. C, **36**, 411-416 (1981)

[6] Forkmann, G.; Heller, W.; Grisebach, H.: Anthocyanin biosynthesis in flowers of Matthiola incana flavanone 3- and flavonoid 3'-hydroxylases. Z. Naturforsch. C, **35**, 691-695 (1980)

[7] Britsch, L.; Grisebach, H.: Purification and characterization of (2S)-flavanone 3-hydroxylase from Petunia hybrida. Eur. J. Biochem., **156**, 569-577 (1986)

[8] Forkmann, G.; Stotz, G.: Selection and characterisation of flavanone 3-hydroxylase mutants of Dahlia, Streptocarpus, Verbena and Zinnia. Planta, **161**, 261-265 (1984)

[9] Dangelmayr, B.; Stotz, G.; Spribille, R.; Forkmann, G.: Relationship between flower development, anthocyanin accumulation and activity of enzymes involved in flavonoid biosynthesis in Matthiola incana R.Br.. Z. Naturforsch. C, **38**, 551-555 (1983)

[10] Charrier, B.; Coronado, C.; Kondorosi, A.; Ratet, P.: Molecular characterization and expression of alfalfa (Medicago sativa L.) flavanone-3-hydroxylase and dihydroflavonol-4-reductase encoding genes. Plant Mol. Biol., **29**, 773-786 (1995)

[11] O'Neill, S.D.; Tong, Y.; Spoerlein, B.; Forkmann, G.; Yoder, J.I.: Molecular genetic analysis of chalcone synthase in Lycopersicon esculentum and an anthocyanin-deficient mutant. Mol. Gen. Genet., **224**, 279-288 (1990)

[12] Lukacin, R.; Britsch, L.: Identification of strictly conserved histidine and arginine residues as part of the active site in Petunia hybrida flavanone 3β-hydroxylase. Eur. J. Biochem., **249**, 748-757 (1997)

[13] Lukacin, R.; Groning, I.; Pieper, U.; Matern, U.: Site-directed mutagenesis of the active site serine290 in flavanone 3β-hydroxylase from Petunia hybrida. Eur. J. Biochem., **267**, 853-860 (2000)

[14] Britsch, L.; Dedio, J.; Saedler, H.; Forkmann, G.: Molecular characterization of flavanone 3 β-hydroxylases. Consensus sequence, comparison with re-

lated enzymes and the role of conserved histidine residues. Eur. J. Biochem., **217**, 745-754 (1993)
[15] Lukacin, R.; Urbanke, C.; Groning, I.; Matern, U.: The monomeric polypeptide comprises the functional flavanone 3β-hydroxylase from Petunia hybrida. FEBS Lett., **467**, 353-358 (2000)
[16] Pelletier, M.K.; Shirley, B.W.: Analysis of flavanone 3-hydroxylase in Arabidopsis seedlings. Coordinate regulation with chalcone synthase and chalcone isomerase. Plant Physiol., **111**, 339-345 (1996)

Pyrimidine-deoxynucleoside 1′-dioxygenase 1.14.11.10

1 Nomenclature

EC number
1.14.11.10

Systematic name
2'-deoxyuridine,2-oxoglutarate:oxygen oxidoreductase (1'-hydroxylating)

Recommended name
pyrimidine-deoxynucleoside 1'-dioxygenase

Synonyms
oxygenase, deoxyuridine-uridine 1'-di-
Additional information (cf. EC 1.14.11.3)

CAS registry number
98865-52-2

2 Source Organism

<1> *Rhodotorula glutinis* [1]

3 Reaction and Specificity

Catalyzed reaction
2'-deoxyuridine + 2-oxoglutarate + O_2 = uracil + 2-deoxyribonolactone + succinate + CO_2

Reaction type
oxidation
oxidative decarboxylation
redox reaction
reduction

Natural substrates and products
S 2'-deoxyuridine + 2-oxoglutarate + O_2 <1> (Reversibility: ? <1> [1]) [1]
P uracil + deoxyribonolactone + succinate + CO_2

Substrates and products
S 2'-deoxy-2'-fluorouridine + 2-oxoglutarate + O_2 <1> (Reversibility: ? <1> [1]) [1]
P uracil + 2-fluororibonolactone + succinate + CO_2

S 2'-deoxyuridine + 2-oxoglutarate + O_2 <1> (Reversibility: ? <1> [1]) [1]
P uracil + deoxyribonolactone + succinate + CO_2
S uridine + 2-oxoglutarate + O_2 <1> (Reversibility: ? <1> [1]) [1]
P ribonolactone + uracil + succinate + CO_2

Cofactors/prosthetic groups
ascorbate <1> (<1> requirement [1]) [1]

Metals, ions
Fe^{2+} <1> (<1> requirement [1]) [1]

Specific activity (U/mg)
5 <1> [1]

pH-Optimum
7.5 <1> (<1> assay at [1]) [1]

Temperature optimum (°C)
25 <1> (<1> assay at [1]) [1]

5 Isolation/Preparation/Mutation/Application

Localization
cytoplasm <1> [1]

Purification
<1> [1]

6 Stability

Storage stability
<1>, -20°C, stable for at least a month in 0.02 M sodium phosphate buffer, pH 7.5, with 0.1 mM EDTA [1]

References

[1] Stubbe, J.: Identification of two α-ketoglutarate-dependent dioxygenases in extracts of Rhodotorula glutinis catalyzing deoxyuridine hydroxylation. J. Biol. Chem., **260**, 9972-9975 (1985)

Hyoscyamine (6S)-dioxygenase 1.14.11.11

1 Nomenclature

EC number
1.14.11.11

Systematic name
L-hyoscyamine,2-oxoglutarate:oxygen oxidoreductase ((6S)-hydroxylating)

Recommended name
hyoscyamine (6S)-dioxygenase

Synonyms
hyoscyamine 6-hydroxylase
hyoscyamine 6β-hydroxylase
oxygenase, hyoscyamine 6β-di-

CAS registry number
103865-33-4

2 Source Organism

<1> *Hyoscyamus niger* [1, 2, 3, 4, 5, 7]
<2> *Atropa belladonna* [3, 6]
<3> *Duboisia leichhardtii* [3]
<4> *Datura fastuosa* [3]
<5> *Hyoscyamus albus* [3]
<6> *Hyoscyamus gyorffi* [3]
<7> *Hyoscyamus pusillus* [3]
<8> *Hyoscyamus muticus* [3]

3 Reaction and Specificity

Catalyzed reaction
L-hyoscyamine + 2-oxoglutarate + O_2 = (6S)-hydroxyhyoscyamine + succinate + CO_2

Reaction type
decarboxylation
epoxidation
hydroxylation

oxidation
redox reaction
reduction

Natural substrates and products

S L-hyoscyamine + 2-oxoglutarate + O_2 <1> (<1>, enzyme is involved in biosynthetic pathway leading to scopolamine [1]; <1>, the enzyme catalyzes two consecutive oxidation reactions leading from hyoscyamine to scopolamine [5]) (Reversibility: ? <1> [1]) [1, 5]
P 6β-hydroxyhyoscyamine + succinate + CO_2 <1> [1]

Substrates and products

S 2-hydroxy-3-phenylpropionyltropine + 2-oxoglutarate + O_2 <1> (<1>, 15% of the activity with L-hyoscyamine [1]) (Reversibility: ? <1> [1]) [1]
P 2-hydroxy-3-phenylpropionyl-6-hydroxytropine + succinate + CO_2 <1> [1]
S 2-oxoglutarate + *p*-hydroxyatropine + O_2 <1> (<1>, 26% of the activity with L-hyoscyamine [1]) (Reversibility: ? <1> [1]) [1]
P ?
S 2-oxoglutarate + trans-cinnamoyltropine + O_2 <1> (<1>, 39% of the activity with L-hyoscyamine [1]) (Reversibility: ? <1> [1]) [1]
P succinate + trans-cinnamoyl-6-hydroxytropine + CO_2 <1> [1]
S 3-hydroxy-3-phenylpropionyltropine + 2-oxoglutarate + O_2 <1> (<1>, 56% of the activity with L-hyoscyamine [1]) (Reversibility: ? <1> [1]) [1]
P 3-hydroxy-3-phenylpropionyl-6-hydroxytropine + succinate + CO_2 <1> [1]
S 6,7-dehydrohyoscyamine + 2-oxoglutarate + O_2 <1> (<1>, 119% of the activity with L-hyoscyamine [1]) (Reversibility: ? <1> [1]) [1]
P scopolamine + succinate + CO_2 <1> [1]
S 6β-hydroxyhyoscyamine <1> (<1>, weak epoxidase activity, 2-5% of the hydroxylase activity [5]) (Reversibility: ? <1> [5]) [5]
P scopolamine
S L-homatropine + 2-oxoglutarate + O_2 <1> (<1>, 81% of the activity with L-hyoscyamine [1]) (Reversibility: ? <1> [1]) [1]
P 6-hydroxyhomatropine + succinate + CO_2 <1> [1]
S L-hyoscyamine + 2-oxoglutarate + O_2 <1-8> (<1>, only the L-isomer of hyoscyamine serves as substrate, D-hyoscyamine is nearly inactive [3]) (Reversibility: ? <1-8> [1-7]) [1-7]
P 6β-hydroxyhyoscyamine + succinate + CO_2 <1-8> [1-7]
S L-norhyoscyamine + 2-oxoglutarate + O_2 <1> (<1>, 81% of the activity with L-hyoscyamine [1]) (Reversibility: ? <1> [1]) [1]
P 6-hydroxynorhyoscyamine + succinate + CO_2 <1> [1]
S apoatropine + 2-oxoglutarate + O_2 <1> (<1>, 45% of the activity with L-hyoscyamine [1]) (Reversibility: ? <1> [1]) [1]
P 6-hydroxyapoatropine + succinate + CO_2 <1> [1]
S isobutyltropine + 2-oxoglutarate + O_2 <1> (<1>, 15% of the activity with L-hyoscyamine [1]) (Reversibility: ? <1> [1]) [1]
P ?

S noratropine-N-acetic acid + 2-oxoglutarate + O_2 <1> (<1>, 17% of the activity with L-hyoscyamine [1]) (Reversibility: ? <1> [1]) [1]
P ?
S phenylacetyltropine + 2-oxoglutarate + O_2 <1> (<1>, 81% of the activity with L-hyoscyamine [1]) (Reversibility: ? <1> [1]) [1]
P 6-hydroxyphenylacetyltropine + succinate + CO_2 <1> [1]
S phenylalanyltropine + 2-oxoglutarate + O_2 <1> (<1>, 8% of the activity with L-hyoscyamine [1]) (Reversibility: ? <1> [1]) [1]
P ?
S Additional information <1> (<1>, high specificity for 2-oxoglutarate in hydroxylation [1]) [1]
P ?

Inhibitors

1,10-phenanthroline <1> (<1>, 0.1 mM, 35% inhibition [1]) [1]
2,2'-dipyridyl <1> (<1>, 0.1 mM, 15% inhibition [1]) [1]
3,4-dihydroxybenzoate <1> [1]
3,4-dihydroxycinnamate <1> [1]
3,4-dihydroxymandelate <1> [1]
3,4-dihydroxyphenylacetate <1> [1]
3,4-dihydroxyphenylpropionate <1> [1]
3-oxoglutaric acid <1> [1]
5,7-dehydrohyoscyamine <1> [1]
Ca^{2+} <1> (<1>, 0.4 mM, 11% inhibition [1]) [1]
Cd^{2+} <1> (<1>, 0.4 mM, complete inhibition [1]) [1]
Co^{2+} <1> (<1>, 0.4 mM, complete inhibition [1]) [1]
Cu^{2+} <1> (<1>, 0.4 mM, complete inhibition [1]) [1]
EDTA <1> (<1>, 0.1 mM, 67% inhibition [1]) [1]
Fe^{3+} <1> (<1>, 0.4 mM, 28% inhibition [1]) [1]
Hg^{2+} <1> (<1>, 0.4 mM, 96% inhibition [1]) [1]
L-homatropine <1> [1]
Mn^{2+} <1> (<1>, 0.4 mM, 95% inhibition [1]) [1]
Ni^{2+} <1> (<1>, 0.4 mM, 96% inhibition [1]) [1]
Tiron <1> (<1>, 0.1 mM, 44% inhibition [1]) [1]
Zn^{2+} <1> (<1>, 0.4 mM, complete inhibition [1]) [1]
apoatropine <1> [1]
bathocuproine <1> (<1>, 0.1 mM, 27% inhibition [1]) [1]
diethyldithiocarbamate <1> (<1>, 0.1 mM, 37% inhibition [1]) [1]
fumarate <1> [1]
nitroblue tetrazolium <1> [1]
noratropine-N-acetic acid <1> [1]
phenylacetyltropine <1> [1]
pyridine 2,3-dicarboxylate <1> (<1>, competitive with respect to 2-oxoglutarate [1]) [1]
pyridine 2,4-dicarboxylate <1> (<1>, competitive with respect to 2-oxoglutarate [1]) [1]

Activating compounds

5,7-dimethyl-5,6,7,8-tetrahydropterine <1> (<1>, partial activation [1]) [1]
acetone <1-8> (<1-8>, 10% v/v, activation [3]) [3]
ascorbate <1-8> (<1>, 4 mM, 5.9fold stimulation [1]; <1-8>, required [3]) [1, 3]
catalase <1-8> (<1-8>, activation [1,3]) [1, 3]
dehydroascorbate <1> (<1>, partial activation [1]) [1]
dithiothreitol <1-8> (<1>, partial activation [1]; <1-8>, activation [3]) [1, 3]
isoascorbate <1> (<1>, partial activation [1]) [1]

Metals, ions

Fe^{2+} <1-8> (<1>, activates [1]; <1-8>, required [3]) [1, 3]

Specific activity (U/mg)

Additional information <1> [1]

K_m-Value (mM)

0.01 <1> (6,7-dehydrohyoscyamine) [1]
0.017 <1> (L-hyoscyamine) [3]
0.035 <1> (L-hyoscyamine) [1]
0.043 <1> (2-oxoglutarate) [1]
0.057 <1> (2-oxoglutarate) [3]

K_i-Value (mM)

0.009 <1> (pyridine 2,4-dicarboxylate) [1]
0.09 <1> (pyridine 3,4-dicarboxylate) [1]

pH-Optimum

7.8 <1> [1]

pH-Range

6-9 <1> (<1>, about 50% of maximal activity at pH 6.0 and pH 9.0 [1]) [1]

4 Enzyme Structure

Molecular weight

41000 <1> (<1>, gel filtration [1]) [1]

Subunits

? <1> (<1>, x * 38999, calculation from nucleotide sequence [2]; <1>, x * 39000 + x * 19000, SDS-PAGE [1]) [1, 2]

5 Isolation/Preparation/Mutation/Application

Source/tissue

anther <2> [6]
pericycle <2> (of root) [6]
pollen mother cell <2> [6]

root <2> (<2> native [6]) [6]
root <1-8> (culture) [1, 2, 3, 6]
Additional information <1, 2> (<1>, mRNA of the hydroxylase is absent in leaves, stems and cultured cells [2]; <2>, no activity in stem, leaf, pistil, petal and sepal tissue [6]) [2, 6]

Purification

<1> (partial [3]) [1, 3]

Cloning

<1> (cDNA from Hyoscyamus niger is simultaneously introduced into Nicotiana tabacum using particle bombardment and expressed under the control of the CaMV 35S promoter [4]; expression in Escherichia coli as a fusion protein to maltose-binding protein [7]) [2, 4, 7]

Engineering

D219H <1> (<1>, inactive mutant enzyme [7]) [7]
D219N <1> (<1>, inactive mutant enzyme [7]) [7]
H217Q <1> (<1>, inactive mutant enzyme [7]) [7]
H66Q <1> (<1>, mutant enzyme has 97% of the activity of the wild-type enzyme [7]) [7]
S274Q <1> (<1>, inactive mutant enzyme [7]) [7]

6 Stability

Organic solvent stability

acetone <1-8> (<1-8>, 10% v/v, activation [3]) [3]

Storage stability

<1>, -20°C, stable for more than 3 months [1]

References

[1] Hashimoto, T.; Yamada, Y.: Purification and characterization of hyoscyamine 6 β-hydroxylase from root cultures of Hyoscyamus niger L. Hydroxylase and epoxidase activities in the enzyme preparation. Eur. J. Biochem., **164**, 277-285 (1987)

[2] Matsuda, J.; Okabe, S.; Hashimoto, T.; Yamada, Y.: Molecular cloning of hyoscyamine 6 β-hydroxylase, a 2-oxoglutarate-dependent dioxygenase, from cultured roots of Hyoscyamus niger. J. Biol. Chem., **266**, 9460-9464 (1991)

[3] Hashimoto, T.; Yamada, Y.: Hyoscyamine 6β-hydroxylase, a 2-oxoglutarate-dependent dioxygenase, an alkaloid-producing root cultures. Plant Physiol., **81**, 619-625 (1986)

[4] Rocha, P.; Stenzel, O.; Parr, A.; Walton, N.; Christou, P.; Drager, B.; Leech, M.J.: Functional expression of tropinone reductase I (trI) and hyoscyamine-6β-hydroxylase (h6h) from Hyoscyamus niger in Nicotiana tabacum. Plant Sci., **162**, 905-913 (2002)

[5] Yun, D.J.; Hashimoto, T.; Yamada, Y.: Transgenic tobacco plants with two consecutive oxidation reactions catalyzed by hyoscyamine 6β-hydroxylase. Biosci. Biotechnol. Biochem., **57**, 502-503 (1993)
[6] Suzuki, K.I.; Yun, D.J.; Chen, X.Y.; Yamada, Y.; Hashimoto, T.: An Atropa belladonna hyoscyamine 6β-hydroxylase gene is differentially expressed in the root pericycle and anthers. Plant Mol. Biol., **40**, 141-152 (1999)
[7] Matsuda, J.; Hashimoto, T.; Yamada, Y.: Analysis of active-site residues in hyoscyamine 6β-hydroxylase. Plant Biotechnol., **14**, 51-57 (1997)

Gibberellin-44 dioxygenase 1.14.11.12

1 Nomenclature

EC number
1.14.11.12

Systematic name
(gibberellin-44),2-oxoglutarate:oxygen oxidoreductase

Recommended name
gibberellin-44 dioxygenase

Synonyms
(gibberellin-44),2-oxoglutarate:oxygen oxidoreductase
GA44 oxidase <1> [1]
gibberellin A44 oxidase
gibberellin-44-dioxygenase
oxygenase, gibberellin A44 di-

CAS registry number
112198-85-3

2 Source Organism

<1> *Spinacia oleracea* [1]

3 Reaction and Specificity

Catalyzed reaction
gibberellin 44 + 2-oxoglutarate + O_2 = gibberellin 19 + succinate + CO_2

Reaction type
oxidation
redox reaction
reduction

Natural substrates and products
S gibberellin 44 + 2-oxoglutarate + O_2 <1> (Reversibility: ? <1> [1]) [1]
P gibberellin 19 + succinate + CO_2 <1> [1]

Substrates and products
S gibberellin 44 + 2-oxoglutarate + O_2 <1> (Reversibility: ? <1> [1]) [1]
P gibberellin 19 + succinate + CO_2 <1> [1]

Metals, ions

Fe^{2+} <1> (<1> required [1]) [1]

Specific activity (U/mg)

0.003 <1> [1]

pH-Optimum

8 <1> [1]

Temperature optimum (°C)

30 <1> (<1> assay at [1]) [1]

4 Enzyme Structure

Molecular weight

381000 <1> (<1> gel filtration, HPLC [1]) [1]

5 Isolation/Preparation/Mutation/Application

Source/tissue

leaf <1> [1]

Purification

<1> [1]

References

[1] Gilmour, S.J.; Bleecker, A.B.; Zeevaart, J.A.D.: Partial purification of gibberelin oxidases from spinach leaves. Plant Physiol., **85**, 87-90 (1987)

Gibberellin 2β-dioxygenase 1.14.11.13

1 Nomenclature

EC number
1.14.11.13

Systematic name
(gibberellin-1),2-oxoglutarate:oxygen oxidoreductase (2β-hydroxylating)

Recommended name
gibberellin 2β-dioxygenase

Synonyms
GA 2-oxidase <2, 5> [4-6]
gibberellin 2β-hydroxylase

CAS registry number
85713-20-8

2 Source Organism

<1> *Arabidopsis sp.* [6]
<2> *Oryza sativa* (tall rice, L. cv. Nipponbare [6]) [6]
<3> *Phaseolus coccineus* (runner bean [6]) [6]
<4> *Phaseolus vulgaris* (var. Canadian Wonder [1,3]) [1, 3]
<5> *Pisum sativum* (garden pea, cv. progress No. 9 [2]; cross between the WT cv. Torsdag (SLN) ans sln line NGB6074 [4]) [2-6]

3 Reaction and Specificity

Catalyzed reaction
gibberellin 1 + 2-oxoglutarate + O_2 = 2β-hydroxygibberellin 1 + succinate + CO_2 (also acts on a number of gibberellins)

Reaction type
oxidation
redox reaction
reduction

Natural substrates and products
S gibberellin 1 + 2-oxoglutarate + O_2 <4> (<5> multifunctional enzyme catalyzing 2β hydroxylation and 2-ketone formation of the C19-gibberellin

substrates giberellin 9, gibberellin 20 gibberellin 1 and gibberellin 4 [5]) (Reversibility: ? <4, 5> [1, 2]) [1, 2, 5]

P gibberellin 8 + succinate + CO_2 <4> [1]

Substrates and products

S [1,2,3-3H3]gibberellin 20 + 2-oxoglutarate + O_2 <4> (Reversibility: ? <4> [3]) [3]

P ?

S [1,2-3H2]gibberellin 1 + 2-oxoglutarate + O_2 <4> (<4> isoenzyme I [3]) (Reversibility: ? <4> [1, 3]) [1, 3]

P ?

S [2,3-3H2]gibberellin 9 + 2-oxoglutarate + O_2 <4> (<4> isoenzyme II [3]) (Reversibility: ? <4> [3]) [3]

P ?

S gibberellin 1 + 2-oxoglutarate + O_2 <2, 4, 5> (<4> marked preference for 3-hydroxylated gibberellins as substrate [1]) (Reversibility: ? <2, 4, 5> [1-6]) [1-6]

P 2β-hydroxygibberellin 1 + succinate + CO_2

S gibberellin 20 + 2-oxoglutarate + O_2 <2, 4, 5> (Reversibility: ? <2, 4, 5> [1, 2, 5, 6]) [1, 2, 5, 6]

P gibberellin 29 + succinate + CO_2 <2> [6]

S gibberellin 4 + 2-oxoglutarate + O_2 <2, 4, 5> (Reversibility: ? <2, 4, 5> [1, 2, 5, 6]) [1, 2, 5, 6]

P gibberellin 34 + succinate + CO_2 <2> [6]

S gibberellin 44 + 2-oxoglutarate + O_2 <2> (Reversibility: ? <2> [6]) [6]

P gibberellin 98 + succinate + CO_2 <2> [6]

S gibberellin 9 + 2-oxoglutarate + O_2 <2, 4, 5> (Reversibility: ? <2, 4, 5> [1, 2, 5, 6]) [1, 2, 5, 6]

P gibberellin 51 + succinate + CO_2 <2> [6]

Inhibitors

N,N-dimethyl succinic acid hydrazide <4> [1]

Zn^{2+} <5> [2]

gibberellin 1 <5> (<5> efficient inhibitor of gibberellin 20 2β-hydroxylation [2]) [2]

gibberellin methyl ester <4> [1]

succinic acid <4> [1]

Cofactors/prosthetic groups

2-oxoglutarate <4, 5> (<5> activity dependent upon [2]) [1-3]

ascorbate <4, 5> (<5> activity dependend upon [2]) [1-3]

Activating compounds

catalase <4> (<4> activity is stimulated by [1]) [1]

Metals, ions

Fe^{2+} <4> (<4> required as cofactor, stimulated by addition of exogenous Fe^{2+} [1]) [1]

Fe^{2+} <4, 5> (<5> activity dependent upon [2]) [1-3]

Specific activity (U/mg)

Additional information <4> (<4> 191 * 10 -12 mol h -1 [1]) [1]

K_m-Value (mM)

0.000027 <4> ([2,3-3H2]gibberellin 9, <4> isoenzyme II [3]) [3]
0.000052 <4> ([1,2-3H2]gibberellin 1, <4> isoenzyme II [3]) [3]
0.00006 <4> ([1,2-3H2]gibberellin 4, <4> isoenzyme I [3]) [3]
0.000069 <5> (gibberellin 1) [2]
0.000085 <4> ([1,2-3H2]gibberellin 1) [1]
0.000103 <4> ([1,2-3H2]gibberellin 1, <4> isoenzyme I [3]) [3]
0.000135 <4> ([1,2-3H2]gibberellin 4, <4> isoenzyme II [3]) [3]
0.000299 <5> (gibberellin 9) [2]
0.000302 <4> ([1,2,3-3H3]gibberellin 20, <4> isoenzyme I [3]) [3]
0.000538 <4> ([2,3-3H2]gibberellin 9, <4> isoenzyme I [3]) [3]
0.00155 <5> (gibberellin 20) [2]
0.0045 <4> ([2,3-3H2]gibberellin 9, <4> isoenzyme II [3]) [3]
0.0118 <4> ([1,2-3H2]gibberellin 1, <4> isoenzyme I [3]) [3]
0.01244 <4> ([1,2,3-3H3]gibberellin 20, <4> isoenzyme II [3]) [3]
0.021 <4> (α-ketoglutarate) [1]

pH-Optimum

5.8 <4> [3]
6-7 <4> (<4> similar optimum for both isoenzymes [3]) [3]
7.4-7.8 <5> [2]

pH-Range

6.1-7.8 <5> [2]

4 Enzyme Structure

Molecular weight

26000 <4> (<4> isoenzyme I, size exclusion column, SDS-PAGE [3]) [3]
35000 <4> (<4> SDS-PAGE [1]) [1]
36000 <4> (<4> gel filtration [1]) [1, 3]
42000 <4> (<4> isoenzyme II, gel filtration, SDS-PAGE [3]) [3]
43000 <5> (<5> major protein, SDS-PAGE [2]) [2]
44000 <5> (<5> gel filtration [2]) [2]
45000 <5> (<5> minor protein, SDS-PAGE [2]) [2]

5 Isolation/Preparation/Mutation/Application

Source/tissue

cotyledon <4, 5> [1-3]
embryo <5> (<5> embryonic axes [2]) [2]
root <5> [5]

seed <2, 4, 5> (<4> mature seeds [3]; <5> mature and immature seeds [2]; <5> mature seeds [4]) [1-6]
seedling <5> [5]

Purification

<4> (two gibberellin 2β-hydroxylases partially purified [3]) [1, 3]
<5> (recombinant enzyme [5]) [2, 5]

Cloning

<1> (3 GA 2-oxidase cDNAs cloned by database screening [6]) [6]
<2> (cDNA isolation of OsGA2ox1, heterologous expressed in Escherichia coli, ectopic expression in transgenic rice [6]) [6]
<3> (GA 2-oxidase cDNA cloned by functional screening [6]) [6]
<5> (SLN gene, 2 cDNAs encoding gibberellin 2-oxidases, PsGA2ox1 isolated by screening of a Lambda-ZAP cDNA library, excised into phagemid form and expressed in Escherichia coli, PsGA2ox2 obtained as PCR product and expressed in Escherichia coli [4]; cDNA library, SLENDER gene encodes gibberellin 2-oxidase [5]; two cDNAs for GA 2-oxidase isolated by functional screening and reverse transcription [6]) [4-6]

6 Stability

pH-Stability

5.8-7.8 <5> (<5> activity is highest at pH 7.0-8.0 and decreases rapidly below pH 7.0, enzyme is unstable when stored below pH 7 or in absence of a thiol reagent [2]) [2]

General stability information

<4>, dialysis of the extracted soluble protein results in complete loss of activity [1]
<5>, enzyme is very acid-labile. EDTA has beneficial effect on enzyme stability if Mg^{2+} is present in storage buffer [2]

Storage stability

<4>, -20°C, enzyme activity lost upon storage, regained by addition of catalase to the reaction mixture [1]

References

[1] Smith, V.A.; MacMillan, J.: Purification and partial characterization of a gibberellin 2β-hydroxylase from Phaseolus vulgaris. J. Plant Growth Regul., **2**, 251-264 (1984)
[2] Smith, V.A.; MacMillan, J.: The partial purification and characterisation of gibbelerin 2β-hydroxylases from seeds of Pisum sativum. Planta, **167**, 9-18 (1986)

[3] Griggs, D.L.; Hedden, P.; Lazarus, C.M.: Partial purification of two gibberellin 2β-hydroxylases from cotyledons of Phaseolus vulgaris. Phytochemistry, **30**, 2507-2512 (1991)
[4] Lester, D.R.; Ross, J.J.; Smith, J.J.; Elliott, R.C.; Reid, J.B.: Gibberellin 2-oxidation and the SLN gene of Pisum sativum. Plant J., **19**, 65-73 (1999)
[5] Martin, D.N.; Proebsting, W.M.; Hedden, P.: The SLENDER gene of pea encodes a gibberellin 2-oxidase. Plant Physiol., **121**, 775-781 (1999)
[6] Sakamoto, T.; Kobayashi, M.; Itoh, H.; Tagiri, A.; Kayano, T.; Tanaka, H.; Iwahori, S.; Matsuoka, M.: Expression of a gibberellin 2-oxidase gene around the shoot apex is related to phase transition in rice. Plant Physiol., **125**, 1508-1516 (2001)

6β-Hydroxyhyoscyamine epoxidase 1.14.11.14

1 Nomenclature

EC number
1.14.11.14

Systematic name
(6S)-6-hydroxyhyoscyamine,2-oxoglutarate oxidoreductase (epoxide-forming)

Recommended name
6β-hydroxyhyoscyamine epoxidase

Synonyms
hydroxyhyoscyamine dioxygenase
hyosOH epoxidase <1> [1]

CAS registry number
121479-53-6

2 Source Organism

<1> *Hyoscyamus niger* [1]

3 Reaction and Specificity

Catalyzed reaction
(6S)-6-hydroxyhyoscyamine + 2-oxoglutarate + O_2 = scopolamine + succinate + CO_2 (requires Fe^{2+} and ascorbate)

Reaction type
oxidation
redox reaction
reduction

Natural substrates and products
S (6S)-6-hydroxyhyoscyamine + 2-oxoglutarate + O_2 <1> (<1> alkaloid metabolism, scopolamine is formed by oxidative transformation of hyoscyamine in several solanaceous species [1]) (Reversibility: ? <1> [1]) [1]
P scopolamine + succinate + CO_2 <1> [1]

Substrates and products

S (6S)-6-hydroxyhyoscyamine + 2-oxoglutarate + O_2 <1> (Reversibility: ? <1> [1]) [1]
P scopolamine + succinate + CO_2 <1> [1]

Inhibitors

$CuSO_4$ <1> (<1> 100% inhibition [1]) [1]
$MgSO_4$ <1> (<1> 24% inhibition [1]) [1]
$MnSO_4$ <1> (<1> 100% inhibition [1]) [1]
$ZnSO_4$ <1> (<1> 100% inhibition [1]) [1]
pyridine 2,4-dicarboxylate <1> [1]
Additional information <1> (<1> no marked effect on enzyme activity by addition of NAD^+, NADH, $NADP^+$, NADPH, ATP + $MgSO_4$, FAD, FMN, pyrroloquinoline quinone, acetyl-CoA, 6,7-dimethyl-5,6,7,8-tetrahydrofolate, phenazine methosulfate, 2,6-dichlorophenolindophenol, cytochrome c and H_2O_2 [1]) [1]

Cofactors/prosthetic groups

2-oxoglutarate <1> [1]
ascorbate <1> [1]

Activating compounds

catalase <1> [1]

Metals, ions

Fe^{2+} <1> (<1> required as co-factor [1]) [1]

Specific activity (U/mg)

0.000198 <1> [1]

K_m-Value (mM)

0.015 <1> ((6S)-6-hydroxyhyoscyamine) [1]

pH-Optimum

7.5 <1> [1]

5 Isolation/Preparation/Mutation/Application

Source/tissue

root <1> [1]

Localization

cytoplasm <1> [1]

Purification

<1> (partial, could not be separated from hyoscyamine 6β-hydroxylase [1]) [1]

6 Stability

Oxidation stability
<1>, requires molecular oxygen [1]

General stability information
<1>, rapid loss of enzyme activity during purification [1]

Storage stability
<1>, -20°C, enzyme is stored in small portions until assayed [1]

References

[1] Hashimoto, T.; Kohno, J.; Yamada, Y.: 6β-hydroxyhyoscyamine epoxidase from cultured roots of Hyoscyamus niger. Phytochemistry, **28**, 1077-1082 (1989)

Gibberellin 3β-dioxygenase 1.14.11.15

1 Nomenclature

EC number
1.14.11.15

Systematic name
(gibberellin-20),2-oxoglutarate:oxygen oxidoreductase (3β-hydroxylating)

Recommended name
gibberellin 3β-dioxygenase

Synonyms
(giberrellin-20),2-oxoglutarate: oxygen oxidoreductase (3β-hydroxylating)
gibberellin 3β dioxygenase
gibberellin 3β-hydroxylase
oxygenase, gibberellin 3β-di-

CAS registry number
116036-68-1

2 Source Organism

<1> *Phaseolus vulgaris* [1, 2, 3]
<2> *Cucurbita maxima* [1, 3, 6]
<3> *Pisum sativum* [4]
<4> *Arabidopsis thaliana* [5]
<5> *Oryza sativa* [7]

3 Reaction and Specificity

Catalyzed reaction
gibberellin 20 + 2-oxoglutarate + O_2 = gibberellin 1 + succinate + CO_2

Reaction type
oxidation
redox reaction
reduction

Natural substrates and products
S gibberellin a20 + 2-oxoglutarate + O_2 <1, 2, 3, 4> (<1,2>, enzyme catalyzes final step to biological active gibberellins [3]; <2>, involved in bio-

synthesis of gibberellin [6]) (Reversibility: ? <1, 2, 3, 4> [3, 4, 5, 6]) [3, 4, 5, 6]

P gibberellin a1 + succinate + CO_2

Substrates and products

S gibberellin a15 + 2-oxoglutarate + O_2 <2> (Reversibility: ? <2> [6]) [6]

P gibberellin a37 + succinate + CO_2 <2> [6]

S gibberellin a20 + 2-oxoglutarate + O_2 <1-5> (Reversibility: ? <1-5> [1-7]) [1-7]

P gibberellin a1 + succinate + CO_2 <1> [2]

S gibberellin a44 + 2-oxoglutarate + O_2 <5> (Reversibility: ? <5> [7]) [7]

P gibberellin a38 + succinate + CO_2 <5> [7]

S gibberellin a5 + 2-oxoglutarate + O_2 <5> (Reversibility: ? <5> [7]) [7]

P gibberellin a3 + succinate + CO_2 <5> [7]

S gibberellin a9 + 2-oxoglutarate + O_2 <1, 2, 5> (Reversibility: <1, 2, 5> [1, 2, 3, 7]) [1, 2, 3, 7]

P gibberellin a4 + succinate + CO_2 <1, 2> [1, 3]

Inhibitors

16-deoxo-2,3-di-dehydro-3-methyl-deoxygibberellin c <1> [3]
16-deoxo-deoxygibberellin c <1, 2> (<2>, slightly [1]) [1]
2,2'-bipyridyl <1> [2]
2,3-didehydro-3-methyl-deoxygibberellin c <1> [3]
2,3-didehydro-3-methyl-gibberellin a9 <1, 2> [3]
2,3-methyl-gibberellin a5 <1> [3]
3-methyl-gibberellin a5 <2> [3]
Cd^{2+} <1> (<1>, 0.5 mM, strong [2]) [2]
Co^{2+} <1> (<1>, 0.5 mM, strong [2]) [2]
Cu^{2+} <1> (<1>, 0.5 mM, strong [2]) [2]
Hg^{2+} <1> (<1>, 0.5 mM, strong [2]) [2]
Mn^{2+} <1> (<1>, 0.5 mM, strong [2]) [2]
Ni^{2+} <1> (<1>, 0.5 mM, strong [2]) [2]
Zn^{2+} <1> (<1>, 0.5 mM, strong [2]) [2]
deoxygibberellin c <1, 2> [1]
gibberellin a15 <1> (<1>, inhibits 3β-hydroxylation of gibberellin a20 [2]) [2]
gibberellin a20 <1> [2]
gibberellin a44 <1> (<1>, inhibits 3β-hydroxylation of gibberellin a20 [2]) [2]
gibberellin a5 <1> (<1>, inhibits 3β-hydroxylation of gibberellin a20 [2]) [2]
gibberellin a9 <1> (<1>, inhibits 3β-hydroxylation of gibberellin a20 [2]) [2]

Cofactors/prosthetic groups

ascorbate <1> (<1>, 5 mM, 10fold stimulation [2]) [2]

Activating compounds

DTT <2> (<2>, stimulates [6]) [6]
ascorbate <2> (<2>, 1 mM and more, 1.7fold stimulation [6]) [6]

bovine serum albumin <1> (<1>, 2 mg/ml, 30% activation of purified enzyme, stimulation by bovine serum albumin and catalase is additive and less pronounced in crude cell extract [2]) [2]
catalase <1> (<1>, 0.5 mg/ml, 30% activation of purified enzyme, stimulation by catalase and bovine serum albumin is additive and less pronounced in crude cell extract [2]) [2]

Metals, ions

Fe^{2+} <1, 2> (<1>, activates, at least 0.2 mM for full activation [2]; <2>, stimulates. Stimulation is negated when added together with DTT. If Fe^{2+} is omitted, the activities are reduced by 25% compared with the maximal value reached at 0.008-0.03 mM $FeSO_4$ [6]) [2, 6]
Fe^{3+} <1> (<1>, activation in the presence of ascorbate [2]) [2]

Specific activity (U/mg)

0.065 <2> [6]
Additional information <1> [2]

K_m-Value (mM)

0.00029 <1> (gibberellin a20) [2]
0.00033 <1> (gibberellin a9) [2]
0.0087 <2> (gibberellin a15) [6]
0.25 <1> (2-oxoglutarate, <1>, with gibberellin a20 as cosubstrate [2]) [2]

pH-Optimum

7.7 <1> [2]

pH-Range

7-9 <1> (<1>, about half-maximal activity at pH 7.0 and pH 9.0 [2]) [2]

4 Enzyme Structure

Molecular weight

42000 <1> (<1>, gel filtration [2]) [2]
58000 <2> (<2>, gel filtration [6]) [6]

Subunits

? <2> (<2>, x * 40000-50000, SDS-PAGE [6]) [6]
monomer <1> (<1>, 1 * 42000, SDS-PAGE [2]) [2]

5 Isolation/Preparation/Mutation/Application

Source/tissue

endosperm <2> [6]
seed <1, 2, 4> (<1,2>, immature [1,2,3]; <4>, expressed predominantly during seed germination [5]; <2>, endosperm [6]) [1, 2, 3, 5, 6]
seedling <3> (<3>, mRNA accumulation is higher in etiolated seedlings than in light-grown seedlings [4]) [4]

Localization

cytoplasm <1, 2> [3]

Purification

<1> (partial [2]) [2]
<2> (partial [6]) [6]

Cloning

<5> (two gibberellin 3β-hydroxylase genes, OsGA3ox1 and OsGA3ox2, expression in Escherichia coli [7]) [7]

References

[1] Saito, T.; Kwak, S.S.; Kamiya, Y.; Yamane, H.; Sakurai, A.; Murofushi, N.; Takahashi, N.: Effects of deoxygibberellin C (DGC) and 16-deoxo-DGC on gibberellin 3β-hydroxylase. Plant Cell Physiol., **32**, 239-245 (1991)
[2] Kwak, S.S.; Kamiya, Y.; Sakurai, A.; Takahashi, N.; Graebe, J.: Partial purification and characterization of gibberellin 3β-hydroxylase from immature seeds of Phaseolus vulgaris. Plant Cell Physiol., **29**, 935-943 (1988)
[3] Saito, T.; Kamiya, Y.; Yamane, H.; Sakurai, A.; Murofushi, N.; Takahashi, N.: Effects of 3-methylgibberellin analogs on gibberellin 3β-hydroxylase and plant growth. Biosci. Biotechnol. Biochem., **56**, 1046-1052 (1992)
[4] Ait-Ali, T.; Frances, S.; Weller, J.L.; Reid, J.B.; Kendrick, R.E.; Kamiya, Y.: Regulation of gibberellin 20-oxidase and gibberellin 3β-hydroxylase transcript accumulation during de-etiolation of pea seedlings. Plant Physiol., **121**, 783-791 (1999)
[5] Yamaguchi, S.; Smith, M.W.; Brown, R.G.S.; Kamiya, Y.; Sun, T.P.: Phytochrome regulation and differential expression of gibberellin 3β-hydroxylase genes in germinating Arabidopsis seeds. Plant Cell, **10**, 2115-2126 (1998)
[6] Lange, T.; Schweimer, A.; Ward, D.A.; Hedden, P.; Graebe, J.E.: Separation and characterization of three 2-oxoglutarate-dependent dioxygenases from Cucurbita maxima L. endosperm involved in gibberellin biosynthesis. Planta, **195**, 98-107 (1994)
[7] Itoh, H.; Ueguchi-Tanaka, M.; Sentoku, N.; Kitano, H.; Matsuoka, M.; Kobayashi, M.: Cloning and functional analysis of two gibberellin 3β-hydroxylase genes that are differently expressed during the growth of rice. Proc. Natl. Acad. Sci. USA, **98**, 8909-8914 (2001)

Peptide-aspartate β-dioxygenase 1.14.11.16

1 Nomenclature

EC number
1.14.11.16

Systematic name
peptide-L-aspartate,2-oxoglutarate:oxygen oxidoreductase (3-hydroxylating)

Recommended name
peptide-aspartate β-dioxygenase

Synonyms
aspartate β-hydroxylase
aspartylpeptide β-dioxygenase

CAS registry number
122544-66-5

2 Source Organism

<1> *Mus musculus* (three different size transcripts 2.8, 4.5 and 6.6 kb, the bigger ones lead to active proteins after expression [1]) [1, 10]
<2> *Homo sapiens* (human osteosarcoma [7]) [2, 4, 7]
<3> *Bos taurus* [3, 5, 6, 8, 9]
<4> *Rattus norvegicus* [10]

3 Reaction and Specificity

Catalyzed reaction
peptide L-aspartate + 2-oxoglutarate + O_2 = peptide 3-hydroxy-L-aspartate + succinate + CO_2

Reaction type
hydroxylation
oxidation
redox reaction
reduction

Natural substrates and products

- **S** peptide L-aspartate + 2-oxoglutarate + O_2 <1, 2> (<2> hydroxylates epidermal growth factor-like domains in transformation-associated proteins [2]) (Reversibility: ? <1, 2> [1, 2]) [1, 2]
- **P** peptide 3-hydroxy-L-aspartate + succinate + CO_2 <1, 2> [1, 2]
- **S** Additional information <1, 2> (<1,2> overexpression may be associated with malignant transformation [1,2]) [1, 2]
- **P** ?

Substrates and products

- **S** peptide L-asparagine + 2-oxoglutarate + O_2 <2, 3> (<2> first epidermal growth factor-like domain of bovine protein S with an asparagine replacing the aspartic acid at position 18 [4];<3> first epidermal growth factor-like domain of human protein S with an asparagine replacing the aspartic acid at position 18 [6,8]) (Reversibility: ? <2, 3> [2, 3, 4, 6, 8]) [2, 3, 4, 6, 8]
- **P** peptide 3-hydroxy-L-asparagine + succinate + CO_2
- **S** peptide L-aspartate + 2-oxoglutarate + O_2 <1, 2, 3, 4> (<2> first epidermal growth factor-like domain of bovine protein S as substrate [2]; <3> first epidermal growth factor-like domain of human protein S as substrate [8,9]; <1,3,4> specific erythro-hydroxylation [9,10]; <1,4> first epidermal growth factor-like domain of human factor IX as substrate [10]; <3> second epidermal growth factor-like domain of bovine protein S [5]) (Reversibility: ? <1, 2, 3, 4> [1, 2, 3, 5, 8, 9, 10]) [1, 2, 3, 5, 8, 9, 10]
- **P** peptide 3-hydroxy-L-aspartate + succinate + CO_2 <1, 2, 3, 4> [1, 2, 3, 5, 8, 9, 10]

Inhibitors

2,2'-dipyridyl <1> (<1> at 1 mM 90% inhibition [10]) [10]
iodoacetamide <3> (<3> at 1 mM less than 5% activity, 2-oxoglutarate and EDTA protects [6]) [6]

Metals, ions

Fe^{2+} <1, 3, 4> (<1> 0.05 mM increases activity 6-fold [10]) [5, 6, 8, 9, 10]

Turnover number (min^{-1})

1 <3> (first epidermal growth factor-like domain, <3> substrate is of human protein S, 56 kDa protein [8]) [8]
1.3 <3> (first epidermal growth factor-like domain, <3> substrate is of human protein S with an asparagine replacing the aspartic acid at position 18, 56 kDa protein [8]) [8]
1.5 <3> (first epidermal growth factor-like domain, <3> substrate is of human protein S, 52 kDa protein [8]) [8]
1.9 <3> (first epidermal growth factor-like domain, <3> substrate is of human protein S with an asparagine replacing the aspartic acid at position 18, 52 kDa protein [8]) [8]

Specific activity (U/mg)

0.59 <3> [8]
14 <3> [6]

K_m-Value (mM)

0.003 <3> (Fe^{2+}) [9]
0.005 <3> (2-oxoglutarate) [9]
0.0083 <3> (Fe^{2+}, <3> 56 kDa protein [8]) [8]
0.0096 <3> (Fe^{2+}, <3> 52 kDa protein [8]) [8]
0.013 <3> (Fe^{2+}, <3> at 0.6 mM 2-oxoglutarate, wild-type [5]) [5]
0.019 <3> (first epidermal growth factor-like domain, <3> substrate is of human protein S with an asparagine replacing the aspartic acid at position 18, native enzyme [6]) [6]
0.021 <3> (2-oxoglutarate, <3> 56 kDa protein [8]) [8]
0.022 <3> (2-oxoglutarate, <3> 52 kDa protein [8]) [8]
0.024 <3> (first epidermal growth factor-like domain, <3> substrate is of human protein S with an asparagine replacing the aspartic acid at position 18, recombinant enzyme [6]) [6]
0.034 <3> (first epidermal growth factor-like domain, <3> substrate is of human protein S with an asparagine replacing the aspartic acid at position 18, 52 kDa protein [8]) [8]
0.035 <3> (first epidermal growth factor-like domain, <3> substrate is of human protein S with an asparagine replacing the aspartic acid at position 18, 56 kDa protein [8]) [8]
0.052 <3> (Fe^{2+}, <3> at 0.6 mM 2-oxoglutarate, H675E-mutant [5]) [5]
0.067 <3> (first epidermal growth factor-like domain, <3> substrate is of human protein S, 52 kDa protein [8]) [8]
0.075 <3> (first epidermal growth factor-like domain, <3> substrate is of human protein S, 56 kDa protein [8]) [8]
0.093 <3> (Fe^{2+}, <3> at 0.6 mM 2-oxoglutarate, H675D-mutant [5]) [5]
0.098 <3> (2-oxoglutarate, <3> at 1 mM Fe^{2+}, H675D-mutant [5]) [5]
0.102 <3> (2-oxoglutarate, <3> at 1 mM Fe^{2+}, wild-type [5]) [5]
0.125 <3> (2-oxoglutarate, <3> at 1 mM Fe^{2+}, H675E-mutant [5]) [5]

pH-Optimum

6.8 <3> [9]

pH-Range

6.5-7.9 <3> (<3> about half-maximal activity at pH 6.5 and 7.9 [9]) [9]

4 Enzyme Structure

Molecular weight

50700 <3> (<3> sedimentation equilibrium [6]) [6]
51030 <3> (<3> MALDI-TOF [6]) [6]
52000 <3> (<3> SDS-PAGE, protein with lower molecular weight [8]) [8]
56000 <2> (<2> transfected enzyme, SDS-PAGE [7]) [7]
56000 <3> (<3> SDS-PAGE, protein with higher molecular weight [8]) [8]

Subunits

monomer <3> (<3> 1 * 50700, sedimentation equilibrium [6]) [6]

5 Isolation/Preparation/Mutation/Application

Source/tissue

adrenal gland <1, 2> [1, 4]
bile <2> (<2> proliferating ducts [2]) [2]
brain <1> [1]
breast cancer cell <2> (<2> highly expressed [4]) [4]
cholangiocarcinoma cell <2> (<2> highly expressed in all cholangiocarcinomas [4]) [4]
colonic carcinoma cell <2> (<2> highly expressed [4]) [4]
fat <1> [1]
heart <1> [1]
hepatoma cell <2> (<2> highly expressed in 4 of 10 hepatocarcinomas, 10-fold activity increase [4]) [4]
kidney <1> [1]
large intestine <1> [1]
liver <1, 2, 3, 4> [1, 4, 8, 9, 10]
lung <1> [1]
ovary <1> [1]
pancreas <1> [1]
skeletal muscle <1> [1]
stomach <1> [1]
testis <1> [1]
thymus <1> [1]

Localization

microsome <4> [10]
rough endoplasmic reticulum <3> [9]
Additional information <1> (<1> L-cell extract [10]) [10]

Purification

<3> (two enzymes, one 52000 Da, the other 56000 Da [8]; partial [9]) [6, 8, 9]

Cloning

<2> (transfected into NIH-3T3 cells [2]; in vitro transcription and translation in the presence of canine pancreas microsomes [7]) [2, 7]
<3> (wild-type and mutant enzymes expressed in Escherichia coli [3]; wild-type and mutants expressed in Escherichia coli BL21 [5]; expressed in wild-type Escherichia coli and in Escherichia coli BL21 [6]) [3, 5, 6]

Engineering

C637A <3> (<3> 38% activity of wild-type [3]) [3]
C644A <3> (<3> 62% activity of wild-type [3]) [3]
C656A <3> (<3> 100% activity of wild-type [3]) [3]
C681A <3> (<3> 60% activity of wild-type [3]) [3]
C696A <3> (<3> 29% activity of wild-type [3]) [3]
G659A <3> (<3> 21% activity of wild-type [5]) [5]
G669A <3> (<3> 90% activity of wild-type [5]) [5]

H667L <3> (<3> 106% activity of wild-type [5]) [5]
H671L <3> (<3> 16% activity of wild-type [5]) [5]
H675D <3> (<3> 20% activity of wild-type [5]) [5]
H675E <3> (<3> 12% activity of wild-type [5]) [5]
H686L <3> (<3> 9% activity of wild-type [5]) [5]
P678V <3> (<3> 90% activity of wild-type [5]) [5]
R682A <3> (<3> 10% activity of wild-type [5]) [5]
R684A <3> (<3> 8% activity of wild-type [5]) [5]
R684K <3> (<3> 87% activity of wild-type [5]) [5]

6 Stability

Storage stability

<3>, 4°C, 50 mM Tris-HCl pH 7.2, 1 mg/ml bovine serum albumin, at least 8 weeks stable [8]

References

[1] Dinchuk, J.E.; Henderson, N.L.; Burn, T.C.; Huber, R.; Ho, S.P.; Link, J.; O'Neil, K.T.; Focht, R.J.; Scully, M.S.; Hollis, J.M.; Hollis, G.F.; Friedman, P.A.: Aspartyl β-hydroxylase (Asph) and an evolutionarily conserved isoform of Asph missing the catalytic domain share exons with junctin. J. Biol. Chem., **275**, 39543-39554 (2000)
[2] Ince, N.; De la Monte, S.M.; Wands, J.R.: Overexpression of human aspartyl (asparaginyl) β-hydroxylase is associated with malignant transformation. Cancer Res., **60**, 1261-1266 (2000)
[3] McGinnis, K.; Ku, G.M.; Fu, J.; Stern, A.M.; Friedman, P.A.: The five cysteine residues located in the active site region of bovine aspartyl (asparaginyl) β-hydroxylase are not essential for catalysis. Biochim. Biophys. Acta, **1387**, 454-456 (1998)
[4] Lavaissiere, L.; Jia, S.; Nishiyama, M.; de la Monte, S.; Stern, A.M.; Wands, J.R.; Friedman, P.A.: Overexpression of human aspartyl(asparaginyl)β-hydroxylase in hepatocellular carcinoma and cholangiocarcinoma. J. Clin. Invest., **98**, 1313-1323 (1996)
[5] McGinnis, K.; Ku, G.M.; VanDusen, W.J.; Fu, J.; Garsky, V.; Stern, A.M.; Friedman, P.A.: Site-directed mutagenesis of residues in a conserved region of bovine aspartyl (asparaginyl) β-hydroxylase: Evidence that histidine 675 has a role in binding Fe^{2+}. Biochemistry, **35**, 3957-3962 (1996)
[6] Jia, S.; McGinnis, K.; VanDusen, W.J.; Burke, C.J.; Kuo, A.; Griffin, P.R.; Sardana, M.K.; Elliston, K.O.; Stern, A.M.; Friedman, P.A.: A fully active catalytic domain of bovine aspartyl (asparaginyl) β-hydroxylase expressed in Escherichia coli: characterization and evidence for the identification of an active-site region in vertebrate a-ketoglutarate-dependent dioxygenases. Proc. Natl. Acad. Sci. USA, **91**, 7227-7231 (1994)

[7] Korioth, F.; Gieffers, C.; Frey, J.: Cloning and characterization of the human gene encoding aspartyl β-hydroxylase. Gene, **150**, 395-399 (1994)
[8] Wang, Q.; VanDusen, W.J.; Petrosky, C.J.; Garsky, V.M.; Stern, A.M.; Friedman, P.A.: Bovine liver aspartyl β-hydroxylase. Purification and characterization. J. Biol. Chem., **266**, 14004-14010 (1991)
[9] Gronke, R.S.; Welsch, D.J.; VanDusen, W.J.; Garsky, V.M.; Sardana, M.K.; Stern, A.M.; Friedman, P.A.: Partial purification and characterization of bovine liver aspartyl β-hydroxylase. J. Biol. Chem., **265**, 8558-8565 (1990)
[10] Gronke, R.S.; VanDusen, W.J.; Garsky, V.M.; Jacobs, J.W.; Sardana, M.K.; Stern, A.M.; Friedman, P.A.: Aspartyl β-hydroxylase: in vitro hydroxylation of a synthetic peptide based on the structure of the first growth factor-like domain of human factor IX. Proc. Natl. Acad. Sci. USA, **86**, 3609-3613 (1989)

Taurine dioxygenase 1.14.11.17

1 Nomenclature

EC number
1.14.11.17

Systematic name
taurine, 2-oxoglutarate:O_2 oxidoreductase (sulfite-forming)

Recommended name
taurine dioxygenase

Synonyms
2-aminoethanesulfonate dioxygenase
SSI3
α-ketoglutarate-dependent taurine dioxygenase

CAS registry number
197809-75-9
297319-14-3
325506-70-5
325506-70-5

2 Source Organism

<1> *Escherichia coli* (strain MC4100 [1]) [1]

3 Reaction and Specificity

Catalyzed reaction
taurine + 2-oxoglutarate + O_2 = sulfite + aminoacetaldehyde + succinate + CO_2

Reaction type
oxidation
redox reaction
reduction

Substrates and products
S 1,3-dioxo-2-isoindolineenthanesulfonic acid + 2-oxoglutarate + O_2 <1> (Reversibility: ? <1> [1]) [1]
P sulfite+ ? + succinate + CO_2

S MOPS + 2-oxoglutarate + O_2 <1> (Reversibility: ? <1> [1]) [1]
P sulfite+ ? + succinate + CO_2
S butanesulfonic acid + 2-oxoglutarate + O_2 <1> (Reversibility: ? <1> [1]) [1]
P sulfite+ butanal + succinate + CO_2
S hexanesulfonic acid + 2-oxoglutarate + O_2 <1> (Reversibility: ? <1> [1]) [1]
P sulfite+ hexanal + succinate + CO_2
S pentanesulfonic acid + 2-oxoglutarate + O_2 <1> (Reversibility: ? <1> [1]) [1]
P sulfite+ pentanal + succinate + CO_2
S taurine + 2-oxoglutarate + O_2 <1> (<1> no substrates are methanesulfonic acid, ethanesulfonic acid, isethionic acid, 2-bromoethanesulfonic acid, L-cysteic acid, sulfosuccinate, 4-aminobenzenesulfonic acid, 2-(4-pyridyl)ethanesulfonic acid, N-phenyltaurine [1]) (Reversibility: ? <1> [1]) [1]
P sulfite+ aminoacetaldehyde + succinate + CO_2
S taurine + α-ketoadipate + O_2 <1> (<1> α-ketoadipate is less active than 2-oxoglutarate, no activity with pyruvate, α-ketobutyrate, α-ketovalerate, α-ketocaproate, α-ketoisovalerate and oxalacetat [1]) (Reversibility: ? <1> [1]) [1]
P sulfite+ aminoacetaldehyde + pentan-1,5-dioic acid + CO_2

Inhibitors

Co^{2+} <1> (<1> inhibits activity by 80-95% at 0.01-0.05 mM [1]) [1]
Cu^{2+} <1> (<1> inhibits activity by 80-95% at 0.01-0.05 mM [1]) [1]
EDTA <1> (<1> complete inactivation [1]) [1]
Zn^{2+} <1> (<1> inhibits activity by 80-95% at 0.01-0.05 mM [1]) [1]

Activating compounds

ascorbate <1> (<1> 50% increase in activity at 0.2-0.8 mM [1]) [1]

Metals, ions

Fe^{2+} <1> (<1> maximal activation between 0.005 and 0.150 mM [1]) [1]
Additional information <1> (<1> Mg^{2+}, Ca^{2+}, Mn^{2+} or Ni^{2+} can not replace iron [1]) [1]

Specific activity (U/mg)

0.012 <1> (<1> strain MC4100 grown in sulfate-free minimal medium containing 0.25 mM taurine as sulfur source [1]) [1]
1.64 <1> (<1> purified enzyme [1]) [1]

K_m-Value (mM)

0.011 <1> (2-oxoglutarate) [1]
0.055 <1> (taurine) [1]
0.145 <1> (MOPS) [1]
0.485 <1> (1,3-dioxo-2-isoindolineenthanesulfonic acid) [1]
0.59 <1> (pentanesulfonic acid) [1]
1.49 <1> (butanesulfonic acid) [1]
1.51 <1> (hexanesulfonic acid) [1]

pH-Optimum
6.9 <1> [1]

4 Enzyme Structure

Molecular weight
32410 <1> (<1> calculation from gene sequence [1]) [1]
37400 <1> (<1> estimated by SDS-PAGE [1]) [1]
81000 <1> (<1> gel filtration on Superose 6 and Superose 12 HR [1]) [1]

Subunits
dimer <1> (<1> 2*37400, SDS-PAGE [1]) [1]

5 Isolation/Preparation/Mutation/Application

Purification
<1> (two-step purification from overexpressing Escherichia coli to apparent homogeneity [1]) [1]

Cloning
<1> [1]

6 Stability

Temperature stability
30 <1> (<1> incubation at 30°C leads to rapid inactivation, effect is enhanced by ascorbate and not due to oxidation of the enzyme-bound ferrous iron [1]) [1]

Storage stability
<1>, -20°C, phosphate buffer, 16% glycerol, 10 weeks, activity increases 4-fold [1]
<1>, -20°C, phosphate buffer, without glycerol, 3 weeks, more than 50% loss of activity [1]

References

[1] Eichhorn, E.; Van der Ploeg, J.R.; Kertesz, M.A.; Leisinger, T.: Characterization of α-ketoglutarate-dependent taurine dioxygenase from Escherichia coli. J. Biol. Chem., **272**, 23031-23036 (1997)

Phytanoyl-CoA dioxygenase 1.14.11.18

1 Nomenclature

EC number
1.14.11.18

Systematic name
phytanoyl-CoA, 2-oxoglutarate:oxygen oxidoreductase (2-hydroxylating)

Recommended name
phytanoyl-CoA dioxygenase

Synonyms
PAHX
PhyH
phytanic acid oxidase
phytanoyl-CoA α-hydroxylase
phytanoyl-CoA hydroxylase

CAS registry number
185402-46-4

2 Source Organism

<1> *Rattus norvegicus* (rat [1, 5, 8, 9, 10]) [1, 5, 8, 9, 10]
<2> *Homo sapiens* (human [2-6, 7, 9, 10]) [2-6, 7, 9, 10]
<3> *monkey* [9]
<4> *Mus musculus* (mouse [9]) [9]

3 Reaction and Specificity

Catalyzed reaction
phytanoyl-CoA + 2-oxoglutarate + O_2 = 2-hydroxyphytanoyl-CoA + succinate + CO_2

Reaction type
oxidation
redox reaction
reduction

Substrates and products

S 3-methylhexadecanoyl-CoA + 2-oxoglutarate + O_2 <1, 2> (Reversibility: ? <1, 2> [10]) [10]

P 2-hydroxy-3-methylhexadecanoyl-CoA + succinate + CO_2

S phytanoyl-CoA + 2-oxoglutarate + O_2 <1, 2> (<2> no activity with octadecanoyl-CoA, lignoceroyl-CoA, 2-methylhexadecanoyl-CoA and 4,8,12-trimethyltridecanoyl-CoA [10]) (Reversibility: ? <1, 2> [1-4, 6-10]) [1-4, 6-10]

P α-hydroxyphytanoyl-CoA + succinate + CO_2

Inhibitors

2-methylhexadecanoyl-CoA <2> (<2> 56% of control activity [10]) [10]
hexadecanoyl-CoA <2> (<2> 74% of control activity [10]) [10]
propyl gallate <2> (<2> no activity with 1 mM, interacts with iron binding [4]) [4]
Additional information <2> (<2> not inhibited by bifonazole, clotrimazole, miconazole, ketoconazole [4]) [4]

Cofactors/prosthetic groups

ATP <1, 2> [10]
GTP <2> [10]
Additional information <2> (<2> UTP, CTP, ITP, AMP, ADP, NAD^+ and FAD can not act as cofactors, ATP and GTP can be replaced by adenosine-5'-O-(3-thiotriphosphate), adenylylimidodiphosphate, adenylyl-(β,γ-methylene)-diphosphonate and guanylyl-imidodiphosphate [10]) [10]

Activating compounds

ascorbate <1, 2> (<1> at 1mM [1,8]; <2> 2.5fold induction at 5 mM [4]) [1, 4, 8]

Metals, ions

Fe^{2+} <1, 2> (<1,2> requirement [1-4,8,10]; <2> optimal activity at 0.5 mM [4]) [1-4, 10]
Mg^{2+} <1> [10]
Additional information <2> (<2> Cu^{2+}, Mn^{2+} or Zn^{2+} can not replace Fe^{2+} [4]) [4]

Specific activity (U/mg)

0.000001 <3> (<3> in kidney cell line COS-1, activity can be induced 4fold by phytanic acid [9]) [9]
0.0000018 <4> (<4> in embryonic carcinoma cell line P19-EC, activity can be induced 4fold by phytanic acid [9]) [9]
0.000002 <2> (<2> in kidney cell line 293, activity can be induced 4fold by phytanic acid [9]) [9]
0.0000022 <2> (<2> in fibroblast homogenate [7]) [7]
0.0000031 <2> (<2> in hepatoma cell line HepG2, no induction by phytanic acid [9]) [9]
0.000024 <1> (<1> in hepatoma cell line FaO, activity can be induced 2.5fold by phytanic acid [9]) [9]

0.000041 <2> (<2> in liver homogenate [2,4,7]) [2, 4, 7]
0.0001 <1> (<1> in purified peroxisomes [1]) [1]
0.00026 <1> (<1> in purified liver peroxisomes [8]) [8]

K_m-Value (mM)
0.049 <2> (2-oxoglutarate) [4]

pH-Optimum
7.5 <1, 2> [4, 8]

pH-Range
6.5-8.5 <2> (<2> at pH 6.5 over 80%, at pH 8.5 50% of control activity [4]) [4]

4 Enzyme Structure

Molecular weight
33000 <1> (<1> purified protein, SDS-PAGE [5]) [5]
35000 <1> (<1> purified protein, SDS-PAGE [8]) [8]
35400 <2> (<2> mature protein after cleavage of presumed leader sequence, calculation from cDNA sequence [5]) [5]
38600 <1> (<1> calculated from amino acid sequence [8]) [8]
41200 <2> (<2> precursor protein with peroxisomal targeting signal type 2, calculation from cDNA sequence [5]) [5]

5 Isolation/Preparation/Mutation/Application

Source/tissue
embryonic cell line <4> (embryonic carcinoma cell line) [9]
hepatoma cell line <1, 2> [9]
kidney <2, 3> [9]
liver <1, 2> [1, 2, 4, 5, 7, 8, 10]
skin fibroblast <2> [6, 7]

Localization
peroxisome <1, 2> [1, 2, 5, 8]

Purification
<1> [5, 8]

Cloning
<1> (in yeast expression vector pEL26 and pEL30 [8]) [8]
<2> (in bacterial expression vector pMALc2 [3]; in mammalian expression vector pcDNA3 [6]; in yeast expression vector [7]; in bacterial expression vector pQE-31 [10]) [3, 6, 7, 10]

References

[1] Mihalik, S.J.; Rainville, A.M.; Watkins, P.A.: Phytanic acid oxidation in rat liver peroxisomes. Production of α-hydroxyphytanoyl-CoA and formate is enhanced by dioxygenase cofactors. Eur. J. Biochem., **232**, 545-551 (1995)

[2] Jansen, G.A.; Mihalik, S.J.; Watkins, P.A.; Moser, H.W.; Jakobs, C.; Denis, C.; Wanders, R.J.A.: Phytanoyl-CoA hydroxylase is present in human liver, located in peroxisomes, and deficient in Zellweger syndrome: direct, unequivocal evidence for the new, revised pathway of phytanic acid α-oxidation in humans. Biochem. Biophys. Res. Commun., **229**, 205-210 (1996)

[3] Mihalik, S.J.; Morrell, J.C.; Kim, D.; Sacksteder, K.A.; Watkins, P.A.; Gould, S.J.: Identification of PAHX, a Refsum disease gene. Nature Genet., **17**, 185-189 (1997)

[4] Jansen, G.A.; Mihalik, S.J.; Watkins, P.A.; Jakobs, C.; Moser, H.W.; Wanders, R.J.A.: Characterization of phytanoyl-Coenzyme A hydroxylase in human liver and activity measurements in patients with peroxisomal disorders. Clin. Chim. Acta, **271**, 203-211 (1998)

[5] Jansen, G.A.; Ofman, R.; Ferdinandusse, S.; Ijlst, L.; Muijsers, A.O.; Skjeldal, O.H.; Stokke, O.; Jakobs, C.; Besley, G.T.N.; Wraith, J.E.; Wanders, R.J.A.: Refsum disease is caused by mutations in the phytanoyl-CoA hydroxylase gene. Nature Genet., **17**, 190-193 (1997)

[6] Chahal, A.; Khan, M.; Pai, S.G.; Barbosa, E.; Singh, I.: Restoration of phytanic acid oxidation in Refsum disease fibroblasts from patients with mutations in the phytanoyl-CoA hydroxylase gene. FEBS Lett., **429**, 119-122 (1998)

[7] Jansen, G.A.; Ferdinandusse, S.; Hogenhout, E.M.; Verhoeven, N.M.; Jakobs, C.; Wanders, R.J.A.: Phytanoyl-CoA hydroxylase deficiency. Enzymological and molecular basis of classical Refsum disease. Adv. Exp. Med. Biol., **466**, 371-376 (1999)

[8] Jansen, G.A.; Ofman, R.; Denis, S.; Ferdinandusse, S.; Hogenhout, E.M.; Jakobs, C.; Wanders, R.J.A.: Phytanoyl-CoA hydroxylase from rat liver: protein purification and cDNA cloning with implications for the subcellular localization of phytanic acid α-oxidation. J. Lipid Res., **40**, 2244-2254 (1999)

[9] Zomer, A.W.M.; Jansen, G.A.; Van der Burg, B.; Verhoeven, N.M.; Jakobs, C.; Van der Saag, P.T.; Wanders, R.J.A.; Poll-The, B.T.: Phytanoyl-CoA hydroxylase activity is induced by phytanic acid. Eur. J. Biochem., **267**, 4063-4067 (2000)

[10] Croes, K.; Foulon, V.; Casteels, M.; Van Veldhoven, P.P.; Mannaerts, G.P.: Phytanoyl-CoA hydroxylase: recognition of 3-methyl-branched acyl-CoAs and requirement for GTP or ATP and Mg^{2+} in addition to its known hydroxylation cofactors. J. Lipid Res., **41**, 629-636 (2000)

Leucocyanidin oxygenase 1.14.11.19

1 Nomenclature

EC number
1.14.11.19

Systematic name
leucocyanidin,2-oxoglutarate:oxygen oxidoreductase

Recommended name
leucocyanidin oxygenase

Synonyms
anthocyanidin synthase <1> [1]
anthocyanin synthase
leucocyanidin dioxygenase

CAS registry number
180984-01-4

2 Source Organism

<1> *Arabidopsis thaliana* [1, 4, 5]
<2> *Perilla frutescens* [2, 3, 5]
<3> *Antirrhinum majus* (snapdragon [3]) [3]
<4> *Zea mays* (maize [3]) [3]
<5> *Petunia hybrida* [3]
<6> *Torenia fournieri* [3]

3 Reaction and Specificity

Catalyzed reaction
leucocyanidin + 2-oxoglutarate + O_2 = cis- and trans-dihydroquercetins + succinate + CO_2 (<1,2> mechanism [1,2])

Reaction type
oxidation
redox reaction
reduction

Substrates and products

S leucocyanidin + 2-oxoglutarate + O_2 <1-6> (<2-6> using (2R,3S,4R)-leucocyanidin or (2R,3S,4R)-leucopelargonidin as substrates [3]; <1,2> more detail to stereochemistry [5]) (Reversibility: ir <1-6> [1-5]) [1-5]

P cis-dihydroquercetin + trans-dihydroquercitin + succinate + CO_2 <1-6> [1-5]

Activating compounds

2-oxoglutarate <2, 4, 5> [2, 3]
ascorbate <4, 5> [3]

Metals, ions

Fe^{2+} <1, 4, 5> (<1> contains [1]; <4,5> dependent on [3]) [1, 3]

Specific activity (U/mg)

Additional information <2> [2]

K_m-Value (mM)

0.038 <2> ((2R,3S,4R)-leucocyanidin) [2]
0.059 <2> (2-oxoglutarate) [2]

5 Isolation/Preparation/Mutation/Application

Source/tissue

whole plant <2> (<2> perilla, red or green [2]) [2]

Purification

<1> (of recombinant enzyme [4]) [4]
<2> (of recombinant enzyme [2]) [2]
<3> (of recombinant enzyme [3]) [3]
<4> (of recombinant enzyme [3]) [3]
<5> (of recombinant enzyme [3]) [3]
<6> (of recombinant enzyme [3]) [3]

Crystallization

<1> [1, 4]

Cloning

<1> (overexpression in Escherichia coli [4]) [4]
<2> (overexpression in Escherichia coli [2,3]) [2, 3]
<3> (overexpression in Escherichia coli [3]) [3]
<4> (overexpression in Escherichia coli [3]) [3]
<5> (overexpression in Escherichia coli [3]) [3]
<6> (overexpression in Escherichia coli [3]) [3]

References

[1] Wilmouth, R.C.; Turnbull, J.J.; Welford, R.W.D.; Clifton, I.J.; Prescott, A.G.; Schofield, C.J.: Structure and mechanism of anthocyanidin synthase from Arabidopsis thaliana. Structure, **10**, 93-103 (2002)
[2] Saito, K.; Kobayashi, M.; Gong, Z.; Tanaka, Y.; Yamazaki, M.: Direct evidence for anthocyanidin synthase as a 2-oxoglutarate-dependent oxygenase: molecular cloning and functional expression of cDNA from a red forma of Perilla frutescens. Plant J., **17**, 181-189 (1999)
[3] Nakajima, J.I.; Tanaka, Y.; Yamazaki, M.; Saito, K.: Reaction mechanism from leucoanthocyanidin to anthocyanidin 3-glucoside, a key reaction for coloring in anthocyanin biosynthesis. J. Biol. Chem., **276**, 25797-25803 (2001)
[4] Turnbull, J.J.; Prescott, A.G.; Schofield, C.J.; Wilmouth, R.C.: Purification, crystallization and preliminary x-ray diffraction of anthocyanidin synthase from Arabidopsis thaliana. Acta Crystallogr. Sect. D, **57**, 425-427 (2001)
[5] Turnbull, J.J.; Sobey, W.J.; Aplin, R.T.; Hassan, A.; Schofield, C.J.; Firmin, J.L.; Prescott, A.G.: Are anthocyanidins the immediate products of anthocyanidin synthase?. Chem. Commun., **2000**, 2473-2474 (2000)

Desacetoxyvindoline 4-hydroxylase 1.14.11.20

1 Nomenclature

EC number
1.14.11.20

Systematic name
desacetoxyvindoline,2-oxoglutarate:oxygen oxidoreductase (4β-hydroxylating)

Recommended name
desacetoxyvindoline 4-hydroxylase

CAS registry number
132084-83-4

2 Source Organism

<1> *Catharanthus roseus* (enzyme is induced after exposure of seedlings to light [1]) [1, 2, 3]
<2> *Catharanthus roseus* [4]
<3> *Catharanthus roseus* (different isoforms in etiolated and light-grown seedlings [5]) [4, 5]
<4> *Catharanthus roseus* [4]

3 Reaction and Specificity

Catalyzed reaction
desacetoxyvindoline + 2-oxoglutarate + O_2 = desacetylvindoline + succinate + CO_2 (<1> ordered Ter Ter mechanism is suggested, deacetylvindoline is the first product released, followed by Co_2 and succinate [3])

Reaction type
oxidation
redox reaction
reduction

Substrates and products
S desacetoxyvindoline + 2-oxoglutarate + O <1> (<1> strictly specific for position 4, no hydroxylation of indole alkaloid substrates with a 2,3-double bond [1]) (Reversibility: ? <1> [1]) [1]
P desacetylvindoline + succinate + CO_2 <1> [1]

Inhibitors

CO <1> (<1> 7.5 mM, 50% inhibition [2]) [2]
desacetylvindoline <1> (<1> product inhibition, noncompetitive vs. 2-oxoglutarate [2]) [2]
succinate <1> (<1> product inhibition, competitive vs. 2-oxoglutarate, noncompetitive vs. desacetoxyvindoline [2]) [2]

Cofactors/prosthetic groups

2-oxoglutarate <1> (<1> absolutely required for activity [1]) [1, 2]

Activating compounds

ascorbic acid <1> (<1> in vitro required for maximal activity [1]) [1, 2]

Metals, ions

Fe^{2+} <1> [2]

Specific activity (U/mg)

0.00517 <1> [2]

K_m-Value (mM)

0.000003 <1> (desacetoxyvindoline) [2]
0.0085 <1> (Fe^{2+}) [2]
0.045 <1> (2-oxoglutarate) [2]
0.045 <1> (O_2) [2]
0.2 <1> (ascorbate) [2]

K_i-Value (mM)

0.115 <1> (deacetylvindoline) [2]
9 <1> (succinate) [2]

pH-Optimum

7.5 <1> [1]

4 Enzyme Structure

Molecular weight

45000 <1> (<1> gel filtration [1]) [1, 2, 3]

Subunits

monomer <1> (<1> 1 * 44700, SDS-PAGE [2]) [2, 3]

5 Isolation/Preparation/Mutation/Application

Source/tissue

fruit <2, 3, 4> (<2,3,4> 5% of activity in leaf [4]) [4]
leaf <1, 2, 3, 4> (<2,3,4> highest activity in leaf [4]) [1, 4]
seedling <3> [5]
stem <2, 3, 4> (<2,3,4> 8% of activity in leaf [4]) [4]

Purification

<1> (Sephadex G-100, Green 19-agarose, hydroxylapatite, 2-oxoglutarate-Sepharose, Mono-Q [2]) [2, 3]

Cloning

<2> (cloning of cDNA, possibly dimorphic allele of a single-copy gene [4]) [4]

<3> (cloning of cDNA, expression in Escherichia coli, possibly dimorphic allele of a single-copy gene [4]) [4, 5]

<4> (cloning of cDNA, possibly dimorphic allele of a single-copy gene [4]) [4]

References

[1] De Carolis, E.; Chan, F.; Balsevich, J.; De Luca, V.: Isolation and characterization of a 2-oxoglutarate dependent dioxygenase involved in the second-to-last step in vindoline biosynthesis. Plant Physiol., **94**, 1323-1329 (1990)

[2] De Carolis, E.; De Luca, V.: Purification, characterization, and kinetic analysis of a 2-oxoglutarate-dependent dioxygenase involved in vindoline biosynthesis from Catharanthus roseus. J. Biol. Chem., **268**, 5504-5511 (1993)

[3] Carolis, E.D.; Luca, V.D.: A novel 2-oxoglutarate-dependent dioxygenase involved in vindoline biosynthesis: Characterization, purification and kinetic properties. Plant Cell Tissue Organ Cult., **38**, 281-287 (1994)

[4] Vazquez-Flota, F.; De Carolis, E.; Alarco, A.M.; De Luca, V.: Molecular cloning and characterization of desacetoxyvindoline-4-hydroxylase, a 2-oxoglutarate dependent-dioxygenase involved in the biosynthesis of vindoline in Catharanthus roseus (L.) G. Don. Plant Mol. Biol., **34**, 935-948 (1997)

[5] Vazquez-Flota, F.A.; De Luca, V.: Developmental and light regulation of desacetoxyvindoline 4-hydroxylase in Catharanthus roseus (L.) G. Don. Evidence of a multilevel regulatory mechanism. Plant Physiol., **117**, 1351-1361 (1998)

Clavaminate synthase 1.14.11.21

1 Nomenclature

EC number
1.14.11.21

Systematic name
deoxyamidinoproclavaminate,2-oxoglutarate:oxygen oxidoreductase (3-hydroxylating)

Recommended name
clavaminate synthase

Synonyms
clavaminate synthase 2
clavaminic acid synthase

CAS registry number
122799-56-8

2 Source Organism

<1> *Streptomyces clavuligerus* [1-5]

3 Reaction and Specificity

Catalyzed reaction
deoxyamidinoproclavaminate + 2-oxoglutarate + O_2 = amidinoproclavaminate + succinate + CO_2 + H_2O
dihydroclavaminate + 2-oxoglutarate + O_2 = clavaminate + CO_2 + 2 H_2O (Contains nonheme iron. Catalyses three separate oxidative reactions in the pathway for the biosynthesis of the β-lactamase inhibitor clavulanate in Streptomyces clavuligerus. The first step (hydroxylation) is separated from the latter two (oxidative cyclization and desaturation) by the action of EC 3.5.3.22, proclavaminate amidinohydrolase. The three reactions are all catalysed at the same nonheme iron site)
proclavaminate + 2-oxoglutarate + O_2 = dihydroclavaminate + CO_2 + 2 H_2O

Natural substrates and products
S deoxyamidinoproclavaminate + 2-oxoglutarate + O_2 <1> (Reversibility: ? <1> [1-5]) [1-5]
P amidinoproclavaminate + succinate + CO_2 + H_2O

S dihydroclavaminate + 2-oxoglutarate + O_2 <1> (Reversibility: ? <1> [1-5]) [1-5]
P clavaminate + CO_2 + H_2O
S proclavaminate + 2-oxoglutarate + O_2 <1> (Reversibility: ? <1> [1-5]) [1-5]
P dihydroclavaminate + CO_2 + H_2O

Substrates and products
S deoxyamidinoproclavaminate + 2-oxoglutarate + O_2 <1> (Reversibility: ? <1> [1-5]) [1-5]
P amidinoproclavaminate + succinate + CO_2 + H_2O
S dihydroclavaminate + 2-oxoglutarate + O_2 <1> (Reversibility: ? <1> [1-5]) [1-5]
P clavaminate + CO_2 + H_2O
S proclavaminate + 2-oxoglutarate + O_2 <1> (Reversibility: ? <1> [1-5]) [1-5]
P dihydroclavaminate + CO_2 + H_2O

References

[1] Salowe, S.P.; Krol, W.J.; Iwatareuyl, D.; Townsend, C.A.: Elucidation of the order of oxidations and identification of an intermediate in the multistep clavaminate synthase reaction. Biochemistry, **30**, 2281-2292 (1991)
[2] Zhou, J.; Gunsior, M.; Bachmann, B.O.; Townsend, C.A.; Solomon, E.I.: Substrate binding to the α-ketoglutarate-dependent non-heme iron enzyme clavaminate synthase 2: Coupling mechanism of oxidative decarboxylation and hydroxylation. J. Am. Chem. Soc., **120**, 13539-13540 (1998)
[3] Zhang, Z.H.; Ren, J.S.; Stammers, D.K.; Baldwin, J.E.; Harlos, K.; Schofield, C.J.: Structural origins of the selectivity of the trifunctional oxygenase clavaminic acid synthase. Nat. Struct. Biol., **7**, 127-133 (2000)
[4] Zhou, J.; Kelly, W.L.; Bachmann, B.O.; Gunsior, M.; Townsend, C.A.; Solomon, E.I.: Spectroscopic studies of substrate interactions with clavaminate synthase 2, a multifunctional α-KG-dependent non-heme iron enzyme: Correlation with mechanisms and reactivities. J. Am. Chem. Soc., **123**, 7388-7398 (2001)
[5] Townsend, C.A.: New reactions in clavulanic acid biosynthesis. Curr. Opin. Chem. Biol., **6**, 583-589 (2002)

Anthranilate 1,2-dioxygenase (deaminating, decarboxylating) 1.14.12.1

1 Nomenclature

EC number
1.14.12.1

Systematic name
anthranilate,NAD(P)H:oxygen oxidoreductase (1,2-hydroxylating, deaminating, decarboxylating)

Recommended name
anthranilate 1,2-dioxygenase (deaminating, decarboxylating)

Synonyms
anthranilate hydroxylase
anthranilic acid hydroxylase
anthranilic hydroxylase
two-component anthranilate 1,2-dioxygenase <2> [7]

CAS registry number
9059-17-0

2 Source Organism

<1> *Acinetobacter calcoaceticus* [3]
<2> *Acinetobacter sp.* (strain ADP1, derived from BD413 [6,7]) [6, 7]
<3> *Pseudomonas aeruginosa* [2]
<4> *Pseudomonas aureofaciens* (wild-type ATCC 15926, ATCC 29574 [4]) [4]
<5> *Pseudomonas fluorescens* (No.23 (ATCC 11250) [1]) [1]
<6> *Pseudomonas sp.* [6, 7]
<7> *Trichosporon cutaneum* [5]

3 Reaction and Specificity

Catalyzed reaction
anthranilate + NAD(P)H + H^+ + O_2 + 2 H_2O = catechol + CO_2 + $NAD(P)^+$ + NH_3 (requires Fe^{2+})

Reaction type
deamination
decarboxylation
oxidation

redox reaction
reduction

Natural substrates and products

S anthranilic acid + NAD(P)H + O_2 + H_2O <1-7> (Reversibility: ? <1-7> [1-7]) [1-7]

P catechol + CO_2 + $NAD(P)^+$ + NH_3

Substrates and products

S 2,3-dihydroxybenzoic acid + NAD(P)H + O_2 + H_2O <5> (Reversibility: ? <5> [1]) [1]

P ?

S anthranilic acid + NAD(P)H + O_2 + H_2O <1-7> (Reversibility: ? <1-7> [1-7]) [1-7]

P catechol + CO_2 + $NAD(P)^+$ + NH_3

S catechol + O_2 + NAD(P)H + H_2O <5> (Reversibility: ? <5> [1]) [1]

P ?

S protocatechuic acid + NAD(P)H + O_2 + H_2O <5> (Reversibility: ? <5> [1]) [1]

P ?

S trans-benzeneglycol + NAD(P)H + O_2 + H_2O <5> (Reversibility: ? <5> [1]) [1]

P ?

S Additional information <5> (<5> salicylic acid and *o*-aminophenol are no substrates [1]) [1]

P ?

Inhibitors

$HgCl_2$ <5> [1]
p-chloromercuribenzoate <5> [1]

Cofactors/prosthetic groups

NADH <5> [1]
NADPH <5> (<5> NAD^+ + $NADP^+$ do not replace NADH or NADPH [1]) [1]

Metals, ions

Fe^{2+} <6> (<6> required [6]) [6]

K_m-Value (mM)

0.12 <4> (anthranilate) [4]

pH-Optimum

7.5 <5> [1]

4 Enzyme Structure

Subunits

trimer <2> (<2> 1 * 19000 + 1 * 39000 + 1 * 54000, DNA-sequence [6]) [6]

5 Isolation/Preparation/Mutation/Application

Purification

<2> (recombinant protein, expressed in Escherichia coli [7]) [7]
<5> (partially [1]) [1]

Cloning

<2> (clustered antABC genes from ADP1 chromosome encodes anthranilate dioxygenase [6,7]; expressed in Escherichia coli [7]) [6, 7]

References

[1] Taniuchi, H.; Hatanaka, M.; Kuno, S.; Hayaishi, O.; Nakajima, M.; Kurihara, N.: Enzymatic formation of catechol from anthranilic acid. J. Biol. Chem., **239**, 2204-2211 (1964)
[2] Roenberg, S.L.; Hegeman, G.D.: Genetics of the mandelate pathway in Pseudomonas aeruginosa. J. Bacteriol., **108**, 1270-1276 (1971)
[3] Wheels, M.L.: Regulation of the synthesis of enzymes of tryptophan dissimilation in Acinetobacter calcoaceticus. Arch. Mikrobiol., **87**, 1-9 (1972)
[4] Salcher, O.; Lingens, F.: Metabolism of tryptophan by Pseudomonas aureofaciens and its relationship to pyrrolnitrin biosynthesis. J. Gen. Microbiol., **121**, 465-471 (1980)
[5] Macheroux, Peter; Kojiro, C.L.; Schopfer, L.M.; Chakraborty, S.; Massey, V.: F NMR studies on fluoroflavins and 8-fluoroflavoproteins. Biochemistry, **29**, 2670-2679 (1990)
[6] Bundy, B.M.; Campbell, A.L.; Neidle, E.L.: Similarities between the antABC-encoded anthranilate dioxygenase and the benABC-encoded benzoate dioxygenase of Acinetobacter sp. strain ADP1. J. Bacteriol., **180**, 4466-4474 (1998)
[7] Eby, D.M.; Beharry, Z.M.; Coulter, E.D.; Kurtz, D.M., Jr.; Neidle, E.L.: Characterization and evolution of anthranilate 1,2-dioxygenase from Acinetobacter sp. strain ADP1. J. Bacteriol., **183**, 109-118 (2001)

Anthranilate 2,3-dioxygenase (deaminating) 1.14.12.2

1 Nomenclature

EC number
1.14.12.2 (transferred to EC 1.14.13.35)

Recommended name
anthranilate 2,3-dioxygenase (deaminating)

Benzene 1,2-dioxygenase 1.14.12.3

1 Nomenclature

EC number
1.14.12.3

Systematic name
benzene,NADH:oxygen oxidoreductase (1,2-hydroxylating)

Recommended name
benzene 1,2-dioxygenase

Synonyms
BDO
benzene dioxygenase
benzene hydroxylase
oxygenase, benzene 1,2-di-

CAS registry number
9075-66-5

2 Source Organism

<1> *Pseudomonas putida* (strain ML2 [2, 6, 7, 8, 9, 10]) [1-4, 6, 7, 8, 9, 10]
<2> *Pseudomonas sp.* [5]

3 Reaction and Specificity

Catalyzed reaction
benzene + NADH + + H^+ + O_2 = cis-cyclohexa-3,5-diene-1,2-diol + NAD^+

Reaction type
oxidation
redox reaction
reduction

Natural substrates and products
S benzene + NADH + O_2 <1, 2> (<1,2>, reaction in benzene catabolism [1-8]) (Reversibility: ? <1, 2> [1-8]) [1-8]
P ?

Substrates and products

- **S** benzene + NADH + O_2 <1, 2> (Reversibility: ? <1, 2> [1-10]) [1-10]
- **P** cis-cyclohexa-3,5-diene-1,2-diol + NAD^+ <1, 2> (<1>, the enzyme oxidizes cyclohexene to a mixture of two products, a monohydroxylated 2-cyclohexen-1-ol product and a dihydroxylated cis-1,2-cyclohexanediol [10]) [1-10]
- **S** benzene + NADPH + O_2 <2> (<2>, activity with NADPH is 10% of the activity with NADH [5]) (Reversibility: ? <2> [5]) [5]
- **P** cis-cyclohexa-3,5-diene-1,2-diol + NAD^+
- **S** trifluoromethylbenzene + NADH + O_2 <1> [7]
- **P** ?

Cofactors/prosthetic groups

FAD <1> (<1>, contains FAD [7]) [7]

NADH <1, 2> [1, 2, 3, 4, 5, 6, 7]

NADPH <2> (<2>, activity with NADPH is 10% of the activity with NADH [5]) [5]

ferredoxin <1> (<1>, the benzene dioxygenase comprises three components: 1. flavoprotein, NADH:ferredoxin oxidoreductase, 2. an intermediate electron-transfer protein, or ferredoxin and 3. a terminal dioxygenase [4, 7, 8]) [4, 7, 8]

Metals, ions

Fe^{2+} <1, 2> (<1>, the benzene dioxygenase comprises three components: 1. flavoprotein, NADH:ferredoxin oxidoreductase, 2. an intermediate electron-transfer protein, or ferredoxin with a [2Fe-2S] cluster and 3. a terminal dioxygenase, containing two [2Fe-2S]iron-sulfur clusters which require two additional Fe^{2+} atoms/molecules for oxygenase activity [4]; <2>, required [5]; <1>, the terminal dioxygenase component contains two Fe^{2+} ions per molecule in addition to two [2Fe-2S] iron-sulfur clusters [7]) [4, 5, 7]

iron <1, 2> (<1>, the intermediate electron-carrying protein possesses one [2Fe-2S] cluster, the terminal dioxygenase possesses 2 [2Fe-2S] clusters [1]; <1>, the terminal dioxygenase protein [2Fe-2S] centres, in the oxidized form the two iron atoms within the centre are high-spin ferric [3]; <1>, the benzene dioxygenase comprises three components: 1. flavoprotein, NADH:ferredoxin oxidoreductase, 2. an intermediate electron-transfer protein, or ferredoxin with a [2Fe-2S] cluster and 3. a terminal dioxygenase, containing two [2Fe-2S]iron-sulfur clusters which require two additional Fe^{2+} atoms/molecules for oxygenase activity [4]; <2>, the terminal dioxygenase component contains 2 atoms of iron and 3 atoms of inorganic sulfur [5]; <1>, the sequence of the ferredoxin component of the benzene dioxygenase contains five Cys residues, four of which are required to coordinate the iron-sulfur cluster [6]; <1>, the terminal dioxygenase component contains two Fe^{2+} ions per molecule in addition to two [2Fe-2S] iron-sulfur clusters. The ferredoxin component contains one [2Fe-2S] cluster [7]; <1>, Rieske-type [2Fe-2S] centres are coordinated by two histidines and two cysteines [9]) [1, 3, 4, 5, 6, 7, 9]

Specific activity (U/mg)
Additional information <2> [5]

K_m-Value (mM)
0.00782 <1> (NADH) [7]
0.0112 <2> (NADH) [5]

4 Enzyme Structure

Molecular weight
11860 <1> (<1>, ferredoxin component of the benzene dioxygenase, fast atom bombardment mass spectrometry [6]) [6]
12000 <1> (<1>, intermediate electron-carrying protein, gel filtration [1]) [1]
12300 <1> (<1>, intermediate electron-carrying protein, meniscus depletion method [1]) [1]
168000 <1> (<1>, gel filtration [2]) [2]
186000 <2> (<2>, terminal dioxygenase component, meniscus depletion method [5]) [5]
215000 <1> (<1>, terminal dioxygenase component, meniscus depletion method [4]) [4]
215300 <1> (<1>, terminal dioxygenase component, meniscus depletion method [1]) [1]

Subunits
? <1> (<1>, α_2, β_2, 2 * 54500 + 2 * 23500, terminal dioxygenase component. 2 * 42000, reductase component. The ferredoxin component has a MW of 12300 Da [7]) [7]
tetramer <1> (<1>, 2 * 23500 + 2 * 54500, SDS-PAGE [2]) [2]
Additional information <1> (<1>, the benzene dioxygenase comprises three components: 1. flavoprotein, NADH:ferredoxin oxidoreductase, 2. an intermediate electron-transfer protein, or ferredoxin and 3. a terminal dioxygenase [4,7]; <1>, 55000 Da is the MW of the α-subunit of the terminal dioxygenase [8]) [4, 7, 8]

5 Isolation/Preparation/Mutation/Application

Localization
soluble <2> [5]

Purification
<1> (iron-sulfur proteins of the benzene dioxygenase system: 1.intermediate electron-carrying protein and terminal dioxygenase [1]) [1, 7]
<2> [5]

Cloning

<1> (genes bedC1 and bedC2 encoding the terminal oxygenase α-subunit and β-subunit, expression in Escherichia coli [8]; expression of the α-subunit and the β-subunit of terminal dioxygenase in Escherichia coli [9]; expression in Escherichia coli JM109 [10]) [8, 9, 10]

Engineering

H119C <1> (<1>, the mutant α-subunit of the terminal dioxygenase is unable to coordinate an EPR-detectable Rieske [2Fe-2S] cluster with the characteristic g factors [8]) [8]
H222M <1> (<1>, in the reconstitution assay with the reductase component, the ferredoxin component and the β-subunit of terminal dioxygenase the mutant α-subunit is unable to reconstitute dioxygenase activity [8]) [8]
H228C <1> (<1>, in the reconstitution assay with the reductase component, the ferredoxin component and the β-subunit of terminal dioxygenase the mutant α-subunit is unable to reconstitute dioxygenase activity [8]) [8]
H98C <1> (<1>, the mutant α-subunit of the terminal dioxygenase is unable to coordinate an EPR-detectable Rieske [2Fe-2S] cluster with the characteristic g factors, detection of a novel EPR spectrum, the intensity of the spectrum is approximately 8% from the wild-type [8]) [8]
Y118S <1> (<1>, the mutant α-subunit of the terminal dioxygenase shows an EPR spectrum of half the intensity of that of the wild-type. In the reconstitution assay with the reductase component, the ferredoxin component and the β-subunit of terminal dioxygenase it shows significantly reduced activities [8]) [8]
Y221A <1> (<1>, in the reconstitution assay with the reductase component, the ferredoxin component and the β-subunit of terminal dioxygenase the mutant α-subunit shows significantly reduced activity [8]) [8]

6 Stability

Storage stability

<1>, -20°C, stable for 6 months to 1 year [7]

References

[1] Crutcher, S.E.; Geary, P.J.: Properties of the iron–sulphur proteins of the benzene dioxygenase system from Pseudomonas putida. Biochem. J., **177**, 393-400 (1979)
[2] Zamanian, M.; Mason, J.R.: Benzene dioxygenase in Pseudomonas putida. Subunit composition and immuno-cross-reactivity with other aromatic dioxygenases. Biochem. J., **244**, 611-616 (1987)
[3] Geary, P.J.; Dickson, D.P.: Moessbauer spectroscopic studies of the terminal dioxygenase protein of benzene dioxygenase from Pseudomonas putida. Biochem. J., **195**, 199-203 (1981)

[4] Geary, P.J.; Saboowalla, F.; Patil, D.; Cammack, R.: An investigation of the iron-sulphur proteins of benzene dioxygenase from Pseudomonas putida by electron-spin-resonance spectroscopy. Biochem. J., **217**, 667-673 (1984)
[5] Axcell, B.C.; Geary, P.J.: Purification and some properties of a soluble benzene-oxidizing system from a strain of Pseudomonas. Biochem. J., **146**, 173-183 (1975)
[6] Morrice, N.; Geary, P.J.; Cammack, R.; Harris, A.; Beg, F.; Aitken, A.: Primary structure of protein B from Pseudomonas putida, member of a new class of 2Fe-2S ferredoxins. FEBS Lett., **231**, 336-340 (1988)
[7] Wackett, L.P.: Benzene dioxygenase from Pseudomonas putida, ML2 (NCIB 12190). Methods Enzymol., **188**, 52-60 (1990)
[8] Mason, J.R.; Butler, C.S.; Cammack, R.; Shergill, J.K.: Structural studies on the catalytic component of benzene dioxygenase from Pseudomonas putida. Biochem. Soc. Trans., **25**, 90-95 (1997)
[9] Shergill, J.K.; Butler, C.S.; White, A.C.; Cammack, R.; Mason, J.R.: EPR, ENDOR and ESEEM studies on recombinant benzene dioxygenase. Biochem. Soc. Trans., **22**, 288S (1994)
[10] Swift, R.J.; Carter, S.F.; Widdowson, D.A.; Mason, J.R.; Leak, D.J.: Expression of benzene dioxygenase from Pseudomonas putida ML2 in cis-1,2-cyclohexanediol-degrading pseudomonads. Appl. Microbiol. Biotechnol., **55**, 721-726 (2001)

3-Hydroxy-2-methylpyridinecarboxylate dioxygenase 1.14.12.4

1 Nomenclature

EC number
1.14.12.4

Systematic name
3-hydroxy-2-methylpyridine-5-carboxylate,NAD(P)H:oxygen oxidoreductase (decyclizing)

Recommended name
3-hydroxy-2-methylpyridinecarboxylate dioxygenase

Synonyms
2-methyl-3-hydroxypyridine 5-carboxylic acid dioxygenase
3-hydroxy-2-methylpyridine carboxylate dioxygenase
methylhydroxypyridine carboxylate dioxygenase
methylhydroxypyridinecarboxylate oxidase

CAS registry number
37256-69-2

2 Source Organism

<1> *Pseudomonas sp.* (MA-1 [2, 3, 4]) [1, 2, 3, 4, 5, 6, 7]

3 Reaction and Specificity

Catalyzed reaction
3-hydroxy-2-methylpyridine-5-carboxylate + NAD(P)H + H^+ + O_2 = 2-(acetamidomethylene)succinate + $NAD(P)^+$ (<1>, ordered mechanism in which 3-hydroxy-2-methylpyridine-5-carboxylate binds first, followed by NADH. The first product NAD^+ is then released, followed by oxygen binding and finally release of the oxygenated and reduced cleavage product 2-(acetamidomethylene)succinate [2]; <1>, reaction proceeds in a concerted fashion via a ternary complex of oxygenase, NADH and 3-hydroxy-2-methylpyridine-5-carboxylate [3]; <1>, the enzyme catalyzes both a classical hydroxylation and a subsequent unique hydrolysis of the hydroxylated substrate to yield the acyclic product [6]; <1>, the binding proceeds in two steps: an enzyme-substrate complex initially formed is followed by a ligand-induced isomerization [7])

Reaction type

oxidation
redox reaction
reduction

Natural substrates and products

S 3-hydroxy-2-methylpyridine-5-carboxylate + NADH + O_2 <1> (<1>, inducible enzyme opens the pyridine ring during the metabolic degradation of vitamin B_6 [3]; <1>, enzyme is involved in degradation of vitamin B_6 [5]) (Reversibility: ? <1> [3, 5]) [3, 5]
P 2-(acetamidomethylene)succinate + NAD^+

Substrates and products

S 3-hydroxy-2-methylpyridine-5-carboxylate + NADH + O_2 <1> (<1>, the nitrogen atom of the 3-hydroxy-2-methylpyridine-5-carboxylate is invariably protonated during the catalytic reaction [5]) (Reversibility: ? <1> [1, 2, 3, 4, 5, 6, 7]) [1, 2, 3, 4, 5, 6, 7]
P 2-(acetamidomethylene)succinate + $NAD(P)^+$ <1> [1, 2, 3]
S 5-hydroxynicotinic acid + NADH + O_2 <1> (Reversibility: ? <1> [4, 5, 6]) [4, 5, 6]
P ?
S 5-pyridoxic acid + NADH + O_2 <1> (Reversibility: ? <1> [3]) [3]
P ?
S N-methyl-5-hydroxynicotinic acid + NADH + O_2 <1> (Reversibility: ? <1> [5]) [5]
P ?

Inhibitors

1-deaza-FAD <1> [1]
5-pyridoxic acid <1> (<1>, competitive with 3-hydroxy-2-methylpyridine-5-carboxylate [3]; <1>, competitive [1]) [1, 3]
6-methylnicotinate <1> [3]
6-methylnicotinic acid <1> (<1>, competitive with 3-hydroxy-2-methylpyridine-5-carboxylate [3]) [3]
NAD^+ <1> (<1>, binds competitively with O_2, but not with NADH [2]) [2]
p-chloromercuribenzoate <1> [3]
p-chloromercuribenzoate <1> (<1>, 0.05 mM, quick and complete inhibition [3]) [3]

Cofactors/prosthetic groups

FAD <1> (<1>, contains 2 mol of FAD per mol of tetrameric enzyme. 412 nM [1]; <1>, contains 2 mol of FAD per mol of enzyme [3]; <1>, contains one FAD per subunit, tetrameric enzyme [4]) [1, 3, 4]
NADH <1> (<1>, interacts with the holoenzyme in a slow catalytically irrelevant manner [1]) [1, 2, 3, 4, 5, 6, 7]

Turnover number (min^{-1})

Additional information <1> [2, 4, 5]

Specific activity (U/mg)
4.03 <1> [7]

K_m-Value (mM)
0.00056 <1> (N-methyl-5-hydroxynicotinic acid, <1>, calculation from stopped-flow data [5]) [5]
0.0011 <1> (N-methyl-5-hydroxynicotinic acid, <1>, calculation from steady-state data [5]) [5]
0.0054 <1> (NADH, <1>, calculation from steady-state data [5]) [5]
0.0059 <1> (O_2, <1>, calculation from stopped-flow data [5]) [5]
0.0117 <1> (O_2, <1>, calculation from steady-state data [5]) [5]
0.065 <1> (5-hydroxynicotinic acid, <1>, recombinant enzyme [4]) [4]
0.068 <1> (5-hydroxynicotinic acid, <1>, wild-type enzyme [4]) [4]
0.126 <1> (O_2, <1>, wild-type enzyme [4]) [4]
0.148 <1> (O_2, <1>, recombinant enzyme [4]) [4]
0.18 <1> (NADH, <1>, recombinant enzyme [4]) [4]
0.205 <1> (NADH, <1>, wild-type enzyme [4]) [4]

K_i-Value (mM)
0.00046 <1> (1-deaza-FAD) [1]
0.023 <1> (5-pyridoxic acid) [1]

pH-Optimum
6.5-8 <1> [3]

4 Enzyme Structure

Molecular weight
160000 <1> (<1>, equilibrium sedimentation [1]) [1]
166000 <1> (<1>, equilibrium sedimentation [3]) [3]

Subunits
tetramer <1> (<1>, 4 * 41700, calculation from nucleotide sequence [4]; <1>, 4 * 43000, SDS-PAGE [1]) [1, 4]

5 Isolation/Preparation/Mutation/Application

Purification
<1> [3, 4, 7]

Crystallization
<1> [3]

Cloning
<1> (expression in Escherichia coli [4]) [4]

6 Stability

Oxidation stability

<1>, the enzyme is very sensitive to oxidation, it loses activity rapidly in absence of mercaptoethanol even at 4°C, it is further stabilized in presence of high concentrations of glycerol or by serum albumin [1]

Storage stability

<1>, -20°C, 50% glycerol, 0.1% 2-mercaptoethanol, 1 month [3]

References

[1] Kishore, G.M.; Snell, E.E.: Interaction of 2-methyl-3-hydroxypyridine-5-carboxylic acid oxygenase with FAD, substrates, and analogues. Spectral and fluorescence investigations. J. Biol. Chem., **256**, 4234-4240 (1981)
[2] Kishore, G.M.; Snell, E.E.: Kinetic investigations on a flavoprotein oxygenase, 2-methyl-3-hydroxypyridine-5-carboxylic acid oxygenase. J. Biol. Chem., **256**, 4228-4233 (1981)
[3] Sparrow, L.G.; Ho, P.P.K.; Sundaram, T.K.; Zach, D.; Nyns, E.J.; Snell, E.E.: The bacterial oxidation of vitamin B_6. VII. Purification, properties, and mechanism of action of an oxygenase which cleaves the 3-hydroxypyridine ring. J. Biol. Chem., **244**, 2590-2600 (1969)
[4] Chaiyen, P.; Ballou, D.P.; Massey, V.: Gene cloning, sequence analysis, and expression of 2-methyl-3-hydroxypyridine-5-carboxylic acid oxygenase. Proc. Natl. Acad. Sci. USA, **94**, 7233-7238 (1997)
[5] Chaiyen, P.; Brissette, P.; Ballou, D.P.; Massey, V.: Reaction of 2-methyl-3-hydroxypyridine-5-carboxylic acid (MHPC) oxygenase with N-methyl-5-hydroxynicotinic acid: studies on the mode of binding, and Protonation Status of the Substrate. Biochemistry, **36**, 13856-13864 (1997)
[6] Chaiyen, P.; Brissette, P.; Ballou, D.P.; Massey, V.: Unusual mechanism of oxygen atom transfer and product rearrangement in the catalytic reaction of 2-methyl-3-hydroxypyridine-5-carboxylic acid oxygenase. Biochemistry, **36**, 8060-8070 (1997)
[7] Chaiyen, P.; Brissette, P.; Ballou, D.P.; Massey, V.: Thermodynamics and reduction kinetics properties of 2-methyl-3-hydroxypyridine-5-carboxylic acid oxygenase. Biochemistry, **36**, 2612-2621 (1997)

5-Pyridoxate dioxygenase 1.14.12.5

1 Nomenclature

EC number
1.14.12.5

Systematic name
5-pyridoxate,NADPH:oxygen oxidoreductase (decyclizing)

Recommended name
5-pyridoxate dioxygenase

Synonyms
5-pyridoxate oxidase
5-pyridoxic-acid oxygenase <1> [2]
compound I oxygenase <2> [1, 2]

CAS registry number
37256-70-5

2 Source Organism

<1> *Arthrobacter sp.* (strain Cr7 [2]) [2]
<2> *Pseudomonas sp.* (strain MA-1 [1,2]) [1, 2]

3 Reaction and Specificity

Catalyzed reaction
3-hydroxy-4-hydroxymethyl-2-methylpyridine-5-carboxylate + NADPH + H^+ + O_2 = 2-(acetamidomethylene)-3-(hydroxymethyl)succinate + $NADP^+$ (a flavoprotein)

Reaction type
oxidation
redox reaction
reduction

Natural substrates and products
S 3-hydroxy-4-hydroxymethyl-2-methylpyridine-5-carboxylate + NADPH + O_2 + H_3O^+ <2> (<2> bacterial vitamin B_6 degradation pathway [1, 2]) (Reversibility: ? <2> [1]) [1, 2]
P 2-(acetamidomethylene)-3-(hydroxymethyl)succinate + $NADP^+$ H_2O <2> [1]

Substrates and products

S 3-hydroxy-2-methylpyridine-5-carboxylic acid + NADH <1> (<1> poorly utilized substrate analoge [2]) (Reversibility: ? <1> [2]) [2]
P 2-[(acetylamino)methylene]succinate+ NAD^+ + H_2O
S 3-hydroxy-4-hydroxymethyl-2-methylpyridine-5-carboxylate + NADPH + O_2 <2> (Reversibility: ? <2> [1]) [1]
P 2-(acetamidomethylene)-3-(hydroxymethyl)succinate + $NADP^+$ + H_2O <2> [1]
S 5-pyridoxic acid + NADPH + O_2 + H3O+ <1, 2> (<2> reaction rate 3% [1]) (Reversibility: ? <1, 2> [1, 2]) [1, 2]
P 2-(acetylaminomethylene)succinate + $NADP^+$ + H_2O
S Additional information <2> (<2> 2,6-dichloroindophenol is no substrate [1]) [1]
P ?

Inhibitors

5-pyridoxic acid <2> [1]
6-methylnicotinic acid <2> [1]
p-chloromercuribenzoate <2> [1]

Cofactors/prosthetic groups

FAD <1, 2> (<2> contains 2 mol of FAD^+ per mol [1]; <1> one FAD per subunit, FMN or riboflavin do not replace FAD as coenzyme [2]) [1, 2]
NADH <1, 2> [2]
NADPH <1, 2> [2]

Specific activity (U/mg)

8.5 <2> [2]

K_m-Value (mM)

0.01 <1> (NADPH) [2]
0.036 <1> (5-pyridoxate) [2]
0.125 <1> (O_2) [2]

K_i-Value (mM)

0.06 <2> (5-pyridoxic acid) [1]
0.2 <2> (6-methylnicotinic acid) [1]

pH-Optimum

6.5-8 <2> (<2> broad pH-optimum in phosphate, diphosphate and Tris buffer [1]) [1]
7-8 <1> [2]
7.5 <1> [2]

pH-Range

5.8-8.6 <1> [2]

4 Enzyme Structure

Molecular weight

35000 <1> (<1> native enzyme, gel filtration [2]) [2]
39200 <1> (<1> native enzyme, sedimentation velocity [2]) [2]
51000 <1> (<1> SDS-PAGE [2]) [2]
166000 <2> (<2> sedimentation equlibrium [1]) [1]

Subunits

monomer <1> (<1> 1 * 51000, SDS-PAGE [2]) [2]
tetramer <2> (<2> 4 * 40000, SDS-PAGE [2]) [2]

5 Isolation/Preparation/Mutation/Application

Localization

cytoplasm <1> [2]

Purification

<1> [2]
<2> [1, 2]

Crystallization

<2> [1]

6 Stability

Oxidation stability

<2>, highly sensitive to oxidation [1]

General stability information

<2>, activity is lost on resolution with acidic ammonium sulfate and can be completely restored with FAD^+, but not with FMN [1]
<2>, highly sensitive to sulfhydryl reagents but not to chelating agents and is stabilized by high concentrations of mercaptoethanol and glycerol [1]

Storage stability

<2>, -20°C, can be stored up to 1 month with loss about 20% of its activity in a mixture of 50% glycerol and 0.1% 2-mercaptoethanol [1]
<2>, 4°C, loses activity rapidly in absence of mercaptoethanol [1]

References

[1] Sparrow, L.G.; Ho, P.P.K.; Sundaram, T.K.; Zach, D.; Nyns, E.J.; Snell, E.E.: The bacterial oxidation of vitamin B6. VII. Purification, properties, and mechanism of action of an oxygenase which cleaves the 3-hydroxypyridine ring. J. Biol. Chem., **244**, 2590-2600 (1969)
[2] Nelson, M.J.K.; Snell, E.E.: Enzymes of vitamin B6 degradation. Purification and properties of 5-pyridoxic-acid oxygenase from Arthrobacter sp. J. Biol. Chem., **261**, 15115-15120 (1986)

2-Hydroxycyclohexanone 2-monooxygenase 1.14.12.6

1 Nomenclature

EC number

1.14.12.6 (transferred to EC 1.14.13.66)

Recommended name

2-hydroxycyclohexanone 2-monooxygenase

Phthalate 4,5-dioxygenase 1.14.12.7

1 Nomenclature

EC number
1.14.12.7

Systematic name
phthalate,NADH:oxygen oxidoreductase (4,5-hydroxylating)

Recommended name
phthalate 4,5-dioxygenase

Synonyms
PDO
oxygenase, phthalate 4,5-di-
phthalate dioxygenase

CAS registry number
63626-44-8

2 Source Organism

<1> *Pseudomonas cepacia* (DB01 [1]; two separate proteins are required for phthalate dioxygenase activity: phthalate dioxygenase and phthalate dioxygenase reductase [1]) [1-7]

3 Reaction and Specificity

Catalyzed reaction
phthalate + NADH + H^+ + O_2 = cis-4,5-dihydroxycyclohexa-1(6),2-diene-1,2-dicarboxylate + NAD^+ + H_2O (<1>, reduced phthalate dioxygenase reductase first interacts with phthylate dioxygenase and reduces the Rieske [2Fe-2S]center. Phthalate and O_2 then bind reversibly and reduced phthalate dioxygenase transfers one more electron to the enzyme [1])

Reaction type
oxidation
redox reaction
reduction

Substrates and products

S 2,3-dicarboxypyridine + NADH + O_2 <1> (<1>, 69% of the activity with *o*-phthalate [4]) [4]

P 5,6-dihydroxy-5,6-dihydropyridine-2,3-dicarboxylate+ NAD^+ + H_2O

S 4-chloro-*o*-phthalate + NADH + O_2 <1> (<1>, 79% of the activity with *o*-phthalate [4]) [4]

P 4-chloro-cis-4,5-dihydroxycyclohexa-1(6),2-diene-1,2-dicarboxylate + NAD^+ + H_2O

S phthalate + NADH + O_2 <1> (<1>, *o*-phthalate [4]) [1-7]

P cis-4,5-dihydroxycyclohexa-1(6),2-diene-1,2-dicarboxylate + NAD^+ + H_2O <1> [1-7]

Inhibitors

1,10-phenanthroline <1> [4]
Cu^{2+} <1> [4]
EDTA <1> [4]
Zn^{2+} <1> [4]
azide <1> (<1>, weak, binds to the six-coordinate resting enzyme to form a new six-coordinate Fe^{2+} species [5]) [5]

Cofactors/prosthetic groups

FMN <1> (<1>, enzyme contains FMN [4]) [4]
NADH <1> (<1>, specific for NADH [1]) [1-7]

Metals, ions

Fe^{2+} <1> (<1>, one Fe^{2+} is bound at a mononuclear site [1]; <1>, mononuclear non-heme Fe^{2+} active site [5]; <1>, mononuclear ferrous center is strictly required for catalytic oxygen activation, change of ligation state is associated with substrate binding [6]) [1, 5]
Fe^{3+} <1> (<1>, stimulates [4]) [4]
iron <1> (<1>, two iron atoms are bound in a Rieske-type [2Fe-2S] center [1]; <1>, the Rieske-type [2Fe-2S]cluster is bound to at least one strongly coordinated nitrogen [2]; <1>, two histidines are coordinated to the [2Fe-2S] Rieske-type clusters [3]; <1>, detailed structural model for the Rieske [2Fe-2S] center [7]; <1>, active form of enzyme contains one mononuclear iron and one [2Fe-2S] center per monomer [4]; <1>, enzyme contains a [2Fe-2S] Rieske cluster [5,6]) [1, 2, 3, 4, 5, 6, 7]
Additional information <1> (<1>, optimal activity at ionic strength of 80 mM [1]) [1]

Specific activity (U/mg)

1.31 <1> [1]

K_m-Value (mM)

0.125 <1> (O_2) [1]
0.3 <1> (phthalate, <1>, 25°C [1]) [1]
0.5 <1> (phthalate, <1>, 4°C [1]) [1]

pH-Optimum

6.8-8.2 <1> [1]

4 Enzyme Structure

Molecular weight
217000 <1> (<1>, gel filtration [4]) [4]

Subunits
tetramer <1> (<1>, 4 * 48000, SDS-PAGE [1,4]) [1, 4]

5 Isolation/Preparation/Mutation/Application

Purification
<1> [1]

6 Stability

pH-Stability
6.8 <1> (<1>, room temperature, 12 h, stable [1]) [1]
8 <1> (<1>, room temperature, 12 h, stable [1]) [1]

Temperature stability
21 <1> (<1>, pH 6.8 or pH 8.0, 12 h stable [1]) [1]
42 <1> (<1>, 1 h, stable [1]) [1]
60 <1> (<1>, 5 min, stable [1]) [1]

Storage stability
<1>, 4°C, pH 8.0, 100 mM HEPES buffer, 20% v/v glycerol, stable for 1 week [1]

References

[1] Batie, C.J.; Ballou, D.P.: Phthalate dioxygenase. Methods Enzymol., **188**, 61-70 (1990)
[2] Cline, J.F.; Hoffman, B.M.; Mims, W.B.; LaHaie, E.; Ballou, D.P.; Fee, J.A.: Evidence for N coordination to Fe in the [2Fe-2S] clusters of Thermus Rieske protein and phthalate dioxygenase from Pseudomonas. J. Biol. Chem., **260**, 3251-3254 (1985)
[3] Gurbiel, R.J.; Batie, C.J.; Sivaraja, M.; True, A.E.; Fee, J.A.; Hoffman, B.M.; Ballou, D.P.: Electron-nuclear double resonance spectroscopy of 15N-enriched phthalate dioxygenase from Pseudomonas cepacia proves that two histidines are coordinated to the [2Fe-2S] Rieske-type clusters. Biochemistry, **28**, 4861-4871 (1989)
[4] Batie, C.J.; LaHaie, E.; Ballou, D.P.: Purification and characterization of phthalate oxygenase and phthalate oxygenase reductase from Pseudomonas cepacia. J. Biol. Chem., **262**, 1510-1518 (1987)

[5] Pavel, E.G.; Martins, L.J.; Ellis, W.R., Jr.; Solomon, E.I.: Magnetic circular dichroism studies of exogenous ligand and substrate binding to the non-heme ferrous active site in phthalate dioxygenase. Chem. Biol., **1**, 173-183 (1994)
[6] Gassner, G.T.; Ballou, D.P.; Landrum, G.A.; Whittaker, J.W.: Magnetic circular dichroism studies on the mononuclear ferrous active site of phthalate dioxygenase from Pseudomonas cepacia show a change of ligation state on substrate binding. Biochemistry, **32**, 4820-4825 (1993)
[7] Tsang, H.T.; Batie, C.J.; Ballou, D.P.; Penner-Hahn, J.E.: X-ray absoprtion spectroscopie of the [2Fe-2S] Rieske cluster in Pseudomonas cepacia phthalate dioxygenase. Determination of core dimensions and iron ligation. Biochemistry, **28**, 7233-7240 (1989)

4-Sulfobenzoate 3,4-dioxygenase 1.14.12.8

1 Nomenclature

EC number
1.14.12.8

Systematic name
4-sulfobenzoate,NADH:oxygen oxidoreductase (3,4-hydroxylating, sulfite-forming)

Recommended name
4-sulfobenzoate 3,4-dioxygenase

Synonyms
4-sulfobenzoate 3,4-dioxygenase system
oxygenase, 4-sulfobenzoate di-

CAS registry number
122933-81-7

2 Source Organism

<1> *Comamonas testosteroni* (T-2 [1, 2, 3]; PSB-4 [3]) [1, 2, 3]

3 Reaction and Specificity

Catalyzed reaction
4-sulfobenzoate + NADH + H^+ + O_2 = 3,4-dihydroxybenzoate + sulfite + NAD^+

Reaction type
oxidation
redox reaction
reduction

Natural substrates and products
S 4-sulfobenzoate + NADH + O_2 <1> (<1> reaction in toluene-*p*-sulfonate degradation pathway, simple multi-component dioxygenase of 2, perhaps 3 components: reductase B coupled to dimeric oxygenase, the third component, reductase C, represents only 15% of total activity [1]) (Reversibility: ? <1> [1]) [1]
P 3,4-dihydroxybenzoate + sulfite + NAD^+ <1> [1]

Substrates and products

S 4-sulfobenzoate + NADH + O_2 <1> (<1> the oxygenase component was expressed at high levels, no expression of a specific reductase activity linked to synthesis of the oxygenase could be detected [2]) (Reversibility: ? <1> [1, 2, 3]) [1, 2, 3, 1, 2, 3]

P 3,4-dihydroxybenzoate + sulfite + NAD^+ <1> [1]

S 4-toluenesulfonate + NADH + O_2 <1> (<1> the oxygenase component was expressed at high levels, no expression of a specific reductase activity linked to synthesis of the oxygenase could be detected [2]; <1> this strain requires the reductase (TsaB) of toluenesulfonate methyl monooxygenase in the incompletely expressed sulfobenzoate dioxygenase system (PsbAC) [3]) (Reversibility: ? <1> [2, 3]) [2, 3]

P ?

S Additional information <1> (<1> poor substrates: 2-sulfo-, 3-sulfobenzoate, benzenesulfonate, 4-methyl-, 4-hydroxy-, 4-amino-, 4-nitro-, 4-chloro-benzenesulfonate, 4-sulfophenyl-acetate,-propionate,-butyrate, no substrates are: benzoate, 4-methyl-, 4-hydroxy-, 4-amino-, 4-nitro-, 4-chloro-benzoate [1]) [1]

P ?

Cofactors/prosthetic groups

FMN <1> (<1> flavoprotein, 1 mol FMN/mol reductase B [1]) [1]
NADH <1> [1]
Additional information <1> (<1> no independent ferredoxin [1]) [1]

Metals, ions

Fe^{2+} <1> (<1> requirement, iron-sulfur protein, 1 mol (2Fe-2S) per mol reductase B, one 2Fe-2S center (Rieske-type) per oxygenase A subunit, atomic absorption spectroscopy [1]) [1, 3]

Specific activity (U/mg)

0.012 <1> (<1> for enzyme system grown on sulfobenzoate [2]) [2]
0.024 <1> (<1> for enzyme system grown on toluenesulfonate [2]) [2]
0.03 <1> (<1> for enzyme system grown on sulfobenzoate [3]) [3]
0.11 <1> (<1> reductase B [1]) [1]
0.8 <1> (<1> oxygenase A [1]) [1]
84 <1> (<1> reductase B, dichlorophenolindophenol [1]) [1]
138 <1> (<1> reductase B, cytochrome c [1]) [1]
318 <1> (<1> reductase B, ferricyanide [1]) [1]

K_m-Value (mM)

0.025-0.032 <1> (4-sulfobenzoate) [1]

pH-Optimum

6 <1> (<1> assay at [1]) [1]

Temperature optimum (°C)

25 <1> (<1> assay at, reductase B [1]) [1]
30 <1> (<1> assay at [1]) [1]

4 Enzyme Structure

Molecular weight

39000 <1> (<1> reductase B, contains 1 mol of FMN and about 2 mol each of iron and inorganic sulfur per mol, Superose gel filtration) [1]
47000 <1> (<1> reductase B, Sephadex G-2000 gel filtration [1]) [1]
50000 <1> (<1> dioxygenase A, SDS-PAGE [3]) [3]
85000 <1> (<1> oxygenase A, native dioxygenase: the enzyme is presumed to be homodimeric, Superose gel filtration [1]) [1]
105000 <1> (<1> oxygenase A, native dioxygenase: the enzyme is presumed to be homodimeric, Sephadex G-2000 gel filtration [1]) [1]
Additional information <1> (<1> simple multi-component dioxygenase of two, perhaps three components: reductase B coupled to dimeric oxygenase, reductase C exhibits only 15% of total activity [1]) [1]

Subunits

dimer <1> (<1> 2 * 50000, oxygenase A, SDS-PAGE [1]) [1]
monomer <1> (<1> 1 * 36000, reductase B, SDS-PAGE [1]) [1]

5 Isolation/Preparation/Mutation/Application

Localization

cytoplasm <1> [1]

Purification

<1> (purification of the oxygenase system [3]) [1, 3]

6 Stability

General stability information

<1>, ammonium sulfate precipitation leads to excessive loss of activity [1]
<1>, chloride-buffers give diminished yields during purification [1]
<1>, glycerol, 30% v/v, stabilizes [1]
<1>, repeated freezing and thawing inactivates [1]

References

[1] Locher, H.H.; Leisinger, T.; Cook, A.M.: 4-Sulphobenzoate 3,4-dioxygenase. Purification and properties of a desulphonative two-component enzyme system from Comamonas testosteroni T-2. Biochem. J., **274**, 833-842 (1991)
[2] Schläfli Oppenberg, H.R.; Chen, G.; Leisinger, T.; Cook, A.M.: Regulation of the degradative pathways from 4-toluenesulphonate and 4-toluenecarboxylate to protocatechuate in Comamonas testosteroni T-2. Microbiology, **141**, 1891-1899 (1995)

[3] Junker, F.; Saller, E.; Schläfli Oppenberg, H.R.; Kroneck, P.M.H.; Leisinger, T.; Cook, A.M.: Degradative pathways for *p*-toluenecarboxylate and *p*-toluenesulfonate and their multicomponent oxygenases in Comamonas testosteroni strains PSB-4 and T-2. Microbiology, **142**, 2419-2427 (1996)

4-Chlorophenylacetate 3,4-dioxygenase 1.14.12.9

1 Nomenclature

EC number
1.14.12.9

Systematic name
4-chlorophenylacetate,NADH:oxygen oxidoreductase (3,4-hydroxylating, dechlorinating)

Recommended name
4-chlorophenylacetate 3,4-dioxygenase

Synonyms
EC 1.13.99.4 (formerly)
oxygenase, 4-chlorophenylacetate 3,4-di-

CAS registry number
105006-00-6

2 Source Organism

<1> *Pseudomonas sp.* (strain CBS 3 [1,2]) [1, 2]

3 Reaction and Specificity

Catalyzed reaction
4-chlorophenylacetate + NADH + H^+ + O_2 + H_2O = 3,4-dihydroxyphenyl acetate + chloride + NAD^+

Reaction type
oxidation
redox reaction
reduction

Substrates and products
S 3-chlorophenylacetate + NADH + O_2 + H_2O <1> (<1>, 10% of the activity of 4-chlorophenylacetate [1]) (Reversibility: ? <1> [1]) [1]
P 3,4-dihydroxyphenylacetate + Cl^- + NAD^+
S 4-bromophenylacetate + NADH + O_2 + H_2O <1> (<1>, 102% of the activity of 4-chlorophenylacetate [1]) (Reversibility: ? <1> [1]) [1]
P 3,4-dihydroxyphenylacetate + Br^- + NAD^+

S 4-chlorophenylacetate + NADH + O_2 + H_2O <1> (Reversibility: ? <1> [1, 2]) [1, 2]
P 3,4-dihydroxyphenylacetate + Cl^- + $NAD(P)^+$
S 4-fluorophenylacetate + NADH + O_2 + H_2O <1> (<1>, 30% of the activity of 4-chlorophenylacetate [1]) (Reversibility: ? <1> [1]) [1]
P 3,4-dihydroxyphenylacetate + F^- + NAD^+
S 4-hydroxyphenylacetate + NADH + O_2 + H_2O <1> (<1>, 4% of the activity of 4-chlorophenylacetate [1]) (Reversibility: ? <1> [1]) [1]
P 3,4-dihydroxyphenylacetate + OH^- + NAD^+
S phenylacetate + NADH + O_2 + H_2O <1> (<1>, 6% of the activity of 4-chlorophenylacetate [1]) (Reversibility: ? <1> [1]) [1]
P ?
S Additional information <1> (<1>, not active as substrate: benzoate, substituted analogues of benzoate, 2-chlorophenylacetate, antipyrin, 4-chlorocinnamate [1]; <1>, the reductase component reduces several redox compunds: 2,6-dichlorophenylindophenol, potassium hexacyanoferrate III, cytochrome c, methylene blue and nitro blue tetrazolium [2]) [1, 2]
P ?

Inhibitors
1,10-phenanthroline <1> (<1>, 1 mM, 34% inhibition [2]) [2]
EDTA <1> (<1>, 1 mM, 32% inhibition [2]) [2]
KCN <1> (<1>, 1 mM, 52% inhibition [2]) [2]
NEM <1> (<1>, 1 mM, 38% inhibition [2]) [2]
NH_4^+ <1> (<1>, inhibition of activity of reductase component [2]) [2]
PCMB <1> (<1>, 0.005 mM, complete inhibition [2]) [2]
iodoacetate <1> (<1>, 0.01 mM, 77% inhibition [2]) [2]

Cofactors/prosthetic groups
FMN <1> (<1>, flavoprotein, contains 1 mol of FMN per mol of reductase, no increase of activity by addition of exogenous FMN [2]) [2]
NADH <1> [1, 2]
NADPH <1> (<1>, 30% as effective as NADH [2]) [2]

Metals, ions
iron <1> (<1>, component A is an iron-sulfur protein, contains 1.6-1.8 mol of iron per subunit and 1.6-1.9 mol of acid-labile sulfide [1]; <1>, reductase component contains an iron-sulfur cluster of the [2Fe-2S]-type, contains 2.1 mol of iron and 1.7 mol of acid-labile sulfide per mol of reductase [2]) [1, 2]

Specific activity (U/mg)
0.144 <1> (<1>, dioxygenase component of of 4-chlorophenylacetate 3,4-dioxygenase [1]) [1]
30 <1> (<1>, reductase component of of 4-chlorophenylacetate 3,4-dioxygenase [2]) [2]

K_m-Value (mM)

0.0023 <1> (cytochrome c) [2]
0.0063 <1> (2,6-dichlorophenolindophenol) [2]
0.032 <1> (NADH) [2]

pH-Optimum

7-7.5 <1> [2]

4 Enzyme Structure

Molecular weight

35000 <1> (<1>, reductase component of 4-chlorophenylacetate 3,4-dioxygenase, gel filtration [2]) [2]
140000 <1> (<1>, dioxygenase component of 4-chlorophenylacetate 3,4-dioxygenase, gel filtration [1]) [1]
144000 <1> (<1>, dioxygenase component of 4-chlorophenylacetate 3,4-dioxygenase, ultracentrifugation [1]) [1]

Subunits

monomer <1> (<1>, 1 * 35000, reductase component of of 4-chlorophenylacetate 3,4-dioxygenase, SDS-PAGE [2]) [2]
trimer <1> (<1>, 3 * 46000-52000, dioxygenase component of of 4-chlorophenylacetate 3,4-dioxygenase, SDS-PAGE [1]) [1]
Additional information <1> (<1>, the enzyme consists of two components, a monomeric reductase and a trimeric dioxygenase [1,2]) [1, 2]

5 Isolation/Preparation/Mutation/Application

Purification

<1> (dioxygenase component [1]; reductase component [2]) [1, 2]

6 Stability

pH-Stability

6-7.5 <1> (<1>, 4°C, 10 days, 50% loss of activity [2]) [2]

Temperature stability

45 <1> (<1>, 3 min, 50% loss of activity [2]) [2]
55 <1> (<1>, 90-100% loss of activity [2]) [2]

Oxidation stability

<1>, reductase component loses activity in presence of molecular oxygen [2]

General stability information

<1>, 75% loss of activity after 16 h dialysis [2]

Storage stability

<1>, 4°C, 20 mM potassium phosphate buffer, pH 6.0-7.5, 10 days, 50% loss of activity [2]

References

[1] Markus, A.; Krekel, D.; Lingens, F.: Purification and some properties of component A of the 4-chlorophenylacetate 3,4-dioxygenase from Pseudomonas species strain CBS. J. Biol. Chem., **261**, 12883-12888 (1986)

[2] Schweizer, D.; Markus, A.; Seez, M.; Ruf, H.H.; Lingens, F.: Purification and some properties of component B of the 4-chlorophenylacetate 3,4-dioxygenase from Pseudomonas species strain CBS 3. J. Biol. Chem., **262**, 9340-9346 (1987)

Benzoate 1,2-dioxygenase 1.14.12.10

1 Nomenclature

EC number
1.14.12.10

Systematic name
benzoate,NADH:oxygen oxidoreductase (1,2-hydroxylating, decarboxylating)

Recommended name
benzoate 1,2-dioxygenase

Synonyms
EC 1.13.99.2 (formerly)
benzoate dioxygenase
benzoate hydroxylase
benzoic hydroxylase
oxygenase, benzoate 1,2-di-

CAS registry number
9059-18-1

2 Source Organism

<1> *Pseudomonas arvilla* (C-1 [1,2]) [1, 2]
<2> *Acinetobacter sp.* (strain ADP1, class IB dioxygenase [3]) [3]
<3> *Rhodococcus sp.* (strain 19070 [4]) [4]
<4> *Pseudomonas putida* (strain mt-2 [5]) [5]

3 Reaction and Specificity

Catalyzed reaction
benzoate + NADH + H^+ + O_2 = catechol + CO_2 + NAD^+ (<1> the enzyme consists of two proteins, a NADH-cytochrome c reductase and an oxygenase [1]; <4> proposed catalytic cycle [5])

Reaction type
oxidation
redox reaction
reduction

Natural substrates and products

S benzoate + O_2 + NADH <1> [2]
P 2-hydro-1,2-dihydroxybenzoate <1> [2]

Substrates and products

S 3-aminobenzoate + O_2 + NADH <1, 3> (<3> activity in E. coli expressing a presumed dioxygenase component from Rhodococcus [4]) (Reversibility: ? <1, 3> [1, 4]) [1, 4]
P 3-amino-2-hydro-1,2-dihydroxybenzoate <1, 3> [1, 4]
S 3-chlorobenzoate + O_2 + NADH <1> (Reversibility: ? <1> [1]) [1]
P 3-chloro-2-hydro-1,2-dihydroxybenzoate <1> [1]
S 3-fluorobenzoate + O_2 + NADH <1> (<1> benzoate analogues with substituents in *meta* position are oxygenated more rapidly than those with *ortho-* or *para* substitution [1]) (Reversibility: ? <1> [1]) [1]
P 3-fluoro-2-hydro-1,2-dihydroxybenzoate <1> [1]
S 3-methylbenzoate + O_2 + NADH <1, 3> (<3> activity in E. coli expressing a presumed dioxygenase component from Rhodococcus [4]) (Reversibility: ? <1, 3> [1, 4]) [1, 4]
P 3-methyl-2-hydro-1,2-dihydroxybenzoate <1, 3> [1, 4]
S 4-fluorobenzoate + O_2 + NADH <1> (Reversibility: ? <1> [1]) [1]
P 4-fluoro-2-hydro-1,2-dihydroxybenzoate <1> [1]
S benzoate + O_2 + NADH <1, 3> (<3> activity in E. coli expressing a presumed dioxygenase component from Rhodococcus [4]) (Reversibility: ? <1, 3> [1, 2, 4]) [1, 2, 4]
P 2-hydro-1,2-dihydroxybenzoate <1> [1, 4]

Cofactors/prosthetic groups

FAD <1, 2, 4> (<1> requirement, benzoate 1,2-dioxygenase system consists of 2 proteins: a 37500 Da NADH-cytochrome c reductase containing 1 FAD and 1 iron-sulfur cluster [2Fe-2S] and an oxygenase which may contain 4 [2Fe-2S] clusters [1]; <4> 1 FAD per enzyme monomer [5]) [1, 3, 5]
NADH <1> (<1> requirement, benzoate 1,2-dioxygenase system consists of 2 proteins: a 37500 Da NADH-cytochrome c reductase containing 1 FAD and 1 iron-sulfur cluster [2Fe-2S] and an oxygenase which may contain 4 [2Fe-2S] clusters [1]) [1]

Metals, ions

iron <1, 2, 3, 4> (<1> iron sulfur protein with [2Fe-2S]clusters and additional iron atoms [2]; <1> α-subunit contains 8.2 mol of iron [2]; <1> 10 mol of iron per mol of enzyme, [2Fe-2S] clusters [1]; <2> reductase component contains a plant type [2Fe-2S] cluster and a presumed mononuclear iron, the terminal dioxygenase contains a Rieske-type [2Fe-2S] cluster [3]; <4> oxygenase component: 2.7 mol/mol of $\alpha\beta$ subunit, each α subunit contains a Rieske [2Fe-2S] cluster and a mononuclear iron site, reductase component: 2.02 mol/monomer [5]) [1, 2, 3, 4, 5]

Turnover number (min^{-1})

22000 <1> (benzoate, <1> in the presence of saturating amounts of NADH-cytochrome c reductase [1]) [1]

Specific activity (U/mg)

1.36 <4> (<4> benzoate oxygenase [5]) [5]
4.9 <1> [1]
790 <4> (<4> benzoate reductase, reduction of $K_3Fe(CN)_6$ [5]) [5]

K_m-Value (mM)

0.0039 <1> (benzoate) [1]
0.0043 <1> (O_2) [1]
0.0048 <1> (NADH) [1]
0.026 <1> (NADH-cytochrome c reductase component) [1]

pH-Optimum

6.7 <1> (<1> oxygenase component [1]) [1]
8.2 <1> (<1> NADH-cytochrome c reductase component [1]) [1]

pH-Range

6-8 <1> (<1> pH 6.0: about 55% activity, pH 8.0: about 60% activity [1]) [1]

Temperature optimum (°C)

24 <1> (<1> assay at [1]) [1]

4 Enzyme Structure

Molecular weight

37500 <1> (<1> NADH-cytochrome c component [1]) [1]
38000 <4> (<4> reductase component [5]) [5]
195000 <4> (<4> oxygenase component, sedimentation analysis [5]) [5]
201000 <1> (<1> sedimentation equilibrium [2]) [2]
270000 <1> (<1> oxygenase component, gel filtration [1]) [1]
273000 <1> (<1> oxygenase component, sedimentation equilibrium [1]) [1]
280000 <1> (<1> oxygenase component, native PAGE [1]) [1]

Subunits

hexamer <1, 4> (<1> $\alpha_3\beta_3$, 3 * 20000 + 3 * 50000, oxygenase component of the enzyme system, SDS-PAGE [2]; <4> $\alpha_3\beta_3$, 3 * 19000 + 3 * 49000, oxygenase component, SDS-PAGE [5]) [2, 5]
monomer <1, 4> (<1> 1 * 38000, reductase component of the enzyme system [2]; <4> 1 * 38000, reductase component, sedimentation analysis, SDS-PAGE [5]) [2, 5]

5 Isolation/Preparation/Mutation/Application

Purification

<1> (oxygenase component [1]) [1]
<4> (benzoate oxgenase and reductase component [5]) [5]

Cloning

<3> (expression of presumed dioxygenase component in Escherichia coli [4]) [4]

References

[1] Yamaguchi, M.; Fujisawa, H.: Purification and characterization of an oxygenase component in benzoate 1,2-dioxygenase system from Pseudomonas arvilla C-1. J. Biol. Chem., **255**, 5058-5063 (1980)
[2] Yamaguchi, M.; Fujisawa, H.: Subunit structure of oxygenase component in benzoate-1,2-dioxygenase system from Pseudomonas arvilla C-1. J. Biol. Chem., **257**, 12497-12502 (1982)
[3] Bundy, B.M.; Campbell, A.L.; Neidle, E.L.: Similarities between the antABC-encoded anthranilate dioxygenase and the benABC-encoded benzoate dioxygenase of Acinetobacter sp. strain ADP1. J. Bacteriol., **180**, 4466-4474 (1998)
[4] Haddad, S.; Eby, D.M.; Neidle, E.L.: Cloning and expression of the benzoate dioxygenase genes from Rhodococcus sp. strain 19070. Appl. Environ. Microbiol., **67**, 2507-2514 (2001)
[5] Wolfe, M.D.; Altier, D.J.; Stubna, A.; Popescu, C.V.; Muenck, E.; Lipscomb, J.D.: Benzoate 1,2-dioxygenase from Pseudomonas putida: single turnover kinetics and regulation of a two-component Rieske dioxygenase. Biochemistry, **41**, 9611-9626 (2002)

Toluene dioxygenase 1.14.12.11

1 Nomenclature

EC number
1.14.12.11

Systematic name
toluene,NADH:oxygen oxidoreductase (1,2-hydroxylating)

Recommended name
toluene dioxygenase

Synonyms
ISPTOL <1> (<1>, oxygenase component of the toluene dioxygenase multienzyme system [1]) [1]
TDO
oxygenase, toluene 2,3-di-

CAS registry number
120038-36-0

2 Source Organism

<1> *Pseudomonas putida* (F1 [5, 7, 9, 10, 12, 13, 14, 15]; UV4 [20, 21]; NCIB11767 [6]; F39D [18]; recombinant enzyme expressed in Escherichia coli JM109(pDTG601) under control of the tac promoter [7]; enzyme from recombinant Escherichia coli [16, 17, 20]) [1, 2, 3, 4, 5, 6, 7, 8, 9, 10, 11, 12, 13, 14, 15, 16, 17, 18, 19, 20, 21, 22]
<2> *Pseudomonas fluorescens* (CFS215 [15]) [15]
<3> *Pseudomonas sp.* (strain W31 [15]) [15]

3 Reaction and Specificity

Catalyzed reaction
toluene + NADH + H^+ + O_2 = (1S,2R)-3-methylcyclohexa-3,5-diene-1,2-diol + NAD^+

Reaction type
oxidation
redox reaction
reduction

Natural substrates and products

- **S** toluene + NADH + O_2 <1> (<1>, initial enzyme of toluene catabolism [6, 16]; <1>, enzyme is involved in meta pathway for catechol degradation [9]) (Reversibility: ? <1> [6, 9, 16]) [6, 9, 16]
- **P** (1S,2R)-3-methylcyclohexa-3,5-diene-1,2-diol + NAD^+

Substrates and products

- **S** (+)-(S)-indan-1-ol + O_2 + NADH <1> (<1>, biotransformation with intact cells [21]) (Reversibility: ? <1> [21]) [21]
- **P** (-)-(R)-indan-1-ol + indan-1-one + (+)-trans-(1S,3S)-1,3-dihydroxyindane + (-)-(3R)-3-hydroxyindan-1-one + NAD^+ <1> [21]
- **S** (+/-)-trans-2-phenyl-1-cyclohexanol + NADH + O_2 <1> (Reversibility: ? <1> [19]) [19]
- **P** 3-(2-hydroxycyclohexanyl)-3,5-cyclohexadiene-1,2-diol + ? <1> [19]
- **S** (-)-(R)-indan-1-ol + NADH + O_2 <1> (<1>, biotransformation with intact cell [21]) (Reversibility: ? <1> [21]) [21]
- **P** trans-(1R,3R)-1,3-dihydroxyindane + (-)-(1R,4R,5S)-1,4,5-trihydroxy-4,5-dihydroindane + NAD^+ <1> [21]
- **S** (1R,2S)-(-)-trans-2-phenyl-1-cyclohexanol + NADH + O_2 <1> (Reversibility: ? <1> [19]) [19]
- **P** 3-(2-hydroxycyclohexanyl)-3,5-cyclohexadiene-1,2-diol + ? <1> [19]
- **S** (R)-1-phenyl-1-ethanol + NADH + O_2 <1> (Reversibility: ? <1> [19]) [19]
- **P** 3-[1(R)-hydroxyethyl]cyclohexa-3,5-diene-1(S),2(R)-diol + ? <1> [19]
- **S** (R)-2-phenylcyclohexanone + NADH + O_2 <1> (Reversibility: ? <1> [19]) [19]
- **P** (4S,4aR,9aR)-4,6,7,8,9,9a-hexahydro-4aH-dibenzofuran-4,5a-diol + ? <1> [19]
- **S** (S)-1-phenyl-1-ethanol + O_2 + NADH + O_2 <1> (Reversibility: ? <1> [19]) [19]
- **P** 3-[1(S)-hydroxyethyl]cyclohexa-3,5-diene-1(S),2(R)-diol + ? <1> [19]
- **S** (S)-2-phenylcyclohexanone + NADH + O_2 <1> (Reversibility: ? <1> [19]) [19]
- **P** (1S,5'S,6'R)-5',6'-dihydroxybicyclohexyl-1',3'-diene-2-one + ? <1> [19]
- **S** (cis)-2-chloro-2-butene + NADH + O_2 <1> (<1>, 12% of the activity with toluene [11]) (Reversibility: ? <1> [11]) [11]
- **P** 2-chloro-2-butene-1-ol + NAD^+ <1> [11]
- **S** (trans)-2-chloro-2-butene + NADH + O_2 <1> (<1>, 4% of the activity with toluene [11]) (Reversibility: ? <1> [11]) [11]
- **P** ?
- **S** 1,1-dichloro-1-propene + NADH + O_2 <1> (<1>, 18% of the activity with toluene [11]) (Reversibility: ? <1> [11]) [11]
- **P** ?
- **S** 1,1-dichloro-1-propene + NADH + O_2 <1> (<1>, 6% of the activity with toluene [11]) (Reversibility: ? <1> [11]) [11]
- **P** 3,3-dichloro-2-propene-1-ol + NAD^+ <1> [11]
- **S** 1-chloro-2-methyl-1-propene + NADH + O_2 <1> (<1>, 13% of the activity with toluene [11]) (Reversibility: ? <1> [11]) [11]

P 2-chloro-2-butene-1-ol + NAD^+ <1> [11]
S 1-phenylcyclohexene + NADH + O_2 <1> (Reversibility: ? <1> [19]) [19]
P (1S,2R)-3-(1-cyclohexenyl)-3,5-cyclohexadiene-1,2-diol + ? <1> [19]
S 2,3-dichloro-1-propene + NADH + O_2 <1> (<1>, 19% of the activity with toluene [11]) (Reversibility: ? <1> [11]) [11]
P 2,3-dichloro-2-propene-1-ol + NAD^+ <1> [11]
S 2-acetoxyindane + O_2 + NADH <1> (<1>, biotransformation with intact cells [21]) (Reversibility: ? <1> [21]) [21]
P indan-2-ol + (-)-cis-(1S,2R)-1,2-dihydroxyindane + (-)-trans-(1R,2R)-1,2-dihydroxyindane + (-)-(2R)-2-hydroxyindan-1-one + NAD^+ <1> [21]
S 2-bromoindane + O_2 + NADH <1> (<1>, biotransformation with intact cells [21]) (Reversibility: ? <1> [21]) [21]
P (-)-cis-(1S,2R)-2-bromoindan-1-ol + (+)-trans-(1S,3S)-1,3-dihydroxy-2-bromoindane + NAD^+ <1> [21]
S 2-carbamoylindane + O_2 + NADH <1> (<1>, biotransformation with intact cells [21]) (Reversibility: ? <1> [21]) [21]
P (-)-cis-(1S,2R)-2-azoindan-1-ol + NAD^+ <1> [21]
S 2-chloroindane + O_2 + NADH <1> (<1>, biotransformation with intact cells [21]) (Reversibility: ? <1> [21]) [21]
P (-)-cis-(1S,2R)-2-chloroindan-1-ol + (+)-trans-(1R,2R)-2-chloroindan-1-ol + NAD^+ <1> [21]
S 2-chloroindane + O_2 + NADH <1> (<1>, biotransformation with intact cells [21]) (Reversibility: ? <1> [21]) [21]
P (-)-trans-(1S,3S)-1,3-dihydroxy-2-chloroindane + NAD^+ <1> [21]
S 2-hexene + NADH + O_2 <1> (Reversibility: ? <1> [11]) [11]
P hexane-2,3-diol + NAD^+ <1> [11]
S 2-indanone + NADH + O_2 <1> (<1>, no reaction with 1-indanone [18]) (Reversibility: ? <1> [18]) [18]
P (S)-2-hydroxy-1-indanone + NAD^+ <1> (<1>, 95% S-enantiomer, product is formed by incorporation of a single atom of molecular oxygen rather than by dioxygenation of enol tautomers of the ketone substrate [18]) [18]
S 2-iodoindane + O_2 + NADH <1> (<1>, biotransformation with intact cells [21]) (Reversibility: ? <1> [21]) [21]
P (-)-cis-(1S,2R)-1,2-dihydroxyindane + (-)-(1R)-1-hydroxyindene + (+)-(1S,3S)-1,3-dihydroxy-2-iodoindane + NAD^+ <1> [2]
S 2-methoxyindane + O_2 + NADH <1> (<1>, biotransformation with intact cells [21]) (Reversibility: ? <1> [21]) [21]
P (-)cis-(1S,2R)-2-methoxyindan-1-ol + (-)-trans-(1S,2R)-2-methoxyindan-1-ol + NAD^+ <1> [21]
S 2-methylindan + O_2 + NADH <1> (<1>, biotransformation with intact cells [21]) (Reversibility: ? <1> [21]) [21]
P (-)-trans-(1R,3R)-1,3-dihydroxy-2-methylindane + (-)-cis-(2S,3R)-3-hydroxy-2-methylindan-1-one + (-)-cis-(1R,2R)-1-hydroxy-2-methylindane + (-)-(2R)-2-methyindan-1-one + NAD^+ <1> [21]
S 3,4-dichloro-1-butene + NADH + O_2 <1> (<1>, 23% of the activity with toluene [11]) (Reversibility: ? <1> [11]) [11]
P 3,4-dichlorobutane-1,2-diol + NAD^+ <1> [11]

S 3-phenylcyclohexene + NADH + O_2 <1> (Reversibility: ? <1> [19]) [19]
P (1S,2R)-3-(cyclohexenyl)-3,5-cyclohexadiene-1,2-diol + ? <1> [19]
S 4-picoline + NADH + O_2 <1> (<1>, E. coli expressed mutant enzyme TDO 2-B38, in which the wild-type stop codon is replaced with a codon encoding threonine, exhibits 5.6-times higher activity towards 4-picoline than the wild-type enzyme [22]) (Reversibility: ? <1> [1]) [22]
P 3-hydroxy-4-picoline + ? <1> [22]
S butyl phenyl sulfide + NADH + O_2 <1> (Reversibility: ? <1> [20]) [20]
P (R)-butyl phenyl sulfoxide + ? <1> (<1>, more than 98% R-enantiomer [20]) [20]
S cis-1,2-dichloroethene + NADH + O_2 <1> (<1>, 12% of the activity with toluene [11]) (Reversibility: ? <1> [11]) [11]
P ?
S cis-1,4-dichloro-2-butene + NADH + O_2 <1> (<1>, 18% of the activity with toluene [11]) (Reversibility: ? <1> [11]) [11]
P 1,4-dichlorobutane-2,3-diol <1> [11]
S cis-1-bromo-1-propene + NADH + O_2 <1> (<1>, 11% of the activity with toluene [11]) (Reversibility: ? <1> [11]) [11]
P ?
S cis-1-chloro-1-propene + NADH + O_2 <1> (<1>, 5% of the activity with toluene [11]) (Reversibility: ? <1> [11]) [11]
P ?
S cis-2-heptene + NADH + O_2 <1> (Reversibility: ? <1> [11]) [11]
P heptane-2,3-diol + NAD^+ <1> [11]
S cis-2-octene + NADH + O_2 <1> (Reversibility: ? <1> [11]) [11]
P octane-2,3-diol + NAD^+ <1> [11]
S cis-2-pentene + NADH + O_2 <1> (<1>, 16% of the activity with toluene [11]) (Reversibility: ? <1> [11]) [11]
P pentane-2,3-diol + NAD^+ <1> [11]
S cis-dibromoethene + NADH + O_2 <1> (<1>, 13% of the activity with toluene [11]) (Reversibility: ? <1> [11]) [11]
P 3-bromo-3-propene-1-ol + NAD^+ <1> [11]
S diphenylmethane + NADH + O_2 <1> (Reversibility: ? <1> [19]) [19]
P (1S,2R)-3-benzyl-3,4-cyclohexadiene-1,2-diol + ?
S ethenyl phenyl sulfide + NADH + O_2 <1> (Reversibility: ? <1> [20]) [20]
P (R)-ethenyl phenyl sulfoxide + ? <1> (<1>, more than 98% R-enantiomer [20]) [20]
S ethyl phenyl sulfide + NADH + O_2 <1> (Reversibility: ? <1> [16]) [16]
P (R)-ethyl phenyl sulfoxide <1> (<1>, more than 98% R-enantiomer [16,20]) [16, 20]
S indan-1-ol + O_2 + NADH <1> (<1>, biotransformation with intacts cells [21]) (Reversibility: ? <1> [21]) [21]
P (-)-cis-(1S,2R)-1,2-dihydroxyindane + (-)-trans-(1R,2R)-1,2-dihydroxyindane + (-)-(2R)-2-hydroxyindan-1-one + NAD^+ <1> [21]
S indan-2-ol + O_2 + NADH <1> (<1>, biotransformation with intact cells [21]) (Reversibility: ? <1> [21]) [21]

P (-)-cis-(1S,2R)-1,2-dihydroxyindane + (-)-trans-(1R,2R)-1,2-dihydroxyindane + NAD^+ <1> [21]

S indan-2-one + O_2 + NADH <1> (<1>, biotransformation with intact cells [21]) (Reversibility: ? <1> [21]) [21]

P indan-2-ol + (-)-cis-(1S,2R)-1,2-dihydroxy-indane + (-)-trans-(1S,2R)-1,2-dihydroxy-indane + NAD^+ <1> [21]

S indane + NADH + O_2 <1> (<1>, monooxygenase reaction of toluene dioxygenase [2]; <1>, biotransformation with intact cells [21]) (Reversibility: ? <1> [2, 21]) [2, 21]

P (-)-(1R)-indanol + NAD^+ <1> (<1>, + indan-1-one [21]; <1>, 84% enantiomeric excess of (-)-(1R)-indanol, 70% of the oxygen in 1-indanol is derived from water [2]) [2, 21]

S indene + O_2 + NADH <1> (<1>, monooxygenase reaction of toluene dioxygenase [2]) (Reversibility: ? <1> [2]) [2]

P (-)cis(1S,2R)-dihydroxyindan (+)-(1S)-indenol + ? <1> (<1>, in addition the enzyme catalyzes the dioxygen addition of the nonaromatic double bond of indene to form cis-1,2-indanediol. The oxygen atom in 1-indenol and cis,1,2-indanediol is derived from molecular oxygen [2]) [2]

S isopropyl phenyl sulfide + NADH + O_2 <1> (Reversibility: ? <1> [20]) [20]

P (R)-isopropyl phenyl sulfoxide + ? <1> (<1>, more than 98% R-enantiomer [20]) [20]

S methoxymethyl phenyl sulfide + NADH + O_2 <1> (Reversibility: ? <1> [20]) [20]

P (R)-methoxymethyl phenyl sulfoxide + ? <1> (<1>, more than 98% R-enantiomer [20]) [20]

S methyl (2-pyridyl) sulfide + NADH + O_2 <1> (Reversibility: ? <1> [20]) [20]

P (R)-methyl (2-pyridyl) sulfoxide + ? <1> (<1>, more than 98% R-enantiomer [20]) [20]

S methyl (2-thienyl) sulfide + NADH + O_2 <1> (Reversibility: ? <1> [20]) [20]

P (R)-methyl (2-thienyl) sulfoxide + ? <1> (<1>, more than 98% R-enantiomer [20]) [20]

S methyl *p*-nitrophenyl sulfide + O_2 <1> (Reversibility: ? <1> [16]) [16]

P methyl *p*-nitrophenyl sulfoxide <1> (<1>, 86% S-enantiomer [16]) [16]

S methyl *p*-tolyl sulfide + O_2 <1> (Reversibility: ? <1> [16]) [16]

P cis-1,2-dihydroxy-3-methyl-6-methylthiocyclohexa-3,5-diene <1> [16]

S methyl phenyl sulfide + NADH + O_2 <1> (Reversibility: ? <1> [16, 20]) [16, 20]

P (R)-methyl phenyl sulfoxide + ? <1> (<1>, more than 98% R-enantiomer [16,20]) [16, 20]

S *p*-methoxyphenyl methyl sulfide + O_2 <1> (Reversibility: ? <1> [16]) [16]

P *p*-methoxyphenyl methyl sulfoxide <1> (<1>, 32% S-enantiomer [16]) [16]

S phenylcyclohexane + NADH + O_2 <1> (Reversibility: ? <1> [19]) [19]

P (1S,2R)-3-(1-cyclohexyl)-3,5-cyclohexadiene-1,2-diol + ? <1> [19]

S propyl phenyl sulfide + NADH + O_2 <1> (Reversibility: ? <1> [20]) [20]
P (R)-propyl phenyl sulfoxide + ? <1> (<1>, more than 98% R-enantiomer [20]) [20]
S toluene + NADH + O_2 <1> (<1>, E. coli expressed mutant enzyme TDO 2-B38, in which the wild-type stop codon is replaced with a codon encoding threonine, exhibits about 20% more activity towards toluene than the wild-type enzyme [22]) (Reversibility: ? <1> [1, 2, 3, 4, 5, 6, 7, 8, 9, 10, 11, 12, 13, 14, 15, 16, 16, 17, 18, 19, 20, 21, 22]) [1, 2, 3, 4, 5, 6, 7, 8, 9, 10, 11, 12, 13, 14, 15, 16, 16, 17, 18, 19, 20, 21, 22]
P (1S,2R)-3-methylcyclohexa-3,5-diene-1,2-diol + NAD^+ <1> [1, 4]
S trans-1,4-dichloro-2-butene + NADH + O_2 <1> (<1>, 18% of the activity with toluene [11]) (Reversibility: ? <1> [11]) [11]
P 1,4-dichloro-2-butanone + NAD^+ <1> [11]
S trans-1-bromo-1-propene + NADH + O_2 <1> (<1>, 3% of the activity with toluene [11]) (Reversibility: ? <1> [11]) [11]
P ?
S trans-1-chloro-1-propene + NADH + O_2 <1> (<1>, 4% of the activity with toluene [11]) (Reversibility: ? <1> [11]) [11]
P ?
S trans-dibromoethene + NADH + O_2 <1> (<1>, 5% of the activity with toluene [11]) (Reversibility: ? <1> [11]) [11]
P ?
S trichloroethylene + O_2 + NADH <1> (<1>, 25% of the activity with toluene [11]) (Reversibility: ? <1> [7, 11]) [7, 11]
P ?
S trichloroethylene + O_2 + NADPH <1> (Reversibility: ? <1> [17]) [17]
P formate + glyoxylate + $NADP^+$ <1> (<1>, formate accounts for 47% of the trichloroethylene oxidized, glyoxylate accounts for 17% of the trichloroethylene oxidized. Both carbon atoms give rise to formic acid [17]) [17]
S Additional information <1> (<1>, the NADH-ferredoxinTOL reductase component catalyzes the NADH-dependent reduction of 2,6-dichloroindophenol, nitroblue tetrazolium and ferricyanide. NADPH is inactive [3]; <1>, enzyme catalyzes monooxygenation and dioxygenation of aliphatic olefins [11]; <1>, the purified α-subunit of oxygenase component is reduced by NADH and catalytic amounts of reductaseTOL and ferredoxinTOL. Reduced α-subunit can not oxidize toluene and catalysis is strictly dependent on the presence of β-subunit [12]; <1>, the oxygen atom in methyl phenyl sulfoxide is derived exclusively from dioxygen [16]; <1>, no reaction with 1-indanone [18]; <1>, screening of substituted arenes containing remote chiral centers as substrates, enantiomers are indiscriminately processed to diastereomeric pairs [19]; <1>, activity with alkyl aryl sulfides [20]) [3, 11, 12, 16, 18, 19, 20]
P ?

Cofactors/prosthetic groups

FAD <1> (<1>, NADH-ferredoxinTOL reductase component is a flavoprotein that contains one mol of FAD per mol of enzyme. K_m: 2.5 nM [3]; <1>, NADH-ferredoxinTOL reductase component contains 1 mol of noncovalently bound FAD per mol of protein [10]) [3, 10]
NADH <1> (<1>, NADH-ferredoxinTOL reductase component [3]) [1, 2, 3, 4, 5, 6, 7, 8, 9, 10, 11, 12, 13, 14, 15, 16, 17, 18, 19, 20, 21, 22]
NADPH <1> (<1>, NADPH is inactive [3]) [17]

Metals, ions

Fe^{2+} <1> (<1>, required [1,4]) [1, 4]
iron <1> (<1>, iron-sulfur protein contains 2 gatom of iron and 2 gatom of acid-labile sulfur per mol of protein [1]; <1>, ferredoxinTOL component of the toluene dioxygenase contains 2 gatoms each of iron and acid-labile sulfur which appear to be organized as a single [2Fe-2S]cluster [8]; <1>, the oxygenase component contains 4-6 iron atoms per holoenzyme, the ferredoxin component contains one [2Fe-s2S] cluster of the Rieske type [10]; <1>, enzyme requires mononuclear iron for activity [14]) [1, 8, 10, 14]

Specific activity (U/mg)

0.0075 <1> [1]
2.46 <1> (<1>, units per mg of α-subunit of oxygenase [12]) [12]
2.58 <1> [13]
Additional information <1> [10]

K_m-Value (mM)

Additional information <1> (<1>, K_m-values of the NADH-ferredoxinTOL reductase component: 0.0046 mM for cytochrome and 0.0105 mM for NADH [3]) [3]

pH-Optimum

6.6-7.5 <1> (<1>, NADH-ferredoxinTOL reductase component [3]) [3]

4 Enzyme Structure

Molecular weight

14700 <1> (<1>, ferredoxinTOL component of the toluene dioxygenase, equilibrium sedimentation [8]) [8]
46500 <1> (<1>, NADH-ferredoxinTOL reductase component, gel filtration [3]) [3]
151000 <1> (<1>, oxygenase component, gel filtration [1]) [1]
Additional information <1> (<1>, the enzyme system consists of monomeric reductaseTOL and ferredoxinTOL and a dimeric terminal oxygenase [1]; <1>, three-component enzyme: 1. reductaseTOL, 2. ferredoxinTOL, 3. a complex iron-sulfur protein [10]) [1, 10]

Subunits

? <1> (<1>, x * 52500 + x * 20800, oxygenase component, SDS-PAGE [1]) [1]
monomer <1> (<1>, 1 * 15500, ferredoxinTOL component of the toluene dioxygenase, SDS-PAGE [8,10]; <1>, 1 * 46000, NADH-ferredoxinTOL reductase component, SDS-PAGE [3,10]) [3, 8, 10]
tetramer <1> (<1>, $\alpha_2\beta_2$, 2 * 46000 + 2 * 15300, oxygenase component of toluene dioxygenase [10]) [10]
Additional information <1> (<1>, the enzyme system consists of monomeric reductaseTOL, a ferredoxinTOL and a dimeric terminal oxygenase [1]) [1]

5 Isolation/Preparation/Mutation/Application

Purification

<1> (oxygenase component, ISPTOL [1,10]; NADH-ferredoxinTOL reductase component of toluene dioxygenase [3,10]; ferredoxinTOL component of the toluene dioxygenase [8,10]; recombinant α-subunit of the oxygenase component [12]; partial [4]; rapid purification of oxygenase component from a polyol-responsive monoclonal antibody [13]; β-subunit of oxygenase component [14]) [1, 3, 4, 8, 10, 12, 13, 14]

Cloning

<1> (genes encoding the three components of the toluene dioxygenase overproduced in Escherichia coli JM109: 1. the reductaseTOL - tolA, 2. the ferredoxinTOL and 3. the two subunits of the terminal dioxygenase - todC1C2 [5]; cloned and expressed in Escherichia coli HB101 [6]; expression of mutant enzyme TDO 2-B38, in which the wild-type stop codon is replaced with a codon encoding Thr [22]) [5, 6, 22]

Engineering

D219A <1> (<1>, mutation at α-subunit of oxygenase component completely abolishes toluene dioxygenase activity, mutation completely eliminates formation of cis-toluene dihydrodiol [14]) [14]
E214A <1> (<1>, mutation at α-subunit of oxygenase component completely abolishes toluene dioxygenase activity, mutation completely eliminates formation of cis-toluene dihydrodiol [14]) [14]
H222A <1> (<1>, mutation at α-subunit of oxygenase component completely abolishes toluene dioxygenase activity, mutation completely eliminates formation of cis-toluene dihydrodiol [14]) [14]
H228A <1> (<1>, mutation at α-subunit of oxygenase component completely abolishes toluene dioxygenase activity, mutation completely eliminates formation of cis-toluene dihydrodiol [14]) [14]
Y221A <1> (<1>, mutation at α-subunit of oxygenase component, 42% of the activity of the wild-type enzyme, formation of cis-toluene dihydrodiol is reduced [14]) [14]
Y266A <1> (<1>, mutation at α-subunit of oxygenase component, 12% of the activity of the wild-type enzyme, formation of cis-toluene dihydrodiol is reduced [14]) [14]

Additional information <1> (<1>, Escherichia coli expressed mutant enzyme TDO 2-B38, in which the wild-type stop codon is replaced with a codon encoding threonine, exhibits 5.6times higher activity towards 4-picoline and about 20% more activity towards toluene than the wild-type enzyme [22]) [22]

Application

synthesis <1> (<1>, screening of substituted arenes containing remote chiral centers as substrates, enantiomers are indiscriminately processed to diastereomeric pairs. Some of these new metabolites are useful as synthons for morphine synthesis [19]) [19]

6 Stability

Storage stability

<1>, -20°C, ferredoxinTOL component of the toluene dioxygenase, stable for over 10 weeks [8]
<1>, -20°C, purified NADH-ferredoxinTOL reductase component is stable for two weeks [3]
<1>, 0-4°C, ferredoxinTOL component of the toluene dioxygenase, stable for up to 72 h [8]
<1>, 0-4°C, purified NADH-ferredoxinTOL reductase component is stable for up to 30 h [3]

References

[1] Subramanian, V.; Liu, T.N.; Yeh, W.K.; Gibson, D.T.: Toluene dioxygenase: purification of an iron-sulfur protein by affinity chromatography. Biochem. Biophys. Res. Commun., **91**, 1131-1139 (1979)
[2] Wackett, L.P.; Kwart, L.D.; Gibson, D.T.: Benzylic monooxygenation catalyzed by toluene dioxygenase from Pseudomonas putida. Biochemistry, **27**, 1360-1367 (1988)
[3] Subramanian, V.; Liu, T.N.; Yeh, W.K.; Narro, M.; Gibson, D.T.: Purification and properties of NADH-ferredoxinTOL reductase. A component of toluene dioxygenase from Pseudomonas putida. J. Biol. Chem., **256**, 2723-2730 (1981)
[4] Yeh, W.K.; Gibson, D.T.; Liu, T.N.: Toluene dioxygenase: a multicomponent enzyme system. Biochem. Biophys. Res. Commun., **78**, 401-410 (1977)
[5] Zylstra, G.J.; Gibson, D.T.: Toluene degradation by Pseudomonas putida F1. Nucleotide sequence of the todC1C2BADE genes and their expression in Escherichia coli. J. Biol. Chem., **264**, 14940-14946 (1989)
[6] Stephens, G.M.; Sidebotham, J.M.; Mann, N.H.; Dalton, H.: Cloning and expression in Escherichia coli of the toluene dioxygenase gene from Pseudomonas putida. FEMS Microbiol. Lett., **57**, 295-300 (1989)

[7] Zylstra, G.J.; Wackett, L.P.; Gibson, D.T.: Trichloroethylene degradation by Escherichia coli containing the cloned Pseudomonas putida F1 toluene dioxygenase genes. Appl. Environ. Microbiol., **55**, 3162-3166 (1989)
[8] Subramanian, V.; Liu, T.N.; Yeh, W.K.; Serdar, C.M.; Wackett, L.P.; Gibson, D.T.: Purification and properties of ferredoxinTOL. A component of toluene dioxygenase from Pseudomonas putida F1. J. Biol. Chem., **260**, 2355-2363 (1985)
[9] Zylstra, G.J.; McCombie, W.R.; Gibson, D.T.; Finette, B.A.: Toluene degradation by Pseudomonas putida F1: genetic organization of the tod operon. Appl. Environ. Microbiol., **54**, 1498-1503 (1988)
[10] Wackett, L.P.: Toluene dioxygenase from Pseudomonas putida F1. Methods Enzymol., **188**, 39-45 (1990)
[11] Lange, C.C.; Wackett, L.P.: Oxidation of aliphatic olefins by toluene dioxygenase: enzyme rates and product identification. J. Bacteriol., **179**, 3858-3865 (1997)
[12] Jiang, H.; Parales, R.E.; Gibson, D.T.: The α subunit of toluene dioxygenase from Pseudomonas putida F1 can accept electrons from reduced FerredoxinTOL but is catalytically inactive in the absence of the β subunit. Appl. Environ. Microbiol., **65**, 315-318 (1999)
[13] Lynch, N.A.; Jiang, H.; Gibson, D.T.: Rapid purification of the oxygenase component of toluene dioxygenase from a polyol-responsive monoclonal antibody. Appl. Environ. Microbiol., **62**, 2133-2137 (1996)
[14] Jiang, H.; Parales, R.E.; Lynch, N.A.; Gibson, D.T.: Site-directed mutagenesis of conserved amino acids in the α subunit of toluene dioxygenase: potential mononuclear non-heme iron coordination sites. J. Bacteriol., **178**, 3133-3139 (1996)
[15] Leahy, J.G.; Olsen, R.H.: Kinetics of toluene degradation by toluene-oxidizing bacteria as a function of oxygen concentration, and the effect of nitrate. FEMS Microbiol. Ecol., **23**, 23-30 (1997)
[16] Lee, K.; Brand, J.M.; Gibson, D.T.: Stereospecific sulfoxidation by toluene and naphthalene dioxygenases. Biochem. Biophys. Res. Commun., **212**, 9-15 (1995)
[17] Li, S.; Wackett, L.P.: Trichloroethylene oxidation by toluene dioxygenase. Biochem. Biophys. Res. Commun., **185**, 443-451 (1992)
[18] Resnick, S.M.; Torok, D.S.; Lee, K.; Brand, J.M.; Gibson, D.T.: Regiospecific and stereoselective hydroxylation of 1-indanone and 2-indanone by naphthalene dioxygenase and toluene dioxygenase. Appl. Environ. Microbiol., **60**, 3323-3328 (1994)
[19] Bui, V.P.; Vidar Hansen, T.; Stenstrom, Y.; Hudlicky, T.; Ribbons, D.W.: A study of substrate specificity of toluene dioxygenase in processing aromatic compounds containing benzylic and/or remote chiral centers. New J. Chem., **25**, 116-124 (2001)
[20] Boyd, D.R.; Sharma, N.D.; Haughey, S.A.; Kennedy, M.A.; McMurray, B.T.; Sheldrake, G.N.; Allen, C.C.R.; Dalton, H.; Sproule, K.: Toluene and naphthalene dioxygenase-catalyzed sulfoxidation of alkyl aryl sulfides. J. Chem. Soc. Perkin Trans., **1**, 1929-1934 (1998)

[21] Bowers, N.I.; Boyd, D.R.; Sharma, N.D.; Goodrich, P.A.; Groocock, M.R.; Blacker, A.J.; Goode, P.; Dalton, H.: Stereoselective benzylic hydroxylation of 2-substituted indanes using toluene dioxygenase as biocatalyst. J. Chem. Soc. Perkin Trans., **1**, 1453-1462 (1999)
[22] Sakamoto, T.; Joern, J.M.; Arisawa, A.; Arnold, F.H.: Laboratory evolution of toluene dioxygenase to accept 4-picoline as a substrate. Appl. Environ. Microbiol., **67**, 3882-3887 (2001)

Naphthalene 1,2-dioxygenase 1.14.12.12

1 Nomenclature

EC number
1.14.12.12

Systematic name
naphthalene,NADH:oxygen oxidoreductase (1,2-hydroxylating)

Recommended name
naphthalene 1,2-dioxygenase

Synonyms
naphthalene dioxygenase
naphthalene oxygenase
oxygenase, naphthalene di-

CAS registry number
9074-04-8

2 Source Organism

<1> *Pseudomonas putida* (wild-type grown on naphthalene [1]; strain 119, unable to grow on naphthalene [1]; strain G7 [9, 22]) [1, 9, 10, 22]
<2> *Pseudomonas sp.* (strain NCIB 9816 [2]; strain NCIB 9816/11, naphthalene dihydrogenase mutant derived from strain NCIB 9816-4 oxidizing naphthalene to cis-(1R,2S)-dihydroxy-1,2-dihydronaphthalene [12, 14, 15, 16, 17, 18, 19, 20, 23, 24, 25]) [2-6, 8, 12, 14, 15, 16, 17, 18, 19, 20, 23, 24, 25]
<3> *Corynebacterium renale* (grown on naphthalene as sole carbon source [7]) [7]
<4> *Pseudomonas fluorescens* (strain NCIMB 40531 [13]) [11, 13]
<5> *Rhodococcus sp.* (strain NCIMB12038 [21, 26]) [21, 26]

3 Reaction and Specificity

Catalyzed reaction
naphthalene + NADH + H^+ + O_2 = (1R,2S)-1,2-dihydronaphthalene-1,2-diol + NAD^+ (<1> proposed mechanism [1]; <2> proposed electron transport chain: NADH, ferredoxixinNAP reductase, ferredoxinNAP, terminal oxygenase ISPNAP [3])

Reaction type

oxidation
redox reaction
reduction

Natural substrates and products

S naphthalene + NADH + O_2 <1, 2> (<1,2> initial oxidative reaction in bacterial naphthalene catabolism [1,6]; <2> three-component dioxygenase, uses two proteins containing three redox centers to transfer electrons to the terminal oxygenase [3]; <2> genes coding for naphthalene dioxygenase: ndoA, ndoB and ndoC [5]) (Reversibility: ? <1, 2> [1, 6]) [1-6]

P (1R,2S)-1,2-dihydronaphthalene-1,2-diol + NAD^+ <1, 2> [1-6]

Substrates and products

S (1S)-indanol + NADH + O_2 <2> (<2> recombinant enzyme [14]) (Reversibility: ? <2> [14]) [14]

P (1S)-indanol + trans-1,3-indandiol <2> (<2> 85.5% indanol, 11.5% 1,3-indandiol, minor products: (1S)-indenol and 1-indanone [14]) [14]

S (R)-1-indanol + NAD^+ + O_2 <2> (<2> activity in E. coli cells expressing recombinant naphthalene dioxygenase [20]) (Reversibility: ? <2> [20]) [20]

P cis-1,3-indandiol + (R)-3-hydroxy-1-indanone + cis-1,2,3-indantriol + NAD^+ <2> (<2> 71% cis-1,3-indandiol, 18.2% (R)-3-hydroxy-1-indanone and 10.8% cis-1,2,3-indantriol [20]) [20]

S (S)-1-indanol + NAD^+ + O_2 <2> (<2> activity in E. coli cells expressing recombinant naphthalene dioxygenase [20]) (Reversibility: ? <2> [20]) [20]

P trans-(1S,3S)-indan-1,3-diol + (R)-3-hydroxy-1-indanone + NAD^+ <2> (<2> 95.5% trans-(1S,3S)-indan-1,3-diol [20]) [20]

S (S)-1-indenol + NADH + O_2 <2> (<2> recombinant enzyme system [20]) (Reversibility: ? <2> [20]) [20]

P syn-2,3-dihydroxy-1-indanol + anti-2,3-dihydroxy-1-indanol <2> [20]

S 1,2-dihydronaphthalene + NADH + O_2 <2> (<2> naphthalene dioxygenase activity in cells induced by salicylate [15]) (Reversibility: ? <2> [15]) [15]

P cis-(1R,2S)-dihydroxy-1,2,3,4-tetrahydronaphthalene + cis-(1R,2S)-dihydroxy-1,2-dihydronaphthalene + NAD^+ <2> [15]

S 2-chlorobiphenyl + NADH + O_2 <1> (<1> recombinant histidine-tagged terminal dioxygenase [22]) (Reversibility: ? <1> [22]) [22]

P 2-chloro-2',3'-dihydrobiphenyl-2,3-diol + NAD^+ <1> [22]

S 2-methoxynaphthalene + NADH + O_2 <2> (<2> naphtalene dioxygenase activity in whole cells and activity in E. coli cells expressing naphtalene dioxygenase [12]) (Reversibility: ? <2> [12]) [12]

P (1R,2S)-dihydroxy-7-methoxy-1,2-dihydronaphthalene + NAD^+ <2> (<2> minor product (1R,2S)-dihydroxy-6-methoxy-1,2-dihydronaphthalene [12]) [12]

S 3-chlorobiphenyl + NADH + O_2 <1> (<1> recombinant histidine-tagged terminal dioxygenase [22]) (Reversibility: ? <1> [22]) [22]

P 3-chloro-2',3'-dihydrobiphenyl-2,3-diol + NAD^+ <1> [22]

S 4-chlorobiphenyl + NADH + O_2 <1> (<1> recombinant histidine-tagged terminal dioxygenase [22]) (Reversibility: ? <1> [22]) [22]

P 4-chloro-2',3'-dihydrobiphenyl-2,3-diol + NAD^+ <1> [22]

S 6,7-dihydro-5H-benzocycloheptene + NADH + O_2 <2> (<2> naphthalene dioxygenase activity in cells induced by salicylate [18]) (Reversibility: ? <2> [18]) [18]

P (1R,2S)-cis-dihydroxybenzocycloheptane + NAD^+ <2> (<2> enantiomeric excess greater than 98% [18]) [18]

S 9,10-dihydroanthracene + NADH + O_2 <2> (<2> naphthalene dioxygenase activity in cells induced by salicylate [17]) (Reversibility: ? <2> [17]) [17]

P cis-(1R,2S)-1,2-dihydroxy-1,2,9,10-tetrahydroanthracene + NAD^+ <2> (<2> more than 95% yield, enantiomeric excess greater than 95%, minor product: 9-hydroxy-9,10-dihydroanthracene [17]) [17]

S 9,10-dihydrophenanthrene + NADH + O_2 <2> (<2> naphthalene dioxygenase activity in cells induced by salicylate [17]) (Reversibility: ? <2> [17]) [17]

P cis-(3S,4R)-3,4-dihydroxy-3,4,9,10-tetrahydrophenanthrene + (S)-9-hydroxy-9,10-dihydrophenanthrene <2> (<2> 70% cis-(3S,4R)-3,4-dihydroxy-3,4,9,10-tetrahydrophenanthrene, enantiomeric excess greater than 95%, 30% (S)-9-hydroxy-9,10-dihydrophenanthrene [17]) [17]

S benzene + NADH + O_2 <2> (<2> 40-50% of O_2 is reduced to H_2O_2 by an uncoupling reaction, only trace amounts of cis-benzene-1,2-dihydrodiol are formed [23]) (Reversibility: ? <2> [23]) [23]

P H_2O_2 + cis-benzene-1,2-dihydrodiol + NAD^+ <2> [23]

S benzocyclobutene + NADH + O_2 <4> (<4> naphtalene dioxygenase activity in intact cells [11]) (Reversibility: ? <4> [11]) [11]

P benzocyclobutene-1-ol + benzocyclobutene-1-one <4> [11]

S dibenzofuran + NADH + O_2 <2> (<2> naphthalene dioxygenase activity in cells induced by salicylate [15]) (Reversibility: ? <2> [16]) [15]

P (1R,2S)-cis-1,2-dihydroxy-1,2-dihydrodibenzofuran + (3S,4R)-cis-3,4-dihydroxy-3,4-dihydrodibenzofuran + NAD^+ <2> (<2> 60-70% (1R,2S)-cis-1,2-dihydroxy-1,2-dihydrodibenzofuran, enantiomeric excess greater than 95%, 30-40% (3S,4R)-cis-3,4-dihydroxy-3,4-dihydrodibenzofuran, enantiomeric excess greater than 95% [16]) [16]

S dibenzothiophene + NADH + O_2 <2> (<2> naphthalene dioxygenase activity in cells induced by salicylate [16]) (Reversibility: ? <2> [16]) [16]

P (1R,2S)-cis-1,2-dihydroxy-dihydrodibenzothiophene + dibenzothiophene sulfoxide <2> (<2> 84-87% (1R,2S)-cis-1,2-dihydroxy-dihydrodibenzothiophene, enantiomeric excess greater than 95%, 15% dibenzothiophene sulfoxide [16]) [16]

S fluorene + NADH + O_2 <2> (<2> naphthalene dioxygenase activity in cells induced by salicylate [15]) (Reversibility: ? <2> [16]) [15]

P (3S,4R)-cis-3,4dihydroxy-3,4-dihydrofluorene + 9-fluorenol + NAD^+ <2> (<2> 80-90% (3S,4R)-cis-3,4dihydroxy-3,4-dihydrofluorene, ennatiomeric excess greater than 95%, 10% 9-fluorenol [16]) [16]

S indan + NADH + O_2 <1> (Reversibility: ? <1> [10]) [10]
P 1-indanol + NAD^+ <1> [10]
S indan + NADH + O_2 <2> (<2> recombinant enzyme, enantiospecific monooxygenation [14]) (Reversibility: ? <2> [14]) [14]
P (1S)-indenol + (1S)-indanol + NAD^+ <2> (<2> 20% indenol, 67.1% indanol, minor products: indene, 1-indanone and (1R,2S)-indandiol [14]) [14]
S indene + NADH + O_2 <2> (<2> recombinant enzyme [14]) (Reversibility: ? <2> [14]) [14]
P (1S)-indenol + (1R,2S)-indandiol + NAD^+ <2> (<2> 57.9% indenol, 29.6% indandiol [14]) [14]
S indole + ? + O_2 <1> (<1> reaction of recombinant E. coli transformed with Pseudomonas sp. DNA [9]) (Reversibility: ? <1> [9]) [9]
P cis-indole-2,3-dihydrodiol <1> [9]
S naphthalene + NAD(P)H + O_2 <1, 5> (<2> 50% activity with NADH compared to NADPH [6]) (Reversibility: ? <1, 5> [1, 21]) [1, 6, 21]
P cis-(1R,2S)-1,2-dihydronaphthalene-1,2-diol + $NAD(P)^+$ <1, 5> (<1> only cis-isomer by bacteria, product rapidly autooxidizes to 1,2-naphthoquinone [1]) [1, 21]
S naphthalene + NADH + O_2 <2, 3> (<2> 3 component enzyme system consisting of ferredoxinNAP reductase, ferredoxinNAP and oxygenase ISPNAP, ferredoxinNAP reductase reduces: 2,6-dichlorophenolindophenol, ferricyanide, nitro blue tetrazolium and cytochrome c, in the presence of FAD ferredoxinNAP reductase transfers electrons to ferredoxin [3]) (Reversibility: ? <2, 3> [2-7]) [2-7, 15]
P cis-(1R,2S)-1,2-dihydronaphthalene-1,2-diol + NAD^+ <2, 3> [2-7, 15]
S styrene + NADH + O_2 <2> (<2> recombinant enzyme system [19]) (Reversibility: ? <2> [19]) [19]
P (R)-1-phenyl-1,2-ethanediol + NAD^+ <2> (<2> 78.6% enantiomeric excess [19]) [19]

Inhibitors

1,10-phenanthroline <2> (<2> 10 mM, 63% inhibition of ferredoxinNAP reductase [3]) [3]
4-chloromercuribenzoate <2> (<2> 0.0005 mM, 94% inhibition of ferredoxinNAP reductase [3]) [3]
H_2O_2 <2> (<2> 0.4 mM, complete inactivation of reduced terminal dioxgenase ISPNAP after 10 min in the absence of ferrous iron [23]) [23]
N-ethylmaleimide <2> (<2> 2 mM, 30% inhibition of ferredoxinNAP reductase [3]) [3]
NaN_3 <2> (<2> 40 mM, 46% inhibition of ferredoxinNAP reductase [3]) [3]
iodoacetate <2> (<2> 10 mM, 50% inhibition of ferredoxinNAP reductase [3]) [3]

Cofactors/prosthetic groups

FAD <1, 2, 5> (<2> component A, ferredoxinNAP reductase is a flavoprotein, enzyme can use both FAD and FMN but exhibits slightly higher affinity for FAD [2,3]; <2> addition enhances ferredoxinNAP reductase activity with all in vitro electron-acceptors, e.g. cytochrome c, 2,6-dichlorophenolindophenol,

Nitroblue tetrazolium and ferricyanide [3]; <2> 1 mol FAD/mol enzyme in flavin-reconstituted protein [3]; <2> stimulates cytochrome c reduction by ferredoxinNAP reductase [6]; <5> stimulates activity of terminal oxygenase ISPNAR [21]) [1-4, 6, 21]
FMN <2, 5> (<2> requirement, ferredoxinNAP reductase is a flavoprotein, enzyme can use both FAD and FMN but exhibits slightly higher affinity for FAD, addition stimulates dioxygenase activity by 53% of FAD-stimulation [3]; <2> stimulates cytochrome c reduction by ferredoxinNAP reductase [6]; <5> stimulates activity of terminal oxygenase ISPNAR [21]) [3, 6, 21]
NADH <1-3, 5> (<2> requirement, the oxygenase accepts two electrons from NADH, the reduction requires component A and C as mediators [2]) [1-4, 6-8, 21]
NADPH <1, 2, 5> (<2> can replace NADH with 39% [3]; <2> less than 50% of the activity with NADH [6]; <2> ferredoxinNAP reductase [3]) [1, 3, 6, 8, 21]
Additional information <2, 3> (<2> spinach ferredoxin cannot replace ferredoxinNAP [4]; <3> cytochrome P-450 is no cofactor [7]) [4, 7]

Activating compounds

ferricyanide <2> (<2> activation, direct reduction by reductaseNAP in the presence of NADH in vitro [3]) [3]

Metals, ions

Fe^{2+} <2, 3, 5> (<2> component A, i.e. NADH-ferredoxinNAP reductase, of the multienzyme system is an iron-containing flavoprotein containing 1.8 g atoms Fe^{2+} and 2 g atoms sulfur [3]; <2> component B, the terminal oxygenase ISPNAP is an iron-sulfur protein, oxidized ISPNAP binds naphthalene without conformational changes that affect its FeS-chromophores, ISPNAP contains 6 g atom Fe^{2+} and 4 g atom acid-labile sulfur per mol enzyme, the enzyme complex is not stimulated by exogenous Fe^{2+} [2]; <3> non-heme iron protein [7]; <2> tightly bound Fe^{2+} [8]; <5> terminal dioxygenase ISPNAR may be a Rieske-type iron-sulfur protein containing 2.4 g atoms of iron and 2.1 g atoms of sulfur per α,β subunit [21]; <2> α subunit of terminal oxygenase ISPNAP contains a Rieske [2Fe-2S] center in one domain and a mononuclear iron in the catalytic domain [25]) [2-4, 6, 7, 8, 21, 25]
Additional information <2> (<2> no significant activation by the addition of Fe^{2+}, Fe^{3+}, Zn^{2+}, Mg^{2+} or Cu^{2+} [6]) [6]

Specific activity (U/mg)

0.006 <5> (<5> activity of terminal dioxygenase ISPNAR [21]) [21]
1.05 <3> [7]
1.37 <2> (<2> in the presence of partially purified components A and C of the 3 component enzyme system plus FAD [2]) [2]
397 <2> (<2> cytochrome c reduction by ferredoxinNAP reductase [3]) [3]
Additional information <4> (<4> 0.000125 mM/min/mg dry cell weight, naphthalene dioxygenase activity in cells grown on succinate in the presence of naphthalene as inducer [13]) [13]

K_m-Value (mM)
1.42 <3> (NADH) [7]
2.9 <3> (naphthalene) [7]

pH-Optimum
6.5 <3> [7]
7.2 <1> (<1> assay at [1]) [1]
7.5 <2> (<2> assay at [2,4]) [2, 4]
Additional information <2> (<2> ferredoxin reductaseNAP, pI: 6.3 [3]) [3]

Temperature optimum (°C)
30 <3> [7]

4 Enzyme Structure

Molecular weight
34900 <2> (<2> ferredoxinNAP reductase, native PAGE [3]) [3]
37000 <2> (<2> ferredoxinNAP reductase, gel filtration [3]) [3]
37100 <2> (<2> ferredoxinNAP reductase, deduced from amino acid sequence [3]) [3]
99000 <3> (<3> gel filtration [7]) [7]
155000 <5> (<5> terminal dioxygenase ISPNAR, gel filtration [21]) [21]
158000 <2> (<2> component B, i.e. oxygenase ISPNAP, gel filtration [2]) [2]
Additional information <1, 2> (<1,2> three component enzyme system consisting of: component A, i.e. ferredoxinNAP reductase, component B, i.e. terminal oxygenase ISPNAP, and component C, i.e. ferredoxinNAP [1-3]) [1-3]

Subunits
? <1> (<1> x * 53000 + x * 25000, histidine-tagged recombinant enzyme, SDS-PAGE [22]) [22]
dimer <3> (<3> 1 * 43000 + 1 * 56000, SDS-PAGE [7]) [7]
monomer <2> (<2> 1 * 36000, ferredoxinNAP reductase, SDS-PAGE [3]) [3]
tetramer <2, 5> (<2> $\alpha_2\beta_2$ 2 * 55000 + 2 * 20000, terminal oxygenase ISPNAP, SDS-PAGE [2]; <5> $\alpha_2\beta_2$, 2 * 55000 + 2 * 23000, SDS-PAGE [21]) [2, 21]
Additional information <5> (<5> X-ray structure data from Pseudomonas sp. 9861-4 suggest an $\alpha_3\beta_3$ hexameric structure [21]) [21]

5 Isolation/Preparation/Mutation/Application

Localization
cytoplasm <2> [2-4, 6]

Purification
<2> (component B i.e. the terminal oxygenase ISPNAP, DEAE-Sephadex, DEAE-cellulose, octyl-Sepharose, presence of 10% ethanol and 10% glycerol is required to maintain stability [2,4]; ferredoxinNAP reductase, Blue Sephar-

ose, DEAE-cellulose [3,4]; affinity chromatography separates the reductase from the two other enzyme components [6]; purification of ferredoxinNAP reductase, ferredoxinNAP and ISPNAP [4,6]) [2-4, 6]
<3> (ammonium sulfate, Sephadex G-75, DEAE-cellulose [7]) [7]
<5> (Blue Sepharose, Q-Sepharose, Superose 12, purification of component B, the terminal oxygenase ISPNAR [21]) [21]

Crystallization
<2> (vapor diffusion, equal volume of enzyme solution, 30 mg/ml, and reservoir solution, 2 M ammonium sulfate, 2-3% dioxane in 50 mM Mes, pH 6.0, mixed on a cover slip, deep red crystals, refinement of X-ray structure at 2.25 A resolution [25]) [25]
<5> (diffraction-quality crystals by hanging-drop method, 2.3 A resolution [26]) [26]

Cloning
<1> (strain G7, fragment of plasmid NAH7 cloned and expressed in Escherichia coli HB101 [5,9]; expression of histidine tagged enzyme in Escherichia coli [22]) [5, 9, 22]
<2> (strain NCIB 9816, nahA-gene containing ndoA, ndoB and ndoC is inserted into plasmid pT7-5, transferred to and expressed in Escherichia coli HB101 [5]; expression of ferredoxinNAP reductase, ferredoxinNAP and ISPNAP in Escherichia coli JM 109 [14]) [5, 14]

Engineering
D362A <2> (<2> mutation in α subunit of terminal dioxgenase ISPNAP, no activity with naphthalene, biphenyl and phenanthrene [24]) [24]
F202L <2> (<2> mutation in α subunit of terminal dioxgenase ISPNAP, no activity with naphthalene, biphenyl and phenanthrene [24]) [24]
F202V <2> (<2> mutation in α subunit of terminal dioxgenase ISPNAP, similar activity with naphthalene, biphenyl and phenanthrene as wild-type [24]) [24]
F352L <2> (<2> mutation in α subunit of terminal dioxgenase ISPNAP, similar activity with naphthalene, biphenyl and phenanthrene as wild-type, major product of biphenyl oxidation: biphenyl-3,4-dihydrodiol instead of biphenyl-2,3-dihydrodiol [24]) [24]
F352V <2> (<2> mutation in α subunit of terminal dioxgenase ISPNAP, similar activity with naphthalene, biphenyl and phenanthrene as wild-type, product of biphenyl oxidation: 96% biphenyl-3,4-dihydrodiol instead of biphenyl-2,3-dihydrodiol [24]) [24]
M366W <2> (<2> mutation in α subunit of terminal dioxgenase ISPNAP, similar activity with naphthalene, biphenyl and phenanthrene as wild-type [24]) [24]
N201A <2> (<2> mutation in α subunit of terminal dioxgenase ISPNAP, similar activity with naphthalene as wild-type, very weak activity with biphenyl [24]) [24]

N201Q <2> (<2> mutation in α subunit of terminal dioxgenase ISPNAP, similar activity with naphthalene, biphenyl and phenanthrene as wild-type [24]) [24]
N201S <2> (<2> mutation in α subunit of terminal dioxgenase ISPNAP, similar activity with naphthalene, biphenyl and phenanthrene as wild-type [24]) [24]
T351N <2> (<2> mutation in α subunit of terminal dioxgenase ISPNAP, similar activity with naphthalene, biphenyl and phenanthrene as wild-type [24]) [24]
T351R <2> (<2> mutation in α subunit of terminal dioxgenase ISPNAP, similar activity with naphthalene, biphenyl and phenanthrene as wild-type [24]) [24]
T351S <2> (<2> mutation in α subunit of terminal dioxgenase ISPNAP, similar activity with naphthalene, biphenyl and phenanthrene as wild-type [24]) [24]
V260A <2> (<2> mutation in α subunit of terminal dioxgenase ISPNAP, similar activity with naphthalene, biphenyl and phenanthrene as wild-type [24]) [24]
V260L <2> (<2> mutation in α subunit of terminal dioxgenase ISPNAP, similar activity with naphthalene, biphenyl and phenanthrene as wild-type [24]) [24]
V260N <2> (<2> mutation in α subunit of terminal dioxgenase ISPNAP, similar activity with naphthalene, biphenyl and phenanthrene as wild-type [24]) [24]
W316A <2> (<2> mutation in α subunit of terminal dioxgenase ISPNAP, similar activity with naphthalene, biphenyl and phenanthrene as wild-type [24]) [24]
W358A <2> (<2> mutation in α subunit of terminal dioxgenase ISPNAP, very low activity with naphthalene and biphenyl [24]) [24]

6 Stability

Temperature stability

25 <2> (<2> room temperature, ferredoxinNAP reductase, $t_{1/2}$: 8 h [3]) [3]

Organic solvent stability

2-methoxyethanol <3> (<3> stable to [7]) [7]

General stability information

<2>, 10% ethanol and 10% glycerol are required for stability [2, 4, 6]
<2>, DTT stabilizes [2, 6]
<2>, dialysis inactivates rapidly, $NaBH_4$ and dithioerythritol restore activity [8]
<2>, dilution inactivates rapidly, $NaBH_4$ and dithioerythritol restore activity [8]
<2>, purification of reductase leads to significant loss of flavin cofactor [3]

Storage stability

<2>, -20°C, ferredoxinNAP reductase, 1 month, minimal loss of activity, prolonged storage leads to precipitation when preparation is heated above 5°C [3]

<2>, 0-5°C, ferredoxinNAP reductase, 5 days, 30% loss of activity [3]

References

[1] Jeffrey, A.M.; Yeh, H.J.C.; Jerina, D.M.; Patel, T.R.; Davey, J.F.; Gibson, D.T.: Initial reactions in the oxidation of naphthalene by Pseudomonas putida. Biochemistry, **14**, 575-584 (1975)

[2] Ensley, B.D.; Gibson, D.T.: Naphthalene dioxygenase: purification and properties of a terminal oxygenase component. J. Bacteriol., **155**, 505-511 (1983)

[3] Haigler, B.E.; Gibson, D.T.: Purification and properties of NADH-ferredoxinNAP reductase, a component of naphthalene dioxygenase from Pseudomonas sp. strain NCIB 9816. J. Bacteriol., **172**, 457-464 (1990)

[4] Ensley, B.D.; Haigler, B.E.: Naphthalene dioxygenase from Pseudomonas NCIB 9816. Methods Enzymol., **188**, 46-52 (1990)

[5] Kurkela, S.; Lehväslaiho, H.; Palva, E.T.; Teeri, T.H.: Cloning, nucleotide sequence and characterization of genes encoding naphthalene dioxygenase of Pseudomonas putida strain NCIB9816. Gene, **73**, 355-362 (1988)

[6] Ensley, B.D.; Gibson, D.T.; Laborde, A.L.: Oxidation of naphthalene by a multicomponent enzyme system from Pseudomonas sp. strain NCIB 9816. J. Bacteriol., **149**, 948-954 (1982)

[7] Dua, R.D.; Meera, S.: Purification and characterisation of naphthalene oxygenase from Corynebacterium renale. Eur. J. Biochem., **120**, 461-465 (1981)

[8] Catterall, F.A.; Williams, P.A.: Some properties of the naphthalene oxygenase from Pseudomonas sp. NCIB 9816. J. Gen. Microbiol., **67**, 117-124 (1971)

[9] Ensley, B.D.; Ratzkin, B.J.; Osslund, T.D.; Simon, M.J.; Wackett, L.P.; Gibson, D.T.: Expression of naphthalene oxidation genes in Escherichia coli results in the biosynthesis of indigo. Science, **222**, 167-169 (1983)

[10] Wackett, L.P.; Kwart, L.D.; Gibson, D.T.: Benzylic monooxygenation catalyzed by toluene dioxygenase from Pseudomonas putida. Biochemistry, **27**, 1360-1367 (1988)

[11] Swanson, E.P: Microbial transformation of benzocyclubutene to benzocyclobutene-1-ol and benzocyclobutene-1-one. Appl. Environ. Microbiol., **58**, 3404-3406 (1992)

[12] Whited, G.M.; Downie, J.C.; Hudlicky, T.; Fearnley, S.P.; Dudding, t.C.; Olivo, H.F.; Parker, D.: Oxidation of 2-methoxynaphthalene by toluene, naphthalene and biphenyl dioxygenases: structure and absolute stereochemistry of metabolites. Bioorg. Med. chem., **2**, 727-734 (1994)

[13] Cidaria, D.; Deidda, F.; Bosetti, A.: A rapid method for naphthalene dioxygenase assay in whole cells of naphthalene cis-dihydrodiol dehydrogenase blocked Pseudomonas fluorescens: screening of potential inducrs of dioxyenase activity. Appl. Microbiol. Biotechnol., **41**, 689-693 (1994)

[14] Gibson, D.T.; Resnick, S.M.; Lee, K.; Brand, J.M.; Torok, D.S.; Wackett, L.P.; Schocken, M.J.; Haigler, B.E.: Desaturation, dioxygenation, and monooxygenation reactions catalyzed by naphthalene dioxygenase from Pseudomonas sp. strain 9816-4. J. Bacteriol., **177**, 2615-2621 (1995)
[15] Torok, D.S.; Resnick, S.M.; Brand, J.M.; Cruden, D.L.; Gibson, D.T.: Desaturation and oxygenation of 1,2-dihydronaphthalene by toluene and naphthalene dioxygenase. J. Bacteriol., **177**, 5799-5805 (1995)
[16] Resnick, S.M.; Gibson, D.T.: Regio- and stereospecific oxidation of fluorene, dibenzofuran, and dibenzothiophene by naphthalene dioxygenase from Pseudomonas sp. strain NCIB 9816-4. Appl. Environ. Microbiol., **62**, 4073-4080 (1996)
[17] Resnick, s.M.; Gibson, D.T.: Regio- and stereospecific oxidation of 9,10-dihydroanthracene and 9,10-dihydrophenanthrene by naphthalene dioxygenase: structure and absolute stereochemistry of metabolites. Appl. Environ. Microbiol., **62**, 3355-3359 (1996)
[18] Resnick, S.M.; Gibson, D.T.: Oxidation of 6,7-dihydro-5H-benzocycloheptene by bacterial strains expressing naphthalene dioxygenase, biphenyl dioxygenase, and toluene dioxygenase yields homochiral mono-ol or cis-diol enantiomers as major products. Appl. Environ. Microbiol., **62**, 1364-1368 (1996)
[19] Lee, K.; Gibson, D.T.: Stereospecific dihydroxylation of the styrene vinyl group by purified naphthalene dioxygenase from Pseudomonas sp. strain NCIB 9816-4. J. Bacteriol., **178**, 3353-3356 (1996)
[20] Lee, K.; Resnick, S.M.; Gibson, D.T.: Stereospecific oxidation of (R)- and (S)-1-indanol by naphthalene dioxygenase from Pseudomonas sp. strain NCIB 9816-4. Appl. Environ. Microbiol., **63**, 2067-2070 (1997)
[21] Larkin, M.J.; Allen, C.C.R.; Kulakov, L.A.; Lipscomb, D.A.: Purification and characterization of a novel naphthalene dioxygenase from Rhodococcus sp. strain NCIMB12038. J. Bacteriol., **181**, 6200-6204 (1999)
[22] Barriault, D.; Sylvestre, M.: Functionality of biphenyl 2,3-dioxygenase components in naphthalene 1,2-dioxygenase. Appl. Microbiol. Biotechnol., **51**, 592-597 (1999)
[23] Lee, K: Benzene-induced uncoupling of naphthalene dioxygenase activity and enzyme inactivation by production of hydrogen peroxide. J. Bacteriol., **181**, 2719-2725 (1999)
[24] Parales, R.E.; Lee, K.; Resnick, S.M.; Jiang, H.; Lessner, D.J.; Gibson, D.T: Substrate specifity of naphthalene dioxgenase: effect of specific amino acids at the active site of the enzyme. J. Bacteriol., **182**, 1641-1649 (2000)
[25] Carredano, E.; Karlsson, A.; Kauppi, B.; Choudhury, D.; Parales, R.E.; Parales, J.V.; Lee, K.; Gibson, D.T.; Eklund, H.; Ramaswamy, S.: Substrate binding site of naphthalene 1,2-dioxygenase: functional implications of indole binding. J. Mol. Biol., **296**, 701-712 (2000)
[26] Malik, Z.A.; Allen, C.C.; Gakhar, L.; Lipscomb, D.A.; Larkin, M.J.; Ramaswamy, S.: Crystallization and preliminary X-ray diffraction analysis of naphthalene dioxygenase from Rhodococcus sp. strain NCIMB 12038. Acta Crystallogr. Sect. D, **58**, 2173-2174 (2002)

2-Chlorobenzoate 1,2-dioxygenase 1.14.12.13

1 Nomenclature

EC number
1.14.12.13

Systematic name
2-chlorobenzoate,NADH:oxygen oxidoreductase (1,2-hydroxylating, dechlorinating, decarboxylating)

Recommended name
2-chlorobenzoate 1,2-dioxygenase

Synonyms
2-halobenzoate 1,2-dioxygenase

CAS registry number
125268-83-9

2 Source Organism

<1> *Pseudomonas cepacia* (2CBS, isolated from water [1-3]; Burkholderia cepacia 2CBS, DSM 9959 [4]) [1-4]
<2> *Pseudomonas sp.* (CPE2 [5]) [5]
<3> *Burkholderia sp.* (TH2, isolated from soil [6]; strain NK8 isolated from soil [7]) [6, 7]

3 Reaction and Specificity

Catalyzed reaction
2-chlorobenzoate + NADH + H^+ + O_2 = catechol + chloride + NAD^+ + CO_2 (requires Fe^{2+}; <1> multi-component dioxygenase system proposed: 1,2-hydroxylating, dehalogenating, decarboxylating [1, 2]; <1> inducible two-component enzyme system: component B is an iron-sulfur flavoprotein (reductase) and red-brown component A (oxygenase) is an iron sulfur protein [2]; <1> two-component nonheme iron dioxygenase system of oxygenase system class IB dioxygenase: NADH:acceptor reductase containing both FAD and cloroplast-type 2Fe-2S and Rieske-type 2Fe-2S-containing oxygenase component [3]; <1> EPR, electron spin echo envelope modulation, electron nuclear double resonance studies and the electrochemical potential are performed. Enzyme contains one ferredoxin-type (reductase) and one Rieske-type (oxy-

genase) 2Fe2S center [4]; <3> broad substrate specificity, ability to transform all 3 monochlorobenzoates almost equally well. Enzyme appears to lack absolute regioselectivty [7])

Reaction type

decarboxylation
dehalogenation
hydroxylation
oxidation
redox reaction
reduction

Natural substrates and products

S 2,5-dichlorobenzoate + NADH + O_2 <2> (<2> involved in the metabolism of 2,5-dichlorobenzoate [5]) (Reversibility: ? <2> [5]) [5]
P 4-chlorocatechol + NAD^+ + chloride + CO_2 <2> [5]
S 2-chlorobenzoate + NADH + O_2 <1-3> (<1> first step of 2-chlorobenzoate degradation via *meta*-cleavage [1,2]; <1> initial step in the degradation of aromatic compounds [3]; <2> involved in the metabolism of 2-chlorobenzoate [5]) (Reversibility: ? <1-3> [1-7]) [1-7]
P catechol + chloride + NAD^+ + CO_2 <1-3> [1-7]

Substrates and products

S 2,5-dichlorobenzoate + NADH + O_2 <2> (Reversibility: ? <2> [5]) [5]
P 4-chlorocatechol + chloride + NAD^+ + CO_2 <2> [5]
S 2-aminobenzoate + NADH + O_2 <1> (Reversibility: ? <1> [2]) [2]
P ? + NAD^+ + CO_2
S 2-bromobenzoate + NADH + O_2 <1, 3> (Reversibility: ? <1, 3> [2, 3, 6]) [2, 3, 6]
P catechol + bromide + NAD^+ + CO_2 <1, 3> [2, 3, 6]
S 2-chlorobenzoate + NADH + O_2 <3> (Reversibility: ? <3> [7]) [7]
P 3-chlorocatechol + NAD^+ + O_2 <3> [7]
S 2-chlorobenzoate + NADH + O_2 <1-3> (<1> the oxygen in the enzymatically catechol is derived from molecular oxygen, the formed byproduct 2,3-dihydroxybenzoate is suggested to be a dead end metabolite [1]; <3> no growth on 2-chlorobenzoate [7]) (Reversibility: ? <1-3> [1-7]) [1-7]
P catechol + chloride + NAD^+ + CO_2 <1-3> [1-7]
S 2-fluorobenzoate + NADH + O_2 <1> (Reversibility: ? <1> [2, 3]) [2, 3]
P catechol + fluoride + NAD^+ + CO_2 <1> [2, 3]
S 2-hydroxybenzoate + NADH + O_2 <1> (Reversibility: ? <1> [2]) [2]
P catechol + ? + NAD^+ + CO_2 <1> [2]
S 2-iodobenzoate + NADH + O_2 <1, 3> (Reversibility: ? <1, 3> [2, 3, 6]) [2, 3, 6]
P catechol + iodide + NAD^+ + CO_2 <1, 3> [2, 3, 6]
S 2-methoxybenzoate + NADH + O_2 <1> (Reversibility: ? <1> [2]) [2]
P catechol + ? + NAD^+ + CO_2 <1> [2]
S 2-methylbenzoate + NADH + O_2 <1, 3> (<3> named *o*-toluate [6]) (Reversibility: ? <1, 3> [2, 6]) [2, 6]
P *o*-cresol + ? + NAD^+ + CO_2 <1, 3> [2, 6]

S 3-chlorobenzoate + NADH + O_2 <3> (Reversibility: ? <3> [6]) [6]
P ? + NAD^+ + CO_2 <3> [6]
S 3-chlorobenzoate + NADH + O_2 <3> (Reversibility: ? <3> [7]) [7]
P 3-chlorocatechol + NAD^+ + CO_2 <3> [7]
S 3-chlorobenzoate + NADH + O_2 <3> (Reversibility: ? <3> [7]) [7]
P 4-chlorocatechol + NAD^+ + CO_2 <3> [7]
S 4-chlorobenzoate + NADH + O_2 <3> (Reversibility: ? <3> [7]) [7]
P 4-chlorocatechol + NAD^+ + CO_2 <3> [7]
S anthranilate + NADH + O_2 <1> (Reversibility: ? <1> [2]) [2]
P catechol + ? + NAD^+ + CO_2 <1> [2]
S benzoate + NADH + O_2 <1, 3> (Reversibility: ? <1, 3> [2, 6]) [2, 6]
P 3,5-cyclohexadiene-1,2-diol-1-carboxylic acid + NAD^+ + CO_2 <1, 3> [2, 6]
S benzoate + NADH + O_2 <3> (<3> 2-chloro-3,5-cyclohexadiene-1,2-diol-1-carboxylic acid as unstable intermediate [7]) (Reversibility: ? <3> [7]) [7]
P catechol + NAD^+ + CO_2 <3> [7]
S Additional information <1> (<1> very broad substrate specificity is listed, benzoate analogs with substituents in ortho position are preferred. Trichloroethylene, dihydroxybenzoates, benzene chlorobenzene, toluene, phenol and chlorophenylacetate isomers do not serve as substrates [2]; <1> very broad substrate specificity [3]) [2, 3]
P ?

Inhibitors

2,2'-dipyridyl <1> (<1> 100% inhibition at 1 mM [2]) [2]
Cu^{2+} <1> (<1> 100% inhibition compared to the activity without any metal [2]) [2]
EDTA <1> (<1> 100% inhibition at 1 mM [2]) [2]
FAD <2> (<2> slight inhibition [5]) [5]
Fe^{2+} <2> (<2> slight inhibition [5]) [5]
KCN <1> (<1> 40% inhibition at 2 mM [2]) [2]
N-ethylmaleimide <1> (<1> 100% inhibition at 2 mM [2]) [2]
Ni^{2+} <1> (<1> 25% inhibition compared to the activity without any metal [2]) [2]
Zn^{2+} <1> (<1> 40% inhibition compared to the activity without any metal [2]) [2]
iodoacetate <1> (<1> 55% inhibition at 2 mM [2]) [2]
o-phenanthroline <1> (<1> 100% inhibition at 0.5 mM [2]) [2]
p-chloromercuribenzoate <1> (<1> 17% inhibition at 0.005 mM [2]) [2]
Additional information <1> (<1> no inhibition with sodium azide [2]) [2]

Cofactors/prosthetic groups

FAD <1-3> (<1> component B contains 0.8 mol per mol of enzyme, 0.002 mM increases activity [2]) [2, 3, 5, 7]
FMN <1> (<1> little increase of activity by addition of 0.002 mM [2]) [2]
NADH <1-3> (<1> can not be replaced by NADPH [1,2]; <2> exogenous [5]) [1-3, 5, 7]

Metals, ions

Fe^{2+} <1-3> (<1> exogenous, required for activity [1,2]; <1> component B contains 1.7 mol and A 5.8 mol of iron per mol of enzyme [2]; <1> iron-sulfur protein [3]; <1> iron atoms are coordinated by 4 cysteines and are bridged by a pair of acid-labile sulfur atoms in the 2Fe2S ferredoxins [4]; <2> required, ability to retain the endogenous Fe^{2+} during the crude extract perparation [5]) [1-5, 7]

sulfide <1, 3> (<1> component B contains 1.7 mol and A 6.0 mol of acid-labile sulfide per mol of enzyme [2]; <1> iron-sulfur protein [3]; <1> iron atoms are coordinated by 4 cysteines and are bridged by a pair of acid-labile sulfur atoms in the 2Fe2S ferredoxins [4]) [2-4, 7]

Specific activity (U/mg)

0.0013 <3> (<3> cells are grown on 3-chlorobenzoate, benzoate as substrate for assay [7]) [7]

0.0024 <3> (<3> recombinant cells are grown on LB/IPTG, benzoate as substrate for assay [7]) [7]

0.0032 <3> (<3> cells are grown on benzoate, benzoate as substrate for assay [7]) [7]

0.006 <3> (<3> with 4-chlorobenzoate as substrate [6]) [6]

0.0085 <1> (<1> crude extract, 3.4 mg protein per ml in the assay [1,2]) [1, 2]

0.018-0.023 <2> (<2> cells grown on 2-clorobenzoate or 2,5-dichlorobenzoate, enzyme activity for the conversion of both substrates [5]) [5]

0.032 <3> (<3> with 2-iodobenzoate as substrate [6]) [6]

0.043 <1> (<1> after anion exchange: fraction A + B + NADH + Fe^{2+} [1]) [1]

0.073 <3> (<3> with 3-chlorobenzoate as substrate [6]) [6]

0.195 <3> (<3> with benzoate as substrate [6]) [6]

0.303 <3> (<3> with 2-bromobenzoate as substrate [6]) [6]

0.458 <3> (<3> with 2-chlorobenzoate as substrate [6]) [6]

0.53 <1> (<1> activity of component A after gel filtration in presence of component B [2]) [2]

4 <1> (<1> activity of component B in presence of component A [2]) [2]

Additional information <2> (<2> activity is similarly influenced by pH, temperature, concentration of oxygen, protein, Fe^{2+}, FAD and NADH in assay medium independent if the cells are grown on 2-chlorobenzoate or 2,5-dichlorobenzoate [5]) [5]

K_m-Value (mM)

0.023 <1> (2-chlorobenzoate, <1> component A [2]) [2]

0.063-0.079 <2> (2,5-dichlorobenzoate, <2> cells grown on 2-chlorobenzoate or 2,5-dichlorobenzoate [5]) [5]

0.078-0.117 <2> (2-chlorobenzoate, <2> cells grown on 2-chlorobenzoate or 2,5-dichlorobenzoate [5]) [5]

0.079 <1> (NADH, <1> component B [2]) [2]

pH-Optimum
5.9 <2> [5]
6.5 <1, 2> (<1,2> assay at [2,5]) [2, 5]
6.6 <1> (<1> assay at [1]) [1]
7.4 <3> (<3> assay at, assay with cells [6]) [6]

pH-Range
6.5-7.5 <3> (<3> assay at [7]) [7]

Temperature optimum (°C)
25 <1, 2> (<1,2> assay at [1,2,5]) [1, 2, 5]
30 <2, 3> (<3> assay at, assay with cells [6]; <3> assay at [7]) [5, 6, 7]

4 Enzyme Structure

Molecular weight
37500-38000 <1> (<1> component B, gel filtration [2]) [2]
200000-220000 <1> (<1> component A, gel filtration [2]) [2]

Subunits
hexamer <1> (<1> 3 * 52000, 3 * 20000, $\alpha_3\beta_3$ structure of component A, SDS-PAGE [2]) [2]
monomer <1> (<1> 1 * 37500, component B, SDS-PAGE [2]; <1> 1 * 37100, NADH:acceptor reductase component, SDS-PAGE [3]) [2, 3]
oligomer <1> (<1> x * 19500, x * 52400, $\alpha_3\beta_3$, small and large subunit of the oxygenase component, SDS-PAGE [3]) [3]

5 Isolation/Preparation/Mutation/Application

Purification
<1> (anion exchange [1]; component A: anion-exchange, gel filtration, component B: anion-exchange, affinity and adsorption chromatography [2]; anion-exchange [3]; gel filtration of oxygenase component [4]) [1-4]

Cloning
<1> (Escherichia coli TG1, JM105, MV1190 and Pseudomonas putida KT2440 are used as host strains [3]) [3]
<3> (several constructs with relevant genes are expressed in Escherichia coli DH5α and S17-1 lambdapir [6]; several gene constructs are expressed in Escherichia coli DH5α, S17-1, S17-lambdapir and HMS174(DE3) [7]) [6, 7]

Engineering
Additional information <1, 3> (<1> mutants are made by treatment with N-methyl-N'-nitro-N-nitrosoguanidine, mutants are unable to grow on 2-chlorobenzoate and catechol [1]; <1> several mutants of Pseudomonas cepacia, putida and sp. are used [3]; <3> interposon mutagenesis, resulting strain TD2 fails to grow on benzoate or 2-chlorobenzoate showing that the cbd gene

is involved in their degradation [6]; <3> disruptants of Burkholderia sp. NK8 are generated by ω cassette interposon mutagenesis: NDBA1, NCAD, NCRD and NBALZ. Mutagenesis shows the involvement of the cbeABCD genes in the oxidation of 2-, 3-, 4-chlorobenzoate and benzoate in NK8 [7]) [1, 3, 6, 7]

6 Stability

General stability information

<1>, sucrose has stabilizing effect [1]

Storage stability

<1>, -20°C component A, 80-85% loss of activity [2]
<1>, -20°C, component A, 20% glycerol, 50% loss of activity [2]
<1>, -20°C, component B, 20% sucrose, 2 weeks, 25% loss of activity [2]
<1>, -80°C or -20°C, component B, 2 weeks, 70% loss of activity [2]
<1>, -80°C or -20°C, component B, 20% glycerol, 2 weeks, 15% loss of activity [2]

References

[1] Fetzner, S.; Muller, R.; Lingens, F.: Degradation of 2-chlorobenzoate by Pseudomonas cepacia 2CBS. Biol. Chem. Hoppe-Seyler, **370**, 1173 - 1182 (1989)
[2] Fetzner, S.; Muller, R.; Lingens, F.: Purification and some properties of 2-halobenzoate 1,2-dioxygenase, a two-component enzyme system from Pseudomonas cepacia 2CBS. J. Bacteriol., **174**, 279-290 (1992)
[3] Haak, B.; Fetzner, S.; Lingens, F.: Cloning, nucleotide sequence, and expression of the plasmid-encoded genes for the two-component 2-halobenzoate 1,2-dioxygenase from Pseudomonas cepacia 2CBS. J. Bacteriol., **177**, 667-675 (1995)
[4] Riedel, A.; Fetzner, S.; Rampp, M.; Lingens, F.; Liebl, U.; Zimmermann, J.L.; Nitschke, W.: EPR, electron spin echo envelope modulation, and electron nuclear double resonance studies of the 2Fe2S centers of the 2-halobenzoate 1,2-dioxygenase from Burkholderia (Pseudomonas) cepacia 2CBS. J. Biol. Chem., **270**, 30869-30873 (1995)
[5] Fava, F.; Baldoni, F.; Marchetti, L.: 2-Chlorobenzoic acid and 2,5-dichlorobenzoic acid metabolism by crude extracts of Pseudomonas sp. CPE2 strain. Lett. Appl. Microbiol., **22**, 275-279 (1996)
[6] Suzuki, K.; Ogawa, N.; Miyashita, K.: Expression of 2-halobenzoate dioxygenase genes (cbdSABC) involved in the degradation of benzoate and 2-halobenzoate in Burkholderia sp. TH2. Gene, **262**, 137-145 (2001)
[7] Francisco, P.B.; Ogawa, N.; Suzuki, K.; Miyashita, K.: The chlorobenzoate dioxygenase genes of Burkholderia sp. strain NK8 involved in the catabolism of chlorobenzoates. Microbiology, **147**, 121-133 (2001)

2-Aminobenzenesulfonate 2,3-dioxygenase 1.14.12.14

1 Nomenclature

EC number
1.14.12.14

Systematic name
2-aminobenzenesulfonate, NADH:oxygen oxidoreductase (2,3-hydroxylating, ammonia-forming)

Recommended name
2-aminobenzenesulfonate 2,3-dioxygenase

Synonyms
2-aminobenzenesulfonate dioxygenase
2-aminosulfobenzene 2,3-dioxygenase
2AS dioxygenase
oxygenase, 2-aminobenzenesulfonate di-

CAS registry number
156621-16-8

2 Source Organism

<1> *Alcaligenes sp.* (mutant of strain O-1 [1]; strain O-1 [2]) [1, 2]

3 Reaction and Specificity

Catalyzed reaction
2-aminobenzenesulfonate + NADH + H^+ + O_2 = 2,3-dihydroxybenzenesulfonate + NH_3 + NAD^+

Reaction type
oxidation
reduction

Natural substrates and products
S 2-aminobenzenesulfonate + NADH + H^+ + O_2 <1> (<1>, initial step in degradation of of 2-aminobenzenesulfonate) [2]

Substrates and products
S 2-aminobenzenesulfonate + NADH + H^+ + O_2 <1> [1]
P 3-sulfocatechol + NH_3 + NAD^+ <1> [1]

5 Isolation/Preparation/Mutation/Application

Source/tissue

culture condition:2-aminobenzenesulfonate-grown cell <1> [2]

References

[1] Junker, F.; Field, J.A.; Bangerter, F.; Ramsteiner, K.; Kohler, H.P.; Joannou, C.L.; Mason, J.R.; Leisinger, T.; Cook, A.M.: Oxygenation and sponataneous deamination of 2-aminobenzenesulphonic acid in Alcaligenes sp. strain O-1 with subsequent meta ring cleavage and spontaneous desulphonation to 2-hydroxymuconic acid. Biochem. J., **300**, 429-436 (1994)

[2] Junker, F.; Leisinger, T.; Cook, A.M.: 3-Sulfocatechol 2,3-dioxygenase and other dioxygenases (EC 1.13.11.2 and EC 1.14.12.) in the degradative pathways of 2-aminobenzenesulfonic, benzenesulfonic and 4-toluenesulfonic acids in Alcaligenes sp. strain O-1. Microbiology, **140**, 1713-1722 (1994)

Terephthalate 1,2-dioxygenase 1.14.12.15

1 Nomenclature

EC number
1.14.12.15

Systematic name
benzene-1,4-dicarboxylate, NADH:oxygen oxidoreductase (1,2-hydroxylating)

Recommended name
terephthalate 1,2-dioxygenase

Synonyms
1,4-dicarboxybenzoate 1,2-dioxygenase
benzene-1,4-dicarboxylate 1,2-dioxygenase
oxygenase, terephthalate, 1,2-di
TER dioxygenase system
TERDOS

CAS registry number
162032-76-0

2 Source Organism

<1> *Comamonas testosteroni* (strain T-2 [1]) [1]

3 Reaction and Specificity

Catalyzed reaction
terephthalate + NADH + H^+ + O_2 = (1R,6S)-dihydroxycyclohexa-2,4-diene-1,4-dicarboxylate + NAD^+

Reaction type
oxidation
reduction

Substrates and products
S 1,4-dicarboxynaphthalene + O_2 <1> [1]
P ?
S 2,5-dicarboxypyridine + O_2 <1> [1]
P ?
S terephthalate + NADH + O_2 <1> [1]
P (1R,2S)-dihydroxy-3,5-cyclohexadiene-1,4-dicarboxylate + NAD^+ <1> [1]

Cofactors/prosthetic groups
NADH <1> [1]

Metals, ions
Iron <1> (<1>, the enzyme is a Rieske [2Fe-2S] protein) [1]

pH-Optimum
7.5-8.5 <1> (<1>, oxygenase component of the TER dioxygenase system) [1]

Temperature optimum (°C)
25-36 <1> (<1>, oxygenase component of the TER dioxygenase system) [1]

4 Enzyme Structure

Molecular weight
126000 <1> (<1>, oxygenase component of the TER dioxygenase system, gel filtration [1]) [1]

Subunits
tetramer <1> (<1>, 2 * 49000, α, + 2 * 18000, β, oxygenase component of the TER dioxygenase system, SDS-PAGE [1]) [1]

5 Isolation/Preparation/Mutation/Application

Source/tissue
culture condition:terephthalate-salt-grown cell <1> [1]

Purification
<1> (oxygenase component of the TER dioxygenase system. The reductase component or components cannot be purified) [1]

References

[1] Schlaefli, H.R.; Weiss, M.A.; Leisinger, T.; Cook, A.M.: Terephthalate 1,2-dioxygenase system from Comamonas testosteroni T-2: purification and some properties of the oxygenase component. J. Bacteriol., **176**, 6644-6652 (1994)

2-Hydroxyquinoline 5,6-dioxygenase 1.14.12.16

1 Nomenclature

EC number
1.14.12.16

Systematic name
quinolin-2-ol, NADH:oxygen oxidoreductase (5,6-hydroxylating)

Recommended name
2-hydroxyquinoline 5,6-dioxygenase

Synonyms
2-hydroxyquinoline 5,6-dioxygenase
2-oxo-1,2-dihydroquinoline 5,6-dioxygenase
oxygenase, 2(1H)-quinolinone 5,6-di-
quinolin-2(1H)-one 5,6-dioxygenase
quinolin-2-ol 5,6-dioxygenase

CAS registry number
172399-50-7 (not distinguished from EC 1.14.13.65)

2 Source Organism

<1> *Comamonas testosteroni* (strain 63) [1]

3 Reaction and Specificity

Catalyzed reaction
quinolin-2-ol + NADH + H^+ + O_2 = 2,5,6-trihydroxy-5,6-dihydroquinoline + NAD^+

Reaction type
oxidation
reduction

Natural substrates and products
S Additional information <1> (<1>, second enzyme in the pathway of quinoline and 3-methylquinoline degradation [1]) [1]

Substrates and products
S 2-oxo-1,2-dihydroquinoline + NADH + O_2 <1> [1]
P ?

S 3-methyl-2-oxo-1,2-dihydroquinoline + NADH + O_2 <1> (<1>, NADH cannot be replaced by NADPH [1]) [1]
P 5,6-dihydro-5,6-dihydroxy-(3-methyl-)2-oxo-1,2-dihydroquinoline + NAD^+ <1> [1]
S 6-hydroxy-2-oxo-1,2-dihydroquinoline + NADH + O_2 <1> [1]
P ?
S 8-hydroxy-2-oxo-1,2-dihydroquinoline + NADH + O_2 <1> [1]
P ?
S 8-hydroxyquinoline + NADH + O_2 <1> [1]
P ?

Inhibitors
1,10-phenanthroline <1> (<1>, 0.5 mM, 82% loss of activity [1]) [1]
4-hydroxymercuribenzoate <1> (<1>, 0.1 mM, complete loss of activity [1]) [1]
Cu^{2+} <1> [1]
EDTA <1> (<1>, inhibits after a prolonged incubation time [1]) [1]
NaCl <1> (<1>, 0.2 M, 50% loss of activity) [1]
acriflavin <1> (<1>, 0.5 mM, 69% loss of activity [1]) [1]
diethyldithiocarbaminate <1> (<1>, inhibits after a prolonged incubation time [1]) [1]
iodoacetate <1> (<1>, 2 mM, complete loss of activity [1]) [1]
quinacrine <1> (<1>, 0.5 mM, 15% loss of activity [1]) [1]

Cofactors/prosthetic groups
NADH <1> (<1>, required, cannot be replaced by NADPH [1]) [1]

Metals, ions
Fe^{2+} <1> (<1>, enhances activity 1.5fold) [1]

Specific activity (U/mg)
0.125 <1> [1]

pH-Optimum
7.3 <1> [1]

5 Isolation/Preparation/Mutation/Application

Purification
<1> (partial) [1]

6 Stability

Temperature stability
30 <1> (<1>, 5 min, 30% loss of activity) [1]
65 <1> (<1>, 5 min, complete loss of activity) [1]

Organic solvent stability

ethanol <1> (5%, decreases enzyme activity to 44%) [1]

General stability information

<1>, dithioerythritol, dithiothreitol and 2-oxo-1,2-dihydroquinoline stabilize [1]

References

[1] Schach, S.; Tshisuaka, B.; Fetzner, S.; Lingens, F.: Quinoline 2-oxidoreductase and 2-oxo-1,2-dihydroquinoline 5,6-dioxygenase from Comamonas testosteroni 63. The first two enzymes in quinoline and 3-methylquinoline degradation. Eur. J. Biochem., **232**, 536-544 (1995)

Nitric oxide dioxygenase 1.14.12.17

1 Nomenclature

EC number
1.14.12.17

Systematic name
nitric oxide, NAD(P)H:oxygen oxidoreductase

Recommended name
nitric oxide dioxygenase

Synonyms
NOD
flavoHb <1> [3]
flavohemoglobin <1> [3]

CAS registry number
214466-78-1

2 Source Organism

<1> *Escherichia coli* [1, 2, 3]

3 Reaction and Specificity

Catalyzed reaction
2 NO + 2 O_2 + NAD(P)H + H^+ = 2 NO_3^- + $NAD(P)^+$ + 2 H^+

Reaction type
oxidation
redox reaction
reduction

Substrates and products
S NO + O_2 + NAD(P)H <1> (Reversibility: ? <1> [1, 3]) [1, 3]
P NO_3^- + $NAD(P)^+$ + H^+

Inhibitors
NaCN <1> (<1> 45% of control activity with 0.025 mM, 2% with 0.25 mM [1]; <1> less than 15% of control activity with 25 mM after NO exposition [2]) [1, 2]
superoxide dismutase <1> [1]

Cofactors/prosthetic groups

FAD <1> (<1> full activity at 0.001 mM [1]; <1> apparent Kd of 40 nM [3]) [1, 2, 3]
NADPH <1> [1, 2]
O_2 <1> [1, 2]

Activating compounds

hemin <1> (<1> maximal activity at 0.001 mM [1]) [1]

Turnover number (min^{-1})

3000 <1> (NADH, <1> per heme at 37°C [3]) [3]
4980 <1> (NADH, <1> per heme at 20°C [3]) [3]
5640 <1> (NO, <1> per heme at 20°C [3]) [3]
40200 <1> (NO, <1> per heme at 37°C [3]) [3]

Specific activity (U/mg)

0.0058 <1> (<1> strain DH5α [2]) [2]
0.015 <1> (<1> strain AB1157 [1]) [1]
0.2095 <1> (<1> strain DH5α after NO-exposition [2]) [2]
0.865 <1> (<1> mutant strain PG118 [1]) [1]

K_m-Value (mM)

0.00011 <1> (NO, <1> at 20°C [3]) [3]
0.00028 <1> (NO, <1> at 37°C [3]) [3]
0.0032 <1> (NADH, <1> at 20°C [3]) [3]
0.0048 <1> (NADH, <1> at 37°C [3]) [3]
0.013 <1> (O_2, <1> with 0.0001 mM NO at 20°C [3]) [3]
0.027 <1> (O_2, <1> with 0.001 mM NO at 20°C [3]) [3]
0.035 <1> (O_2, <1> with 0.0001 mM NO at 37°C [3]) [3]
0.1 <1> (O_2, <1> with 0.001 mM NO at 37°C [3]) [3]
0.18 <1> (NADPH, <1> at 20°C [3]) [3]

4 Enzyme Structure

Molecular weight

43000 <1> (SDS-PAGE [1]) [1]

5 Isolation/Preparation/Mutation/Application

Purification

<1> [1]

6 Stability

General stability information

<1>, stabilizing of activity during gel filtration by adding 10 mM NaN_3 [1]

References

[1] Gardner, P.R.; Gardner, A.M.; Martin, L.A.; Salzman, A.L.: Nitric oxide dioxygenase: an enzymic function for flavohemoglobin. Proc. Natl. Acad. Sci. USA, **95**, 10378-10383 (1998)
[2] Gardner, P.R.; Costantino, G.; Salzman, A.L.: Constitutive and adaptive detoxification of nitric oxide in Escherichia coli. Role of nitric-oxide dioxygenase in the protection of aconitase. J. Biol. Chem., **273**, 26528-26533 (1998)
[3] Gardner, A.M.; Martin, L.A.; Gardner, P.R.; Dou, Y.; Olson, J.S.: Steady-state and transient kinetics of Escherichia coli nitric-oxide dioxygenase (flavohemoglobin). The B10 tyrosine hydroxyl is essential for dioxygen binding and catalysis. J. Biol. Chem., **275**, 12581-12589 (2000)

Biphenyl 2,3-dioxygenase 1.14.12.18

1 Nomenclature

EC number
1.14.12.18

Systematic name
biphenyl,NADH:oxygen oxidoreductase (2,3-hydroxylating)

Recommended name
biphenyl 2,3-dioxygenase

Synonyms
BDO
BPDO
BPH dox
BPO
biphenyl 2,3-dioxygenase
biphenyl dioxygenase

CAS registry number
103289-55-0

2 Source Organism

<1> *Sphingomonas yanoikuyae* (strain B8/36, former name Beijerinckia sp. [1, 7]) [1, 7]
<2> *Pseudomonas sp.* (strain LB400 [2, 3, 4, 5, 9, 10, 13, 14]; other name Burkholderia sp. [13, 14]) [2, 3, 4, 5, 9, 10, 13, 14]
<3> *Pseudomonas pseudoalcaligenes* (strain KF707 [2]) [2]
<4> *Comamonas testosteroni* (strain B-356 [6, 10, 12]) [6, 10, 12]
<5> *Rhodococcus globerulus* (strain P6 [8, 11]) [8, 11]

3 Reaction and Specificity

Catalyzed reaction
biphenyl + NADH + + H^+ + O_2 = (2R,3S)-3-phenylcyclohexa-3,5-diene-1,2-diol + NAD^+ (<2> with chlorinated biphenyls the product is 2,3-dihydroxybiphenyl and HCl [4])

Reaction type

oxidation
redox reaction
reduction

Substrates and products

S 2,2'-dibromobiphenyl + NADH + O_2 <2> (Reversibility: ? <2> [14]) [14]
P 2,3-dihydroxy-2'-bromobiphenyl + NAD^+ + HBr
S 2,2'-dichlorobiphenyl + NADH + O_2 <2> (Reversibility: ? <2> [14]) [14]
P 2,3-dihydroxy-2'-chlorobiphenyl + NAD^+ + HCl
S 2,2'-dichlorobiphenyl + NADH + O_2 <2, 4, 5> (<5> no substrate [8]) (Reversibility: ? <2, 4, 5> [3, 4, 10, 11]) [3, 4, 10, 11]
P 5,6-dihydroxy-1-phenylcyclohexa-1,3-diene + 2,3-dihydroxy-2'-chlorobiphenyl + NAD^+ + HCl
S 2,2'-difluorobiphenyl + NADH + O_2 <2> (Reversibility: ? <2> [14]) [14]
P 2,3-dihydroxy-2'-fluorobiphenyl + 5,6-dihydroxy-2,2'-difluorobiphenyl + NAD^+ + HF
S 2,2'-dihydroxybiphenyl + NADH + O_2 <2> (Reversibility: ? <2> [14]) [14]
P 2,3,2'-trihydroxybiphenyl + trihydroxybiphenyl + NAD^+
S 2,2'-dinitrobiphenyl + NADH + O_2 <2> (Reversibility: ? <2> [14]) [14]
P 2,3-dihydroxy-2'-nitrobiphenyl + NAD^+ + NO_2
S 2,3'-dichlorobiphenyl + NADH + O_2 <2> (Reversibility: ? <2> [3]) [3]
P 5,6-dihydroxy-1-phenylcyclohexa-1,3-diene + 2,3-dihydroxy-3'-chlorobiphenyl + 5',6'-dihydroxy-1'-phenylcyclohexa-1',3'-diene + NAD^+ + HCl
S 2,4'-dichlorobiphenyl + NADH + O_2 <2> (Reversibility: ? <2> [4]) [4]
P 2,3-dihydroxy-4'-chlorobiphenyl + NAD^+ + HCl
S 2,5,2',5'-tetrachlorobiphenyl + NADH + O_2 <2> (<4> no substrate [10]; <5> no substrate [11]) (Reversibility: ? <2> [3, 10]) [3, 10, 11]
P cis-3,4-dihydroxy-2,5-dichloro-1-[2',5'-dichlorophenyl]-cyclohexa-1,5-diene + 3,4-dihydroxy-1-phenylcyclohexa-1,5-diene + NAD^+ + HCl
S 2,5,2'-trichlorobiphenyl + NADH + O_2 <2> (Reversibility: ? <2> [3]) [3]
P cis-3,4-dihydroxy-2,5-dichloro-1-[2'-chlorophenyl]-cyclohexa-1,5-diene + 2',3'-dihydroxy-2,5-dichlorobiphenyl + 5',6'-dihydrodiol + NAD^+ + HCl
S 2,5,3'-trichlorobiphenyl + NADH + O_2 <2> (Reversibility: ? <2> [3, 13]) [3, 13]
P 3,4-dihydroxy-1-phenylcyclohexa-1,5-diene + 5',6'-dihydroxy-1'-phenylcyclohexa-1',3'-diene + 2,3-catechol + NAD^+ + HCl
S 2,5-dichlorobiphenyl + NADH + O_2 <2, 4, 5> (<4> no 3,4-dihydrodiol as product [10]) (Reversibility: ? <2, 4, 5> [3, 10, 11, 13]) [3, 10, 11, 13]
P cis-2',3'-dihydroxy-1'-(2,5-dichlorophenyl)-cyclohexa-4',6'-diene + 3,4-dihydroxy-1-phenylcyclohexa-1,5-diene + NAD^+ + HCl
S 2-chlorobiphenyl + NADH + O_2 <2, 5> (<5> no catechol formation [8]) (Reversibility: ? <2, 5> [3, 8, 13]) [3, 8, 13]
P cis-2',3'-dihydroxy-1'-(2-chlorophenyl)-cyclohexa-4',6'-diene + catechol + NAD^+
S 3,3'-dichlorobiphenyl + NADH + O_2 <2, 4, 5> (<2> poor substrate [3]) (Reversibility: ? <2, 4, 5> [3, 10, 11]) [3, 10, 11]

P 5,6-dihydroxy-1-phenylcyclohexa-1,3-diene + 4,5-dihydroxy-1-phenylcyclohexa-1,2-diene + NAD^+ + HCl
S 3,4'-dichlorobiphenyl + NAD(P)H + O_2 <5> (Reversibility: ? <5> [8]) [8]
P 5,6-dihydroxy-3,4'-dichlorobiphenyl + $NAD(P)^+$
S 3-chlorobiphenyl + NADH + O_2 <2, 5> (Reversibility: ? <2, 5> [3, 8, 13]) [3, 8, 13]
P cis-2',3'-dihydroxy-1'-(3-chlorophenyl)-cyclohexa-4',6'-diene + NAD^+
S 4,4'-dichlorobiphenyl + NAD(P)H + O_2 <5> (Reversibility: ? <5> [8]) [8]
P 2,3-dihydroxy-4,4'-dichlorobiphenyl + $NAD(P)^+$
S 4-chlorobiphenyl + NADH + O_2 <2, 5> (Reversibility: ? <2, 5> [4, 8]) [4, 8]
P 2',3'-dihydrodiol-4-chlorobiphenyl + NAD^+
S 6,7-dihydro-5H-benzocycloheptene + NAD(P)H + O_2 <1> (Reversibility: ? <1> [7]) [7]
P (-)-cis-(1R,2S)-dihydroxybenzocycloheptane + $NAD(P)^+$
S biphenyl + NAD(P)H + O_2 <1> (Reversibility: ? <1> [1]) [1]
P (+)-cis-(1S,2R)-dihydroxy-3-phenylcyclohexa-3,5-diene (cis-biphenyl dihydrodiol) + $NAD(P)^+$
S biphenyl + NAD(P)H + O_2 <4, 5> (Reversibility: ? <4, 5> [10, 11]) [10, 11]
P 2,3-dihydro-dihydroxybiphenyl + $NAD(P)^+$
S biphenyl + NAD(P)H + O_2 <2> (Reversibility: ? <2> [3, 5, 9, 13]) [3, 5, 9, 13]
P (+)-cis-(2R,3S)-dihydroxy-3-phenylcyclohexa-3,5-diene (cis-biphenyl dihydrodiol) + $NAD(P)^+$
S carbazole + NADH + O_2 <1> (<1> formation of an unstable cis-carbazole-3,4-dihydrodiol is proposed [1]; <2> no products with carbazole detected [5]) (Reversibility: ? <1> [1]) [1, 5]
P 3-hydroxycarbazole + NAD^+
S dibenzo-*p*-dioxine + NADH + O_2 <2> (Reversibility: ? <2> [14]) [14]
P 2,3,2'-trihydroxy-diphenylether + dibenzo-*p*-dioxine-dihydrodiol + NAD^+
S dibenzofurane + NADH + O_2 <2> (Reversibility: ? <2> [14]) [14]
P monohydroxydibenzofuran + 2,3,2'-trihydroxybiphenyl + dibenzofuran-1,2-dihydrodiol + dibenzofuran-3,4-dihydrodiol + NAD^+
S naphthalene + NADPH <2> (<2> poor substrate [5]) (Reversibility: ? <2> [5]) [5]
P cis-naphthalene 1,2-dihydrodiol + $NADP^+$
S Additional information <2, 3> (<2,3> overview on polychlorinated biphenyls [2,3,13]; <2> no substrates are benzene, toluene, 2,5-dichlorotoluene, carbazole and dibenzothiophene [5]) [2, 3, 5, 13]
P ?

Cofactors/prosthetic groups

ferredoxin <2, 4> (<4> part of a multicomponent enzyme [6]) [5, 6, 9]

Metals, ions

Fe^{2+} <2, 4> (<2> omission of ferrous ammonium sulfate reduces specific activity of purified ISP 7.4fold [5]; <4> contains a [2Fe-2S] Rieske-type center carrying a mononuclear Fe^{2+} [6]) [5, 6, 9]

Specific activity (U/mg)

0.15 <2> (<2> referred to iron sulfur protein (terminal oxygenase) after purification [3]) [3]
4.9 <4> (<4> recombinant protein after purification [12]) [12]

K_m-Value (mM)

0.0026 <4> (2,2'-dichlorobiphenyl, <4> recombinant, purified protein [12]) [12]
0.0033 <4> (3,3'-dichlorobiphenyl, <4> recombinant, purified protein [12]) [12]
0.0062 <4> (biphenyl, <4> recombinant, purified protein [12]) [12]
0.028 <4> (O_2, <4> recombinant, purified protein [12]) [12]
0.058 <2> (NADH, <2> recombinant, purified NADH: ferredoxin oxidoreductase component [9]) [9]
0.094 <4> (biphenyl, <4> parental protein [6]) [6]
0.156 <2> (NADPH, <2> recombinant, purified NADH: ferredoxin oxidoreductase component [9]) [9]
1.1 <4> (biphenyl, <4> recombinant His-tagged protein [6]) [6]

pH-Optimum

5.5-6 <4> [6]
7.2 <2> (<2> NADH: ferredoxin oxidoreductase component [9]) [9]

Temperature optimum (°C)

32 <2> (<2> NADH: ferredoxin oxidoreductase component [9]) [9]
40 <4> [6]

4 Enzyme Structure

Molecular weight

25200 <4> (<4> β subunit with His-tag, SDS-PAGE [6]) [6]
27300 <2> (<2> β subunit of ISP, SDS-PAGE [5]) [5]
41500 <2> (<2> NADH: ferredoxin component, gel filtration with Superose 6 [9]) [9]
43600 <2> (<2> NADH: ferredoxin component, SDS-PAGE [9]) [9]
44000 <4> (<4> native, recombinant and purified α subunit with His-tag, HPLC gel filtration [6]) [6]
53000 <2> (<2> α subunit of ISP, SDS-PAGE [5]) [5]
53600 <4> (<4> α subunit with His-tag, SDS-PAGE [6]) [6]
186000 <4> (<4> native protein with His-tag, HPLC gel filtration [6]) [6]
209000 <2> (<2> native protein, gel filtration with Superose 12 [5]) [5]
234000 <4> (<4> native protein, HPLC gel filtration [6]) [6]

Subunits

heterohexamer <2, 4> (<2> $\alpha_3\beta_3$, 3*53000 + 3*27300, measurement of molecular weight and Stokes' radius with gel filtration [5]; <4> $\alpha_3\beta_3$, 3*53600 + 3*25200, SDS-PAGE and gel filtration [6]) [5, 6]

5 Isolation/Preparation/Mutation/Application

Purification

<2> (purification of the iron sulfur protein (ISP) of biphenyl 2,3-dioxygenase [5]; purification of the NADH: ferredoxin oxidoreductase component [9]) [5, 9]

<4> (His-tagged ISP, expressed in Escherichia coli [6]) [6]

<4> (purification from heterologous expression in Pseudomonas putida KT2442 [12]) [12]

<5> (His-tagged ISP, expressed in Escherichia coli and Pseudomonas putida [11]) [11]

Cloning

<2> (in Escherichia coli, strain BL21(DE3) [4,14]; in Escherichia coli, strain M15 [10]) [4, 10, 14]

<4> (in Escherichia coli, strains M15 and SG13009 [6]; in Pseudomonas putida strain KT2442 [12]) [6, 12]

<5> (in Pseudomonas putida strain KT2442 [8,11]; in Escherichia coli strain M15 and SG13009 [11]) [8, 11]

Engineering

T335A/F336T/N338T/I341T <2> (<2> conversion of sequence to corresponding sequence of Pseudomonas pseudoalcaligenes strain KF707 [2]) [2]

T375N <4> (<4> conversion of sequence to corresponding sequence of Pseudomonas sp. strain LB400 [10]) [10]

6 Stability

Temperature stability

5-55 <2> (<2> NADH: ferredoxin oxidoreductase component: no activity at 5 or 55°C, 15.2% of control activity at 50°C [9]) [9]

Oxidation stability

<4>, 20 min preincubation of His-tagged enzyme with 5 mM dithiothreitol on ice can restore activity of older preparations [6]

Storage stability

<2>, -75°C, 4 months, 10% loss of activity [9]

<4>, -70°C, months [6]

References

[1] Resnick, S.M.; Torok, D.S.; Gibson, D.T.: Oxidation of carbazole to 3-hydroxycarbazole by naphthalene 1,2-dioxygenase and biphenyl 2,3-dioxygenase. FEMS Microbiol. Lett., **113**, 297-302 (1993)

[2] Erickson, B.D.; Mondello, F.J.: Enhanced biodegradation of polychlorated biphenyls after site-directed mutagenesis of a biphenyl dioxygenase gene. Appl. Environ. Microbiol., **59**, 3858-3862 (1993)

[3] Haddock, J.D.; Horton, J.R.; Gibson, D.T.: Dihydroxylation and dechlorination of chlorinated biphenyls by purified biphenyl 2,3-dioxygenase from Pseudomonas sp. strain LB400. J. Bacteriol., **177**, 20-26 (1995)

[4] Seeger, M.; Timmis, K.N.; Hofer, B.: Degradation of chlorobiphenyls catalyzed by the bph-encoded biphenyl-2,3-dioxygenase and biphenyl-2,3-dihydrodiol-2,3-dehydrogenase of Pseudomonas sp. LB400. FEMS Microbiol. Lett., **133**, 259-264 (1995)

[5] Haddock, J.D.; Gibson, D.T.: Purification and characterization of the oxygenase component of biphenyl 2,3-dioxygenase from Pseudomonas sp. strain LB400. J. Bacteriol., **177**, 5834-5839 (1995)

[6] Hurtubise, Y.; Barriault, D.; Sylvestre, M.: Characterization of active recombinant His-tagged oxygenase component of Comamonas testosteroni B-356 biphenyl dioxygenase. J. Biol. Chem., **271**, 8152-8156 (1996)

[7] Resnick, S.M.; Gibson, D.T.: Oxidation of 6,7-dihydro-5H-benzocycloheptene by bacterial strains expressing naphthalene dioxygenase, biphenyl dioxygenase, and toluene dioxygenase yields homochiral mono-ol or cis-diol enantiomers as major products. Appl. Environ. Microbiol., **62**, 1364-1368 (1996)

[8] McKay, D.B.; Seeger, M.; Zielinski, M.; Hofer, B.; Timmis, K.N.: Heterologous expression of biphenyl dioxygenase-encoding genes from a gram-positive broad-spectrum polychlorinated biphenyl degrader and characterization of chlorobiphenyl oxidation by the gene products. J. Bacteriol., **179**, 1924-1930 (1997)

[9] Broadus, R.M.; Haddock, J.D.: Purification and characterization of the NADH:ferredoxinBPH oxidoreductase component of biphenyl 2,3-dioxygenase from Pseudomonas sp. strain LB400. Arch. Microbiol., **170**, 106-112 (1998)

[10] Hurtubise, Y.; Barriault, D.; Sylvestre, M.: Involvement of the terminal oxygenase β subunit in the biphenyl dioxygenase reactivity pattern toward chlorobiphenyls. J. Bacteriol., **180**, 5828-5835 (1998)

[11] Chebrou, H.; Hurtubise, Y.; Barriault, D.; Sylvestre, M.: Heterologous expression and characterization of the purified oxygenase component of Rhodococcus globerulus P6 biphenyl dioxygenase and of chimeras derived from it. J. Bacteriol., **181**, 4805-4811 (1999)

[12] Imbeault, N.Y.; Powlowski, J.B.; Colbert, C.L.; Bolin, J.T.; Eltis, L.D.: Steady-state kinetic characterization and crystallization of a polychlorinated biphenyl-transforming dioxygenase. J. Biol. Chem., **275**, 12430-12437 (2000)

[13] Arnett, C.M.; Parales, J.V.; Haddock, J.D.: Influence of chlorine substituents on rates of oxidation of chlorinated biphenyls by the biphenyl dioxygenase of Burkholderia sp. strain LB400. Appl. Environ. Microbiol., **66**, 2928-2933 (2000)
[14] Seeger, M.; Camara, B.; Hofer, B.: Dehalogenation, denitration, dehydroxylation, and angular attack on substituted biphenyls and related compounds by a biphenyl dioxygenase. J. Bacteriol., **183**, 3548-3555 (2001)

Salicylate 1-monooxygenase 1.14.13.1

1 Nomenclature

EC number
1.14.13.1

Systematic name
salicylate,NADH:oxygen oxidoreductase (1-hydroxylating, decarboxylating)

Recommended name
salicylate 1-monooxygenase

Synonyms
oxygenase, salicylate 1-mono-
salicylate 1-hydroxylase
salicylate hydroxylase (decarboxylating)
salicylate monooxygenase
salicylic hydroxylase

CAS registry number
9059-28-3

2 Source Organism

<1> *Trichosporon cutaneum* [16]
<2> *Pseudomonas putida* (UUC-1, strain is capable of utilizing salicylate at high concentrations [20]; strain BS202-P1, grown on phenanthrene [21]) [1, 2, 5-8, 10, 11, 13, 14, 17-21]
<3> *Pseudomonas sp.* (ATCC 29351 [12]) [3, 4, 9, 12]
<4> *Pseudomonas cepacia* [15]

3 Reaction and Specificity

Catalyzed reaction
salicylate + NADH + H^+ + O_2 = catechol + NAD^+ + H_2O + CO_2 (a flavoprotein, FAD)

Reaction type
oxidation
oxidative decarboxylation
redox reaction
reduction

Substrates and products

S 1-hydroxy-2-naphthoate + NADH + O_2 <2> (Reversibility: ? <2> [21]) [21]

P 1,2-dihydroxynaphthalene + CO_2 + H_2O+ NAD^+

S 2,3-dihydroxybenzoate + NADH + O_2 <1-3> (Reversibility: ? <1-3> [1, 3, 4, 9, 16, 21]) [1, 3, 4, 9, 16, 21]

P pyrogallol + CO_2 + H_2O + NAD^+ <1, 3> [3, 16]

S 2,4-dihydroxybenzoate + NAD(P)H + O_2 <1> (Reversibility: ? <1> [16]) [16]

P benzene-1,2,4-triol + CO_2 + H_2O + $NAD(P)^+$ <1> [16]

S 2,5-dihydroxybenzoate + NADH + O_2 <1-3> (<1> or NADPH [16]) (Reversibility: ? <1-3> [1, 4, 9, 16, 21]) [1, 4, 9, 16, 21]

P benzene-1,2,5-triol + CO_2 + H_2O + NAD^+ <1> [16]

S 2,6-dihydroxybenzoate + NADH + O_2 <1-3> (Reversibility: ? <1-3> [1, 3, 4, 9, 16, 21]) [1, 3, 4, 9, 16, 21]

P pyrogallol + CO_2 + H_2O + NAD^+ <1, 3> [3, 16]

S 4-aminosalicylate + NADH + O_2 <1, 2> (Reversibility: ? <1, 2> [16, 21]) [16, 21]

P 4-aminocatechol + CO_2 + NAD^+ + H_2O

S 4-chlorosalicylate + NADH + O_2 <1, 2> (Reversibility: ? <1, 2> [16, 21]) [16, 21]

P 1,2-dihydroxy-4-chlorobenzene + CO_2 + NAD^+ + H_2O

S 5-aminosalicylate + NADH + O_2 <1> (Reversibility: ? <1> [16]) [16]

P 5-aminocatechol + CO_2 + NAD^+ + H_2O

S 5-chlorosalicylate + NADH + O_2 <1> (Reversibility: ? <1> [16]) [16]

P 1,2-dihydroxy-5-chlorobenzene + CO_2 + NAD^+ + H_2O

S 5-fluorosalicylate + NADH + O_2 <1> (Reversibility: ? <1> [16]) [16]

P 1,2-dihydroxy-5-fluorobenzene + CO_2 + NAD^+ + H_2O

S 5-methoxysalicylate + NADH + O_2 <1> (Reversibility: ? <1> [16]) [16]

P 1,2-dihydroxy-5-methoxybenzene + CO_2 + NAD^+ + H_2O

S 5-methylsalicylate + NADH + O_2 <1, 2> (Reversibility: ? <1, 2> [16, 21]) [16, 21]

P 1,2-dihydroxy-5-methylbenzene + CO_2 + NAD^+ + H_2O

S *m*-hydroxybenzoate + NADH + O_2 <3> (<3> 6% of the reaction with salicylate [4,9]) (Reversibility: ? <3> [4, 9]) [4, 9]

P 1,3-dihydroxybenzene + CO_2 + NAD^+ + H_2O

S *o*-iodophenol + NADH + O_2 <2> (Reversibility: ? <2> [5, 6]) [5, 6]

P catechol + iodide + NAD^+

S *o*-nitrophenol + NADH + O_2 <2> (Reversibility: ? <2> [6]) [6]

P catechol + nitrite + NAD^+

S *p*-aminosalicylate + NADH + O_2 <2, 3> (Reversibility: ? <2, 3> [1, 4, 9]) [1, 4, 9]

P 1,2-dihydroxy-4-aminobenzene + NAD^+ + CO_2 + H_2O

S salicylaldehyde + NADH + O_2 <2> (<2> mechanism [10]) (Reversibility: ? <2> [10]) [10]

P catechol + formate + NAD^+

S salicylate + NADH + O_2 <1-4> (Reversibility: ? <1-4> [1-21]) [1-21]

P catechol + NAD^+ + H_2O + CO_2

S Additional information <2-4> (<2> mechanism [1]; <2> enzyme catalyzes formation of catechol from substrate analogues such as *o*-nitro-, *o*-amino-, *o*-iodo-, *o*-bromo- and *o*-chlorophenol by removing the ortho substituted groups [6]; <2> *o*-fluorophenol is not converted to catechol, though NADH oxidation is observed [6]; <2> enzyme catalyses hydroxylation and dehalogenation of *o*-halogenophenols and also denitrification of *o*-nitrophenol with unusual stoichiometry [6]; <2> by chemical treatment of the enzyme with dicarbonyl reagents, such as glyoxal, the original oxygenase activity is converted to the salicylate-dependent NADH-dehydrogenase activity with free FAD as electron acceptor [14]; <2, 3> mechanism [1, 2, 4, 5, 7, 8]; <4> apoenzyme-flavin interaction [15]) [1, 2, 4-8, 14, 15]

P ?

Inhibitors

1,10-phenanthroline <3> [3]
Br^- <3> [9]
Cl^- <1, 3> [4, 9, 16]
$CuSO_4$ <3> [3]
F^- <3> [4, 9]
I^- <3> [4, 9]
NO_3^- <3> [9]
Na_2MoO_4 <3> (<3> slight [3]) [3]
SCN^- <3> [4, 9]
ascorbic acid <3> (<3> slight [3]) [3]
benzoate <3> (<3> competitive inhibitor [9]) [9]
p-chloromercuribenzoate <3> [3, 9]
trinitrobenzenesulfonic acid <2> (<2> irreversible inactivation, modification of a lysine residue results in loss of NADH-dehydrogenase activity suggesting its role in the NADH-binding site of the enzyme [17]) [17]
Additional information <2> (<2> chemical modification of one arginine residue with glyoxal causes the enzyme to act as dehydrogenase, but not as oxygenase [14]) [14]

Cofactors/prosthetic groups

FAD <1-4> (<1-4> flavoprotein [1-4, 9, 11, 13, 15, 16]; <3> 1 mol of FAD loosly bound to 1 mol of enzyme, MW 57000 [3]; <3> 2 FAD per enzyme molecule, 2 subunits, total MW 91000 [4]; <4> enzyme with dimeric structure binds 1 FAD per monomer [15]; <4> nature of flavin binding [15]; <2> ^{13}C-NMR, ^{15}N-NMR and ^{31}P-NMR investigation of interaction between FAD and the apoprotein [11]) [1-4, 9, 11, 15, 16, 21]
NADH <1-4> [1-16, 21]
NADPH <1-3> (<2> 1% of the activity with NADH [1]; <1> 60% of the activity with NADH [16]; <2> 50% of the activity with NADH [21]) [1, 4, 16, 21]

Activating compounds

flavin 1,N^6-ethenoadenine dinucleotide <4> (<4> weaker binding to the apoenzyme than FAD [15]) [15]

Metals, ions

Additional information <1> (<1> no metal ion requirement [16]) [16]

Specific activity (U/mg)

5 <2> [21]
10.57 <3> [4, 9]
13.5 <3> [3]
37 <1> [16]
Additional information <2, 3> [2, 12, 13]

K_m-Value (mM)

0.0016 <2> (salicylate, <2> cosubstrate NADH [6]; <2> native enzyme [17]) [6, 17]
0.0018 <2> (salicylate, <2> modified enzyme [17]) [17]
0.0019 <3> (salicylate) [3]
0.0026 <3> (NADH) [3]
0.0027 <3> (salicylate) [9]
0.0027 <3> (salicylate, <3> cosubstrate NADH [4]) [4]
0.0037 <2> (NADH, <2> native enzyme [17]) [17]
0.004 <2> (1-hydroxy-2-naphthoate) [21]
0.011 <3> (2,6-dihydroxybenzoate, <3> cosubstrate NADH [4]) [4, 9]
0.015 <3> (*p*-aminosalicylate, <3> cosubstrate NADH [4]) [4, 9]
0.0167 <3> (NADH) [9]
0.017 <3> (NADH, <3> cosubstrate salicylate [4]) [4, 9]
0.028 <3> (2,3-dihydroxybenzoate, <3> cosubstrate NADH [4]) [4, 9]
0.034 <2> (*o*-nitrophenol, <3> cosubstrate NADH [4]) [6]
0.044 <3> (NADH, <3> cosubstrate *p*-aminosalicylate [4]) [4]
0.049 <2> (NADH, <2> modified enzyme [17]) [17]
0.065 <3> (2,5-dihydroxybenzoate, <3> cosubstrate NADH [4]) [4, 9]
0.087 <2> (salicylaldehyde) [10]
0.091 <3> (NADH, <3> cosubstrate 2,3-dihydroxybenzoate [4]) [4]
0.1 <2, 3> (NADPH, <3> cosubstrate salicylate [4]; <2> cosubstrates O_2 and salicylate [6]) [4, 6]
0.118 <2> (NADH, <2> cosubstrate salicylaldehyde [10]) [10]
0.13 <2> (O_2, <2> cosubstrate *o*-nitrophenol [6]) [6]
0.14 <3> (NADH, <3> cosubstrate 2,4-dihydroxybenzoate [4]) [4]
0.143 <3> (2,4-dihydroxybenzoate, <3> cosubstrate NADH [4]) [4]
0.196 <2> (O_2, <2> cosubstrates salicylate and NADH [10]) [10]
0.23 <3> (NADH, <3> cosubstrate 2,5-dihydroxybenzoate or 2,6-dihydroxybenzoate [4]) [4]
Additional information <1-3> (<2> FAD, salicylate and NADH, comparison of wild-type, recombinant and mutant enzyme [18]) [6, 8-10, 16]

K_i-Value (mM)
3.1 <3> (benzoate) [9]
60 <3> (Cl^-) [4, 9]

pH-Optimum
7-7.5 <2> [21]
7-8.5 <3> [9]
7.5 <1> [16]
7.5-8 <3> [3]
7.8 <2> (<2> substrate salicylaldehyde [10]) [10]
8 <2> (<2> substrate *o*-iodophenol [6]) [6]

pH-Range
6-8 <2> (<2> 65% of maximal activity at pH 6, 96% of maximal activity at pH 8 [21]) [21]
6-10 <3> (<3> about 50% of activity maximum at pH 6 and 10 [3]) [3]
6.5-8.5 <1> (<1> about 75% of activity maximum at pH 6.5 and 8.5 [16]) [16]

Temperature optimum (°C)
20 <3> (<3> assay at [3]) [3]
25-27 <3> (<3> assay at [4]) [4]
30 <2> [21]

Temperature range (°C)
20-50 <2> (<2> 62% of maximal activity at 20°C, 70% of maximal activity at 50°C, no activity at 55°C [21]) [21]

4 Enzyme Structure

Molecular weight
43000 <2> (<2> SDS-PAGE [20]) [20]
43000-48700 <3> (<3> gel filtration, sedimentation equilibrium, calculation of MW per flavin [4,9]) [4, 9]
43400-45300 <1> (<1> gel filtration, SDS-PAGE [16]) [16]
45000 <2> (<2> SDS-PAGE,gel filtration [13]; <2> SDS-PAGE [21]; <2> recombinant native and mutant protein, expressed in E. coli, SDS-PAGE [18]) [13, 18, 21]
57200 <3> (<3> calculation from diffusion and sedimentation data [3]) [3]

Subunits
dimer <3> (<3> 2 * 43000-48700, sedimentation equilibrium [4, 9]) [4, 9]
monomer <1, 2> (<2> 1 * 52000, SDS-PAGE under dissociating conditions [7]; <2> 1 * 45000, SDS-PAGE [13]; <1> 1 * 45300 SDS-PAGE [16]) [7, 13, 16]

5 Isolation/Preparation/Mutation/Application

Purification

<1> [16]
<2> (cells grown on 1-hydroxy-2-naphthoate [21]) [2, 13, 21]
<3> (affinity chromatography [12]) [3, 4, 9, 12]

Crystallization

<2> [2]
<2> (apoenzyme is crystallized by dialysis method, using ammonium sulfate as the precipitant [18]) [18]

Cloning

<2> (expression in Escherichia coli [18]) [18]

Engineering

K163E <2> (<2> site directed mutagenesis, Lys163 is involved in the NADH-binding site [18]) [18]
K163G <2> (<2> site directed mutagenesis, Lys163 is involved in the NADH-binding site [18]) [18]
K163R <2> (<2> site directed mutagenesis, Lys163 is involved in the NADH-binding site [18]) [18]

Application

analysis <2> (<2> biosensor system for determining salicylate in body fluids [20]) [20]

6 Stability

pH-Stability

8 <3> (<3> highest stability [3]) [3]

Temperature stability

25 <3> (<3> apoenzyme unstable above [3]) [3]
38 <3> (<3> 30 min, complete inactivation in absence of FAD [3]; <3> 2 h, 10% loss of activity in the presence of FAD [3]) [3]

General stability information

<1>, dithiothreitol, 0.5 mM, stabilizes [16]
<2>, glycerol, 10%, stabilizes [13]

Storage stability

<1>, 4°C, 0.5 mM dithiothreitol, 10% loss of activity after 1 week [16]
<2>, -20°C, 20 mM KH_2PO_4 buffer pH 7.5 [21]
<2>, -20°C, stable for a few weeks [2]

References

[1] Suzuki, K.; Takemori, S.; Katagiri, M.: Mechanism of the salicylate hydroxylase reaction. IV. Fluorometric analysis of the complex formation. Biochim. Biophys. Acta, **191**, 77-85 (1969)
[2] Takemori, S.; Yasuda, H.; Mihara, K.; Suzuki, K.; Katagiri, M.: Mechanism of the salicylate hydroxylase reaction. II. The enzyme-substrate complex. Biochim. Biophys. Acta, **191**, 58-68 (1969)
[3] Yamamoto, S.; Katagiri, M.; Maeno, H.; Hayaishi, O.: Salicylate hydroxylase, a monooxygenase requiring flavin adenine dinucleotide. J. Biol. Chem., **240**, 3408-3413 (1965)
[4] Kamin, H.; White-Stevens, R.H.; Presswood, R.P.: Salicylate hydroxylase. Methods Enzymol., **53**, 527-543 (1978)
[5] Suzuki, K.; Gomi, T.; Itagaki, E.: Intermediate and mechanism of hydroxylation of *o*-iodophenol by salicylate hydroxylase. J. Biochem., **109**, 791-797 (1991)
[6] Suzuki, K.; Gomi, T.; Kaidoh, T.; Itagaki, E.: Hydroxylation of *o*-halogenophenol and *o*-nitrophenol by salicylate hydroxylase. J. Biochem., **109**, 348-353 (1991)
[7] Takemori, S.; Hon-Nami, K.; Kawahara, F.; Katagiri, M.: Mechanism of the salicylate 1-monooxygenase reaction. VI. The monomeric nature of the enzyme. Biochim. Biophys. Acta, **342**, 137-144 (1974)
[8] Takemori, S.; Nakamura, M.; Suzuki, K.; Katagiri, M.; Nakamura, T.: Mechanism of the salicylate hydroxylase reaction. V. Kinetic analyses. Biochim. Biophys. Acta, **284**, 382-393 (1972)
[9] White-Stevens, R.H.; Kamin, H.: Studies of a flavoprotein, salicylate hydroxylase. I. Preparation, properties, and the uncoupling of oxygen reduction from hydroxylation. J. Biol. Chem., **247**, 2358-2370 (1972)
[10] Suzuki, K.; Katagiri, M.: Mechanism of salicylate hydroxylase-catalyzed decarboxylation. Biochim. Biophys. Acta, **657**, 530-534 (1981)
[11] Vervoort, J.; Van Berkel, W.J.H.; Muller, F.; Moonen, C.T.W.: NMR studies on *p*-hydroxybenzoate hydroxylase from Pseudomonas fluorescens and salicylate hydroxylase from Pseudomonas putida. Eur. J. Biochem., **200**, 731-738 (1991)
[12] You, K.S.; Roe, C.R.: Affinity chromatography of Pseudomonas salicylate hydroxylase. Anal. Biochem., **114**, 177-185 (1981)
[13] You, I.S.; Murray, R.I.; Jollie, D.; Gunsalus, I.C.: Purification and characterization of salicylate hydroxylase from Pseudomonas putida PpG7. Biochem. Biophys. Res. Commun., **169**, 1049-1054 (1990)
[14] Suzuki, K.; Ohnishi, K.: Functional modification of an arginine residue on salicylate hydroxylase. Biochim. Biophys. Acta, **1040**, 327-336 (1990)
[15] Wang, L.H.; Tu, S.C.; Lusk, R.C.: Apoenzyme of Pseudomonas cepacia salicylate hydroxylase. Preparation, fluorescence property, and nature of flavin binding. J. Biol. Chem., **259**, 1136-1142 (1984)

[16] Sze, I.S.Y.; Dagley, S.: Properties of salicylate hydroxylase and hydroxyquinol 1,2-dioxygenase purified from Trichosporon cutaneum. J. Bacteriol., **159**, 353-359 (1984)
[17] Suzuki, K.; Mizuguchi, M.; Gomi, T.; Itagaki, E.: Identification of a lysine residue in the NADH-binding site of salicylate hydroxylase from Pseudomonas putida S-1. J. Biochem., **117**, 579-585 (1995)
[18] Suzuki, K.; Asao, E.; Nakamura, Y.; Nakamura, M.; Ohnishi, K.; Fukuda, S.: Overexpression of salicylate hydroxylase and the crucial role of Lys163 as its NADH binding site. J. Biochem., **128**, 293-299 (2000)
[19] Yabuuchi, T.; Suzuki, K.; Sato, T.; Ohnishi, K.; Itagaki, E.; Morimoto, Y.: Crystallization and preliminary X-ray analysis of salicylate hydroxylase from Pseudomonas putida S-1. J. Biochem., **119**, 829-831 (1996)
[20] Banat, I.M.; Marchant, A.; Nigam, P.; Gaston, S.J.S.; Kelly, B.A.; Marchant, R.: Production, partial characterization, and potential diagnostic use of salicylate hydroxylase from Pseudomonas putida UUC-1. Enzyme Microb. Technol., **16**, 665-670 (1994)
[21] Balashova, N.V.; Stolz, A.; Knackmuss, H.J.; Kosheleva, I.A.; Naumov, A.V.; Boronin, A.M.: Purification and characterization of a salicylate hydroxylase involved in 1-hydroxy-2-naphthoic acid hydroxylation from the naphthalene and phenanthrene-degrading bacterial strain Pseudomonas putida BS202-P1. Biodegradation, **12**, 179-188 (2001)

4-Hydroxybenzoate 3-monooxygenase 1.14.13.2

1 Nomenclature

EC number
1.14.13.2

Systematic name
4-hydroxybenzoate,NADPH:oxygen oxidoreductase (3-hydroxylating)

Recommended name
4-hydroxybenzoate 3-monooxygenase

Synonyms
4-HBA-3-hydroxylase
4-hydroxybenzoate 3-hydroxylase
4-hydroxybenzoate 3-monooxygenase
4-hydroxybenzoate monooxygenase
4-hydroxybenzoic hydroxylase
PHBAD
PHBH
PHBHase
POHBase
oxygenase, 4-hydroxybenzoate 3-mono-
p-hydroxybenzoate hydroxylase
p-hydroxybenzoate-3-hydroxylase
p-hydroxybenzoic acid hydrolase
p-hydroxybenzoic acid hydroxylase
p-hydroxybenzoic hydroxylase
para-hydroxybenzoate hydroxylase

CAS registry number
9059-23-8

2 Source Organism

<1> *Pseudomonas fluorescens* [1, 2, 3, 4, 5, 6, 8, 10, 12, 13, 15, 17, 16, 24, 25, 26, 27, 29, 32]
<2> *Pseudomonas desmolytica* [6, 7, 14]
<3> *Pseudomonas putida* (A 3.12 [6]; M-6 [6]) [6]
<4> *Pseudomonas aeruginosa* [9, 11, 20, 26, 33, 35, 36, 37, 38, 39]
<5> *Pseudomonas mendocina* (KR1 [18]) [18]
<6> *Rhodococcus opacus* (557, 420, and 1G [19]) [19]
<7> *Rhodococcus rhodochrous* (172 [19]) [19]

<8> *Rhodococcus sp.* (400 [19]) [19]
<9> *Rhodococcus rhodnii* (135 [19]) [19]
<10> *Acinetobacter calcoaceticus* [21]
<11> *Klebsiella pneumoniae* (enzyme is expressed in mutant strain MAO4, but not in wild-type strain [22]) [22]
<12> *Moraxella sp.* (strain GU2 [23]) [23]
<13> *Pseudomonas sp.* (CBS3 [28]) [28, 30]
<14> *Comamonas testosteroni* (Kh 122-3S [30]) [30, 31]
<15> *Acinetobacter sp.* [30]
<16> *Rhodococcus erythropolis* [34]

3 Reaction and Specificity

Catalyzed reaction

4-hydroxybenzoate + NADPH + H^+ + O_2 = protocatechuate + $NADP^+$ + H_2O (<1>, bi uni uni uni ping-pong mechanism [6])

Reaction type

oxidation
redox reaction
reduction

Natural substrates and products

S 4-hydroxybenzoate + NADPH + O_2 <1, 11, 12, 14> (<1>, inducible enzyme [1, 4, 5]; <1>, first step in the bacterial metabolism when 4-hydroxybenzoate is used as growth substrate [8]; <5>, enzyme of the toluene-4-monooxygenase catabolic pathway [18]; <11>, enzyme is expressed at basal level, presence of 4-hydroxybenzoate enhances activity [22]; <12>, degradation of 4-hydroxybenzoate [23]; <1>, enzyme catalyzes an intermediate step in the degradation of aromatic compounds in soil microorganisms [27]; <13, 14, 15>, high degree of homology observed between the enzyme from Comamonas and of Pseudomonas and Acinetobacter indicates the common evolutionary origin of the enzyme in the divergent pathways of 4-hydroxybenzoate among these soil bacteria of different genera [30]) (Reversibility: ? <1, 11, 12, 14> [1, 4, 5, 8, 18, 22, 23, 27, 30]) [1, 4, 5, 8, 18, 22, 23, 27, 30]

P ?

Substrates and products

S 2,4-dihydroxybenzoate + NADPH + O_2 <1, 4, 10, 13> (<1>, slow reaction, formation of at least 3 intermediates, a spectral intermediate that is believed to be an oxygenated form of the enzyme-bound flavin prosthetic group [3]; <1>, 1.5% of the activity with 4-hydroxybenzoate [4]; <1>, about 1% of the activity with 4-hydroxybenzoate [6]; <10>, 3.1% of the activity with 4-hydroxybenzoate [21]; <13>, 8% of the activity with 4-hydroxybenzoate [28]) (Reversibility: ? <1, 4, 10, 13> [2, 3, 4, 6, 20, 21, 28]) [2, 3, 4, 6, 20, 21, 28]

P 2,3,4-trihydroxybenzoate + 2,4,5-trihydroxybenzoate + $NADP^+$ + H_2O <4> [20]

S 2-chloro-4-hydroxybenzoate + NADH + O_2 <13> (<13>, 40% of the activity with 4-hydroxybenzoate [28]) (Reversibility: ? <13> [28]) [28]

P ?

S 2-fluoro-4-hydroxybenzoate + NADH + O_2 <13> (<13>, 50% of the activity with 4-hydroxybenzoate [28]) (Reversibility: ? <13> [28]) [28]

P ?

S 3-bromo-4-hydroxybenzoate + NADPH + O_2 <1> (<1>, 3.2% of the activity with 4-hydroxybenzoate [4]) (Reversibility: ? <1> [4]) [4]

P ?

S 3-chloro-4-hydroxybenzoate + NADPH + O_2 <10> (<10>, 6.5% of the activity with 4-hydroxybenzoate [21]) (Reversibility: ? <10> [21]) [21]

P ?

S 3-fluoro-4-hydroxybenzoate + NADPH + O_2 <1> (<1>, about 1% of the activity with 4-hydroxybenzoate [6]) (Reversibility: ? <1> [6]) [6]

P ?

S 4-aminobenzoate + NADPH + O_2 <1> (<1>, about 1% of the activity with 4-hydroxybenzoate [6]) (Reversibility: ? <1> [6]) [6]

P ?

S 4-hydroxybenzoate + NADH + O_2 <6, 7, 8, 9, 12, 13> (Reversibility: ? <6, 7, 8, 9, 12, 13> [19, 23, 28]) [19, 23, 28]

P protocatechuate + NAD^+ + H_2O <6, 7, 8, 9> [19]

S 4-hydroxybenzoate + NADPH + O_2 <1-16> (<1>, at least three intermediates [12]) (Reversibility: ? <1-16> [1-34]) [1-34]

P protocatechuate + $NADP^+$ + H_2O <1, 4> [1, 2, 35]

S 4-hydroxybenzoate + NADPH + ferricyanide <1> (Reversibility: ? <1> [1]) [1]

P protocatechuate + $NADP^+$ + ferrocyanide <1> [1]

S 4-mercaptobenzoate + NADPH + O_2 <1, 4> (<1>, 50% of the activity with 4-hydroxybenzoate [6]) (Reversibility: ? <1, 4> [6, 9]) [6, 9]

P 4,4'-dithiobisbenzoate + ? <1> [6]

S 4-toluate + NADPH + O_2 <1> (<1>, 0.29% of the activity with 4-hydroxybenzoate [4]) (Reversibility: ? <1> [4]) [4]

P ?

S benzene sulfonate + + NADPH + O_2 <1> (<1>, 0.34% of the activity with 4-hydroxybenzoate [4]) (Reversibility: ? <1> [4]) [4]

P ?

S Additional information <1, 4, 11> (<1>, under anaerobic conditions, the enzyme can catalyze a reduction of FAD by NADPH provided that 4-hydroxybenzoate is present [4]; <1>, mechanism of oxygen insertion [12]; <11>, a Tyr seems to be involved in substrate activation [22]; <1>, Arg42 is involved in binding of the 2'-phosphoadenosine moiety of NADPH [27]; <4>, mutant enzyme Y385F hydroxylates 3,4-dihydroxybenzoate to form gallic acid [35]) [4, 12, 22, 27, 35]

P ?

Inhibitors

(-)-epigallocatechin-3-O-gallate <4> (<4>, non-competitive, binds to the enzyme in the proximity of the FAD binding site via formation of three hydrogen bonds [26]) [26]
2,4-dihydroxybenzoate <10> (<10>, competitive with 4-hydroxybenzoate [21]) [21]
3,4-dihydroxybenzoate <1, 10> (<1>, at high concentration [2]; <10>, competitive with 4-hydroxybenzoate [21]) [2, 21]
3-chloro-4-hydroxybenzoate <10> (<10>, mixed-type [21]) [21]
3-hydroxybenzoate <1, 12> (<1>, above 1 mM [4]; <12>, 2 mM, 26% inhibition [23]) [4, 23]
4-aminobenzoate <1, 10> (<1>, above 1 mM [4]; <10>, competitive [21]) [2, 4, 21]
4-aminosalicylate <4> (<4>, competitive [9]) [9]
4-fluorobenzoate <1> (<1>, slight [2]; <1>, above 1 mM [4]) [2, 4]
4-hydroxy-3-nitrobenzoic acid <12> (<12>, 2 mM, 23% inhibition [23]) [23]
4-hydroxybenzaldehyde <1, 12> (<1>, slight [2]; <12>, 2 mM, 27% inhibition [23]) [2, 23]
4-hydroxybenzoate <1, 10, 12> (<1>, at high concentrations [2]; <1, 10>, above 1 mM [4,21]; <1>, mixed-type [6]; <12>, above 0.1 mM [23]) [2, 4, 6, 21, 23]
4-hydroxybutyrate <2> (<2>, 92 mM, pH 8, very slight inhibition [7]) [7]
4-hydroxycinnamate <1> [2]
4-hydroxyphenylacetic acid <12> (<12>, 2 mM, 16% inhibition [23]) [23]
4-nitrophenol <12> (<12>, 2 mM, 39% inhibition [23]) [23]
6-aminonicotinate <4> (<4>, competitive [9]) [9]
6-hydroxynicotinate <1, 4> (<1>, at high concentrations [2]) [2, 9]
Br^- <1, 2> (<1, 2>, competitive with respect to NADPH [5, 7]; <2>, mixed type inhibition with respect to 4-hydroxybenzoate [7]) [5, 7]
CNS^- <1, 4> (<1>, competitive with respect to NADPH [5]) [5, 9]
Cl^- <1, 2, 4, 10, 13> (<1, 2>, competitive with respect to NADPH [5, 7]; <2>, mixed type inhibition with respect to 4-hydroxybenzoate [7]; <10>, non-competitive with NADPH [21]; <13>, competitive with respect to NADH [28]) [5, 6, 7, 9, 21, 28]
F^- <1, 2> (<1>, competitive with respect to NADPH [5]; <2>, 19 mM, weak [7]) [5, 7]
Fe^{2+} <1> [4]
Hg^{2+} <12> (<12>, 0.1 mM $HgCl_2$, complete inhibition [23]) [23]
I^- <1, 2, 4> (<1,2>, competitive with respect to NADPH [5,7]; <2>, mixed type inhibition with respect to 4-hydroxybenzoate [7]) [5, 6, 7, 9]
N-iodosuccinimide <14> (<14>, reversed by dithiothreitol [31]) [31]
N_3^- <2, 4> [7, 9]
PCMB <1, 12> (<12>, 0.015 mM, more than 70% inhibition, partially restored by addition of 0.14 M 2-mercaptoethanol [23]) [16, 23]
SO_4^{2-} <2> [7]
Zn^{2+} <1> (<1>, 0.2 mM, reversed by 10 mM glutathione [4]) [4]
acetate <2> (<2>, 92 mM, pH 8, very slight inhibition [7]) [7]

benzoate <1, 10, 12> (<1>, slight [2]; <10>, competitive with 4-hydroxybenzoate [21]; <12>, 2 mM, 48% inhibition [23]; <1>, above 1 mM [4]) [2, 4, 21, 23]
borate <2> [7]
citrate <2> (<2>, 92 mM, pH 8, very slight inhibition [7]) [7]
coumaric acid <12> (<12>, 2 mM, 30% inhibition [23]) [23]
diethyl dicarbonate <1, 11> (<1>, inhibition of wild-type enzyme, no inhibition of mutant enzyme H162R [29]) [22, 29]
formate <2> (<2>, 92 mM, pH 8, very slight inhibition [7]) [7]
fumarate <2> (<2>, 92 mM, pH 8, very slight inhibition [7]) [7]
iodoacetamide <14> (<14>, reversed by dithiothreitol [31]) [31]
maleate <2> (<2>, 92 mM, pH 8, very slight inhibition [7]) [7]
n-dodecyl gallate <4> (<4>, non-competitive [26]) [26]
o-iodosobenzoate <14> (<14>, reversed by dithiothreitol [31]) [31]
p-hydroxy-3-iodomethylbenzoate <14> (<14>, 1 mM, irreversible crosslinking to the substrate binding site [31]) [31]
phenylglyoxal <2> (<2>, pseudo-first order kinetics, incorporation into the substrate-binding site [14]) [14]
phosphate <2> [7]
propionate <2> (<2>, 92 mM, pH 8, very slight inhibition [7]) [7]
protocatechuate <1> (<1>, above 1 mM [4]) [4]
salicylate <1> (<1>, slight [2]; <1>, above 1 mM [4]) [2, 4]
tartrate <2> (<2>, 92 mM, pH 8, very slight inhibition [7]) [7]
Additional information <1> (<1>, one of the five sulfhydryl groups reacts rapidly and specifically with NEM, without inactivation of the enzyme [16]; <1>, multisubunit inhibition by ether derivatives [24]) [16, 24]

Cofactors/prosthetic groups

1-deaza-FAD <4> (<4>, when 1-deaza-FAD is used as cofactor, the enzyme carries out each step in catalysis except the transfer of oxygen to 4-hydroxybenzoate [9]) [9]
6-hydroxy-FAD <4> (<4>, when 6-hydroxy-FAD is used as cofactor, the enzyme has a lower turnover rate than the native enzyme [9]) [9]
FAD <1, 4, 6, 7, 8, 9, 12, 13> (<1,4>, contains 1 mol of FAD per mol of enzyme [1,6,11]; <1>, contains approximately 0.8 mol FAD per mol of enzyme [4]; <4>, contains 1 mol of FAD per subunit [9]; <6>, K_m for strain IG: 185 nM [19]; <6>, K_m for strain 420: 220 nM [19]; <6>, K_m for strain 557: 190 nM [19]; <7>, K_m: 420 nM [19]; <8>, K_m: 440 nM [19]; <9>, K_m: 225 nM [19]; <4>, movement of the flavin occurs before reduction [20]; <12>, contains 0.12 mol FAD per mol of enzyme [23]; <1>, flavin motion in 4-hydroxybenzoate hydroxylase is important for efficient reduction [25]; <13>, K_m: 150 nM [28]; <4>, two flavin conformations in the enzyme: the in-position and the out-position. Substrate hydroxylation occurs while the flavin in the enzyme is in the in-conformation. Flavin must move to the out-conformation for proper formation of the charge-transfer complex between NADPH and FAD that is necessary for rapid flavin reduction [38]; <4>, thermodynamic

and kinetic constants of the enzyme reconstituted with 8-substituted flavins [38]) [1, 4, 6, 9, 11, 19, 20, 23, 25, 28, 32, 38]
NADH <6, 7, 8, 9, 12, 13> (<12,13>, activity with NADH is higher than activity with NADPH [23,28]) [19, 23, 28]
NADPH <1, 2, 3, 4, 5, 6, 7, 8, 9> (<1,10>, no activity with NADH [4,6,21]; <12>, activity with NADH is higher than activity with NADPH [23]) [1, 2, 3, 4, 5, 6, 7, 8, 9, 10, 11, 12, 13, 14, 15, 16, 17, 18, 19]
arabinoflavin adenine dinucleotide <1> (<1>, like native enzyme the arabinoflavin adenine dinucleotide containing 4-hydroxybenzoate hydroxylase preferentially binds the phenolate form of the substrate. The oxidative part of the catalytic cycle of a FAD-containing 4-hydroxybenzoate hydroxylase differs from the native enzyme. Partial uncoupling of hydroxylation results in the formation of about 0.3 mol of 3,4-dihydroxybenzoate and 0.7 mol of H_2O_2 per mol of NADPH oxidized [25]) [25]
Additional information <13> (<13>, helix H_2 is involved in determining the coenzyme specificity [28]) [28]

Activating compounds

2,4-dihydroxybenzoate <1> (<1>, increases the rate of NADP oxidation by 4-hydroxybenzoate, no hydroxylation during subsequent reoxidation by O_2 [2]) [2]
3,4-dihydroxybenzoate <1> (<1>, increases the rate of NADP oxidation by 4-hydroxybenzoate, no hydroxylation during subsequent reoxidation by O_2 [2]) [2]
5-hydroxypicolinate <4> (<4>, stimulates rapid oxidation of NADPH [9]) [9]
benzoate <1> (<1>, increases the rate of NADP oxidation by 4-hydroxybenzoate, no hydroxylation during subsequent reoxidation by O_2 [2]) [2]

Turnover number (min^{-1})

21.6 <4> (4-hydroxybenzoate, <4>, mutant enzyme Y385F [35]) [35]
25.2 <4> (4-hydroxybenzoate, <4>, mutant enzyme Y201F [35]) [35]
342 <4> (4-hydroxybenzoate, <4>, wild-type enzyme [35]) [35]
396 <13> (NADPH, <13>, 25°C [28]) [28]
408 <13> (NADH, <13>, 25°C [28]) [28]
444 <13> (4-hydroxybenzoate, <13>, 35°C [28]) [28]
516 <13> (NADPH, <13>, 35°C [28]) [28]
696 <13> (NADH, <13>, 35°C [28]) [28]
798 <13> (NADPH, <13>, 45°C [28]) [28]
1014 <13> (NADH, <13>, 45°C [28]) [28]
2200 <1> (4-hydroxybenzoate) [4]
Additional information <1, 4, 6, 7, 8, 9> [9, 19, 27, 29]

Specific activity (U/mg)

5.76 <11> [22]
10.2 <12> [23]
25 <9> [19]
38 <10> [21]
47.3 <1> [1]

47.5 <1> [6]
52.2 <1> [15]
Additional information <4, 13> [11, 28]

K_m-Value (mM)

0.003 <6> (4-hydroxybenzoate, <6>, strain 557 [19]) [19]
0.0033 <8> (4-hydroxybenzoate) [19]
0.0072 <7> (4-hydroxybenzoate) [19]
0.0073 <9> (4-hydroxybenzoate) [19]
0.0079 <6> (4-hydroxybenzoate, <6>, strain IG [19]) [19]
0.0081 <6> (4-hydroxybenzoate, <6>, strain 420 [19]) [19]
0.0095 <12> (4-hydroxybenzoate, <12>, reaction with NADH [23]) [23]
0.0109 <4> (4-hydroxybenzoate) [11]
0.0122 <6> (NADH, <6>, strain 557 [19]) [19]
0.0139 <8> (NADH) [19]
0.014 <1> (2,4-dihydroxybenzoate) [3]
0.0167 <7> (NADH) [19]
0.0195 <6> (NADH, <6>, strain IG [19]) [19]
0.02 <1> (4-hydroxybenzoate, <1>, wild-type enzyme [29]) [29]
0.02 <6> (NADH, <6>, strain 420 [19]) [19]
0.0213 <1> (4-hydroxybenzoate) [4]
0.022 <11> (4-hydroxybenzoate) [22]
0.0226 <7> (NADPH) [19]
0.0227 <1> (NADPH, <1>, reaction with 4-hydroxybenzoate [4]) [4]
0.0232 <4> (NADPH) [11]
0.0238 <6> (NADPH, <6>, strain IG [19]) [19]
0.025 <1> (4-hydroxybenzoate, <1>, wild-type enzyme [27]) [27]
0.025 <11> (NADPH) [22]
0.027 <1> ($NADP^+$, <1>, reaction with 2,4-dihydroxybenzoate [3]) [3]
0.03 <1> (4-hydroxybenzoate, <1>, mutant enzyme H162R, H162Y, H162K, R269K, R269Y, R269N, R269S and R269T [29]) [29]
0.03 <1> (NADPH, <1>, wild-type enzyme [29]) [29]
0.0335 <6> (NADPH, <6>, strain 557 [19]) [19]
0.037 <12> (NADH) [23]
0.037 <11> (O_2) [22]
0.0371 <4> (O_2) [11]
0.0398 <9> (NADH) [19]
0.04 <1> (4-hydroxybenzoate, <1>, mutant enzyme H162N, H162S, H162T and H162D [29]) [29]
0.041 <10> (4-hydroxybenzoate) [21]
0.05 <1> (NADPH, <1>, mutant enzyme H162R [29]) [29]
0.05 <1> (NADPH, <1>, wild-type enzyme [27]) [27]
0.05 <1> (O_2, <1>, reaction with 2,4-dihydroxybenzoate [3]) [3]
0.06 <1> (4-hydroxybenzoate, <1>, mutant enzyme R269D [29]) [29]
0.0658 <6> (NADPH, <6>, strain 420 [19]) [19]
0.07 <1> (NADPH, <1>, mutant enzyme H162Y and R269K [29]) [29]
0.075 <1> (4-hydroxybenzoate, <1>, mutant enzyme R42S [27]) [27]

0.075 <13> (4-hydroxybenzoate, <13>, 35°C [28]) [28]
0.08 <13> (NADH, <13>, 25°C [28]) [28]
0.11 <1> (4-hydroxybenzoate, <1>, mutant enzyme R42K [27]) [27]
0.12 <13> (NADH, <13>, 35°C [28]) [28]
0.14 <13> (NADPH, <13>, 25°C [28]) [28]
0.155 <9> (NADPH) [19]
0.16 <13> (NADPH, <13>, 35°C [28]) [28]
0.18 <10> (3-chloro-4-hydroxybenzoate) [21]
0.18 <13> (NADH, <13>, 45°C [28]) [28]
0.2 <1, 13> (NADPH, <13>,45°C [28]; <1>, mutant enzyme H162K [29]) [28, 29]
0.25 <10> (2,4-dihydroxybenzoate) [21]
0.32 <1> (NADPH, <1>, mutant enzyme R269S [29]) [29]
Additional information <1, 2> (<1>, K_m value for NADPH above 0.5 mM: mutant enzymes R269D, R269T, R269N, R269Y, H162D, H162T, H162S, H162N [29]) [7, 10, 29]

K_i-Value (mM)

0.014 <4> ((-)-epigallocatechin-3-O-gallate) [26]
0.0181 <4> (n-dodecyl gallate) [26]
0.05 <10> (4-aminobenzoate) [21]
0.31 <10> (2,4-dihydroxybenzoate) [21]
0.53 <10> (benzoate) [21]
0.55 <10> (3,4-dihydroxybenzoate) [21]
5.9 <1> (Cl^-) [5]
6 <1> (CNS^-) [5]
9.7 <1> (I^-) [5]
50 <13> (Cl^-) [28]
65 <1> (Br^-) [5]
150 <1> (F^-) [5]

pH-Optimum

7 <7> [19]
7-7.3 <12> (<12>, reaction with NADH [23]) [23]
7.2 <6> (<6>, strain 400 and 557 [19]) [19]
7.4 <6, 8> (<6>, strain IG [19]) [19]
7.5-7.8 <12> (<12>, reaction with NADPH [23]) [23]
7.5-8.5 <1> [10]
7.7 <9> [19]
8 <10, 11, 13> [21, 22, 28]
8-8.1 <4> (<4>, HEPES or Tris-SO_4^{2-} buffer [11]) [11]

Temperature optimum (°C)

25-30 <11> [22]
45 <13> [28]

4 Enzyme Structure

Molecular weight

65000 <1> (<1>, gel filtration [1,6]) [1, 6]
75000 <4> (<4>, gel filtration [11]) [11]
76000-82000 <1> (<1>, gel filtration [15]) [15]
78000 <1> (<1>, gel filtration [4]) [4]
80000 <11> (<11>, gel filtration [22]) [22]
83000-90000 <1> (<1>, equilibrium sedimentation [15]) [15]
83600 <1> (<1>, low speed sedimentation without reaching equilibrium [4]) [4]
85000 <12> (<12>, gel filtration [23]) [23]
89000 <10> (<10>, gel filtration [21]) [21]
90000 <13> (<13>, gel filtration [28]) [28]

Subunits

? <14> (<14>, x * 71000, SDS-PAGE [31]) [31]
dimer <1, 4, 11, 12, 13> (<11>, 2 * 40000, SDS-PAGE [22]; <1>, 1 * 43000-45000, SDS-PAGE [15]; <13>, 2 * 43632, calculation from nucleotide sequence [28]; <13>, 2 * 44000, SDS-PAGE [28]; <4,12>, 2 * 45000, SDS-PAGE [11,23]; <4>, 2 * 45100 [9]) [9, 11, 15, 22, 23, 28]
tetramer <6, 7, 8, 9> (<7,8>, 4 * 45000, SDS-PAGE [19]; <6>, 4 * 46000, strain 557, SDS-PAGE [19]; <6>, 4 * 47500, strain 420, SDS-PAGE [19]; <9>, 4 * 48500, SDS-PAGE [19]; <6>, 4 * 51000, strain IG, SDS-PAGE [19]) [19]

5 Isolation/Preparation/Mutation/Application

Purification

<1> (microheterogeneity of the highly purified enzyme. Different form of enzyme molecules are due to the partial oxidation of Cys116 in the sequence of the enzyme [13]; recombinat enzyme, cloned in Escherichia coli [25]) [1, 6, 13, 15, 25, 29]
<4> [9, 11]
<9> [19]
<10> (recombinant enzyme [21]) [21]
<11> [22]
<12> [23]
<13> [28]
<14> [30, 31]

Crystallization

<1> (enzyme bound in a crystal is able to convert 4-hydroxybenzoate [17]; <1>, crystals of a arabinoflavin adenine dinucleotide-containing 4-hydroxybenzoate hydroxylase in complex with 4-hydroxybenzoate are obtained using the hanging drop method [25]; <1>, crystal of the enzyme complexed with 4-hydroxybenzoate are obtained using the hanging-drop method [27]; crystal-

ization of mutant enzymes H162R and R269T by hanging drop vapor diffusion method [29]; <1>, crystal structure of wild-type *p*-hydroxybenzoate hydroxylase complexed with 4-aminobenzoate, 2,4-dihydroxybenzoate, and 2-hydroxy-4-aminobenzoate and of the Tyr222Ala mutant complexed with 2-hydroxy-4-aminobenzoate [32]) [1, 4, 6, 8, 17, 25, 27, 29, 32]
<4> (crystal structure of mutant enzyme Y201F, Y385F, and N300D [33]) [9, 33]
<14> [31]

Cloning

<1> (mutant enzymes R42S and R42K expressed in transformed Escherichia coli TG2 cells [27]) [27]
<4> (expression in Escherichia coli, mutant enzyme S212A [20]; expression in Escherichia coli [26,38]; expression of mutant enzymes Y201F and Y385F in Escherichia coli [35]; plasmid mutagenesis for high-level expression of 4-hydroxybenzoate hydroxylase [37]; mutant enzymes K297M, N300D and Y385F [39]) [20, 26, 35, 37, 38, 39]
<10> (overexpression in Escherichia coli [21]) [21]
<13> [28]

Engineering

H162D <1> (<1>, no reliable turnover rate due to impaired NADPH binding [29]) [29]
H162K <1> (<1>, less efficient than wild-type enzyme due to a clear increase in the apparent K_m-value for NADPH [29]) [29]
H162N <1> (<1>, no reliable turnover rate due to impaired NADPH binding [29]) [29]
H162R <1> (<1>, rather efficient enzyme with similar catalytic properties as wild-type enzyme [29]) [29]
H162S <1> (<1>, no reliable turnover rate due to impaired NADPH binding [29]) [29]
H162T <1> (<1>, no reliable turnover rate due to impaired NADPH binding [29]) [29]
H162Y <1> (<1>, rather efficient enzyme with similar catalytic properties as wild-type enzyme [29]) [29]
K297M <4> (<4>, decreased positive charge in active site, about 35fold slower hydroxylation rate than the wild-type enzyme. Substitution of 8-Cl-FAD in the mutant gives about 1.8fold increase in hydroxylation rate compared to the wild-type enzyme [39]) [39]
N300D <4> (<4>, mutation has profound effect on enzyme structure. The side chain of Asp300 moves away from the flavin, disrupting the interaction of the carboxamide group with the flavin O(2) atom, and the α-helix H10 that begins at residue 297 is displaced, altering its dipole interaction with the flavin ring [33]; <4>, 330fold reduced reduction rate of the flavin of the enzyme by NADPH compared to wild-type enzyme, redox potential of the flavin is 20-40 mV lower than that of the wild-type enzyme. The mutation interferes with the orientation of pyridine nucleotide and flavin during reduction, stabilizes flavin C(4a) intermediates, prevents substrate ionization, and alters the rates

and strengths of ligand binding [36]; <4>, decreased positive charge in active site, about 35fold slower hydroxylation rate than the wild-type enzyme, Substitution of 8-Cl-FAD in the mutant gives about 1.8fold increase in hydroxylation rate compared to the wild-type enzyme [39]) [33, 36, 39]
R269D <1> (<1>, no reliable turnover rate due to impaired NADPH binding [29]) [29]
R269K <1> (<1>, rather efficient enzyme with similar catalytic properties as wild-type enzyme [29]) [29]
R269N <1> (<1>, no reliable turnover rate due to impaired NADPH binding [29]) [29]
R269S <1> (<1>, less efficient than wild-type enzyme due to a clear increase in the apparent K_m-value for NADPH [29]) [29]
R269T <1> (<1>, no reliable turnover rate due to impaired NADPH binding [29]) [29]
R269Y <1> (<1>, no reliable turnover rate due to impaired NADPH binding [29]) [29]
R42K <1> (<1>, low activity results from impaired binding of NADPH [27]) [27]
R42S <1> (<1>, low activity results from impaired binding of NADPH [27]) [27]
S212A <4> (<4>, the turnover of the substrate 2,4-dihydroxybenzoate is 1.5-fold faster than the rate observed with the wild-type [20]) [20]
Y201F <4> (<4>, crystals differ from the wild-type enzyme at two surface positions, 228 and 249 [33]; <4>, less than 6% of the activity of the wild-type enzyme. Reduction of FAD by NADPH is slower by 10fold, when the mutant enzyme-4-hydroxybenzoate complex reacts with oxygen, a long-lived flavin-C(4a)-hydroperoxide is observed, which slowly eliminates H_2O_2 with very little hydroxylation [35]) [33, 35]
Y385F <4> (<4>, crystals differ from the wild-type enzyme at two surface positions, 228 and 249 [33]; <4>, less than 6% of the activity of the wild-type enzyme. Reduction of FAD by NADPH is slower by 100fold, the mutant enzyme reacts with oxygen to form 25% oxidized enzyme and 75% flavin hydroperoxide, which successfully hydroxylates the substrate. The mutant also hydroxylates the product 3,4-dihydroxybenzoate to form gallic acid [35]; <4>, mutant enzyme with a disrupted hydrogen-bonding network, substitution of 8-Cl-FAD in the mutant gives about 1.5fold increase in hydroxylation rate compared to the wild-type enzyme [39]) [33, 35, 39]

6 Stability

pH-Stability

5.5-6.5 <1> (<1>, optimal stability [10]) [10]
6.4-8 <1> (<1>, at 0°C or at 25°C, 50 h, stable [1]) [1]
7-7.5 <8> (<8>, optimal stability [19]) [19]
9-9.5 <6> (<6>, optimal stability [19]) [19]

Temperature stability

4 <10, 11> (<10>, 24 h, 40% loss of activity without stabilizer, 10% loss of activity in presence of 4-hydroxybenzoate, FAD and EDTA [21]; <11>, 50 mM phosphate buffer, 50 mM Tris-HCl, pH 8.0, 10% glycerol, 10 mM EDTA, loss of activity after 72 h [22]) [21, 22]
21 <11> (<11>, room temperature, without glycerol and EDTA, complete inactivation after 20 h [22]) [22]
25 <1, 4> (<1>, 50 h, stable [6]; <1>, pH 6.4-8.0, 50 h, stable [6]; <4>, pH 5.5-8.5, stable [11]) [6, 11]
40 <1, 4> (<1>, pH 5.5-7.5, stable [10]; <4>, pH 8.0, stable [11]) [10, 11]
50 <6, 7, 8, 9, 10> (<6>, 50% loss of activity after 9 min, strain IG [19]; <7>, 50% loss of activity after 50 min [19]; <8>, 50% loss of activity after 37 min [19]; <6>, 50% loss of activity after 15 min, strain 420 [19]; <9>, 50% loss of activity after 180 min [19]; <6>, 50% loss of activity after 45 min, strain 557 [19]; <10>, 10 min, 60% loss of activity, in presence of 0.1 mM 4-hydroxybenzoate the enzyme retains 80% of ist activity [21]) [19, 21]
52 <12> (<12>, pH 7.5, 20 mM potassium phosphate buffer, 20 min, stable [23]) [23]
60 <1, 12> (<1>, pH 5.5-6.5, rather stable [10]; <12>, in presence of 1 mM 4-hydroxybenzoate, 40% loss of activity after 20 min [23]) [10, 23]
65 <12> (<12>, 1 mM 4-hydroxybenzoate, 95% loss of activity [23]) [23]

General stability information

<1>, stable to freezing and thawing [6]
<1>, very stable, even in absence of stabilizing agents [1]
<3>, extremely unstable, undergoing rapid inactivation unless protected by substrates and other stabilizing agents [6]
<6, 7, 8, 9>, enzyme rapidly loses activity in dilute solutions, below 2 mg/ml, without stabilizer [19]

Storage stability

<1>, 4°C, as ammonium sulfate paste, indefinitely stable [6]
<4>, 0°C-4°C, as a precipitate under a solution of 50 mM potassium phosphate and 0.5 mM EDTA, pH 6.5-7.0, with 70% saturated ammonium sulfate, idefinitely stable [9]
<10>, 4°C, 24 h, 40% loss of activity without stabilizer, 10% loss of activity in presence of 4-hydroxybenzoate, FAD and EDTA [21]
<11>, 4°C, 50 mM phosphate buffer, 50 mM Tris-HCl, pH 8.0, 10% glycerol, 10 mM EDTA, loss of activity after 72 h [22]
<12>, -70°C, 5 mM potassium phosphate, pH 7.5, 5% glycerol, stable for several months [23]

References

[1] Howell, L.G.; Spector, T.; Massey, V.: Purification and properties of *p*-hydroxybenzoate hydroxylase from Pseudomonas fluorescens. J. Biol. Chem., **247**, 4340-4350 (1972)

[2] Spector, T.; Massey, V.: Studies on the effector specificity of *p*-hydroxybenzoate hydroxylase from Pseudomonas fluorescens. J. Biol. Chem., **247**, 4679-4687 (1972)
[3] Spector, T.; Massey, V.: *p*-Hydroxybenzoate hydroxylase from Pseudomonas fluorescens. Evidence for an oxygenated flavin intermediate. J. Biol. Chem., **247**, 5632-5636 (1972)
[4] Hosokawa, K.; Stanier, R.Y.: Crystallization and properties of *p*-hydroxybenzoate hydroxylase from Pseudomonas putida. J. Biol. Chem., **241**, 2453-2460 (1966)
[5] Steennis, P.J.; Cordes, M.M.; Hilkens, J.G.H.; Mueller, F.: On the interaction of *para*-hydroxybenzoate hydroxylase from Pseudomonas fluorescens with halogen ions. FEBS Lett., **36**, 177-180 (1973)
[6] Husain, M.; Schopfer, L.M.; Massey, V.: *p*-Hydroxybenzoate hydroxylase and melilotate hydroxylase. Methods Enzymol., **53**, 543-558 (1978)
[7] Shoun, H.; Arima, K.; Beppu, T.: Inhibition of *p*-hydroxybenzoate hydroxylase by anions: possible existence of two anion-binding sites in the site for reduced nicotinamide adenine dinucleotide phosphate. J. Biochem., **93**, 169-176 (1983)
[8] Drenth, J.; Hol, W.G.J.; Wierenga, R.K.: Crystallization and preliminary x-ray investigation of *p*-hydroxybenzoate hydroxylase from Pseudomonas fluorescens. J. Biol. Chem., **250**, 5268-5269 (1975)
[9] Entsch, B.: Hydroxybenzoate hydroxylase. Methods Enzymol., **188**, 138-147 (1990)
[10] van Berkel, W.J.H.; Mueller, F.: The temperature and pH dependence of some properties of *p*-hydroxybenzoate hydroxylase from Pseudomonas fluorescens. Eur. J. Biochem., **179**, 307-314 (1989)
[11] Entsch, B.; Ballou, D.P.: Purification, properties, and oxygen reactivity of *p*-hydroxybenzoate hydroxylase from Pseudomonas aeruginosa [published erratum appears in Biochim Biophys Acta 1990 Mar 29;1038(1):139]. Biochim. Biophys. Acta, **999**, 313-322 (1989)
[12] Entsch, B.; Ballou, D.P.; Massey, V.: Flavin-oxygen derivatives involved in hydroxylation by *p*-hydroxybenzoate hydroxylase. J. Biol. Chem., **251**, 2550-2563 (1976)
[13] van Berkel, W.J.H.; Mueller, F.: The elucidation of the microheterogeneity of highly purified *p*-hydroxybenzoate hydroxylase from Pseudomonas fluorescens by various biochemical techniques. Eur. J. Biochem., **167**, 35-46 (1987)
[14] Shoun, H.; Beppu, T.; Arima, K.: An essential arginine residue at the substrate-binding site of *p*-hydroxybenzoate hydroxylase. J. Biol. Chem., **255**, 9319-9324 (1980)
[15] Mueller, F.; Voordouw, G.; van Berkel, W.J.H.; Steennis, P.J.; Visser, S.; van Rooijen, P.J.: A study of *p*-hydroxybenzoate hydroxylase from Pseudomonas fluorescens. Improved purification, relative molecular mass, and amino acid composition. Eur. J. Biochem., **101**, 235-244 (1979)
[16] van Berkel, W.J.H.; Weijer, W.J.; Mueller, F.; Jekel, P.A.; Beintema, J.J.: Chemical modification of sulfhydryl groups in *p*-hydroxybenzoate hydroxylase from Pseudomonas fluorescens. Involvement in catalysis and assignment in the sequence. Eur. J. Biochem., **145**, 245-256 (1984)

[17] Schreuder, H.A.; van der Laan, J.M.; Hol, W.G.J.; Drenth, J.: Crystal structure of *p*-hydroxybenzoate hydroxylase complexed with its reaction product 3,4-dihydroxybenzoate. J. Mol. Biol., **199**, 637-648 (1988)
[18] Whited, G.M.; Gibson, D.T.: Separation and partial characterization of the enzymes of the toluene-4-monooxygenase catabolic pathway in Pseudomonas mendocina KR1. J. Bacteriol., **173**, 3017-3020 (1991)
[19] Jadan, A.P.; van Berkel, W.J.H.; Golovleva, L.A.; Golovlev, E.L.: Purification and properties of *p*-hydroxybenzoate hydroxylases from Rhodococcus strains. Biochemistry (Moscow), **66**, 898-903 (2001)
[20] Moran, G.R.; Entsch, B.; Palfey, B.A.; Ballou, D.P.: Mechanistic insights into *p*-hydroxybenzoate hydroxylase from studies of the mutant Ser212Ala. Biochemistry, **38**, 6292-6299 (1999)
[21] Fernandez, J.; Dimarco, A.A.; Ornston, L.N.; Harayama, S.: Purification and characterization of Acinetobacter calcoaceticus 4-hydroxybenzoate 3-hydroxylase after its overexpression in Escherichia coli. J. Biochem., **117**, 1261-1266 (1995)
[22] Suarez, M.; Martin, M.; Ferrer, E.; Garrido-Pertierra, A.: Purification and characterization of 4-hydroxybenzoate 3-hydroxylase from a Klebsiella pneumoniae mutant strain. Arch. Microbiol., **164**, 70-77 (1995)
[23] Sterjiades, R.: Properties of NADH/NADPH-dependent *p*-hydroxybenzoate hydroxylase from Moraxella sp. Biotechnol. Appl. Biochem., **17**, 77-90 (1993)
[24] Salituro, F.G.; Demeter, D.A.; Weintraub, H.J.R.; Lippert, B.J.; Resvick, R.J.; McDonald, I.A.: Multisubstrate inhibition of 4-hydroxybenzoate 3-monooxygenase. J. Med. Chem., **37**, 4076-4078 (1994)
[25] van Berkel, W.J.H.; Eppink, M.H.M.; Schreuder, H.A.: Crystal structure of *p*-hydroxybenzoate hydroxylase reconstituted with the modified FAD present in alcohol oxidase from methylotrophic yeasts: evidence for an arabinoflavin. Protein Sci., **3**, 2245-2253 (1994)
[26] Abe, I.; Kashiwagi, K.; Noguchi, H.: Antioxidative galloyl esters as enzyme inhibitors of *p*-hydroxybenzoate hydroxylase. FEBS Lett., **483**, 131-134 (2000)
[27] Eppink, M.H.M.; Schreuder, H.A.; Van Berkel, W.J.H.: Lys42 and Ser42 variants of *p*-hydroxybenzoate hydroxylase from Pseudomonas fluorescens reveal that Arg42 is essential for NADPH binding. Eur. J. Biochem., **253**, 194-201 (1998)
[28] Seibold, B.; Matthes, M.; Eppink, M.H.M.; Lingens, F.; Van Berkel, W.J.H.; Mueller, R.: 4-Hydroxybenzoate hydroxylase from Pseudomonas sp. CBS3. Purification, characterization, gene cloning, sequence analysis and assignment of structural features determining the coenzyme specificity. Eur. J. Biochem., **239**, 469-478 (1996)
[29] Eppink, M.H.M.; Schreuder, H.A.; Van Berkel, W.J.H.: Interdomain binding of NADPH in *p*-hydroxybenzoate hydroxylase as suggested by kinetic, crystallographic and modeling studies of histidine 162 and arginine 269 variants. J. Biol. Chem., **273**, 21031-21039 (1998)

[30] Chen, R.; Scott, R., Jr.; Tsugita, A.; Hosokawa, K.: Purification and characterization of *p*-hydroxybenzoate 3-hydroxylase from Comamonas testosteroni. Res. Commun. Biochem. Cell Mol. Biol., **1**, 51-60 (1997)
[31] Chen, R.; Oki, H.; Scott, R.P.; Yamaguchi, H.; Kusunoki, M.; Matsuura, Y.; Chaen, H.; Tsugita, A.; Hosokawa, K.: Crystallization and further characterization of *meta*-hydroxybenzoate 4-hydroxylase from Comamonas testosteroni. Res. Commun. Biochem. Cell Mol. Biol., **2**, 253-274 (1998)
[32] Schreuder, H.A.; Mattevi, A.; Obmolova, G.; Kalk, K.H.; Hol, W.G.J.; van der Bolt, F.J.T.; van Berkel, W.J.H.: Crystal structures of wild-type *p*-hydroxybenzoate hydroxylase complexed with 4-aminobenzoate, 2,4-dihydroxybenzoate, and 2-hydroxy-4-aminobenzoate and of the Tyr222Ala mutant complexed with 2-hydroxy-4-aminobenzoate. Evidence for a proton channel and a new binding mode of the flavin ring. Biochemistry, **33**, 10161-10170 (1994)
[33] Lah, M.S.; Palfey, B.A.; Schreuder, H.A.; Ludwig, M.L.: Crystal structures of mutant Pseudomonas aeruginosa *p*-hydroxybenzoate hydroxylases: The Tyr201Phe, Tyr385Phe, and Asn300Asp variants. Biochemistry, **33**, 1555-1564 (1994)
[34] Suemori, A.; Nakajima, K.; Kurane, R.; Nakamura, Y.: Comparison of chemical inactivation of salicylate 5-hydroxylase, *m*-hydroxybenzoate 6-hydroxylase, and *p*-hydroxybenzoate 3-hydroxylase from Rhodococcus erythropolis. Seimei Kogaku Kogyo Gijutsu Kenkyusho Kenkyu Hokoku, **2**, 27-30 (1994)
[35] Entsch, B.; Palfey, B.A.; Ballou, D.P.; Massey, V.: Catalytic function of tyrosine residues in *para*-hydroxybenzoate hydroxylase as determined by the study of site-directed mutants. J. Biol. Chem., **266**, 17341-17349 (1991)
[36] Palfey, B.A.; Entsch, B.; Ballou, D.P.; Massey, V.: Changes in the catalytic properties of *p*-hydroxybenzoate hydroxylase caused by the mutation Asn300Asp. Biochemistry, **33**, 1545-1554 (1994)
[37] Moran, G.R.; Entsch, B.: Plasmid mutagenesis by PCR for high-level expression of *para*-hydroxybenzoate hydroxylase. Protein Expr. Purif., **6**, 164-168 (1995)
[38] Ortiz-Maldonado, M.; Gatti, D.; Ballou, D.P.; Massey, V.: Structure-function correlation of the reaction of reduced nicotinamide analogues with *p*-hydroxybenzoate hydroxylase substituted with a series of 8-substituted flavins. Biochemistry, **38**, 16636-16647 (1999)
[39] Ortiz-Maldonado, M.; Aeschliman, S.M.; Ballou, D.P.; Masey, V.: Synergistic interactions of multiple mutations on catalysis during the hydroxylation reaction of *p*-hydroxybenzoate hydroxylase: studies of the Lys297Met, Asn300Asp, and Tyr385Phe mutants reconstituted with 8-Cl flavin. Biochemistry, **40**, 8705-8716 (2001)

4-Hydroxyphenylacetate 3-monooxygenase 1.14.13.3

1 Nomenclature

EC number
1.14.13.3

Systematic name
4-hydroxyphenylacetate,NADH:oxygen oxidoreductase (3-hydroxylating)

Recommended name
4-hydroxyphenylacetate 3-monooxygenase

Synonyms
4 HPA 3-hydroxyylase
4-hydroxyphenylacetate 3-hydroxylase
4-hydroxyphenylacetic acid 3-hydroxylase
p-hydroxyphenylacetate 3-hydroxylase
p-hydroxyphenylacetate hydroxylase
p-hydroxyphenylacetic 3-hydroxylase

CAS registry number
37256-71-6

2 Source Organism

<-1> no activity in *Escherichia coli K-12* (DH1 CECT 416, ATCC 27325 [9]) [8, 9]
<1> *Pseudomonas ovalis* [1]
<2> *Pseudomonas putida* (P23X1, NCIB 9865 [3, 7]; mutant P23X1 [7]; strain U, CECT 4848 [10]) [2, 3, 7, 10, 12, 15, 16]
<3> *Escherichia coli* (strain C [4]; strain W, ATCC 11105 [8,9,11,14]) [4, 8, 9, 11, 14, 15]
<4> *Acinetobacter sp.* (wild-type strain 3B-1 and mutant 3B-3 [5]) [5]
<5> *Arthrobacter sp.* (PRL W15 [6]) [6]
<6> *Klebsiella pneumoniae* (M5a1 strain UN [13]) [13, 14]
<7> *Acinetobacter baumannii* [15]

3 Reaction and Specificity

Catalyzed reaction

4-hydroxyphenylacetate + NADH + H^+ + O_2 = 3,4-dihydroxyphenylacetate + NAD^+ + H_2O (flavoprotein; <2,4> enzymes of 4-hydroxyphenylacetic acid *meta*-cleavage pathway are sythesized following at least two sequential inductive events [5,7]; <2> genes encoding the first four enzymes of the 4-hydroxyphenylacetic acid *meta*-pathway exist as two operons, one containing the hydroxylase gene while the other contains the other three [7]; <3> two enzymes suggested to be required for hydroxylase activity, member of new family of two-component aromatic hydroxylases. The hydroxylase, a flavoprotein, is encoded by hpaB and the coupling protein by hpaC gene [8]; <2> family of external flavoprotein monooxygenases. The activity is induced by the substrates [10]; <3> new type of FADH-utilizing monooxygenase [11]; <2> two-protein enzyme: the flavoprotein catalyzes the non-productive oxidation of NADH without substrate hydroxylation and the coupling protein is an absolute requirement for productive hydroxylation. Reaction mechanism and formation of 3 flavin-oxygen intermediates are studied [12]; <6> two protein component enzyme: flavin-containing hydroxylase and coupling protein that appears to be an effector protein, both needed for productive hydroxylation. Enzyme synthesis is induced by 3- or 4-hydroxyphenylacetate [13]; <3> member of novel two-component flavin-diffusible monooxygenase family, both components required for hydroxylation. The genes are located on the same operon. Physical interaction between HpaB and HpaC is not required, enhancement of activity by direct interaction is not excluded. Contains reductase component [14]; <7> different from other hydroxylases. Two-protein component enzyme: component C_1 is a flavoprotein and C_2 the hydroxylase component. Mechanism: 4-hydroxyphenylacetate binds to C1, then the enzyme-bound flavin is reduced by NADH, this reduced flavin is transferred to C_2 and the reoxidation of the flavin occurs concurrently with the hydroxylation of the substrate [15]; <2> dehydration of the C4a-hydroxyflavin is the rate-determining step in catalysis [16])

Reaction type

hydroxylation
oxidation
redox reaction
reduction

Natural substrates and products

S 4-hydroxyphenylacetate + NADH + O_2 <2-7> (<2> key enzyme in pathway of degradation of phenylalanine, tyrosine and other aromatic amines [2]; <2,3> degradation of 4-hydroxyphenylacetate by inducible *meta*-cleavage pathway [3-5,7]; <5> degradation of tyrosine with 3,4-dihydroxyphenylacetate as key intermediate [6]; <3> inducible chromosomally encoded *meta*-cleavage pathway of aromatic degradation [8]; <3> 4-hydroxyphenylacetic acid degradative pathway [9]; <2> degradative pathway involved in the assimilation of different aromatic compounds [10]; <6> *meta*-cleavage pathway [13]; <3> initial reaction in the degradation of 4-hydroxyphenylacetate [14]; <7> biodegradation of aromatic compounds [15]) (Reversibility: ? <2-7> [2-10, 12-15]) [2-10, 12-15]

P 3,4-dihydroxyphenylacetate + NAD^+ + H_2O <2-7> [2-10, 12-15]

Substrates and products

S 2,4-dihydroxybenzoate <2> (Reversibility: ? <2> [16]) [16]

P ?

S 2,5-dihydroxyphenylacetate + ? <3> (<3> 155% of 4-hydroxyphenylacetate activity [9]) (Reversibility: ? <3> [9]) [9]

P ?

S 2-phenylbutyric acid + H_2 <2> (<2> 86% of 4-hydroxyphenylacetate activity [10]) (Reversibility: ? <2> [10]) [10]

P ?

S 2-phenylpropionic acid + ? <2> (<2> 77% of 4-hydroxyphenylacetate activity [10]) (Reversibility: ? <2> [10]) [10]

P ?

S 3,4-dihydroxybenzoate + ? <6> (<6> 30% of 4-hydroxyphenylacetate activity [13]) (Reversibility: ? <6> [13]) [13]

P ?

S 3,4-dihydroxyphenylacetate + ? <3, 6> (<3> 65% of 4-hydroxyphenylacetate [9]; <6> 50% of 4-hydroxyphenylacetate activity [13]) (Reversibility: ? <3, 6> [9, 13]) [9, 13]

P ?

S 3-(4-hydroxyphenyl)propionate + ? <2> (<2> 83% of 4-hydroxyphenylacetate activity [10]) (Reversibility: ? <2> [10, 12, 16]) [10, 12, 16]

P 3-(3,4-dihydroxy)phenylpropionate + ? <2> [10, 12, 16]

S 3-hydroxyphenylacetate + NAD(P)H + O_2 <5, 7> (<7> 98% of 4-hydroxyphenylacetate activity [15]) (Reversibility: ? <5, 7> [6, 15]) [6, 15]

P 3,4-hydroxyphenylacetate + $NAD(P)^+$ + H_2O <5> [6]

S 3-hydroxyphenylacetate + NADH + O_2 <2, 3, 6> (<3> 82% activity of 4-hydroxyphenylacetate [9]) (Reversibility: ? <2, 3, 6> [2, 9, 13]) [2, 9, 13]

P 3,4-hydroxyphenylacetate + NAD^+ + H_2O <2, 3, 6> [2, 9, 13]

S 4-aminobenzoate + ? <2> (Reversibility: ? <2> [16]) [16]

P 4-amino-3-hydroxybenzoate + ? <2> [16]

S 4-aminophenylacetate + ? <2> (Reversibility: ? <2> [12, 16]) [12, 16]

P 4-amino-3-hydroxyphenylacetat + ? <2> [12, 16]

S 4-fluorophenylacetate + NADH + O_2 <2> (Reversibility: ? <2> [2]) [2]

P 4-fluoro-3-hydroxyphenylacetate + NAD^+ + H_2O <2> [2]

S 4-hydroxybenzoate + ? <6> (<6> 15% of 4-hydroxyphenylacetate activity [13]) (Reversibility: ? <6> [13]) [13]

P 3,4-dihydroxybenzoate + ? <6> [13]

S 4-hydroxyphenylacetate + NAD(P)H + O_2 <5> (<5> enzyme is induced by 4-hydroxyphenylacetate, no constitutive synthesis [6]) (Reversibility: ir <5> [6]) [6]

P 3,4-dihydroxyphenylacetate + $NAD(P)^+$ + H_2O <5> [6]

S 4-hydroxyphenylacetate + NADH + O_2 <1-4, 6, 7> (<2-4> enzyme is induced by 4-hydroxyphenylacetate, no constitutive synthesis [3-5,7]; <3> O_2 can be replaced by FADH [11]) (Reversibility: ? <1, 2, 6, 7> [1-3, 8-16]; ir <2-4> [4, 5, 7]) [1-5, 7-16]

P 3,4-dihydroxyphenylacetate + NAD^+ + H_2O <1-4, 6, 7> [1-5, 7-16]

S L-(3,4-dihydroxy)phenylalanine + ? <6> (<6> 10% of 4-hydroxyphenylacetate [13]) (Reversibility: ? <6> [13]) [13]

P ? <6> [13]

S L-tyrosine + ? <6> (<6> 8% of 4-hydroxyphenylacetate activity [13]) (Reversibility: ? <6> [13]) [13]

P L-(3,4-dihydroxy)phenylalanine + ? <6> [13]

S catechol + ? <6> (<6> 30% of 4-hydroxyphenylacetate activity [13]) (Reversibility: ? <6> [13]) [13]

P ?

S *p*-cresol + ? <3> (<3> 51% of 4-hydroxyphenylacetate activity [9]) (Reversibility: ? <3> [9]) [9]

P ?

S phenol + ? <3> (<3> for HpaB [14]) (Reversibility: ? <3> [8, 14]) [8, 14]

P catechol + ? <3> [8, 14]

S phenylacetic acid + ? <2, 7> (<2> 100% of 4-hydroxyphenylacetate activity [10]; <7> 97% of 4-hydroxyphenylacetate activity [15]) (Reversibility: ? <2, 7> [10, 15]) [10, 15]

P ?

S Additional information <2, 3, 6, 7> (<2> strict substrate specificity [3]; <3> 3- and 4-hydroxyphenylacetate are both catalyzed by 3- or 4-hydroxyphenylacetate hydroxylase by the same route with 3,4-dihydroxyphenylacetate as the first common intermediate, cells grown on one compound are fully induced for the catabolism of the other [4]; <3> rather broad substrate specificity, acting on mono- and dihydric phenols. Several substrates are below 50% of activity [9]; <2> broad substrate range. Several other substrates show activities below 50%. The length of the acetyl moiety is very important in substrate recognition [10]; <3> hydroxylation of aromatic compounds [11]; <6> dihydroxylated compounds are transformed, no reaction with phenylacetate, 4-chlorobenzoate or 2-hydroxyphenylacetate [13]; <3> hydroxylates phenol derivatives [14]; <7> a variety of aromatic compounds that contain a hydroxyl group in *para*-position can be hydroxylated. 4-hydroxyphenylacetate hydroxylase is an effector for C_1 and substrate for C_2 [15]; <2> substrates become hydroxylated at a position ortho to the hydroxyl group, 4-chlorophenylacetate, 4-fluorobenzoate and benzoate are not hydroxylated [16]) [3, 4, 9-11, 13-16]

P ?

Inhibitors

4-hydroxyphenylacetate <2> (<2> substrate inhibition by concentrations greater than 0.1 mM [16]) [16]

Cibacron blue F3GA <2> (<2> dye [2]) [2, 10]

Co^{2+} <2> [10]

Cu^{2+} <2> [2, 10]
EDTA <5> (<5> inhibits non-purified enzyme, ammonium sulfate precipitate [6]) [6]
FAD <7> (<7> inhibitory effect at more than 0.15 mM [15]) [15]
FMN <7> (<7> inhibitory effect at more than 0.15 mM [15]) [15]
Hg^{2+} <2> [2, 10]
N-ethylmaleimide <2> (<2> 48% inhibition at 1 mM [2]) [2, 10]
Zn^{2+} <2> [10]
iodoacetate <2> [10]
o-phenanthroline <1> (<1> complete inhibition at 1 mM [1]) [1]
p-chloromercuribenzoate <2> (<2> 92% inhibition at 1 mM [2]) [2]
Additional information <2> (<2> not inhibited by Mn^{2+}, Mg^{2+}, Fe^{3+} and Fe^{2+} [2]; <2> SH-groups in the active site of the enzyme suggested [10]) [2, 10]

Cofactors/prosthetic groups

FAD <2, 3, 6, 7> (<2> flavoprotein monooxygenase, absolute requirement [2,12]; <3> as redox chromophore [8]; <3> no stimulation [9]; <2> required for maximum activity, external flavoprotein monooxygenase [10]; <3> reduction of FAD is rate limiting [11]; <3,6> stimulates [13,14]; <7> alternative to FMN for C_1 [15]; <2> tightly associated with small component, native cofactor [15]) [2, 8-15]
$FADH_2$ <3, 7> (<3> used as substrate and cofactor, enzyme binds FADH in absence of 4-hydroxyphenlyacetate and protects it from rapid autoxidation by O_2 [11]) [11, 15]
FMN <3, 7> (<3> used as substrate by HpaC protein [14]; <7> native cofactor for C_1 component [15]) [14, 15]
$FMNH_2$ <7> [15]
NADH <2, 3, 6, 7> (<1,3> can not be replaced by NADPH [1,4]; <2> absolute requirement [2]; <3> dependent, not NADPH dependent [9]; <2> unable to use NADPH [10,13]; <3> electron donor [14]) [2, 4, 8-10, 13-16]
NADPH <3> [15]
riboflavin <3, 7> (<3> used as substrate by HpaC protein [14]; <7> component C_1 [15]) [14, 15]
Additional information <5> (<5> similar enzyme which oxidizes both NADH and NADPH [6]) [6]

Activating compounds

carbohydrates <2> (<2> stimulate synthesis of enzyme insignificantly [2]) [2]
Additional information <2> (<2> not activated by β-mercaptoethanol, glutathione and dithiothreiotol [2]) [2]

Metals, ions

Br^- <2> (<2> retards the oxidative half-reaction [12]) [12]
Cl^- <2> (<2> retards the oxidative half-reaction [12]) [12]
F^- <2> (<2> retards the oxidative half-reaction [12]) [12]
I^- <2> (<2> retards the oxidative half-reaction [12]) [12]
N_3^- <2> (<2> retards the oxidative half-reaction [12]) [12]

Turnover number (min^{-1})
50 <2> (4-hydroxyphenylacetate) [16]

Specific activity (U/mg)
0.026 <3> (<3> with 3-hydroxyphenylacetate as substrate [4]) [4]
0.028 <3> (<3> with 4-hydroxyphenylacetate as substrate [4]) [4]
0.14 <3> (<3> activity of HpaB protein in presence of NADH and HpaC protein [14]) [14]
0.231 <3> (<3> for 4-hydroxyphenylacetate oxidation [11]) [11]
0.46 <3> [14]
6.5 <2> [2]
8.89 <7> (<7> C_2 component [15]) [15]
197.3 <2> [10]
201 <7> (<7> C_1 component [15]) [15]
Additional information <2-5> (<2> maximal in enzymes of the late log phase [2]; <2,4> no functional enzyme and activity detected for mutant, wild-type shows activity only if exposed to 4-hydroxyphenylacetate [5,7]; <5> activity is reduced by about 40% if 3-hydroxyphenylacetate is used as substrate instead of 4-hydroxyphenylacetate [6]; <3> 120000 Da enzyme encoded by hpaB shows low hydroxylase activity which is increased in the presence of FAD and the coupling protein HpaC [8]) [2, 3, 5-8]

K_m-Value (mM)
0.0021 <3> (FMN) [14]
0.0026 <3> (riboflavin) [14]
0.0031 <3> (FAD) [14]
0.0042 <2> (FAD) [10]
0.012 <7> (NADH, <7> apparent, with FMN [15]) [15]
0.014 <7> (4-hydroxyphenylacetate, <7> apparent, with FAD [15]) [15]
0.019 <7> (4-hydroxyphenylacetate, <7> apparent, with FMN [15]) [15]
0.028 <7> (NADH, <7> apparent, with FAD [15]) [15]
0.038 <2> (4-hydroxyphenylacetate) [10]
0.04 <3> (NADH) [14]
0.041 <2> (NADH) [10]
0.059 <2> (NADH) [2]
0.16 <2> (phenylacetic acid) [10]
0.2 <2> (4-hydroxyphenylacetate) [2]

pH-Optimum
7 <3> (<3> assay at [11]; <3> optimal for HpaC protein [14]) [11, 14]
7.2 <1> (<1> assay at [1]) [1]
7.5 <2, 5, 7> (<5,7> assay at [2,6,15]) [2, 6, 10, 15]
7.8 <3> (<3> assay at [4,14]) [4, 14]
8 <3> (<3> assay at [8]) [8]

pH-Range
5-8 <3> (<3> more than 80% of activity [14]) [14]

Temperature optimum (°C)

22 <3> (<3> assay at [4,14]) [4, 14]
24 <3> (<3> assay at [11]) [11]
25 <7> (<7> assay at [15]) [15]
28 <5> (<5> assay at [6]) [6]
30 <1-3> (<1-3> assay at [1,2,8]) [1, 2, 8]
40 <2> [10]
45 <2> [2]

4 Enzyme Structure

Molecular weight

65000 <2> (<2> gel filtration [10]) [10]
73000 <7> (<7> component C1, gel filtration [15]) [15]
91000 <2> (<2> gel filtration [2]) [2]
120000 <3> (<3> gel filtration [8]) [8]
209000 <7> (<7> component C2, gel filtration [15]) [15]

Subunits

? <3, 6> (<3> x * 18679, SDS-PAGE, enzyme is needed for activity [8]; <6> x * 58781 flavoprotein HpaA, x * 18680, coupling protein, estimation from gene sequence [13,14]) [8, 13-14]
dimer <2, 3, 7> (<2> 2 * x, SDS-PAGE [2]; <3> 2 * 58781, homodimer, encodes the hydroxylase [8]; <3> 2 * 59000, SDS-PAGE [9]; <2> 2 * 32500, α_2 oligomeric structure, SDS-PAGE [10]; <7> x * 32000, small component, SDS-PAGE [15]; <2> 2 * 30750, small component [15]; <3> 2 * 19000 small component, 2 * 59000 large component [15]) [2, 8-10, 15]
monomer <2> (<2> 1 * 38500, large component [15]) [15]
tetramer <7> (<7> homotetramer, 4 * 50000, large component, SDS-PAGE [15]) [15]

5 Isolation/Preparation/Mutation/Application

Purification

<1> (protamine sulfate and ammonium sulfate [1]) [1]
<2> (ammonium sulfate precipitation, gel filtration, affinity chromatography [2]; ammonium sulfate precipitation, ion-exchange, gel filtration [10]) [2, 10, 12, 15]
<3> (affinity chromatography, gel filtration [8]; difficult because of the low stability of the enzyme in solution [9]; ammonium sulfate precipitation, ion exchange [11]; ammonium sulfate precipitation, ion-exchange, gel filtration. Affinity chromatography for the expressed cholin-binding domain containing HpaB protein [14]) [8, 9, 11, 14, 15]
<5> (ultracentrifugation, ammonium sulfate precipitation [6]) [6]
<6> (partially [13]) [13]
<7> (protamine sulfate precipitation, ion-exchange, gel filtration [15]) [15]

Cloning

<3> (expressed in Escherichia coli DH1 [8]; expression in Escherichia coli K-12 [9]; hpaB gene is expressed in Escherichia coli BL21(DE3) [11]; expressed in Escherichia coli DH1 and TG1 [14]) [8, 9, 11, 14]
<6> (HpaA and HpaH are expressed in Escherichia coli W-21, CC118 and YS1 as well as Klebsiella pneumonia mutant strain AG813 [13]) [13]

Engineering

Additional information <2-4, 6> (<2> several mutants are created, e.g. strains with enzyme defectives for analysis of the 4-hydroxyphenylacetate pathway [3]; <3> creation of 4-hydroxyphenylacetate-negative mutants by minimal salt medium with ethylmethanesulfonate [4]; <4> 3B-3 mutant deficient in the metabolism of 4-hydroxyphenylacetic acid are created by exposure to N-methyl-N'-nitro-N-nitrosoguanidine [5]; <2> hydroxylase deficient mutant P23X6 created by exposure to ethylmethane sulfonate [7]; <3> W21, 4-hydroxyphenylacetate deficient mutant [9]; <6> mutant AG813, defective in the 4-hydroxyphenylacetate hydroxylase [13]) [3-5, 7, 9, 13]

6 Stability

Temperature stability

4 <2> (<2> 15-20 days [2]) [2]

General stability information

<2>, unstable [2]
<3>, HpaC consists of low stability [14]
<3>, rather unstable in solution [9]

Storage stability

<2>, -20°C, stable for at least 5 months [10]
<2>, 4°C, 15-20 days [2]
<3>, -20°C, HpaC, small protein, very stable, no significant loss of activity during 2 months [14]
<3>, 4°C, 20 mM phosphate buffer, pH 8.0, concentrated with polyethylene glycol 20000 [8]

References

[1] Adachi, K.; Takeda, Y.; Senoh, S.; Kita, H.: Metabolism of *p*-hydroxyphenylacetic acid in Pseudomonas ovalis. Biochim. Biophys. Acta, **93**, 483-493 (1964)
[2] Raju, S.G.; Kamath, A.V.; Vaidyanathan, C.S.: Purification and properties of 4-hydroxyphenylacetic acid 3-hydroxylase from Pseudomonas putida. Biochem. Biophys. Res. Commun., **154**, 537-543 (1988)
[3] Barbour, M.G.; Bayly, R.C.: Control of *meta*-cleavage degradation of 4-hydroxyphenylacetate in Pseudomonas putida. J. Bacteriol., **147**, 844-850 (1981)

[4] Cooper, R.A.; Skinner, M.A.: Catabolism of 3- and 4-hydroxyphenylacetate by the 3,4-dihydroxyphenylacetate pathway in Escherichia coli. J. Bacteriol., **143**, 302-306 (1980)

[5] Barbour, M.G.; Bayly, R.C.: Regulation of the 4-hydroxyphenylacetic acid *meta*-cleavage pathway in an Acinetobacter sp. Biochem. Biophys. Res. Commun., **79**, 663-670 (1977)

[6] Blakley, E.R.: The catabolism of L-tyrosine by an Arthrobacter sp.. Can. J. Microbiol., **23**, 1128-1139 (1977)

[7] Barbour, M.G.; Bayly, R.C.: Regulation of the *meta*-cleavage of 4-hydroxyphenylacetic acid by Pseudomonas putida. Biochem. Biophys. Res. Commun., **76**, 565-571 (1977)

[8] Prieto, M.A.; Garcia, J.L.: Molecular characterization of 4-hydroxyphenylacetate 3-hydroxylase of Escherichia coli. A two-protein component enzyme. J. Biol. Chem., **269**, 22823-22829 (1994)

[9] Prieto, M.A.; Perez-Aranda, A.; Garcia, J.L.: Characterization of an Escherichia coli aromatic hydroxylase with a broad substrate range. J. Bacteriol., **175**, 2162-2167 (1993)

[10] Fernandez-Medarde, A.; Luengo, J.M.: Purification and characterization of the 4-hydroxyphenylacetic acid-3-hydroxylase from Pseudomonas putida U. FEMS Microbiol. Lett., **157**, 47-53 (1997)

[11] Xun, L.; Sandvik, E.R.: Characterization of 4-hydroxyphenylacetate 3-hydroxylase (HpaB) of Escherichia coli as a reduced flavin adenine dinucleotide-utilizing monooxygenase. Appl. Environ. Microbiol., **66**, 481-486 (2000)

[12] Arunachalam, U.; Massey, V.; Miller, S.M.: Mechanism of *p*-hydroxyphenylacetate-3-hydroxylase. A two-protein enzyme. J. Biol. Chem., **269**, 150-155 (1994)

[13] Gibello, A.; Suarez, M.; Allende, J.L.; Martin, M.: Molecular cloning and analysis of the genes encoding the 4-hydroxyphenylacetate hydroxylase from Klebsiella pneumoniae. Arch. Microbiol., **167**, 160-166 (1997)

[14] Galan, B.; Diaz, E.; Prieto, M.A.; Garcia, J.L.: Functional analysis of the small component of the 4-hydroxyphenylacetate 3-monooxygenase of Escherichia coli W: a prototype of a new flavin:NAD(P)H reductase subfamily. J. Bacteriol., **182**, 627-636 (2000)

[15] Chaiyen, P.; Suadee, C.; Wilairat, P.: A novel two-protein component flavoprotein hydroxylase. *p*-Hydroxyphenylacetate hydroxylase from Acinetobacter baumannii. Eur. J. Biochem., **268**, 5550-5561 (2001)

[16] Arunachalam, U.; Massey, V.: Studies on the oxidative half-reaction of *p*-hydroxyphenylacetate 3-hydroxylase. J. Biol. Chem., **269**, 11795-11801 (1994)

Melilotate 3-monooxygenase 1.14.13.4

1 Nomenclature

EC number
1.14.13.4

Systematic name
3-(2-hydroxyphenyl)propanoate,NADH:oxygen oxidoreductase (3-hydroxylating)

Recommended name
melilotate 3-monooxygenase

Synonyms
2-hydroxyphenylpropionate hydroxylase
2-hydroxyphenylpropionic hydroxylase
melilotate hydroxylase
melilotic hydroxylase
oxygenase, melilotate 3-mono-

CAS registry number
37256-72-7

2 Source Organism

<1> *Arthrobacter sp.* [1, 2, 4]
<2> *Pseudomonas sp.* [1, 3, 5-9]

3 Reaction and Specificity

Catalyzed reaction
3-(2-hydroxyphenyl)propanoate + NADH + H^+ + O_2 = 3-(2,3-dihydroxyphenyl)propanoate + NAD^+ + H_2O (<1,2> mechanism [1,5-7])

Reaction type
oxidation
redox reaction
reduction

Natural substrates and products
S 3-(2-hydroxyphenyl)propanoate + NADH + O_2 <1> (<1> i.e. melilotate, role in metabolism of coumarin [4]) (Reversibility: ir <1> [4]) [4]
P 3-(2,3-dihydroxyphenyl)propanoate + NAD^+ + H_2O <1> [4]

Substrates and products

S 3-(2-hydroxyphenyl)propanoate + NADH + O_2 <1, 2> (<1> i.e. melilotate, reversibility could not be demonstrated [4]) (Reversibility: ? <1, 2> [1-3, 5-9]; ir <1> [4]) [1-9]

P 3-(2,3-dihydroxyphenyl)propanoate + NAD^+ + H_2O <1, 2> (<1> reversibility could not be demonstrated [4]) [1-9]

S Additional information <1> (<1> high specificity, low activity with 3-hydroxyphenylpropanoate with 20% of melilotate reduction, phenylpropanoate with 1-2% of meliotate reduction [4]) [4]

P ?

Inhibitors

2,2'-dipyridyl <1> [4]
$FeCl_3$ <2> [3]
$FeSO_4$ <1> [4]
KCN <1> [4]
N-ethylmaleimide <1> [4]
p-chloromercuribenzoate <1> [4]
Additional information <1> (<1> not inhibitory: iodoacetic acid, iodoacetamide [4]) [4]

Cofactors/prosthetic groups

FAD <1, 2> (<1, 2> flavoprotein [1-5]; <1, 2> FAD: prosthetic group [1-3]; <2> 1 mol FAD per protein of MW 65000 [3]) [1-5]
NADH <1, 2> (<2> uses the re-face of the flavin ring [8]; <2> A-stereospecificity [9]) [1-9]

Turnover number (min^{-1})

232-269 <2> (3-(2-hydroxyphenyl)propanoate) [7]

Specific activity (U/mg)

12.5 <1> [2]
35.3 <2> [3]

K_m-Value (mM)

0.0013 <2> (melilotate, <2> overview [1]) [1, 5]
0.0027 <2> (3-(2-hydroxyphenyl)propanoate) [7]
0.0047 <2> (NADH, <2> overview [1]) [1, 5]
0.02 <2> (O_2, <2> stopped flow [7]) [7]
0.05 <2> (O_2, <2> overview [1]) [1, 5]
0.086 <1> (melilotate) [4]
0.091 <1> (NADH) [4]

pH-Optimum

7-7.5 <1> [4]

pH-Range

5-9 <1> (<1> at pH 5.0 and 9.0: about 30% of maximum activity [4]) [4]

Temperature optimum (°C)

25 <2> (<2> assay at [3]) [3]
30 <1> (<1> assay at [2, 4]) [2, 4]

4 Enzyme Structure

Molecular weight

65000 <1> (<1> sucrose density gradient centrifugation, gel filtration [2]) [1, 2]
238000-250000 <2> (<2> gel filtration, ultracentrifugation [3]) [1, 3]

Subunits

tetramer <2> (<2> 4 * 65000, SDS-PAGE [3]) [1, 3]

5 Isolation/Preparation/Mutation/Application

Purification

<1> (partial [4]) [2, 4]
<2> [3]

Renaturation

<1> (activity can be restored with FAD [4]) [4]

6 Stability

Oxidation stability

<2>, enzyme is rapidly reduced by irradiation with visible light in presence of EDTA or by dithionite [3]

Storage stability

<1>, -20°C, stable for 1 week [4]
<1>, -20°C, stable for 3 weeks, highly purified enzyme, concentrated by dialysis [2]
<1>, -70°C, stable for 1 week [2]

References

[1] Flashner, M.S.; Massey, V.: Flavoprotein oxygenases. Mol. Mech. Oxygen Activ. (Hayaishi, O., ed.) Academic Press, New York, 245-283 (1974)
[2] Levy, C.C.: Melilotate hydroxylase. Purification of the enzyme and the nature of the prosthetic group. J. Biol. Chem., **242**, 747-753 (1967)
[3] Strickland, S.; Massey, V.: The purification and properties of the flavoprotein melilotate hydroxylase. J. Biol. Chem., **248**, 2944-2952 (1973)
[4] Levy, C.C.; Frost, P.: The metabolism of coumarin by a microorganism. V. Melilotate hydroxylase. J. Biol. Chem., **241**, 997-1003 (1966)
[5] Strickland, S.; Massey, V.: The mechanism of action of the flavoprotein melilotate hydroxylase. J. Biol. Chem., **248**, 2953-2962 (1973)

[6] Strickland, S.; Schopfer, L.M.; Massey, V.: Kinetic and mechanistic studies on the reaction of melilotate hydroxylase with deuterated melilotate. Biochemistry, **14**, 2230-2235 (1975)
[7] Schopfer, L.M.; Massey, V.: Kinetic and mechanistic studies on the reduction of melilotate hydroxylase by reduced pyridine nucleotides. J. Biol. Chem., **254**, 10634-10643 (1979)
[8] Manstein, D.J.; Pai, E.F.: Absolute stereochemistry of flavins in enzyme-catalyzed reactions. Biochemistry, **25**, 6807-6816 (1986)
[9] You, K.S.; Arnold, L.J.; Kaplan, N.O.: The stereospecificity of bacterial external flavoprotein monooxygenases for nicotinamide adenine dinucleotide. Arch. Biochem. Biophys., **180**, 550-554 (1977)

Imidazoleacetate 4-monooxygenase 1.14.13.5

1 Nomenclature

EC number
1.14.13.5

Systematic name
4-imidazoleacetate,NADH:oxygen oxidoreductase (5-hydroxylating)

Recommended name
imidazoleacetate 4-monooxygenase

Synonyms
EC 1.14.1.5 (formerly)
imidazoleacetate hydroxylase
imidazoleacetic hydroxylase
imidazoleacetic monooxygenase

CAS registry number
9029-65-6

2 Source Organism

<1> *Pseudomonas sp.* [1-6]

3 Reaction and Specificity

Catalyzed reaction
4-imidazoleacetate + NADH + H^+ + O_2 = 5-hydroxy-4-imidazoleacetate + NAD^+ + H_2O

Reaction type
oxidation
redox reaction
reduction

Natural substrates and products
S 4-imidazoleacetate + NADH + O_2 <1> (<1>, the enzyme is part of the histidine catabolic pathway in which imidazaloneacetate is converted to aspartic acid by way of formiminoaspartic acid [2,6]) (Reversibility: ? <1> [1-6]) [1-6]
P 5-hydroxy-4-imidazoleacetate + NAD^+ + H_2O <1> [1-6]

Substrates and products

S 4-imidazoleacetate + NAD(P)H + O_2 + H^+ <1> (<1>, highly specific for 4-imidazoleacetate [1]; <1>, electrons can be trasferred from NADH to 2,6-dichloroindophenol in the presence of the enzyme [4]) (Reversibility: ? <1> [1-5]) [1-5]

P 5-hydroxy-4-imidazoleacetate + $NAD(P)^+$ + H_2O <1> [1-5]

S 5-hydroxyindoleacetate + NAD(P)H + O_2 + H^+ <1> (<1>, 9.4% relative activity to 4-imidazoleacetate [1]) (Reversibility: ? <1> [1]) [1]

P ? + $NAD(P)^+$ + H_2O <1> [1]

S N-methylimidazoleacetate + NAD(P)H + O_2 + H^+ <1> (<1>, 7.6% relative activity to 4-imidazoleacetate [1]) (Reversibility: ? <1> [1]) [1]

P 5-hydroxy-N-methylimidazoleacetate + $NAD(P)^+$ + H_2O <1> [1]

S imidazolelactate + NADH + O_2 + H^+ <1> (<1>, 4% relative activity to 4-imidazoleacetate [3]) (Reversibility: ? <1> [3, 4]) [3, 4]

P 5-hydroxyimidazolelactate + NAD^+ + H_2O <1> [3, 4]

S imidazolepropionate + NAD(P)H + O_2 + H^+ <1> (<1>, 9.5% relative activity to 4-imidazoleacetate [1]; <1>, 10% relative activity to 4-imidazoleacetate [3]) (Reversibility: ? <1> [1, 3, 4]) [1, 3, 4]

P 5-hydroxyimidazolepropionate + $NAD(P)^+$ + H_2O <1> [1, 3, 4]

S imidazolepyruvate + O_2 + NAD(P)H + H^+ <1> (<1>, 2.8% relative activity to 4-imidazoleacetate [1]) (Reversibility: ? <1> [1]) [1]

P 5-hydroxyimidazolepyruvate + $NAD(P)^+$ + H_2O <1> [1]

S indoleacetate + NAD(P)H + O_2 + H^+ <1> (<1>, 7.3% relative activity to 4-imidazoleacetate [1]) (Reversibility: ? <1> [1]) [1]

P 5-hydroxyindoleacetate + $NAD(P)^+$ + H_2O <1> [1]

S sodium diethyldithiocarbamate + NADH + O_2 <1> (Reversibility: ? <1> [3, 4]) [3, 4]

P ? + NAD^+ + H_2O <1> (<1>, reaction product not identified [3,4]) [3, 4]

Inhibitors

5-hydroxyindoleacetate <1> (<1>, 11% inhibition at 4.5 mM [1]) [1]
N-methylimidazoleacetate <1> (<1>, 11% inhibition at 4.5 mM [1]) [1]
dihydroxymandelate <1> (<1>, 53% inhibition at 4.5 mM [1]) [1]
dihydroxyphenylacetate <1> (<1>, 66% inhibition at 4.5 mM [1]) [1]
ethyl-mercuric chloride <1> [5]
homogentisate <1> (<1>, 50% inhibition at 4.5 mM [1]) [1]
homovanillic acid <1> (<1>, 5% inhibition at 4.5 mM [1]) [1]
imidazolepyruvate <1> (<1>, 56% inhibition at 4.5 mM [1]) [1]
indoleacetate <1> (<1>, 22% inhibition at 4.5 mM [1]; <1>, 30% inhibition at 10 mM [4]) [1, 4]
m-hydroxyphenylacetate <1> (<1>, 79% inhibition at 4.5 mM [1]) [1]
mercury compounds <1> (<1>, maximum inactivation at a ratio about 2 mol of $AgNO_3$ per mol of enzyme [5]) [3-5]
methyl-mercuric chloride <1> [5]
nicotinate <1> (<1>, 65% inhibition at 20 mM [4]) [4]
p-chloromercuribenzoate <1> (<1>, maximum inactivation at a ratio about 2 mol of *p*-chloromercuribenzoate per mole of enzyme [5]) [2-5]

p-hydroxybenzoate <1> (<1>, 38% inhibition at 4.5 mM [1]) [1]
p-hydroxyphenylacetate <1> (<1>, 47% inhibition at 4.5 mM [1]) [1]
phenyl-mercuric acetate <1> [5]
phenylacetate <1> (<1>, 25% inhibition at 4.5 mM [1]; <1>, 79% inhibition at 20 mM [4]) [1, 4]
silver compounds <1> [3-5]
sodium mersalyl <1> [5]
vanillyl mandelic acid <1> (<1>, 9% inhibition at 4.5 mM [1]) [1]

Cofactors/prosthetic groups

FAD <1> (<1>, FAD is the only prosthetic group [2]; <1>, 1 mol per mol enzyme [2-6]; <1>, cannot be replaced by FMN or riboflavin [4,6]) [2-6]
NADH <1> [1-6]
NADPH <1> (<1>, less efficient than NADH [2, 3]; <1>, β-NADPH is approximately 13% as active as β-NADH [3, 4]; <1>, α-NADH cannot replace β-NADH at 0.2 mM [4]) [2-4]
Additional information <1> (<1>, the enzyme contains 2 mol of titratable suflhydryl groups, one of which is essential, possibly as a substrate-binding site, for the oxygenation of imidazoleacetate [5]) [5]

Turnover number (min^{-1})

2260 <1> (imidazoleacetate) [4]

Specific activity (U/mg)

25 <1> [3, 4, 6]
Additional information <1> [1]

K_m-Value (mM)

0.002 <1> (NADH) [3]
0.002-0.01 <1> (NADH) [3, 4]
0.01 <1> (NADH, <1>, spectrophotometric method [4]) [4]
0.01 <1> (sodium diethyldithiocarbamate, <1>, polarographic method [4]) [4]
0.015 <1> (NADH, <1>, substrate: sodium diethyldithiocarbamate, polarographic method [4]) [4]
0.02 <1> (O_2, <1>, substrate: imidazoleacetate or sodium diethyldithiocarbamate, polarographic method [4]) [3, 4]
0.03 <1> (NADH, <1>, polarographic method [4]) [4]
0.3 <1> (imidazoleacetate, <1>, spectrophotometric method [4]) [3, 4]
0.48 <1> (imidazoleacetate) [1]
0.5 <1> (imidazoleacetate, <1>, polarographic method [4]) [4]
23 <1> (5-hydroxyindoleacetate) [1]
44 <1> (N-methylimidazoleacetate) [1]
64 <1> (indoleacetate) [1]

pH-Optimum

9 <1> [3, 4]

4 Enzyme Structure

Molecular weight

87000-90000 <1> (<1>, sedimentation equilibrium, gel filtration [2-6]) [2-6]

5 Isolation/Preparation/Mutation/Application

Purification

<1> (partial, using ammonium sulfate precipitation and column chromatography on DEAE-Sephadex A-50 [1]; using protamine sulfate treatment, ammonium sulfate fractionation, chromatography on TEAE-cellulose column, calcium phosphate gel treatment and chromatography on DEAE-Sephadex column [3, 4, 6];) [1, 3, 4, 6]

Crystallization

<1> [2-6]

6 Stability

Temperature stability

20 <1> (<1>, 2% loss of activity, 0.1 M potassium phosphate, pH 7.2, 5 min [4]) [4]

30 <1> (<1>, 50% loss of activity, 0.1 M potassium phosphate, pH 7.2, 5 min [3,4]) [3, 4]

40 <1> (<1>, 95% loss of activity, 0.1 M potassium phosphate, pH 7.2, 5 min [4]) [4]

Storage stability

<1>, -20°C, potassium phosphate buffer, pH 7.2, no loss of activity for at least 1 month [3, 4]

<1>, -20°C, stable for at least 6 months [1]

<1>, 4°C, 5-100 mM potassium phosphate buffer, pH 7.2, almost full activity retained for several days [4]

References

[1] Watanabe, T.; Kambe, H.; Imamura, I.; Taguchi, Y.; Tamura, T.; Wada, H.: Method for enzymatic determination of imidazole acetic acid. Anal. Biochem., **130**, 321-327 (1983)

[2] Flashner, M.S.; Massey, V.: Flavoprotein oxygenases. Mol. Mech. Oxygen Activ. (Hayaishi, O., ed.) Academic Press, New York, 245-283 (1974)

[3] Nozaki, M.: Imidazolacetate monooxygenase (Pseudomonas). Methods Enzymol., **17B**, 773-777 (1971)

[4] Maki, Y.; Yamamoto, S.; Nozaki, M.; Hayaishi, O.: Studies on monooxygenases. II. Crystallization and some properties of imidazole acetate monooxygenase. J. Biol. Chem., **244**, 2942-2950 (1969)
[5] Okamoto, H.; Nozaki, M.; Hayaishi, O.: A role of sulfhydryl groups in imidazoleacetate monooxygenase. Biochem. Biophys. Res. Commun., **32**, 30-36 (1968)
[6] Maki, Y.; Yamamoto, S.; Nozaki, M.; Hayaishi, O.: Crystallization of imidazolacetate monooxygenase and its characterization as flavoprotein. Biochem. Biophys. Res. Commun., **25**, 609-614 (1966)

Orcinol 2-monooxygenase 1.14.13.6

1 Nomenclature

EC number
1.14.13.6

Systematic name
orcinol,NADH:oxygen oxidoreductase (2-hydroxylating)

Recommended name
orcinol 2-monooxygenase

Synonyms
orcinol hydroxylase
oxygenase, orcinol 2-mono

CAS registry number
37217-34-8

2 Source Organism

<1> *Aspergillus niger* [1]
<2> *Pseudomonas fluorescens* [2]
<3> *Pseudomonas putida* (01 [2-6, 8]; ORC [7]) [2-7, 8, 9]

3 Reaction and Specificity

Catalyzed reaction
orcinol + NADH + H^+ + O_2 = 2,3,5-trihydroxytoluene + NAD^+ + H_2O

Reaction type
oxidation
redox reaction
reduction

Natural substrates and products
S orcinol + NADH + O_2 <1> (Reversibility: ? <1> [1]) [1]
P 2,3,5-trihydroxytoluene + NAD^+ + H_2O <1> [1]

Substrates and products
S 3-chlorophenol + NADH + O_2 <3> (<3> 33% of activity with orcinol [7]) (Reversibility: ? <3> [7]) [7]
P ?

S 3-cresol + NADH + O_2 <3> (<2, 3> nonsubstrate effector, which increases NADH oxidase activity without being hydroxylated [2, 8, 9]; <3> no substrate [4, 8]; <3> 20 to 70% coupled in various assays [5]; <3> 11-14% of the activity with orcinol [7]) (Reversibility: ? <3> [5, 7]) [5, 7]
P 3-methylcatechol + NAD + H_2O <3> [5, 7]
S 3-ethylphenol + NADH + O_2 <3> (<3> 30% of activity with orcinol [7]; <3> no substrate, mimic orcinol as effector [8]) (Reversibility: ? <3> [7]) [7]
P ?
S 3-trifluoromethylphenol + NADH + O_2 <3> (<3> 94% of activity with orcinol [7]) (Reversibility: ? <3> [7]) [7]
P ?
S 4-bromoresorcinol + NADH + O_2 <3> (Reversibility: ? <3> [8]) [8]
P ?
S 4-methylresorcinol + NADH + O_2 <3> (Reversibility: ? <3> [8]) [8]
P ?
S orcinol + NADH + O_2 <1-3> (<2,3> highly specific for orcinol, i.e. 5-methyl-1,3-benzenediol [2,4]) (Reversibility: ? <1-3> [1-9]) [1-9]
P 2,3,5-trihydroxytoluene + NAD^+ + H_2O <1-3> [1-9]
S resorcinol + NADH + O_2 <2, 3> (<3> no substrate [4]; <2> behaves both as a substrate and a nonsubstrate effector [2]; <2,3> hydroxylated to a limited extent [2,9]; <3> the enzyme processes resorcinol to hydroxylated product 66% of the time [5]; <3> 23% of activity with orcinol [7]) (Reversibility: ? <2, 3> [2, 5-9]) [2, 5-9]
P hydroxyquinol + NAD + H_2O <2, 3> [2, 5-9]
S Additional information <3> (<3> reduced orcinol hydroxylase is able to transfer reducing equivalents to a variety of electron acceptors, other than oxygen. These include free FAD, ferricyanide, cytochrome c, acetylpyridine-NAD and tetrazolium salts [8]) [7, 8]
P ?

Inhibitors

1,3-dimethoxybenzene <3> (<3> 45% inhibition at 1.7 mM [8]) [8]
2-hydroxy-4-methoxybenzoate <3> (<3> 68% inhibition at 1.7 mM [8]) [8]
2-methylresorcinol <3> (<3> 27% inhibition at 1.7 mM [8]) [8]
2-nitroorcinol <3> (<3> 64% inhibition at 1.7 mM [8]) [8]
3,4-dimethylphenol <3> (<3> 54% inhibition at 1.7 mM [8]) [8]
3,5-dihydroxybenzoate <3> (<3> 59% inhibition at 1.7 mM [8]) [8]
3,5-dimethylphenol <3> (<3> 68% inhibition at 1.7 mM [8]) [8]
3-hydroxybenzaldehyde <3> (<3> 23% inhibition at 1.7 mM [8]) [8]

Cofactors/prosthetic groups

3-acetyl pyridine nucleotide <2, 3> (<2,3> reduced [2,8,9]) [2, 8, 9]
FAD <2, 3> (<2,3> flavoprotein [2,3,7-9]; <3> 1 mol of FAD per mol of protein [2,3,7,8]) [2, 3, 7-9]
NADH <1-3> (<2,3> the best donor [2]; <3> 4R stereospecificity with respect to dihydronicotinamide oxidation with the substrates: orcinol, resorcinol and *m*-cresol [5]; <3> mixed type 4R,4S stereospecificity with respect to

dihydronicotinamide oxidation with the substrates *m*-cresol and resorcinol, 4R chirality with orcinol [6]; <3> 4R stereospecificity for the transfer of hydride, substrate: orcinol [9]) [1-3, 5-9]
NADPH <2, 3> [2, 7-9]

Activating compounds
m-cresol <2, 3> (<2,3> increases activity [2,8]) [2, 8]

Turnover number (min^{-1})
1560 <3> (orcinol, <3> crystalline enzyme [8]) [8]

Specific activity (U/mg)
2.8 <3> (<3> apoenzyme without flavin addition [8]) [8]
3.5 <3> (<3> apoenzyme with addition of FMN [8]) [8]
11.7 <3> [7]
12.9 <3> [3]
14.7 <3> (<3> apoenzyme with addition of FAD [8]) [8]
18 <3> (<3> holoenzyme [8]) [8]
24 <3> [8]
37 <3> [5]
Additional information <1, 3> (<1> at different time intervals [1]) [1, 4]

K_m-Value (mM)
0.026 <3> (orcinol) [5]
0.03 <3> (orcinol, <3> cofactor: NADH [8]) [8]
0.05 <3> (orcinol, <3> cofactor: 3-acetylpyridine NADH [8]) [8]
0.07 <3> (O_2) [8]
0.085 <3> (orcinol, <3> cofactor: NADPH [8]) [8]
0.13 <3> (NADH, <3> substrate: orcinol [8]) [8]
0.17 <3> (3-cresol) [5]
0.17 <3> (NADH) [5]
0.19 <3> (resorcinol) [8]
0.22 <3> (resorcinol) [5]
0.65 <3> (3-acetylpyridine NADH, <3> substrate: orcinol [8]) [8]
2.5 <3> (NADH, <3> substrate: resorcinol [8]) [8]
Additional information <3> [8]

pH-Optimum
6.8 <3> (<3> assay at [3, 5, 7, 8]) [3, 5, 7, 8]
Additional information <3> (<3> difficult to evaluate, because the values do not take into account the nonenzymic oxidation rate of the product of the reaction [8]) [8]

Temperature optimum (°C)
30 <3> (<3> assay at [7]) [7]

4 Enzyme Structure

Molecular weight

60000-70000 <3> (<3> gel filtration [2]) [2]
63000 <3> (<3> gel filtration [8]) [8]
65000 <3> (<3> sedimentation equilibrium [8]) [8]
68000 <3> (<3> gel filtration [7]) [7]
70000 <3> (<3> ultracentrifugation [3]) [3]

Subunits

monomer <3> (<3> 1 * 60000-70000 [2]; <3> 1 * 70000, SDS-PAGE [7]; <3> 1 * 68000, SDS-PAGE [8]) [2, 7, 8]

5 Isolation/Preparation/Mutation/Application

Purification

<3> (using protamine sulfate treatment, column chromatography on DEAE-cellulose, Sephadex G-100, Sephadex G-75, a second DEAE-cellulose column chromatography and crystallization [3]; ORC, using protamine sulfate treatment, column chromatography on DEAE-cellulose, Sephadex G-100, Sephadex G-75 and a second DEAE-cellulose column chromatography [7]; using protamine sulfate treatment, DEAE-cellulose column chromatography, ammonium sulfate treatment and column chromatography on Sephadex G-100 followed by ammonium sulfate treatment and column chromatography on hydroxylapatite [8]) [3, 7, 8]

Renaturation

<3> (most of the activity of the apoenzyme is reconstituted by the addition of FAD, FMN is a poor substitute for FAD [8]) [8]

Crystallization

<3> (in 2-3 days by addition of ammonium sulfate until turbidity [3]) [2, 3, 8]

6 Stability

pH-Stability

7 <3> (<3> highest stability [8]) [8]

General stability information

<3>, thiol reagents, e.g. 2-mercaptoethanol, stabilize [8]

Storage stability

<3>, 4°C, stabilized by thiol reagents, e.g. 2-mercaptoethanol, some days [8]

References

[1] Sahasrabudhe, S.R.; Lala, D.; Modi, V.V.: Degradation of orcinol by Aspergillus niger. Can. J. Microbiol., **32**, 535-538 (1986)
[2] Flashner, M.S.; Massey, V.: Flavoprotein oxygenases. Orcinol hydroxylase. Mol. Mech. Oxygen Activ. (Hayaishi, O., ed.) Academic Press, New York, 245-283 (1974)
[3] Otha, Y.; Ribbons, D.W.: Crystallization of orcinol hydroxylase from Pseudomonas putida. FEBS Lett., **11**, 189-192 (1970)
[4] Ribbons, D.W.; Ohta, Y.; Higgins, I.J.: Specificity of a catabolic pathway–a lesson learned from indirect assays. J. Bacteriol., **106**, 702-703 (1971)
[5] Ryerson, C.C.; Walsh, C.: The stereochemistry of NADH utilization by the flavoenzyme monooxygenase orcinol hydroxylase. J. Biol. Chem., **254**, 4349-4351 (1979)
[6] Ribbons, D.W.; Ohta, Y.; Higgins, I.J.: Electron transport in a flavoprotein, orcinol hydroxylase. Mol. Basis of Electron Transp. (Proc. Miami Winter Symp., Schultz, J., Cameron, B.F., eds.), **4**, 251-274 (1972)
[7] Ohta, Y.; Ribbons, D.W.: Bacterial metabolism of resorcinylic compounds: purification and properties of orcinol hydroxylase and resorcinol hydroxylase from Pseudomonas putida ORC. Eur. J. Biochem., **61**, 259-269 (1976)
[8] Ohta, Y.; Higgins, I.J.; Ribbons, D.W.: Metabolism of resorcinylic compounds by bacteria. Purification and properties of orcinol hydroxylase from Pseudomonas putida 01. J. Biol. Chem., **250**, 3814-3825 (1975)
[9] Higgins, I.J.; Ribbons, D.W.: Stereospecificity of hydride transfer in the flavin mono-oxygenase orcinol hydroxylase. Biochem. J., **127**, 65P (1972)

Phenol 2-monooxygenase 1.14.13.7

1 Nomenclature

EC number
1.14.13.7

Systematic name
phenol,NADPH:oxygen oxidoreductase (2-hydroxylating)

Recommended name
phenol 2-monooxygenase

Synonyms
oxygenase, phenol 2-mono-
phenol hydroxylase
phenol *o*-hydroxylase

CAS registry number
37256-84-1

2 Source Organism

<1> *Trichosporon cutaneum* (yeast [1, 4, 7, 10, 19]) [1, 3-10, 13, 15, 17-19, 20, 22-25]
<2> *Brevibacterium fuscum* [2]
<3> *Candida tropicalis* [11]
<4> *Rhodococcus sp.* (strain P1 [12]) [12]
<5> *Pseudomonas sp.* (strain CF600 [14]) [14]
<6> *Ralstonia eutropha* (strain E2, formerly Alcaligenes sp. [16]) [16]
<7> *Acinetobacter radioresistens* [21]

3 Reaction and Specificity

Catalyzed reaction
phenol + NADPH + H^+ + O_2 = catechol + $NADP^+$ + H_2O

Reaction type
oxidation
redox reaction
reduction

Natural substrates and products

S phenol + NADPH + O_2 <3, 4> (<3> enzyme of phenol degradation pathway [11]; <4> first step of phenol degradation [12]) (Reversibility: ? <3, 4> [11, 12]) [11, 12]

P ?

Substrates and products

S 2,3,4-trifluorophenol + O_2 + NADPH <1> (Reversibility: ? <1> [19]) [19]

P ?

S 2,3,5,6-tetrafluorophenol + O_2 + NADPH <1> (Reversibility: ? <1> [19]) [19]

P ?

S 2,3,5,6-tetrafluorophenol + O_2 + NADPH <1> (Reversibility: ? <1> [6]) [6]

P 3,4,6-trifluoro-2-benzoquinone + $NADP^+$ + F^- <1> [6]

S 2,3,5-trifluorophenol + O_2 + NADPH <1> (Reversibility: ? <1> [19]) [19]

P ?

S 2,3,6-trifluorophenol + O_2 + NADPH <1> (Reversibility: ? <1> [19]) [19]

P ?

S 2,3-difluorophenol + O_2 + NADPH <1> (Reversibility: ? <1> [19]) [19]

P ?

S 2,4-difluorophenol + O_2 + NADPH <1> (Reversibility: ? <1> [19]) [19]

P ?

S 2,5-difluorophenol + O_2 + NADPH <1> (Reversibility: ? <1> [19]) [19]

P ?

S 2,6-difluorophenol + O_2 + NADPH <1> (Reversibility: ? <1> [19]) [19]

P 3

S 2-amino-3-methylphenol + O_2 + NADPH <2> (Reversibility: ? <> []) [2]

P ?

S 2-aminophenol + O_2 + NADPH <1-4> (Reversibility: ? <1-4> [1, 2, 11, 12]) [1, 2, 11, 12]

P ?

S 2-chlorophenol + O_2 + NADPH <1-4> (Reversibility: ? <1-4> [1, 2, 7, 11, 12]) [1, 2, 7, 11, 12]

P ?

S 2-fluorophenol + O_2 + NADPH <1> (Reversibility: ? <1> [1, 7, 13, 19]) [1, 7, 13, 19]

P ?

S 2-methylyphenol + O_2 + NADPH <1, 2, 3, 4> (<2, 3> i.e. *o*-cresol [2,11]) (Reversibility: ? <1, 2, 3, 4> [2, 7, 11, 12]) [2, 7, 11, 12]

P ?

S 3,4,5-trifluorophenol + O_2 + NADPH <1> (Reversibility: ? <1> [19]) [19]

P ?

S 3,4-difluorophenol + O_2 + NADPH <1> (Reversibility: ? <1> [19]) [19]

P ?

S 3,5-difluorophenol + O_2 + NADPH <1> (Reversibility: ? <1> [19]) [19]

P ?

S 3-aminophenol + O_2 + NADPH <1-4> (Reversibility: ? <1-4> [1, 2, 6, 11, 12]) [1, 2, 6, 11, 12]

P ?

S 3-chloro-4-fluorophenol + O_2 + NADPH <1> (Reversibility: ? <1> [19]) [19]

P ?

S 3-chlorophenol + O_2 + NADPH <1-4> (Reversibility: ? <1-4> [1, 2, 6, 7, 11, 12, 13]) [1, 2, 6, 7, 11, 12, 13]

P ?

S 3-fluorophenol + O_2 + NADPH <1> (<1> below pH 6.5 3-fluorophenol is preferentially hydroxylated at the C_6 ortho position, at increasing pH the C_2 *ortho*-hydroxylation becomes more predominant [20]) (Reversibility: ? <1> [1, 7, 13, 19, 20]) [1, 7, 13, 19, 20]

P ?

S 3-hydroxyphenol + O_2 + NADPH <1> (Reversibility: ? <1> [6]) [6]

P ?

S 3-methylyphenol + O_2 + NADPH <1, 2, 3, 4> (<1, 2, 3> i.e. *m*-cresol [2, 25, 11]) (Reversibility: ? <1, 2, 3, 4> [2, 6, 7, 11, 13, 25, 12]) [2, 6, 7, 11, 13, 25, 12]

P ?

S 4-aminophenol + O_2 + NADPH <1-4> (Reversibility: ? <1-4> [1, 2, 11, 12]) [1, 2, 11, 12]

P ?

S 4-chloro-3-fluorophenol + O_2 + NADPH <1> (Reversibility: ? <1> [19]) [19]

P ?

S 4-chlorocatechol + O_2 + NADPH <3> (Reversibility: ? <3> [11]) [11]

P ?

S 4-chlorophenol + O_2 + NADPH <1-4> (Reversibility: ? <1-4> [1, 2, 7, 11, 12, 13, 17]) [1, 2, 7, 11, 12, 13, 17]

P ?

S 4-fluorophenol + O_2 + NADPH <1> (Reversibility: ? <1> [1, 7, 13, 17, 19]) [1, 7, 13, 17, 19]

P ?

S 4-hydroxyphenol + O_2 + NADPH <1> (Reversibility: ? <1> [6]) [6]

P ?

S 4-methylyphenol + O_2 + NADPH <1, 2, 3, 4> (<3> i.e. *p*-cresol [11]) (Reversibility: ? <1, 2, 3, 4> [2, 7, 11, 12, 13, 17]) [2, 7, 11, 12, 13, 17]

P ?

S catechol + O_2 + NADPH <1, 3> (<4> not [12]) (Reversibility: ? <1, 3> [7, 11, 13]) [7, 11, 13]

P ?

S metol + O_2 + NADPH <3> (Reversibility: ? <3> [11]) [11]

P ?

S orcinol + O_2 + NADPH <2, 3> (<3> weak [11]) (Reversibility: ? <2, 3> [2, 11]) [2, 11]

P ?

S pentafluorophenol + O_2 + NADPH <1> (Reversibility: ? <1> [19]) [19]
P ?
S phenol + NADPH + O_2 <1-6> (<1> reaction mechanism [5]; <7> cytochrome c, 2,6-dichlorophenolindophenol, potassium ferricyanide and nitro blue tetrazolium can act as electron acceptors in vitro [21]) (Reversibility: ? <1-6> [1-7, 10-13, 14, 16, 17, 21, 25]) [1-7, 10-13, 14, 16, 17, 21, 25]
P catechol + $NADP^+$ + H_2O <1, 3> [4, 11, 17, 21]
S phloroglucinol + O_2 + NADPH <2-4> (Reversibility: ? <2-4> [2, 11, 12]) [2, 11, 12]
P ?
S pyrogallol + O_2 + NADPH <4> (Reversibility: ? <4> [12]) [12]
P ?
S quinol + O_2 + NADPH <1, 3, 4> (Reversibility: ? <1, 3, 4> [1, 11, 12]) [1, 11, 12]
P 1,2,4-trihydroxybenzene + $NADP^+$ + H_2O
S resorcinol + NADPH + O_2 <1-4> (<1> reaction mechanism [18]) (Reversibility: ? <1-4> [2, 7, 11, 12, 13, 17, 18, 25]) [2, 7, 11, 12, 13, 17, 18, 25]
P ?
S thiophenol + O_2 + NADPH <1> (Reversibility: ? <1> [6]) [6]
P ?
S Additional information <1-4> (<1> broad specificity, reaction results in the formation of the corresponding *o*-diols [1]; <3> broad specificity [11]; <2> not: *p*-hydroxybenzoic acid [2]; <2, 3> not: salicylic acid [2, 11]; <2> not: *p*-hydroxyphenylacetic acid [2]; <2> not: 2,4-, 2,5- and 2,6-dimethylphenols [2]; <4> not: 3-nitrophenol, 4-nitrophenol [12]; <1> overview of possible reaction products of fluorinated phenols [19]) [1, 2, 11, 12, 19]
P ?

Inhibitors

1,10-phenanthroline <1> (<1> slight inhibition at 0.0005 to 0.001 mM [1]) [1]
2-fluorophenol <1> (<1> substrate inhibition [13]) [13]
3-chlorophenol <1> (<1> substrate inhibition [13]) [13]
3-fluorophenol <1> (<1> substrate inhibition [13]) [13]
4-chlorophenol <1> (<1> substrate inhibition [13]) [13]
4-fluorophenol <1> (<1> substrate inhibition [13]) [13]
5,5'-dithiobis(2-nitrobenzoate) <1> [3]
$AgNO_3$ <1> (<1> 0.01 mM [1]) [1]
Br^- <1> (<1> 50% inhibition at 0.072 M [8]) [8]
CN^- <1> (<1> 50% inhibition at 0.004 M [8]) [8]
Cl^- <1> (<1> 90% inhibition at 0.1 M, irreversible [1]; <1> 50% inhibition at 0.022 M [8]) [1, 5, 8]
$CuSO_4$ <1> (<1> 0.01 mM [1]) [1]
F^- <1> (<1> 50% inhibition at 0.01 M [8]) [8]
$FeSo_4$ <1> (<1> 0.1 mM [1]) [1]
H_2O_2 <1> (<1> 71% inhibition at 0.1 M [1]) [1]

$HgCl_2$ <1> (<1> 0.01 mM [1]) [1]
I^- <1> (<1> 50% inhibition at 0.05 M [8]) [8]
Mg^{2+} <1> [8]
NO_3^- <1> (<1> 50% inhibition at 0.035 M [8]) [8]
SDS <1> [1]
Triton X-100 <1> [1]
acetate <1> (<1> 50% inhibition at 0.123 M [8]) [8]
ammonium sulfate <1> [1]
ascorbate <1> (<1> 52% inhibition at 50 mM [1]) [1]
azide <1> [5]
catechol <1> (<1> substrate inhibition [13]) [13]
copper-chelating agents <2> (<1> not [1]) [2]
dithiothreitol <5> (<5> dithiothreitol acts as H_2O_2 generator and inhibits the oxygenase component of the enzyme, catalase protects the loss of activity [14]) [14]
ethylene glycol <1> [8]
formaldehyde <1> [9]
glutardialdehyde <1> [9]
glutathione <2> [2]
guanidinium chloride <1> (<1> 70-80% inhibition at 0.1 M [1]) [1]
p-chloromercuribenzoate <1> (<1> inhibition is reversed by dithiothreitol [1]) [1]
p-hydroxymercuribenzoate <1> [3]
peroxidase <1> [1]
phenol <1> (<1> excess phenol inhibits [9, 17]; <1> substrate inhibition [10, 13]) [9, 10, 13, 17]
phosphate <1> (<1> 60-70% inhibition at 5 mM [9]) [9]
potassium ethylxanthate <2> (<2> 1 mM [2]) [2]
pyridoxal phosphate <1> (<1> reversible, 50% loss of activity in 2 min at 0.5 mM [9]) [9]
pyridoxamine phosphate <1> (<1> slight [9]) [9]
resorcinol <1> (<1> substrate inhibition [13]) [13]
sodium borohydride <1> [1]
sodium diethyldithiocarbamate <2> (<2> 1 mM [2]) [2]
sodium dithionite <1> [1]
trichloroacetate <1> (<1> complete inhibition at 0.1 M [8]) [8]
urea <1> (<1> 70-80% inhibition at 2 M [1]) [1]

Cofactors/prosthetic groups

FAD <1, 3> (<1> flavoprotein [1]; <1> 1 mol FAD per mol of enzyme [1]; <1> 2 FAD groups per enzyme molecule [3,4]; <3> activates [11]; <1> 1 FAD per monomer [22]; <1> 2 FAD per dimer and 3 FAD per tetramer after removing FAD and reconstituting the apoenzyme with the cofactor [23]) [1, 3, 4, 8, 9, 11, 15, 21, 22, 23]
NADH <2> (<2> less active than NADPH [2]) [2, 21]
NADPH <1-4> [1-13, 15, 21]

Activating compounds

PEG 400 <1> (<1> slight increase [13]) [13]
dithiothreitol <1> (<1> 20% increase of activity at 1 mM [1]) [1]
thiophenol <1> (<1> binds to the enzyme and stimulates NADPH oxidation [6]) [6]

Metals, ions

Cu^{2+} <2> (<2> participation of Cu^{2+} in reaction [2]) [2]
Fe^{2+} <7> (<7> iron-sulfur cluster of the type 2Fe-2S [21]) [21]
anions <1> (<1> effect of anions on attachment of flavin [8]) [8]
monovalent anions <1> (<1> effect on mechanism [5]) [5]
Additional information <1> (<1> does not contain heme, non-heme iron or copper [1]) [1]

Turnover number (min^{-1})

125-690 <1> (NADPH, <1> value depends on phenolic substrate [7]) [7]
270-790 <1> (various phenolic substrates) [7]
720 <1> (phenol) [22]
Additional information <1, 7> (<1> various phenolic substrates [19]; <7> with various electron acceptors [21]) [19, 21]

Specific activity (U/mg)

0.78-1.4 <5> [14]
0.84 <1> (<1> tetrameric form [23]) [23]
1.5 <1> (<1> reconstituted tetrameric form [23]) [23]
2.3 <1> (<1> dimeric form [23]) [23]
2.6 <1> (<1> reconstituted dimeric form [23]) [23]
3.4 <1> [8]
5 <1> [19]
5.4 <1> [22]
5.5 <1> [10]
6 <1> [20]
7 <1> [22]
8.3 <1> [1]
429.5 <7> [21]

K_m-Value (mM)

0.0015 <3> (resorcinol, <3> crude extract [11]) [11]
0.003 <1> (phenol) [17]
0.005 <1> (2-fluorophenol) [7]
0.005 <3> (phenol, <3> crude extract [11]) [11]
0.008 <1> (3-fluorophenol) [7]
0.012 <1> (catechol) [7]
0.017 <1> (4-fluorophenol) [7]
0.018 <1> (phenol) [1]
0.03 <1> (NADPH, <1> + 3-fluorophenol [7]) [7]
0.032 <1> (resorcinol) [7]
0.039 <1> (4-chlorophenol) [7]
0.04 <1> (NADPH, <1> + 2-fluorophenol [7]) [7]

0.05 <1> (NADPH, <1> + phenol or resorcinol [7]) [7]
0.053 <1> (O_2) [1]
0.055 <1> (3-chlorophenol) [7]
0.071 <1> (NADPH) [1]
0.09 <1> (NADPH, <1> + catechol [7]) [7]
0.1 <1> (NADPH, <1> + 4-fluorophenol [7]) [7]
0.5 <1> (NADPH, <1> + 4-chlorophenol [7]) [7]
0.6 <1> (NADPH, <1> + 3-chlorophenol [7]) [7]
0.7 <1> (NADPH, <1> + 2-chlorophenol or 3-methylphenol [7]) [7]
0.8 <1> (NADPH, <1> + 4-methylphenol [7]) [7]
1.3 <1> (NADPH, <1> + 2-methylphenol [7]) [7]
1.7 <1> (NADPH, <1> in the absence of a phenolic substrate [7]) [7]
Additional information <1, 7> (<1> K_m value is 6-20times higher when phenol derivative is added before NADPH [13]; <7> K_m values with various electron acceptors [21]) [13, 21]

K_i-Value (mM)
0.17 <1> (phenol) [10]
0.5 <1> (azide) [5]
0.72 <1> (phenol) [13]
1.305 <1> (phenol) [17]
1.8 <1> (resorcinol) [13]
2.3 <1> (3-chlorophenol) [13]
2.3 <1> (4-fluorophenol) [13]
2.6 <1> (4-chlorophenol) [13]
6.3 <1> (3-fluorophenol) [13]
8.6 <1> (Cl^-) [5]
12.5 <1> (catechol) [13]
30.8 <1> (2-fluorophenol) [13]

pH-Optimum
7.2-7.6 <1> (<1> in phosphate buffer [1]) [1]
7.5 <2> [2]
7.6 <1> [9]
7.6-8 <3> [11]
7.9 <4> [12]
8.2 <1> (<1> in Tris-Cl buffer [1]) [1]
Additional information <1> (<1> effect of pH on oxidative half-reaction [6]) [6]

pH-Range
7-8.7 <3> (<3> about 50% of activity maximum at pH 7.0 and 8.7 [11]) [11]

Temperature optimum (°C)
20 <4> [12]
30 <3> (<3> enzyme from resorcinol grown cells [11]) [11]
40 <3> (<3> enzyme from phenol-induced cells [11]) [11]

Temperature range (°C)

10-35 <4> (<4> 10°C: about 80% of activity maximum, 35°C: about 50% of activity maximum [12]) [12]

4 Enzyme Structure

Molecular weight

38800 <7> (<7> gel filtration [21]) [21]
41000 <7> (<7> SDS-PAGE [21]) [21]
67000 <1> (<1> SDS-PAGE [22]) [22]
148000 <1> (<1> gel filtration [1, 3]) [1, 3]
302000 <1> (<1> gel filtration [23]) [23]

Subunits

dimer <1, 5> (<1> 2 * 76000, SDS-PAGE [3]; <5> SDS-PAGE shows three polypeptides with molecular masses of 13000, 39000 and 60000, gel filtration experiments are consistent with the existence of a dimer [14]; <1> homodimer, each subunit consists of 3 domains [15]) [3, 10, 14, 15]
tetramer <1> (<1> 4 * 76000, non-denaturing PAGE, after expression in E.coli [23]) [23]

5 Isolation/Preparation/Mutation/Application

Purification

<1> (homogeneity [1]; 90% pure [19, 20]) [1, 4, 10, 15, 19, 20, 22, 23, 24, 25]
<2> (partial [2]) [2]
<5> (purification of the oxygenase component [14]) [14]
<7> [21]

Crystallization

<1> (complexed with FAD a phenol, hanging drop vapor diffusion method [15]) [15]

Cloning

<1> [15]
<1> [15, 17, 22, 23, 24, 25]
<6> [16]

Engineering

D54N <1> (<1> slower reaction than wild type enzyme, higher dissociation constant for binding of phenol than wild type enzyme [24]) [24]
P364S <1> (<1> only 13% of the FAD is utilized to hydroxylate the substrate phenol, when resorcinol is used as substrate, the reaction is not significantly different from the reaction of the wild type enzyme [25]) [25]
R281M <1> (<1> slower reaction than wild type enzyme, binds the FAD cofactor more weakly than wild type enzyme [24]) [24]
Y298F <1> (<1> binds phenol more weakly than wild type enzyme [24]) [24]

6 Stability

pH-Stability

5 <1> (<1> stable for at least 7 days [1]) [1]

Temperature stability

4 <3> (<3> stable for some hours [11]) [11]

General stability information

<1>, chloride destabilizes [1]
<1>, phosphate stabilizes [1]
<2>, dilution causes considerable loss of activity and cannot be prevented by addition of proteins such as egg or serum albumin or substances of high molecular weight such as Carbowax-4000 or polyvinylpyrrolidone [2]
<2>, partially purified enzyme loses considerable activity upon dialysis or aging, addition of boiled extract prepared from crude extract fully restores activity [2]
<3>, longer exposure to ultrasound drastically reduces enzyme activity [11]

Storage stability

<1>, -20°C, 4-6 weeks stable [1]
<1>, -70°C, 6 months stable [10]
<1>, anion exchanger-immobilized enzyme is stable for several months at 4°C in 0.01 M buffers at pH 7.6 [4]

References

[1] Neujahr, H.Y.; Gaal, A.: Phenol hydroxylase from yeast. Purification and properties of the enzyme from Trichosporon cutaneum. Eur. J. Biochem., **35**, 386-400 (1973)
[2] Nakagawa, H.; Takeda, Y.: Phenol hydroxylase. Biochim. Biophys. Acta, **62**, 423-426 (1962)
[3] Neujahr, H.Y.; Gaal, A.: Phenol hydroxylase from yeast. Sulfhydryl groups in phenol hydroxylase from Trichosporon cutaneum. Eur. J. Biochem., **58**, 351-357 (1975)
[4] Kjellen, K.G.; Neujahr, H.Y.: Immobilization of phenol hydroxylase. Biotechnol. Bioeng., **21**, 715-719 (1979)
[5] Detmer, K.; Massey, V.: Effect of monovalent anions on the mechanism of phenol hydroxylase. J. Biol. Chem., **259**, 11265-11272 (1984)
[6] Detmer, K.; Massey, V.: Effect of substrate and pH on the oxidative half-reaction of phenol hydroxylase. J. Biol. Chem., **260**, 5998-6005 (1985)
[7] Neujahr, H.Y.; Kjellen, K.G.: Phenol hydroxylase from yeast. Reaction with phenol derivatives. J. Biol. Chem., **253**, 8835-8841 (1978)
[8] Neujahr, H.Y.: Effect of anions, chaotropes, and phenol on the attachment of flavin adenine dinucleotide to phenol hydroxylase. Biochemistry, **22**, 580-584 (1983)

[9] Neujahr, H.Y.; Kjellen, K.G.: Phenol hydroxylase from yeast: a lysyl residue essential for binding of reduced nicotinamide adenine dinucleotide phosphate. Biochemistry, **19**, 4967-4972 (1980)
[10] Selitz, T.; Neujahr, H.Y.: Phenol hydroxylase from yeast. A model for phenol binding and an improved purification procedure. Eur. J. Biochem., **170**, 343-349 (1987)
[11] Krug, M.; Straube, G.: Degradation of phenolic compounds by the yeast Candida tropicalis HP 15. II. Some properties of the first two enzymes of the degradation pathway. J. Basic Microbiol., **26**, 271-281 (1986)
[12] Straube, G.: Phenol hydroxylase from Rhodococcus sp. P 1. J. Basic Microbiol., **27**, 229-232 (1987)
[13] Mörtberg, M.; Neujahr, H.Y.: In situ and in vitro kinetics of phenol hydroxylase. Biochem. Biophys. Res. Commun., **146**, 41-46 (1987)
[14] Cadieux, E.; Vrajmasu, V.; Achim, C.; Powlowski, J.; Muenck, E.: Biochemical, Mossbauer, and EPR studies of the diiron cluster of phenol hydroxylase from Pseudomonas sp. strain CF 600. Biochemistry, **41**, 10680-10691 (2002)
[15] Enroth, C.; Neujahr, H.; Schneider, G.; Lindqvist, Y.: The crystal structure of phenol hydroxylase in complex with FAD and phenol provides evidence for a concerted conformational change in the enzyme and its cofactor during catalysis. Structure, **6**, 605-617 (1998)
[16] Hino, S.; Watanabe, K.; Takahashi, N.: Phenol hydroxylase cloned from Ralstonia eutropha strain E2 exhibits novel kinetic properties. Microbiology, **144**, 1765-1772 (1998)
[17] Kaelin, M.; Neujahr, H.Y.; Weissmahr, R.N.; Sejlitz, T.; Joehl, R.; Fiechter, A.; Reiser, J.: Phenol hydroxylase from Trichosporon cutaneum: gene cloning, sequence analysis, and functional expression in Escherichia coli. J. Bacteriol., **174**, 7112-7120 (1992)
[18] Maeda-Yorita, K.; Massey, V.: On the reaction mechanism of phenol hydroxylase. New information obtained by correlation of fluorescence and absorbance stopped flow studies. J. Biol. Chem., **268**, 4134-4144 (1993)
[19] Peelen, S.; Rietjens, I.M.C.M.; Boersma, M.G.; Vervoort, J.: Conversion of phenol derivatives to hydroxylated products by phenol hydroxylase from Trichosporon cutaneum. A comparison of regioselectivity and rate of conversion with calculated molecular orbital substrate characteristics. Eur. J. Biochem., **227**, 284-291 (1995)
[20] Peelen, S.; Rietjens, I.M.C.M.; van Berkel, W.J.H.; van Workum, W.A.T.; Vervoort, J.: Fluorine-19 NMR study on the pH-dependent regioselectivity and rate of the *ortho*-hydroxylation of 3-fluorophenol by phenol hydroxylase from Trichosporon cutaneum. Implications for the reaction mechanism. Eur. J. Biochem., **218**, 345-353 (1993)
[21] Pessione, E.; Divari, S.; Griva, E.; Cavaletto, M.; Rossi, G.L.; Gilardi, G.; Giunta, C.: Phenol hydroxylase from Acinetobacter radioresistens is a multicomponent enzyme. Purification and characterization of the reductase moiety. Eur. J. Biochem., **265**, 549-555 (1999)
[22] Waters, S.; Neujahr, H.Y.: A fermentor culture for production of recombinant phenol hydroxylase. Protein Expr. Purif., **5**, 534-540 (1994)

[23] Waters, S.; Neujahr, H.Y.: Sources and nature of heterogeneity in recombinant phenol hydroxylase derived from the basidiomycetous soil yeast Trichosporon cutaneum. Biotechnol. Appl. Biochem., **25**, 235-242 (1997)
[24] Xu, D.; Ballou, D.P.; Massey, V.: Studies of the mechanism of phenol hydroxylase: mutants Tyr289Phe, Asp54Asn, and Arg281Met. Biochemistry, **40**, 12369-12378 (2001)
[25] Xu, D.; Enroth, C.; Lindqvist, Y.; Ballou, D.P.; Massey, V.: Studies of the mechanism of phenol hydroxylase: Effect of mutation of proline 364 to serine. Biochemistry, **41**, 13627-13636 (2002)

Dimethylaniline monooxygenase (N-oxide-forming) 1.14.13.8

1 Nomenclature

EC number
1.14.13.8

Systematic name
N,N-dimethylaniline,NADPH:oxygen oxidoreductase (N-oxide-forming)

Recommended name
dimethylaniline monooxygenase (N-oxide-forming)

Synonyms
DMA oxidase
FAD-containing monooxygenase
FMO
FMO 1A1
FMO 1B1
FMO 1C1
FMO 1D1
FMO 1E1
FMO-I
FMO-II
FMO1
FMO2
FMO3
FMO5
N,N-dimethylaniline monooxygenase
dimethylaniline N-oxidase
dimethylaniline oxidase
flavin monooxygenase
flavin-containing monooxygenase
mixed-function amine oxidase
oxygenase, dimethylaniline mono- (N-oxide-forming)

CAS registry number
148848-55-9
37256-73-8

2 Source Organism

<1> *Sus scrofa* [1-4, 6-8, 10, 13, 14, 15, 16, 21, 23]
<2> *Mesocricetus auratus* [6, 14]

<3> *Mus musculus* [5, 13]
<4> *Oryctolagus cuniculus* [5, 9, 11, 13]
<5> *Rattus norvegicus* [6, 13, 19, 29]
<6> *Cavia porcellus* [12, 20]
<7> *Squalus acanthias* (dogfish shark [17]) [17]
<8> *Carcharhinus falciformis* (silky shark [17]) [17]
<9> *Homo sapiens* (diet of Brussels sprouts [18]) [18, 20, 24, 26, 27, 28]
<10> *Macaca fascicularis* (macacque [22]) [22]
<11> *Scophthalmus maximus* (turbot [25]) [25]

3 Reaction and Specificity

Catalyzed reaction

N,N-dimethylaniline + NADPH + H^+ + O_2 = N,N-dimethylaniline N-oxide + $NADP^+$ + H_2O (<1, 2> reaction mechanism of S-oxygenation of N-substituted thioureas [14]; <1> ordered ter-bi mechanism with an irreversible step between the second and third substrate, NADPH is added first, followed by O_2 and the oxidizable organic substrate last [4])

Reaction type

oxidation
redox reaction
reduction

Natural substrates and products

S 1,1-dimethylhydrazine + NADPH + O_2 <1, 2, 5> (<1, 2, 5> possibly, and other 1,1-disubstituted hydrazines [6]) (Reversibility: ? <1, 2, 5> [6]) [6]
P formaldehyde + $CH_3N_2H_3$ + $NADP^+$ <1, 2, 5> [6]

Substrates and products

S 1,1-dimethylhydrazine + NADPH + O_2 <1, 2, 5> (Reversibility: ? <1, 2, 5> [6]) [6]
P formaldehyde + $CH_3N_2H_3$ + $NADP^+$ <1, 2, 5> [6]
S 1,2,3,4-tetrahydroisoquinoline + NADPH + O_2 <1, 5> (Reversibility: ? <1, 5> [19, 21]) [19, 21]
P ?
S 1,2-dimethylhydrazine + NADPH + O_2 <2, 5> (Reversibility: ? <2, 5> [6]) [6]
P ?
S 1,2-dimethylphenylhydrazine + NADPH + O_2 <2, 5> (Reversibility: ? <2, 5> [6]) [6]
P ?
S 1-butanethiol + NADPH + O_2 <1, 3-5> (Reversibility: ? <1, 3-5> [13]) [13]
P ?
S 1-methyl-1-phenylhydrazine + NADPH + O_2 <1, 2, 5> (Reversibility: ? <1, 2, 5> [6]) [6]

P ?
S 1-methyl-1-phenylhydrazine + NADPH + O_2 <2, 5> (Reversibility: ? <2, 5> [6]) [6]
P ?
S 1-methyl-2-benzylhydrazine + NADPH + O_2 <2, 5> (Reversibility: ? <2, 5> [6]) [6]
P ?
S 1-methyl-2-thioimidazole + NADPH + O_2 <1> (Reversibility: ? <1> [15]) [15]
P ?
S 1-methyl-4-phenyl-1,2,3,6-tetrahydropyridine + NADPH + O_2 <1, 5> (Reversibility: ? <1, 5> [16, 19, 21]) [16, 19, 21]
P ?
S 1-methyl-6,7-dihydroxytetrahydroisoquinoline + NADPH + O_2 <1> (Reversibility: ? <1> [21]) [21]
P ?
S 10-([N,N-dimethylaminopentyl]-2-trifluoromethyl)phenothiazine + NADPH + O_2 <9> (Reversibility: ? <9> [18]) [18]
P ?
S 2-mercaptobenzimidazole + NADPH + O_2 <1, 3-5> (Reversibility: ? <1, 3-5> [13, 14]) [13, 14]
P ?
S N,N-dimethylaniline + NADPH + O_2 <1-7, 11> (Reversibility: ? <1-7, 11> [1-15, 17, 21, 25]) [1-15, 17, 21, 25]
P N,N-dimethylaniline N-oxide + $NADP^+$ + H_2O
S N-aminohomopiperidine + NADPH + O_2 <1, 2, 5> (Reversibility: ? <1, 2, 5> [6]) [6]
P ?
S N-aminohomopiperidine + NADPH + O_2 <2, 5> (Reversibility: ? <2, 5> [6]) [6]
P ?
S N-aminomorpholine + NADPH + O_2 <2, 5> (Reversibility: ? <2, 5> [6]) [6]
P ?
S N-aminopiperidine + NADPH + O_2 <2, 5> (Reversibility: ? <2, 5> [6]) [6]
P ?
S N-aminopiperidine + NADPH + O_2 <1-3, 5> (Reversibility: ? <1-3, 5> [6, 13]) [6, 13]
P tetrazene + $NADP^+$ + H_2O + ? <1, 2, 5> [6]
S N-aminopyrrolidone + NADPH + O_2 <2, 5> (Reversibility: ? <2, 5> [6]) [6]
P ?
S N-methyl-1,2,3,4-tetrahydroisoquinoline + NADPH + O_2 <5> (Reversibility: ? <5> [19]) [19]
P ?
S R(-)-deprenyl + NADPH + O_2 <1> (<1> inhibitor of monoaminoxidase B [16]) (Reversibility: ? <1> [16]) [16]

P ?
S α-naphthylthiourea + NADPH + O_2 <1, 3-5> (Reversibility: ? <1, 3-5> [13]) [13]
P ?
S aminopyrine + NADPH + O_2 <1, 3-5> (Reversibility: ? <1, 3-5> [13]) [13]
P ?
S benzylhydrazine + NADPH + O_2 <2, 5> (Reversibility: ? <2, 5> [6]) [6]
P ?
S β-ethylphenylhydrazine + NADPH + O_2 <2, 5> (Reversibility: ? <2, 5> [6]) [6]
P ?
S butylhydrazine + NADPH + O_2 <2, 5> (Reversibility: ? <2, 5> [6]) [6]
P ?
S chlorpromazine + NADPH + O_2 <6> (Reversibility: ? <6> [12]) [12]
P ?
S cysteamine + NADPH + O_2 <1, 3-6> (Reversibility: ? <1, 3-6> [12, 13, 19]) [12, 13, 19]
P ?
S dibenzylamine + NADPH + O_2 <1, 3-5> (Reversibility: ? <1, 3-5> [13]) [13]
P ?
S ephedrine + NADPH + O_2 <1, 3-5> (Reversibility: ? <1, 3-5> [13]) [13]
P ?
S ethylene sulfide + NADPH + O_2 <1, 3-5> (Reversibility: ? <1, 3-5> [13]) [13]
P ?
S ethylenthiourea + NADPH + O_2 <1> (Reversibility: ? <1> [14]) [14]
P ?
S imipramine + NADPH + O_2 <1, 3-5> (Reversibility: ? <1, 3-5> [13, 23]) [13, 23]
P ?
S isopropylhydrazine + NADPH + O_2 <2, 5> (Reversibility: ? <2, 5> [6]) [6]
P ?
S methamphetamine + NADPH + O_2 <1, 3-6> (Reversibility: ? <1, 3-6> [12, 13]) [12, 13]
P ?
S methimazole + NADPH + O_2 <1, 3-7, 9> (<5, 6, 9> recombinant protein expressed in E. coli [20, 29]; <9> FMO3 5000 times more efficient than FMO5) (Reversibility: ? <1, 3-7, 9> [3, 12-15, 17, 20, 23, 24, 29]) [3, 12-15, 17, 20, 23, 24, 29]
P N-methylmethimidazole-2-sulfinic acid + 2 $NADP^+$ + 2 H_2O <1> [3]
S methyl *p*-tolyl sulfide + NADPH + O_2 <10> (Reversibility: ? <10> [22]) [22]
P methyl *p*-tolyl sulfoxide + $NADP^+$ + H_2O <10> (<10> stereochemistry: product 49% R-enantiomer [22]) [22]
S methylphenylsulfide + NADPH + O_2 <1, 3-5> (Reversibility: ? <1, 3-5> [13]) [13]

P ?

S n-decylamine + NADPH + O_2 <3, 4> (<3, 4> lung enzyme active, liver enzyme not [5]) (Reversibility: ? <3, 4> [5]) [5]

P 1-nitrosodecane + $NADP^+$ * H_2O

S n-octylamine + NADPH + O_2 <3, 4, 5, 6, 9> (<3, 4> lung enzyme active, liver enzyme not [5]; <5> not oxidized [19]; <6, 9> recombinant protein expressed in E. coli [20]) (Reversibility: ? <3, 4, 6, 9> [5, 20]) [5, 19, 20]

P 1-nitrosooctane + $NADP^+$ + H_2O

S *p*-chloro-N-methylaniline + NADPH + O_2 <1, 3-5> (Reversibility: ? <1, 3-5> [13]) [13]

P ?

S pargyline + NADPH + O_2 <1> (<1> inhibitor of monoaminoxidase B [16]) (Reversibility: ? <1> [16]) [16]

P ?

S phenylhydrazine + NADPH + O_2 <2, 5> (Reversibility: ? <2, 5> [6]) [6]

P ?

S phenylthiourea + NADPH + O_2 <1, 3-6> (Reversibility: ? <1, 3-6> [12, 13, 14]) [12, 13, 14]

P ?

S procarbazine + NADPH + O_2 <2, 5> (Reversibility: ? <2, 5> [6]) [6]

P ?

S ranitidine + NADPH + O_2 <9> (Reversibility: ? <9> [28]) [28]

P ?

S secondary amine + NADPH + O_2 <1> (Reversibility: ? <1> [2]) [2]

P secondary nitrone + $NADP^+$ + H_2O <1> (<1> first oxidation to the N-hydroxy amine and then to the corresponding nitrone [2]) [2]

S tamoxifen + NADPH + O_2 <1> (Reversibility: ? <1> [23]) [23]

P ?

S tertiary amine + NADPH + O_2 <1> (Reversibility: ? <1> [2]) [2]

P tertiary N-oxide + $NADP^+$ + H_2O

S thioacetamide + NADPH + O_2 <1, 3-5> (Reversibility: ? <1, 3-5> [13, 29]) [13, 29]

P ?

S thiobenzamide + NADPH + O_2 <3-5> (Reversibility: ? <3-5> [13]) [13]

P ?

S thiocarbanilide + NADPH + O_2 <1> (Reversibility: ? <1> [14]) [14]

P ?

S thiourea + NADPH + O_2 <1, 3-6> (Reversibility: ? <1, 3-6> [5, 12, 13, 14, 19]) [5, 12, 13, 14, 19]

P ?

S trifluoperazine + NADPH + O_2 <1> (Reversibility: ? <1> [23]) [23]

P ?

S trimethylamine + NADPH + O_2 <1, 3-5, 9> (Reversibility: ? <1, 3-5, 9> [13, 24, 29]) [13, 24, 29]

P ?

S Additional information <1-6, 9> (<1> reaction can be functionally separated into 2 partial reactions: 1. a reduced pyridine nucleotide and oxygen-dependent N-oxide synthase, 2. an N-oxide dealkylase [1]; <1> study of reductive half-reaction [7]; <1> study of oxidative half-reaction [8]; <1, 2> S-oxygenation of N-substituted thioureas [14]; <1> catalyzes NADPH- and O_2-dependent N-oxidation of N-substituted amines and hydrazines and the S-oxidation of thioureylenes and thiols [2, 3]; <6> specificity of FMO-I and FMO-II [12]; <3, 4, 5> overview on specificity [13]; <9> modulation of activity by site directed mutagenesis [26]; <9> overview on substrate specifities and requirements of FMO1, FMO3 [27]; <9> comparison of FMO3 and FMO5 [28]) [1-3, 7, 8, 12-14, 26, 27]

P ?

Inhibitors

(E)-2-[2-(4-(dimethylamino)phenyl)vinyl]benzoic acid <1> (<1> i.e. DS2CO, 2 mM, mechanism [23]) [23]
(E)-3-[2-(4-(dimethylamino)phenyl)vinyl]benzoic acid <1> (<1> i.e. DS3CO; 2 mM, 80-90% inhibition, mechansim [23]) [23]
2-diethylaminoethyl-2,2-diphenylpentanoate <1> (<1> SFK-525A, inhibition of dealkylation [1]) [1]
CO <1> (<1> inhibition of dealkylation [1]) [1]
$MgCl_2$ <6, 9> (<6,9> 100 mM, 100% inhibition within 6 min [20]) [20]
SDS <1> [3]
anionic detergents <1> [3]
deprenyl <1> (<1> strong, oxidative activity toward 1-methyl-4-phenyl-1,2,3,6-tetrahydropyridine, MPTP [16]) [16]
fatty acids <1> [3]
indole-3-carbinol <9> (<9> and its acid condensation products, strong [18]) [18]
methimazole <7, 11> (<7, 11> oxidation of dimethylaniline [17, 25]; <7> not competitive [17]) [17, 25]
pargyline <1> (<1> strong, oxidative activity toward 1-methyl-4-phenyl-1,2,3,6-tetrahydropyridine, MPTP [16]) [16]
sodium cholate <6, 9> (<6, 9> 1%, time-dependent sensitivity, maximum 65-100% inhibition [20]) [20]
stearate <1> [3]
thiobenzamide <7> (<7> oxidation of dimethylaniline, competitive [17]) [17]
trimethylamine <7, 11> (<7> 1 mM, oxidation of dimethylaniline, competitive [17]) [17, 25]
Additional information <11> (<11> not: piperonyl butoxide [25]) [25]

Cofactors/prosthetic groups

FAD <1-6> (<1-6> flavoprotein [3, 4, 6-9, 12-14]; <4, 5> FAD binding domain [9,29]; <6> 1 FAD per enzyme [12]) [3, 4, 6-9, 12-14, 29]
NADH <1> (<1> can partially replace NADPH [1]; <1> concentration of NADPH required for half-maximal velocity is one-tenth of that for NADH [2]) [1-3]
NADPH <1, 2, 4, 5> (<4, 5> $NADP^+$ binding domain [9, 29]) [1-3, 6, 9, 29]

Activating compounds

guanidines <1> (<1> stimulate NADPH- and O_2-dependent oxidation of tertiary amines and sulfur-containing substrates with alkyl side-chains of less than 5 carbons [3]) [3]
lipophilic primary alkylamines <1> (<1> stimulate NADPH- and O_2-dependent oxidation of tertiary amines and sulfur-containing substrates with alkyl side-chain of less than 5 carbons [3]) [3]
n-octylamine <1, 3-5> (<1> stimulation of tert-amine oxidation, not sec-amine oxidation [3]; <3-5> allosteric activation of pig liver and mouse lung enzymes, not mouse, rabbit or rat liver enzymes [13]; <1> thyroid enzyme only active in presence of n-octylamine [15]) [3, 13-15]
tertiary amines <1> (<1> self-activation with alkyl side chains of 7 or more [3]) [3]

Specific activity (U/mg)

0.41 <6> (<6> thiobenzamide S-oxidase activity, FMO-II [12]) [12]
0.436 <6> (<6> thiobenzamide S-oxidase activity, FMO-I [12]) [12]
0.52 <4> (<4> thiourea [5]) [5]
1.43 <1> [3]
2.4 <3> (<3> thiourea [5]) [5]
6.2 <1> (<1> 1-methyl-4-phenyl-1,2,3,6-tetrahydropyridine [16]) [16]

K_m-Value (mM)

0.006 <5> (1-methyl-4-phenyl-1,2,3,6-tetrahydropyridine) [19]
0.008 <5> (N-methyl-1,2,3,4-tetrahydroisoquinoline) [19]
0.012 <1> (pargyline) [16]
0.014 <1> (deprenyl) [16]
0.018 <1> (dimethylaniline) [21]
0.02 <1> (dimethylaniline) [6]
0.027 <5> (thiourea) [19]
0.028 <1> (NAPDH, <1> with dimethylaniline [21]) [21]
0.03 <1> (N-aminopiperidine) [6]
0.038 <1> (1-methyl-4-phenyl-1,2,3,6-tetrahydropyridine) [16, 21]
0.043 <1> (O_2, <1> with dimethylaniline [21]) [21]
0.044 <7> (dimethylaniline) [17]
0.049 <1> (O_2, <1> with dimethylaniline [21]) [21]
0.05 <1> (NADPH, <1> with 1-methyl-4-phenyl-1,2,3,6-tetrahydropyridine [21]) [21]
0.059 <5> (1,2,3,4-tetrahydroisoquinoline) [19]
0.08 <1> (1-methyl-1-phenylhydrazine) [6]
0.1 <1> (N-aminopyrrolidine) [6]
0.17 <1> (N-aminohomopiperidine) [6]
0.38 <1> (1,1-dimethylhydrazine) [6]
0.61 <1> (N-aminomorpholine) [6]
1.2 <5> (cysteamine) [19]
2 <1> (1-methyl-2-benzylhydrazine) [6]
3 <1> (phenylhydrazine) [6]
3.3 <1> (β-ethylphenylhydrazine) [6]

5.7 <1> (procarbazine) [6]
6.9 <1> (butylhydrazine) [6]
7 <1> (benzylhydrazine) [6]
8.3 <1> (isopropylhydrazine) [6]
12 <1> (1,2-dimethylhydrazine) [6]
15 <1> (n-propylhydrazine) [6]
35 <1> (methylhydrazine) [6]
40 <1> (ethylhydrazine) [6]
Additional information <1-5> (<1, 2> 2 K_m values for oxygenation of thiocarbamides: 1. K_{m1} for oxygenation to sulfenic acid, 2. K_{m2} for oxygenation of sulfenic acid to sulfinic acid [14]) [13, 14]

K_i-Value (mM)

0.008-0.013 <9> (indole-3-carbinol) [18]
0.009 <1> (pargyline, <1> oxidative activity toward 1-methyl-4-phenyl-1,2,3,6-tetrahydropyridine, MPTP [16]) [16]
0.014 <1> (deprenyl, <1> oxidative activity toward 1-methyl-4-phenyl-1,2,3,6-tetrahydropyridine, MPTP [16]) [16]

pH-Optimum

7.6 <1> (<1> N,N-dimethylaniline, immobilized enzyme [2]) [2]
8.4 <1> [4]
8.4 <2> (<2> demethylation of 1,1-dimethylhydrazine [6]) [6]
8.5 <1> (<1> hydrazine oxidation [6]) [6]
8.8 <3> [13]
8.8 <11> [25]
8.8-9 <3, 4> (<3,4> liver, thiobenzamide S-oxidation [5]) [5]
9 <3> [13]
9.6 <7> [17]
9.8 <3-5> (<3,4> thiobenzamide S-oxidation [5]) [5, 13]

pH-Range

7.5-9 <1> (<1> pH 7.5: about 40% of activity maximum, pH 9.0: about 95% of activity maximum [6]) [6]
7.6-9.6 <3, 4> (<3,4> about 50% of activity maximum at pH 7.6 and 9.6, liver, thiobenzamide S-oxidation [5]) [5]
8.4-10.4 <3, 4> (<3,4> about 50% of activity maximum at pH 8.4 and 10.4, lung, thiobenzamide S-oxidation [5]) [5]
Additional information <3-5> [13]

Temperature optimum (°C)

25 <7> [17]
25 <1> [21]
25 <11> [25]
25-28 <1> (<1> N,N-dimethylaniline, immobilized enzyme [2]) [2]
32 <1> [21]

4 Enzyme Structure

Molecular weight

50000 <7, 8> (<7,8> Western blot, anti-FMO 2 antisera [17]) [17]
54000 <6> (<6> SDS-PAGE, FMO-I [12]) [12]
55000 <11> (<11> Western blot, anti-FMO1 and anti-FMO$_2$ (mammalian) antisera [25]) [25]
55000-60000 <4> (<4> SDS-PAGE, amino acid composition [9]) [9]
56000 <1> (<1> SDS-PAGE, gel filtration [21]) [21]
56000 <5> (<5> SDS-PAGE [29]) [29]
56000 <6> (<6> SDS-PAGE, FMO-II [12]) [12]
56000-59000 <3, 4> (<3, 4> SDS-PAGE [5]) [5]
58000 <6> (<6> recombinant protein expressed in E. coli [20]) [20]
58950 <1> (<1> calculation from amino acid sequence [10]) [10]
59000 <10> (<10> SDS-PAGE [22]) [22]
60000 <5> (<5> Western blot, anti-rat liver FMO antisera [19]) [19]
60000 <9> (<9> recombinant protein expressed in E. coli [20]) [20]
64000 <1> (<1> SDS-PAGE [4]) [4]

Subunits

? <1, 6> (<1> x * 64000, SDS-PAGE, active enzyme exists as aggregating units of the monomer, amino acid composition [4]; <6> x * 54000, FMO-I, SDS-PAGE [12]; x * 56000, FMO-II, SDS-PAGE [12]; <1> x * 58952, calculation from amino acid sequence [10]) [4, 10, 12]
monomer <1> (<1> 1 * 56000, gel filtration [21]) [21]
octamer <1> (<1> 8 * 65000, SDS-PAGE [3]) [3]

Posttranslational modification

glycoprotein <1> (<1> amino acid analysis [21]) [21]

5 Isolation/Preparation/Mutation/Application

Source/tissue

brain <5> [19]
kidney <3-5> [13]
liver <1-6, 9, 10> [1-4, 6-8, 10-14, 16, 17, 20-26, 28, 29]
lung <3-5> [5, 9, 11, 13]
thyroid gland <1> [15]
urine <9> [18]

Localization

microsome <1, 7> [16, 17, 19, 21, 28, 29]

Purification

<1> (<1> amino acid composition [21]) [3, 7, 21]
<3> [5]
<4> [5]

<5> (<5> partial [19]) [19, 29]
<6> (<6> 2 forms: FMO-I and FMO-II [12]) [12]
<8> (<8> partial [17]) [17]
<10> (<10> N-terminal amino acid sequence [22]) [22]

Cloning
<9> (missense mutations causing fish-odour syndrome [24]) [24]
<1, 4> (cDNA data [10,11]) [9, 10, 11]

6 Stability

Temperature stability
15-45 <7> (<7> 15°C, 76% loss of activity, 45°C 99% loss of activity [17]) [17]
38 <1> (<1> pH 7.6, half-life of free enzyme: 10 min, half-life of immobilized enzyme 5 h [2]) [2]

Storage stability
<1>, -15°C, several months with little or no loss of activity [3]
<1>, glass-bead immobilized enzyme: 0-4°C, 0.025 M phosphate buffer, several months with little or no loss of activity [2]

References

[1] Ziegler, D.M.; Pettit, F.H.: Microsomal oxidases. I. The isolation and dialkylarylamine oxygenase activity of pork liver microsomes. Biochemistry, **5**, 2932-2938 (1966)
[2] Poulsen, L.L.; Sofer, S.S.; Ziegler, D.M.: Properties and applications of an immobilized mixed-function hepatic drug oxidase. Methods Enzymol., **44**, 849-856 (1976)
[3] Ziegler, D.M.; Poulsen, L.L.: Hepatic microsomal mixed-function amine oxidase. Methods Enzymol., **52**, 142-151 (1978)
[4] Poulsen, L.L.; Ziegler, D.M.: The liver microsomal FAD-containing monooxygenase. Spectral characterization and kinetic studies. J. Biol. Chem., **254**, 6449-6455 (1979)
[5] Tynes, R.E.; Sabourin, P.J.; Hodgson, E.: Identification of distinct hepatic and pulmonary forms of microsomal flavin-containing monooxygenase in the mouse and rabbit. Biochem. Biophys. Res. Commun., **126**, 1069-1075 (1985)
[6] Prough, R.A.; Freeman, P.C.; Hines, R.N.: The oxidation of hydrazine derivatives catalyzed by the purified liver microsomal FAD-containing monooxygenase. J. Biol. Chem., **256**, 4178-4184 (1981)
[7] Beaty, N.B.; Ballou, D.P.: The reductive half-reaction of liver microsomal FAD-containing monooxygenase. J. Biol. Chem., **256**, 4611-4618 (1981)
[8] Beaty, N.B.; Ballou, D.P.: The oxidative half-reaction of liver microsomal FAD-containing monooxygenase. J. Biol. Chem., **256**, 4619-4625 (1981)

[9] Guan, S.; Falick, A.M.; Williams, D.E.; Cashman, J.R.: Evidence for complex formation between rabbit lung flavin-containing monooxygenase and calreticulin. Biochemistry, **30**, 9892-9900 (1991)
[10] Gasser, R.; Tynes, R.E.; Lawton, M.P.; Korsmeyer, K. K.; Ziegler, D.M.; Philpot, R.M.: The flavin-containing monooxygenase expressed in pig liver: primary sequence, distribution, and evidence for a single gene. Biochemistry, **29**, 119-124 (1990)
[11] Lawton, M.P.; Gasser, R.; Tynes, R.E.; Hodgson, E.; Philpot, R.M.: The flavin-containing monooxygenase enzymes expressed in rabbit liver and lung are products of related but distinctly different genes. J. Biol. Chem., **265**, 5855-5861 (1990)
[12] Yamada, H.; Yuno, K.; Oguri, K.; Yoshimura, H.: Multiplicity of liver microsomal flavin-containing monooxygenase in the guinea pig: its purification and characterization. Arch. Biochem. Biophys., **280**, 305-312 (1990)
[13] Tynes, R.E.; Hodgson, E.: Catalytic activity and substrate specificity of the flavin-containing monooxygenase in microsomal systems: characterization of the hepatic, pulmonary and renal enzymes of the mouse, rabbit, and rat. Arch. Biochem. Biophys., **240**, 77-93 (1985)
[14] Poulsen, L.L.; Hyslop, R.M.; Ziegler, D.M.: S-Oxygenation of N-substituted thioureas catalyzed by the pig liver microsomal FAD-containing monooxygenase. Arch. Biochem. Biophys., **198**, 78-88 (1979)
[15] Paterson, J.R.; Hood, H.T.; Skellern, G.G.: The role of porcine thyroid peroxidase and FAD-containing monooxygenase in the metabolism of 1-methyl-2-thioimidazole (methimazole). Biochem. Biophys. Res. Commun., **116**, 449-455 (1983)
[16] Wu, R.F.; Ichikawa, Y.: An essential lysyl residue (Lys208) in the substrate-binding site of porcine FAD-containing monooxygenase. Eur. J. Biochem., **229**, 749-753 (1995)
[17] Schlenk, D.; Li-Schlenk, R.: Characterization of liver flavin-containing monooxygenase of the dogfish shark (Squalus acanthias) and partial purification of liver flavin-containing monooxygenase of the silky shark (Carcharhinus falciformis). Comp. Biochem. Physiol. B, **109**, 655-664. (1994)
[18] Cashman, J.R.; Xiong, Y.; Lin, J.; Verhagen, H.; van Poppel, G.; van Bladeren, P.J.; Larsen-Su, S.; Williams, D.: In vitro and in vivo inhibition of human flavin-containing monooxygenase form 3 (FMO3) in the presence of dietary indoles. Biochem. Pharmacol., **58**, 1047-1055 (1999)
[19] Kawaji, A.; Miki, T.; Takabatake, E.: Partial purification and substrate specificity of flavin-containing monooxygenase from rat brain microsomes. Biol. Pharm. Bull., **18**, 1657-1659 (1995)
[20] Overby, L.H.; Buckpitt, A.R.; Lawton, M.P.; Atta-Asafo-Adjei, E.; Schulze, J.; Philpot, R.M.: Characterization of flavin-containing monooxygenase 5 (FMO5) cloned from human and guinea pig: evidence that the unique catalytic properties of FMO5 are not confined to the rabbit ortholog. Arch. Biochem. Biophys., **317**, 275-284 (1995)

[21] Wu, R.F.; Ichikawa, Y.: Characteristic properties and kinetic analysis with neurotoxins of porcine FAD-containing monooxygenase. Biochim. Biophys. Acta, **1208**, 204-210 (1994)
[22] Sadeque, A.J.M.; Thummel, K.E.; Rettie, A.E.: Purification of macacque liver flavin-containing monooxygenase: A form of the enzyme related immunochemically to an isoenzyme expressed selectively in adult human liver. Biochim. Biophys. Acta, **1162**, 127-134 (1993)
[23] Clement, B.; Weide, M.; Ziegler, D.M.: Inhibition of purified and membrane-bound flavin-containing monooxygenase 1 by (N,N-dimethylamino)stilbene carboxylates. Chem. Res. Toxicol., **9**, 599-604. (1996)
[24] Dolphin, C.T.; Janmohamed, A.; Smith, R.L.; Shephard, E.A.; Phillips, I.R.: Compound heterozygosity for missense mutations in the flavin-containing monooxygenase 3 (FMO3) gene in patients with fish-odour syndrome. Pharmacogenetics, **10**, 799-807 (2000)
[25] Peters, L.D.; Livingstone, D.R.; Shenin-Johnson, S.; Hines, R.N.; Schlenk, D.: Characterization of hepatic flavin monooxygenase from the marine teleost turbot (Scophthalmus maximus L.). Xenobiotica, **25**, 121-131. (1995)
[26] Adali, O.; Carver, G.C.; Philpot, R.M.: The effect of arginine-428 mutation on modulation of activity of human liver flavin monooxygenase 3 (FMO3) by imipramine and chlorpromazine. Exp. Toxicol. Pathol., **51**, 271-276 (1999)
[27] Cashman, J.R.: Human flavin-containing monooxygenase: substrate specificity and role in drug metabolism. Curr. Drug Metab., **1**, 181-191 (2000)
[28] Overby, L.H.; Carver, G.C.; Philpot, R.M.: Quantitation and kinetic properties of hepatic microsomal and recombinant flavin-containing monooxygenases 3 and 5 from humans. Chem. Biol. Interact., **106**, 29-45 (1997)
[29] Lattard, V.; Buronfosse, T.; Lachuer, J.; Longin-Sauvageon, C.; Moulin, C.; Benoit, E.: Cloning, sequencing, tissue distribution, and heterologous expression of rat flavin-containing monooxygenase 3. Arch. Biochem. Biophys., **391**, 30-40 (2001)

Kynurenine 3-monooxygenase 1.14.13.9

1 Nomenclature

EC number
1.14.13.9

Systematic name
L-kynurenine,NADPH:oxygen oxidoreductase (3-hydroxylating)

Recommended name
kynurenine 3-monooxygenase

Synonyms
EC 1.14.1.2 (formerly)
EC 1.99.1.5 (formerly)
L-kynurenine-3-hydroxylase
kynurenine 3-hydroxylase
kynurenine hydroxylase
oxygenase, kynurenine 3-mono-

CAS registry number
9029-61-2

2 Source Organism

<1> *Schistocerca gregaria* [9]
<2> *Ephestia kuehniella* (wild-type and 3 mutants [10]) [10]
<3> *Rattus norvegicus* (Osborne-Mendel strain [2]; male wistar rat [13]; female wistar rat [19]) [1-5, 8, 12, 13, 14, 16, 17, 18, 19, 21, 23]
<4> *Saccharomyces cerevisiae* (wild type and different sterol mutants [6]; mutant E105 has an enzymatic defect of kynurenine 3-hydroxylase [7]) [6, 7]
<5> *Meriones unguiculatus* (female mongolian gerbil [8]) [8, 16]
<6> *Saccharomyces carlsbergensis* [11]
<7> *Macaca mulatta* (male rhesus macaques [8]) [8]
<8> *Homo sapiens* [8, 20, 22]
<9> *Mus musculus* (mouse, strain D57Bl/6J [8]) [8]
<10> *Sus scrofa* (pig [15]) [15]

3 Reaction and Specificity

Catalyzed reaction

L-kynurenine + NADPH + H^+ + O_2 = 3-hydroxy-L-kynurenine + $NADP^+$ + H_2O

Reaction type

oxidation
redox reaction
reduction

Natural substrates and products

S L-kynurenine + NADPH + O_2 <3, 4, 5> (<4> rate limiting step in the pyridine nucleotide biosynthesis from tryptophan [7]; <5> increased activity in injured brain regions following cerebral ischemia [8]; <3> increased activity in the spinal cord with experimental allergic encephalopathy [19]; <3> key enzyme in the kynurenine pathway of tryptophan degradation [12]) (Reversibility: ? <3, 4, 5> [2, 7, 8, 12, 17, 19]) [2, 7, 8, 12, 17, 19]

P ? + NAD

Substrates and products

S L-kynurenine + NADPH + O_2 <1-6> (<3> no reaction with D-isomer [2]; <4> NADH is less effective [6]) (Reversibility: r <3> [18]; ? <1-6> [1-11, 19, 20]) [1-11, 18, 19, 20]

P 3-hydroxy-L-kynurenine + $NADP^+$ + H_2O <3> [2, 18]

S *o*-hydroxybenzoyl-DL-alanine + NADPH + O_2 <3> (<3> about 25% of the activity with L-kynurenine [2]) (Reversibility: ? <3> [2]) [2]

P ? + NAD

S Additional information <3> (<3> catalyzes NADH- and NADPH-linked reductions of low molecular weight acceptors such as 2,6-dichlorophenolindophenol and ferricyanide [3]) [3]

P ?

Inhibitors

(R,S)-2-amino-oxo-4-(3',4'-dichlorophenyl)butanoic acid <3> (<3> FCE 28833, 50% inhibition at 0.2 microM, blocks not only the cerebral but also the peripheral enzyme [13]) [13]

(*m*-nitrobenzoyl)-alanine <3, 5> (<3,5> mNBA, leads to an increase of L-kynurenine and kynurenic acid concentrations in the brain cortex after application in vivo [16]; <3> various pyrrolo[3,2-c]quinoline derivates cause enzyme inhibition [17]; 50% inhibition at 0.9 micromolar [23]) [16, 17, 23]

2-benzyl-4-(3,4-dichlorophenyl)-4-oxo-butanoic acid <3> [12]

2-hydroxy-4-(3,4-dichlorophenyl)-4-oxobutanoic acid <3> [12]

2-oxo acid derivatives <6> (<6> of the 3 branched chain amino acids [11]) [11]

2-oxoglutarate <6> (<6> mixed type inhibitor [11]) [11]

3,4-dimethoxy-N-[4-(3-nitrophenyl)thiazol-2-yl]benzenesulfonamide <3, 5> (<3> Ro-61-8048, 50% inhibition at 37 nM [14]; <3> leads to an increase of L-kynurenine and kynurenic acid concentrations in the brain cortex after application in vivo [16]; <5> leads to an increase of L-kynurenine and kynurenic acid concentrations in the brain cortex after application in vivo [16]; <3> inhibition after oral or intraperitoneal administration [21]) [14, 16, 21]
4-amino-N-[4-[2-fluoro-5-(trifluoromethyl)phenyl]thiazol-2-yl]benzenesulfonamide <3> (<3> 50% inhibition at 19nM [14]) [14]
7-chloro-3-methyl-1H-pyrrolo[3,2-c]quinoline-4-carboxylic acid <3> (<3> relatively potent and selective inhibitor [17]) [17]
CN^- <2, 3> (<2> high concentration [10]; <3> inhibition at 0.01 M [2]) [2, 10]
Cl^- <3> (<3> 70% inhibition with 0.1 M NaCl or KCl, competitive with respect to NADPH and non-competitive with respect to L-kynurenine [18]) [18]
EDTA <5> (<5> 68% inhibition at 2 mM [8]) [8]
KCl <2> (<2> high concentration [10]) [10]
L-tryptophan <5> (<5> 35% inhibition at 2 mM and 22% inhibition at 0.2 mM [8]) [8]
NADH <4> (<4> above 3 mM [7]) [7]
α-ketoisocaproate <6> (<6> non competitive inhibition [11]) [11]
anthranilic acid <5> (<5> 22% inhibition at 2 mM and 11% inhibition at 0.2 mM [8]) [8]
kynurenic acid <5> (<5> 13% inhibition at 2 mM and 5% inhibition at 0.2 mM [8]) [8]
p-chloromercuribenzoate <4> (<4> weak, 50% inhibition at 0.4 M [7]) [7]
pyridoxal <3> (<3> less potent than pyridoxal phosphate [18]) [18]
pyridoxal 5'-phosphate <3> (<3> non competitive inhibition [18]) [18]
pyruvate <6> (<6> mixed type inhibitor [11]) [11]
xanthommatin <2> [10]
xanthurenic acid <5> (<5> 48% inhibition at 2 mM and 13% inhibition at 0.2 mM [8]) [8]
Additional information <3, 8> (<3> inhibition by N-(4-phenylthiazol-2-yl)-benzenesulfonamides with various modifications [14]; <8> inhibition by various 4-aryl-2-hyroxy-4-oxobut-2-enoic acids and esters at 10 micromolar [20]; <8> inhibition by various 2-amino-4-aryl-4-oxobut-2-enoic acids and esters at 10 micromolar [20]) [14, 20]

Cofactors/prosthetic groups

FAD <3> (<3> flavoprotein [1-3]; <3> 4 mol of FAD per subunit [3];<3> one molecule non-covalently bound FAD per molecule of enzyme [18]) [1-3, 18]
NADH <2, 3, 4> (<2,3> no activity [2,10]; <4> less effective than NADPH [7]; <3> less effective than NADPH [18]) [3, 7, 10]
NADPH <2-4, 6> [1-7, 10, 11, 18]

Activating compounds

azide <3> (<3> 0.005 M [2]) [2]
thiocyanate <3> (<3> 0.01 M [2]) [2]

Metals, ions

Br^- <3> (<3> strong stimulation [2]) [2]
CN^- <3> (<3> stimulates [2]) [2]
Cl^- <3> (<3> strong stimulation at 0.02 M [2]) [2]
F^- <3> (<3> stimulates [2]) [2]
I^- <3> (<3> stimulates [2]) [2]

Specific activity (U/mg)

0.028 <3> (<3> NADPH oxidase activity [18]) [18]
0.073 <3> [5]
0.139 <3> [4]
2 <3> (<3> fusion protein with glutathione-S-transferase [18]) [18]
47 <3> (<3> native [1]) [1]
Additional information <3, 4, 5, 7, 8, 9> (<3> 20000000 cpm per micromole [1]; <3> specific activities of apoenzyme and holoenzyme with different electron acceptors [3]; <4> enzyme activity at different temperatures in different strains [6]; <4> enzyme activity with different cofactors and under anaerobic conditions [7]; <5> specific activity in various organs [8]; <3> specific activity in organs of rats with experimental allergic encephalopathy [19]) [1, 3, 6, 7, 8, 19]

K_m-Value (mM)

0.00012 <3> (FAD) [3]
0.0003 <3> (FAD) [1]
0.016 <4> (NADPH, <4> under aerobic conditions [7]) [7]
0.0164 <5> (L-kynurenine) [8]
0.017 <4> (NADPH, <4> under unaerobic conditions [7]) [7]
0.023 <3> (NADPH) [2]
0.025 <3> (kynurenine) [2]
0.04 <4> (L-kynurenine, <4> under aerobic conditions [7]) [7]
0.045 <4> (L-kynurenine, <4> under unaerobic conditions [7]) [7]
0.15 <2> (NADPH) [10]
0.33 <2> (kynurenine) [10]

K_i-Value (mM)

0.0000048 <3> (3,4-dimethoxy-N-[4-(3-nitrophenyl)thiazol-2-yl]benzenesulfonamide, <3> competitive inhibitor [14]) [14]
0.000095 <3> ((R,S)-2-amino-oxo-4-(3'-4'-dichlorophenyl)butanoic acid, <3> competitive inhibitor [14]) [14]
0.19 <3> (pyridoxal 5'-phosphate) [18]
1 <3> (pyridoxal) [18]
4.2 <6> (α-ketoisocaproate, <6> inhibition for L-kynurenine [11]) [11]
8.3 <6> (α-ketoisocaproate, <6> inhibition for NADPH [11]) [11]

pH-Optimum

7.5 <3> (<3> fusion protein with glutathione-S-transferase [18]) [18]
8 <3, 4> [2, 7]
8-8.5 <5> [8]
8.2 <2> [10]

pH-Range

7-9 <3> (<3> about 50% of maximal activity at pH 7 and 9 [18]) [18]
7-9.3 <3> (<3> about 50% of maximal activity at pH 7 and 9.3 [2]) [2]
7.1-8.9 <4> (<4> about 50% of maximal activity at pH 7.1 and 8.9 [7]) [7]

Temperature optimum (°C)

23 <3> (<3> assay at [2]) [2]
25 <3> (<3> assay at [5]) [5]
30 <3> (<3> assay at [3]) [3]

Temperature range (°C)

7-10 <5> (<5> 50% activity at 7 and 10°C [8]) [8]

4 Enzyme Structure

Molecular weight

49000 <10> (<10> SDS-PAGE [15]) [15]
55000 <3> (<3> SDS-PAGE [18]) [18]
55760 <3> (<3> calculated from primary sequence [18]) [18]
55760 <8> (<8> calculated from primary sequence [22]) [22]
200000 <3> (<3> or more, gel filtration, SDS-PAGE [4]) [4]
345000 <3> (<3> gel filtration [3]) [3]

Subunits

dimer <3> (<3> 2 * 160000 in 0.2% Triton-X100, SDS-PAGE [3]) [3]

5 Isolation/Preparation/Mutation/Application

Source/tissue

brain <3, 5, 7, 8> (<3> neurons and astrocytes [19]) [8, 19]
eye <1> [9]
integument <1> [9]
intestine <5> [8]
kidney <3, 5> [8, 14]
liver <3, 5, 10> [1-5, 8, 15, 17, 18]
lung <5> [8]
monocyte <8> [20]
ovary <2> [10]
spinal cord <3> [19]
spleen <5> [8]

Localization

cytoplasmic vesicle <3> [19]
mitochondrial outer membrane <3, 6, 10> [1, 4, 5, 11, 15, 18]
mitochondrion <1-4, 6, 10> [1-7, 9-11, 15, 17, 18]
mitochondrion <4> (promitochondrion) [6]

Purification

<3> (fusion protein with gluthatione-S-transferase [18]) [18]
<3> (partial [2]; homogeneity [3,4]; affinity chromatography [5]) [2-5]
<6> (partial [11]) [11]
<10> (partial [15]) [15]

Cloning

<3> (fusion protein with glutathione-S-transferase [18]) [18]
<8> (kinetic properties similar to the native liver enzyme [22]) [22]

6 Stability

Temperature stability

4 <5> (<5> no loss of enzymatic activity after 24 hours [8]) [8]
25 <5> (<5> 22% loss of enzymatic activity after 24 hours [8]) [8]
37 <3> (<3> stable for four hours [18]) [18]
100 <3> (<3> no enzymatic activity after boiling [1]) [1]

General stability information

<3>, enzyme is protected against inactivation in the solubilized state by the presence of DTT, Triton X-100, and FAD in 0.1 M Tris-acetate buffer at pH 8.1 [4]
<3>, solubilized enzyme is appreciably stabilized in 0.05 M Tris-acetate buffer, pH 8.1, 1 mM DTT, 0.2 M mannitol, 0.2% Triton X-100, 0.005 mM FAD [3]

Storage stability

<3>, -80°C stable for 6 months [18]
<5>, -70°C no loss of enzymatic activity after 24 hours [8]

References

[1] Okamoto, H.; Hayaishi, O.: Flavin adenine dinucleotide requirement for kynurenine hydroxylase of rat liver mitochondria. Biochem. Biophys. Res. Commun., **29**, 394-399 (1967)
[2] Saito, Y.; Hayaishi, O.; Rothenberg, S.: Studies on oxygenases: Enzymatic formation of 3-hydroxy-L-kynurenine from L-kynurenine. J. Biol. Chem., **229**, 921-934 (1957)
[3] Nisimoto, Y.; Takeuchi, F.; Shibata, Y.: Molecular properties of L-kynurenine 3-hydroxylase from rat liver mitochondria. J. Biochem., **81**, 1413-1425 (1977)
[4] Nisimoto, Y.; Takeuchi, F.; Shibata, Y.: Isolation of L-kynurenine 3-hydroxylase from the mitochondrial outer membrane of rat liver. J. Biochem., **78**, 573-581 (1975)
[5] Nishimoto, Y.; Takeuchi, F.; Shibata, Y.: Purification of L-kynurenine 3-hydroxylase by affinity chromatography. J. Chromatogr., **169**, 357-364 (1979)

[6] McLean-Bowen, C.A.; Parks, L.W.: Corresponding changes in kynurenine hydroxylase activity, membrane fluidity, and sterol composition in Saccharomyces cerevisiae mitochondria. J. Bacteriol., **143**, 1325-1333 (1981)
[7] Schott, H.H.; Staudinger, H.: The regulatory function of L-kynurenine 3-hydroxylase (EC 1.14.1.2) for the biosynthesis of pyridine nucleotides in anaerobically and aerobically grown Saccharomyces cerevisiae. Hoppe-Seyler's Z. Physiol. Chem., **352**, 1654-1658 (1971)
[8] Saito, K.; Quearry, B.J.; Saito, M.; Nowak, T.S.Jr.; Markey S.P.; Heyes, M.P.: Kynurenine 3-hydroxylase in brain: species activity differences and effect of gerbil cerebral ischemia. Arch. Biochem. Biophys., **307**, 104-109 (1993)
[9] Pinamonti, S.; Chiarelli-Alvisi, G.: Kynurenine-3-hydroxylase of Schistocera gregaria. Insect Biochem., **4**, 9-16 (1974)
[10] Stratakis, E.: Subcellular localization and properties of kynurenine 3-hydroxylase from eggs of Ephestia kuhniella Z. J. Comp. Physiol., **141**, 451-456 (1981)
[11] Shin, M.; Sano, K.; Umezawa, C.: Inhibition of L-kynurenine 3-hydroxylase from Saccharomyces carlsbergensis by α-keto acid derivatives of branched chain amino acids. J. Nutr. Sci. Vitaminol., **28**, 191-201 (1982)
[12] Giordani A.; Pevarello P.; Cini, M.; Bormetti, R.; Greco, F.; Toma, S.; Speciale, C.; Varasi, M.: 4-phenyl-4-oxo-butanoic acid derivatives inhibitors of kynurenine 3-hydroxylase. Bioorg. Med. Chem. Lett., **8**, 2907-2012 (1998)
[13] Speciale, C.; Cini, M.; Wu, H.Q.; Salvati, P.; Schwarcz, R.; Molinari, A.; Calabresi, M.; Varasi, M.: Kynurenic acid-enhancing and anti-ischemic effects of the potent kynurenine 3-hydroxylase inhibitor FCE28833 in rodents. Adv. Exp. Med. Biol., **398**, 221-227 (1996)
[14] Rover, S.; Cesura, A.M.; Huguenin, P.; Kettler, R.; Szente, A.: Synthesis and biochemical evaluation of N-(4-phenylthiazol-2-yl)benzenesulfonamides as high-affinity inhibitors of kynurenine 3-hydroxylase. J. Med. Chem., **40**, 4378-4385 (1997)
[15] Uemura, T.; Hirai, K.: Purification of L-kynurenine 3-monooxygenase from mitochondrial outer membrane of pig liver. Adv. Exp. Med. Biol., **467**, 619-623 (1999)
[16] Cozzi, A.; Carpenedo, R.; Moroni, F.: Kynurenine hydroxylase inhibitors reduce ischemic brain damage: studies with (*m*-nitrobenzoyl)-alanine (mNBA) and 3,4-dimethoxy-[-N-4-(nitrophenyl)thiazol-2yl]-benzenesulfonamide (Ro 61-8048) in models of focal or global brain ischemia. J. Cereb. Blood Flow Metab., **19**, 771-777 (1999)
[17] Heidempergher, F.; Pevarello, P.; Pillan, A.; Pinciroli, V.; Torre, A.D.; Speciale, C.; Marconi, M.; Cini, M.; Toma, S.; Greco, F.; Varasi, M.: Pyrrolo[3,2-c]quinoline derivates: a new class of kynurenine-3-hydroxylase inhibitors. Farmaco, **54**, 152-160 (1999)
[18] Breton, J.; Avanzi, N.; Magagnin, S.; Covini, N.; Magistrelli, G.; Cozzi, L.; Isacchi, A.: Functional characterization and mechanism of action of recombinant human kynurenine 3-hydroxylase. Eur. J. Biochem., **267**, 1092-1099 (2000)
[19] Chiarugi, A.; Cozzi, A.; Massacesik L.; Moroni, F.: Kynurenine 3-momooxygenase activity and neurotoxic kynurenine metabolites increase in the

spinal cord of rats with experimental allergic encephalomyelitis. Neurosci., **102**, 687-695 (2001)
[20] Drysdale, M.J.; Hind, S.L.; Jansen, M.; Reinhard, J.F.Jr: Synthesis and SAR of 4-aryl-2-hydroxy-4-oxobut-2-enoic acids and esters and 2-amino-4-aryl-4-oxobut-2-enoic acids and esters: Potent inhibitors of kynurenine-3-hydroxylase as potential neuroprotective agents. J. Med. Chem., **43**, 123-127 (2000)
[21] Urenjak, J.; Obrenowitch, T.P.: Kynurenine 3-hydroxylase inhibition in rats: Effects on extracellular kynurenic acid concentration and N-methyl-D-aspartate-induced depolarisation in the striatum. J. Neurochem., **75**, 2427-2433 (2000)
[22] Alberati-Giani, D.; Cesuura, A.M.; Broger, C.; Warren, W.D.; Rover, S.; Malherbe, P.: Cloning and functional expression of human kynurenine 3-monooxygenase. FEBS Lett., **410**, 407-412 (1997)
[23] Pellicciari, R.; Natalini, B.; Costantino, G.; Mahmoud, M.R.; Mattoli, L.; Sadeghpour, B.M.; Moroni, F.; Chiarugi, A.; Carpenedo, R.: Modulation of the kynurenine pathway in search for new neruoprotective agents. Synthesis and preliminary evaluation of (*m*-nitrobenzoys)alanine, a potent inhibitor of kynurenine-3-hydroxylase. J. Med. Chem., **37**, 647-655 (1994)

2,6-Dihydroxypyridine 3-monooxygenase 1.14.13.10

1 Nomenclature

EC number
1.14.13.10

Systematic name
2,6-dihydroxypyridine,NADH:oxygen oxidoreductase (3-hydroxylating)

Recommended name
2,6-dihydroxypyridine 3-monooxygenase

Synonyms
2,6-dihydroxypyridine oxidase

CAS registry number
39279-38-4

2 Source Organism

<1> *Arthrobacter nicotivorans* (formerly Arthrobacter oxidans [1-3]) [1-3]

3 Reaction and Specificity

Catalyzed reaction
2,6-dihydroxypyridine + NADH + H^+ + O_2 = 2,3,6-trihydroxypyridine + NAD^+ + H_2O

Reaction type
hydroxylation
oxidation
redox reaction
reduction

Natural substrates and products
S 2,6-dihydroxypyridine + NADH + O_2 <1> (Reversibility: ? <1> [1-3]) [1-3]
P 2,3,6-trihydroxypyridine + NAD^+ + H_2O

Substrates and products
S 2,6-dihydroxypyridine + NADH + O_2 <1> (Reversibility: ? <1> [1-3]) [1-3]
P 2,3,6-trihydroxypyridine + NAD^+ + H_2O <1> (<1> product dimerizes spontaneously in the presence of oxygen to form a blue pigment [1]; <1> pyridine ring cleavage follows in a second reaction [2]) [1-3]

S 2,6-dihydroxypyridine + NADH + electron acceptor <1> (<1> methylene blue and 2,6-dichlorophenol-indophenol can act as alternative electron acceptors [2]) (Reversibility: ? <1> [2]) [2]
P 2,3,6-trihydroxypyridine + NAD^+ + reduced electron acceptor <1> [3]
S Additional information <1> (<1> 2-hydroxypyridine, 3-hydroxypyridine, 4-hydroxypyridine, 2,5-dihydroxypyridine, 2,6-dipicolinic acid, *p*-hydroxybenzoic acid, nicotinic acid and nicotine are no substrates for this enzyme [2]) [2]
P ?

Inhibitors
2,3-dihydroxypyridine <1> (<1> 100% inhibition at 0.1 mM, irreversible inhibitor [1]) [1]
2,6-dihydroxynicotinamide <1> (<1> 55% inhibition at 0.1 mM, reversible inhibitor [1]) [1]
2,6-dimethoxypyridine <1> (<1> 100% inhibition at 0.1 mM, irreversible inhibitor [1]) [1]
2-hydroxypyridine <1> (<1> 48% inhibition at 0.1 mM, reversible inhibitor [1]) [1]
Cu^{2+} <1> (<1> 50% inhibition at 0.017 mM [2]) [2]
Hg^{2+} <1> (<1> 50% inhibition at 0.005 mM [2]) [2]
NaN_3 <1> (<1> 50% inhibition at 2 mM [2]) [2]
Zn^{2+} <1> (<1> 50% inhibition at 0.7 mM [2]) [2]
p-chloromercuribenzoate <1> (<1> 50% inhibition at 0.01 mM [2]) [2]
resorcine <1> (<1> 51% inhibition at 0.1 mM, reversible inhibitor [1]) [1]

Cofactors/prosthetic groups
FAD <1> (<1> 2 mol FAD per dimer [1]) [1, 2]
NADH <1> (<1> no reaction with NADPH [1]) [1, 2]
NADPH <1> (<1> less effective than NADH [2]) [2]

Activating compounds
CO <1> (<1> 35% activation [2]) [2]
EDTA <1> (<1> 38% activation at 0.05 mM [2]) [2]
KCN <1> (<1> 18% activation at 0.5 mM [2]) [2]

Specific activity (U/mg)
2.16 <1> [2]
90 <1> [1]

K_m-Value (mM)
0.0083 <1> (2,6-dihydroxypyridine, <1> pH 8, 20°C [1]) [1]

pH-Optimum
8 <1> [1, 2]

Temperature optimum (°C)
20 <1> [1]
30 <1> [2]

4 Enzyme Structure

Molecular weight

89000 <1> (<1> sucrose density gradient centrifugation [2]) [2]
90000 <1> (<1> gel filtration [1]) [1]

Subunits

homodimer <1> [1]

5 Isolation/Preparation/Mutation/Application

Localization

soluble <1> [2]

Purification

<1> (partial [2]) [1, 2]

Renaturation

<1> (after removal of FAD through precipitation with 50% $(NH_4)_2SO_4$ at pH 2.0, incubation with 0.1 mM FAD leeds to 90% activity recovery [1]) [1]

Cloning

<1> (expressed in Escherichia coli XL-1 blue [1]) [1]

6 Stability

pH-Stability

7 <1> (<1> most stable in 50 mM potassium phosphate buffer [2]) [2]

Temperature stability

4 <1> (<1> stable [1]) [1]
15 <1> (<1> 30% loss of activity in 10 min [2]) [2]
30 <1> (<1> complete loss of activity in 10 min [2]) [2]
52 <1> (<1> inactivation [1]) [1]

Organic solvent stability

acetone <1> (<1> complete loss of activity [2]) [2]
ethanol <1> (<1> stabilizes during preparative manipulations [2]) [2]
glycerol <1> (<1> stabilizes during preparative manipulations [2]) [2]

General stability information

<1>, dialysis against 1.5 M guanidinium hydrochloride removes FAD, no renaturation possible [1]

References

[1] Baitsch, D.; Sandu, C.; Brandsch, R.; Igloi, G.L.: Gene cluster on pAO1 of Arthrobacter nicotinovorans involved in degradation of the plant alkaloid nicotine: cloning, purification, and characterization of 2,6-dihydroxypyridine 3-hydroxylase. J. Bacteriol., **183**, 5262-5267 (2001)
[2] Holmes, P.E.; Rittenberg, S.C.: The bacterial oxidation of nicotine. VII. Partial purification and properties of 2,6-dihydroxypyridine oxidase. J. Biol. Chem., **247**, 7622-7627 (1972)
[3] Holmes, P.E.; Rittenberg, S.C.; Knackmuss, H.J.: The bacterial oxidation of nicotine. 8. Synthesis of 2,3,6-trihydroxypyridine and accumulation and partial characterization of the product of 2,6-dihydroxypyridine oxidation. J. Biol. Chem., **247**, 7628-7633 (1972)

trans-Cinnamate 4-monooxygenase 1.14.13.11

1 Nomenclature

EC number
1.14.13.11

Systematic name
trans-cinnamate,NADPH:oxygen oxidoreductase (4-hydroxylating)

Recommended name
trans-cinnamate 4-monooxygenase

Synonyms
C4H
CA4H
CA4Hase
CYP 73
cytochrome P450 73
P450C4H
cinnamate 4-hydroxylase
cinnamate 4-monooxygenase
cinnamate hydroxylase
cinnamic 4-hydroxylase
cinnamic acid 4-hydroxylase
cinnamic acid 4-monooxygenase
cinnamic acid *p*-hydroxylase
cytochrome P_{450} cinnamate 4-hydroxylase
hydroxylase, cinnamate 4-
oxygenase, cinnamate 4-mono-
t-cinnamic acid hydroxylase
trans-cinnamate 4-hydroxylase
trans-cinnamic acid 4-hydroxylase

CAS registry number
9077-75-2

2 Source Organism

<1> *Pisum sativum* [1, 10]
<2> *Sorghum sp.* (Sorghum-Sudangrass hybrid [2]) [2]
<3> *Petroselinum sp.* (Petroselinum hortense [3]; Petroselinum crispum [23]) [3, 12, 23]
<4> *Cucumis sativus* (cv. Venlo Pickling [13]) [7, 8, 13]

<5> *Glycine max* [5]
<6> *Phaseolus mungo* [6]
<7> *Helianthus tuberosus* (expressed by an optimized yeast system [20]) [9, 14, 16, 17, 18, 20, 27]
<8> *Brassica napo-brassica* [11]
<9> *Populus sp.* (Populus trichocarpa * Populus deltoides [21]) [21]
<10> *Vicia sativa* [15]
<11> *Catharanthus roseus* [19]
<12> *Medicago sativa* [22]
<13> *Phaseolus vulgaris* (L. cv. immuna [24]) [4, 24, 26]
<14> *Arabidopsis thaliana* [25]
<15> *Nicotiana tabacum* [26]

3 Reaction and Specificity

Catalyzed reaction
trans-cinnamate + NADPH + H^+ + O_2 = 4-hydroxycinnamate + $NADP^+$ + H_2O

Reaction type
oxidation
redox reaction
reduction

Natural substrates and products
S trans-cinnamate + NADPH + O_2 <1, 3, 4, 7, 8, 9, 12, 13, 14, 15> (<4>, blue-light treatment of etiolated seedlings mediates a transient and concurrent increase in activity of cotyledons but not of hypocotyl [7]; <7>, the enzyme is regulated by the NADPH/$NADP^+$ ratio [9]; <1>, enzyme is stimulated by light [10]; <8>, enzyme is involved in the biosynthesis of lignin precursors [11]; <3>, enzyme catalyzes an important step in the biosynthesis of flavonoids, lignin and other plant phenolics [12]; <7>, enzyme catalyzes the first oxidative step of the phenylpropanoid pathway [17, 18, 26]; <9>, the enzyme plays a central role in phenylpropanoid metabolism and lignin biosynthesis and possibly anchors a phenylpropanoid enzyme complex to the endoplasmic reticulum [21]; <12, 13>, second enzyme of the phenylpropanoid pathway [22, 26]; <14>, enzyme of the phenylpropanoid pathway [25]; <15>, enzyme may function in lignification [26]) (Reversibility: ? <1, 3, 4, 7, 8, 9, 12, 13, 14, 15> [7, 9, 10, 11, 12, 17, 18, 21, 22, 25, 26]) [7, 9, 10, 11, 12, 17, 18, 21, 22, 25, 26]
P 4-hydroxycinnamate + $NADP^+$ + H_2O

Substrates and products
S 7-ethoxycoumarin + O_2 + NADPH <7> (<7>, demethylase [20]) (Reversibility: ? <7> [20]) [20]
P umbelliferone + ?
S 7-methoxycoumarin + O_2 + NADPH <7> (<7>, demethylase [20]) (Reversibility: ? <7> [20]) [20]

P umbelliferone + ?
S chlorotoluron + O_2 + NADPH <7> (<7>, methylhydroxylase activity [20]) (Reversibility: ? <7> [10]) [20]
P ?
S *p*-chloro-N-methylaniline + O_2 + NADPH <7> (<7>, N-demethylase activity [20]) (Reversibility: ? <7> [20]) [20]
P *p*-chloroaniline + ?
S trans-cinnamate + NADPH + O_2 <1-15> (<1>, tetrahydrofolate required for reaction [1]) (Reversibility: ? <1-15> [1-27]) [1-27]
P 4-hydroxycinnamate + $NADP^+$ + H_2O <1-15> [1-27]

Inhibitors

1,4-naphthoquinone <2> [2]
1-aminobenzotriazole <7, 10> (<7>, autocatalytic inactivation [14]) [14, 15, 20]
11-dodecynoic acid <10> (<10>, slight [15]) [15]
2,4-dinitrophenol <4> (<4>, 0.1 mM, 83% loss of activity [8]) [8]
2-isopropyl-4-pentenamide <7> [20]
2-mercaptoethanol <4> (<4>, 2.0 mM [8]) [8]
3(2,4-dichlorophenoxy)-1-propyne <7> [20]
3-(2,4-dichlorophenoxy)-1-propyne <10> (<10>, mechanism-based inhibitor [15]) [15]
3-phenoxy-1-propyne <10> (<10>, mechanism-based inhibitor [15]) [15]
CO <4> [8]
$CuSO_4$ <4> (<4>, 1 mM, complete loss of activity [8]) [8]
EDTA <4> (<4>, 1 mM, 18% loss of activity [8]) [8]
FAD <4> (<4>, 1 mM, 73% loss of activity [8]) [8]
FMN <4> (<4>, 1 mM, 75% loss of activity [8]) [8]
KCN <4> (<4>, 10 mM, 21% loss of activity [8]) [8]
KCl <4> (<4>, 200 mM, 39% loss of activity. 500 mM, 69% loss of activity [8]) [8]
$MnCl_2$ <4> (<4>, 1 mM, 27% loss of activity [8]) [8]
$NADP^+$ <7> (<7>, competitive [9]) [9]
antimycin A <4> (<4>, 0.005 mM, 16% loss of activity [8]) [8]
ascorbic acid <4> (<4>, 2.0 mM [8]) [8]
benzoquinone <2> [2]
chlorogenic acid <4> (<4>, 0.1 mM, 9% loss of activity [8]) [8]
cis-cinnamate <3> (<3>, competitive [12]) [12]
dithiothreitol <4> (<4>, 2.0 mM [8]) [8]
gallic acid <4> (<4>, 0.1 mM, 20% loss of activity [8]) [8]
menadione <2, 4> (<4>, 0.1 mM, 92% loss of activity [8]) [2, 8]
phenoxy-1-propyne <7> [20]
sodium azide <4> (<4>, 10 mM, 64% loss of activity [8]) [8]
Additional information <4> (<4>, no inhibition by *p*-coumaric acid [8]; <4>, a specific and reversible macromolecular inhibitor from dark-grown hypocotyl of Cucumis sativus [13]) [8, 13]

Cofactors/prosthetic groups

NADH <4, 7> (<4>, NADH has a synergistic effect on NADPH-supported hydroxylation, at both nonsaturating, 0.1 mM, and near-saturating, 1 mM, concentrations of NADPH [8]; <7>, NADH alone can not supply the enzyme with electrons. NADH, 0.5 mM, results in 51% stimulation when NADPH is present at a saturating level, 0.5 mM [9]) [8, 9]

NADPH <1-15> (<1>, no activity with NADH [1]; <4>, NADH has a synergistic effect on NADPH-supported hydroxylation, at both nonsaturating, 0.1 mM, and near-saturating, 1 mM, concentrations of NADPH [8]; <7>, NADH alone can not supply the enzyme with electrons. NADH, 0.5 mM, results in 51% stimulation when NADPH is present at a saturating level, 0.5 mM [9]) [1-27]

cytochrome P_{450} <2, 4, 5, 7, 9, 13, 14> (<14>, cytochrome P_{450}-dependent monooxygenase [25]; <2>, required [2]; <5>, activity is dependent on NADPH:cytochrome P-450 reductase [5]; <4>, involved in hydroxylation [8]; <7>, electrons are transferred from NADPH, the preferential electron donor of the system, to cytochrome P-450 via NADPH-cytochrome P-450 reductase [9]) [2, 5, 8, 9, 14, 16, 17, 18, 20, 21, 24, 25, 27]

Activating compounds

2-mercaptoethanol <4> (<4>, 0.1 mM, stimulates [8]) [8]
ascorbic acid <4> (<4>, 0.1 mM, stimulates [8]) [8]
dithiothreitol <4> (<4>, 0.1 mM, stimulates [8]) [8]
glutathione <4> (<4>, 0.1 mM, stimulates [8]) [8]
lipid <3> (<3>, lipid component required [3]) [3]

Metals, ions

KCN <4> (<4>, 1 mM, increase to 125% of the activity [8]) [8]

Turnover number (min^{-1})

0.014 <7> (chlorotoluron) [20]
3.4 <7> (7-ethoxycoumarin) [20]
11.4 <7> (7-methoxycoumarin) [20]
142 <7> (*p*-chloro-N-methylaniline) [20]
297 <7> (trans-cinnamate) [20]
Additional information <7> [18]

Specific activity (U/mg)

Additional information <5, 6> (<6>, simple and sensitive assay method, based on the migration of tritium during the enzyme-catalyzed hydroxylation [6]) [5, 6]

K_m-Value (mM)

0.0023 <2> (trans-cinnamate) [2]
0.0025 <4> (trans-cinnamate) [8]
0.005 <7> (trans-cinnamate) [20]
0.013 <7> (NADPH) [9]
0.021 <2> (O_2) [2]
0.555 <7> (7-ethoxycoumarin) [20]

1.919 <7> (*p*-chloro-N-methylaniline) [20]
4 <7> (7-methoxycoumarin) [20]
Additional information <7> [18]

K_i-Value (mM)

0.34 <3> (cis-cinnamate) [12]

pH-Optimum

7.5 <1, 4> [1, 8]

pH-Range

5.5-8.8 <4> (<4>, about 50% of maximal activity at pH 5.5 and at pH 8.8 [8]) [8]

4 Enzyme Structure

Subunits

? <5, 7, 13> (<13>, x * 57000, SDS-PAGE [24]; <7>, x * 57000, SDS-PAGE [16]; <5>, x * 58000, SDS-PAGE [5]) [5, 16, 24]

5 Isolation/Preparation/Mutation/Application

Source/tissue

cell suspension culture <3, 13> [12, 24]
cotyledon <4> [7, 8]
disc <8> (<8>, activity increases during aging of disks from very low initial values [11]) [11]
hypocotyl <4, 6> (<6>, etiolated [6]) [6, 7, 8]
leaf <9> [21]
root <14> [25]
seedling <1, 2, 4> [1, 2, 7, 10, 13]
tuber <7> [9, 14, 16, 17]
xylem <9, 14> (<9>, highest level of activity [21]; <14>, cell undergoing lignification [25]) [21, 25]

Localization

endoplasmic reticulum <4, 7, 9> [8, 9, 21]
microsome <1, 2, 3, 6, 7, 10, 12, 13> (<1>, more than 95% of the activity [1]; <13>, membrane [24]) [1, 2, 3, 6, 10, 14, 15, 16, 18, 20, 22, 23, 24]
mitochondrion <1> (<1>, less than 5% of the activity [1]) [1]

Purification

<5> [5]
<7> [16]
<13> [24]

Cloning

<3> (expression in Saccharomyces cerevisiae [23]) [23]
<7> (expression in Saccharomyces cerevisiae [18]) [17, 18, 27]
<9> (cinnamate 4-hydroxylase and cinnamate 4-hydroxylase fused to the FLAG epitope expressed in yeast. A chimeric cinnamate 4-hydroxylase/green fluorescent protein gene is engineered and stable expressed in Arabidopsis [21]) [21]
<11> (translational fusion with P_{450} reductase in Escherichia coli [19]) [19]
<12> (generation of transgenic tobacco lines with altered activity levels of cinnamic acid 4-hydroxylase by sense or antisense expression of an alfalafa cDNA [22]) [22]
<13> (successful expression requires removal of the intron, yeast does not process the intron at all [4]; transformation of Nicotiana tabacum with a truncated enzyme from Phaseolus vulgaris [26]) [4, 26]
<14> [25]

6 Stability

Storage stability

<1>, -20°C or 4°C, 0.1 M phosphate buffer, homogenate loses 2/3 of its activity in 24 h [1]

References

[1] Russell, D.W.; Conn, E.E.: The cinnamic acid 4-hydroxylase of pea seedlings. Arch. Biochem. Biophys., **122**, 256-258 (1967)
[2] Potts, J.R.M.; Weklych, R.; Conn, E.E.: The 4-hydroxylation of cinnamic acid by sorghum microsomes and the requirement for cytochrome P-450. J. Biol. Chem., **249**, 5019-5026 (1974)
[3] Bueche, T.; Sandermann, H.: Lipid dependence of plant microsomal cinnamic acid 4-hydroxylase. Arch. Biochem. Biophys., **158**, 445-447 (1973)
[4] Nedelkina, S.; Jupe, S.C.; Blee, K.A.; Schalk, M.; Werck-Reichhart, D.; Bolwell, G.P.: Novel characteristics and regulation of a divergent cinnamate 4-hydroxylase (CYP73A15) from French bean: engineering expression in yeast. Plant Mol. Biol., **39**, 1079-1090 (1999)
[5] Kochs, G.; Werck-Reichhart, D.; Grisebach, H.: Further characterization of cytochrome P_{450} involved in phytoalexin synthesis in soybean: cytochrome P_{450} cinnamate 4-hydroxylase and 3,9-dihydroxypterocarpan 6a-hydroxylase. Arch. Biochem. Biophys., **293**, 187-194 (1992)
[6] Suzuki, K.: trans-Cinnamic acid 4-hydroxylase of Phaseolus mungo. A new assay method. Anal. Biochem., **88**, 468-474 (1978)
[7] Billett, E.E.; Smith, H.: Control of phenylalanine ammonia-lyase and cinnamic acid 4-hydroxylase in gherkin tissue. Phytochemistry, **19**, 1035-1041 (1980)

[8] Billett, E.E.; Smith, H.: Cinnamic acid 4-hydroxylase from Gherkin tissue. Phytochemistry, **17**, 1511-1516 (1978)
[9] Beneviste, I.; Salaun, J.P.; Durst, F.: Wounding-induced cinnamic acid hydroxylase in Jerusalem artichoke tuber. Phytochemistry, **16**, 69-73 (1977)
[10] Beneviste, I.; Salaun, J.P.; Durst, F.: Phytochrome-mediated regulation of a monooxygenase hydroxylating cinnamic acid in etiolated pea seedlings. Phytochemistry, **17**, 359-363 (1978)
[11] Rhodes, M.J.C.; Hill, A.C.R.; Wooltorton, L.S.C.: Activity of enzymes involved in lignin biosynthesis in swede root disks. Phytochemistry, **15**, 707-710 (1976)
[12] Pfaendler, R.; Scheel, D.; Sandermann, H.; Grisebach, H.: Stereospecificity of plant microsomal cinnamic acid 4-hydroxylase. Arch. Biochem. Biophys., **178**, 315-316 (1977)
[13] Billett, E.E.; Wallace, W.; Smith, H.: A specific and reversible macromolecular inhibitor of phenylalanine ammonia-lyase and cinnamic acid-4-hydroxylase in gherkins. Biochim. Biophys. Acta, **524**, 219-230 (1978)
[14] Reichhart, D.; Simon, A.; Durst, F.: Autocatalytic inactivation of plant cytochrome P-450 enzymes: selective inactivation of cinnamic acid 4-hydroxylase from Helianthus tuberosus by 1-aminobenzotriazole. Arch. Biochem. Biophys., **216**, 522-529 (1982)
[15] Salauen, J.P.; Simon, A.; Durst, F.; Reich, N.O.; Ortiz de Montellano, P.R.: Differential inactivation of plant lauric acid ω- and in-chain-hydroxylases by terminally unsaturated fatty acids. Arch. Biochem. Biophys., **260**, 540-545 (1988)
[16] Gabriac, B.; Werck-Reichhart, D.; Teutsch, H.; Durst, F.: Purification and immunocharacterization of a plant cytochrome P_{450}: the cinnamic acid 4-hydroxylase. Arch. Biochem. Biophys., **288**, 302-309 (1991)
[17] Teutsch, H.G.; Hasenfratz, M.P.; Lesot, A.; Stoltz, C.; Garnier, J.M.; Jeltsch, J.M.; Durst, F.; Werck-Reichhart, D.: Isolation and sequence of a cDNA encoding the Jerusalem artichoke cinnamate 4-hydroxylase, a major plant cytochrome P_{450} involved in the general phenylpropanoid pathway. Proc. Natl. Acad. Sci. USA, **90**, 4102-4106 (1993)
[18] Urban, P.; Werck-Reichhart, D.; Teutsch, H.G.; Durst, F.; Regnier, S.; Kazmaier, M.; Pompon, D.: Characterization of recombinant plant cinnamate 4-hydroxylase produced in yeast. Kinetic and spectral properties of the major plant P_{450} of the phenylpropanoid pathway. Eur. J. Biochem., **222**, 843-850 (1994)
[19] Hotze, M.; Schroeder, G.; Schroeder, J.: Cinnamate 4-hydroxylase from Catharanthus roseus, and a strategy for the functional expression of plant cytochrome P_{450} proteins as translational fusions with P_{450} reductase in Escherichia coli. FEBS Lett., **374**, 345-350. (1995)
[20] Pierrel, M.A.; Batard, Y.; Kazmaier, M.; Mignotte-Vieux, C.; Durst, F.; Werck-Reichhart, D.: Catalytic properties of the plant cytochrome P_{450} CYP73 expressed in yeast. Substrate specificity of a cinnamate hydroxylase. Eur. J. Biochem., **224**, 835-844 (1994)

[21] Ro, D.K.; Mah, N.; Ellis, B.E.; Douglas, C.J.: Functional characterization and subcellular localization of poplar (Populus trichocarpa.times. Populus deltoides) cinnamate 4-hydroxylase. Plant Physiol., **126**, 317-329 (2001)
[22] Blount, J.W.; Korth, K.L.; Masoud, S.A.; Rasmussen, S.; Lamb, C.; Dixon, R.A.: Altering expression of cinnamic acid 4-hydroxylase in transgenic plants provides evidence for a feedback loop at the entry point into the phenylpropanoid pathway. Plant Physiol., **122**, 107-116 (2000)
[23] Koopmann, E.; Logemann, E.; Hahlrock, K.: Regulation and functional expression of cinamate 4-hydroxylase from parsley. Plant Physiol., **119**, 49-55 (1999)
[24] Rodgers, M.W.; Zimmerlin, A.; Werck-Reichhart, D.; Bolwell, G.P.: Microsomally associated heme proteins from French bean: characterization of the cytochrome P_{450} cinnamate-4-hydroxylase and two peroxidases. Arch. Biochem. Biophys., **304**, 74-80. (1993)
[25] Bell-Lelong, D.A.; Cusumano, J.C.; Meyer, K.; Chapple, C.: Cinnamate-4-hydroxylase expression in Arabidopsis. Regulation in response to development and the environment. Plant Physiol., **113**, 729-738. (1997)
[26] Blee, K.; Choi, J.W.; O'Connell, A.P.; Jupe, S.C.; Schuch, W.; Lewis, N.G.; Bolwell, G.P.: Antisense and sense expression of cDNA coding for CYP73A15, a class II cinnamate 4-hydroxylase, leads to a delayed and reduced production of lignin in tobacco. Phytochemistry, **57**, 1159-1166. (2001)
[27] Teutsch, H.G.; Hasenfratz, M.P.; Lesot, A.; Stoltz, C.; Garnier, J.M.; Jeltsch, J.M.; Durst, F.; Werck-Reichhart, D.: Isolation and sequence of a cDNA encoding the Jerusalem artichoke cinnamate 4-hydroxylase, a major plant cytochrome P_{450} involved in the general phenylpropanoid pathway. Proc. Natl. Acad. Sci. USA, **90**, 4102-4106 (1993)

Benzoate 4-monooxygenase 1.14.13.12

1 Nomenclature

EC number
1.14.13.12

Systematic name
benzoate,NADPH:oxygen oxidoreductase (4-hydroxylating)

Recommended name
benzoate 4-monooxygenase

Synonyms
benzoate 4-hydroxylase
benzoate-4-hydroxylase
benzoate-*p*-hydroxylase
benzoic 4-hydroxylase
benzoic acid 4-hydroxylase
hydroxylase, benzoate 4-
oxygenase, benzoate 4-mono-

CAS registry number
39391-25-8

2 Source Organism

<1> *Rhodotorula graminis* [1, 4]
<2> *Pseudomonas sp.* [2]
<3> *Aspergillus niger* (strain UBC 814 [3]; strain N204 [7]) [3, 5, 7, 8]
<4> *Rhodotorula minuta* (red yeast [6]) [6]

3 Reaction and Specificity

Catalyzed reaction
benzoate + NADPH + H^+ + O_2 = 4-hydroxybenzoate + $NADP^+$ + H_2O

Reaction type
oxidation
redox reaction
reduction

Natural substrates and products

S benzoate + NADPH + O_2 <3> (<3> enzyme is involved in degradation of chlorinated benzoic acid derivatives e.g. 2-chlorobenzoate and 3-chlorobenzoate [5]) (Reversibility: ? <3> [5]) [5]
P 4-hydroxybenzoate + $NADP^+$ + H_2O <3> [5]

Substrates and products

S 2-chlorobenzoate + NADPH + O_2 <3> (Reversibility: ? <3> [8]) [8]
P 2-chloro-4-hydroxybenzoate + $NADP^+$ + H_2O <3> [8]
S 2-fluorobenzoate + NADPH + O_2 <3> (Reversibility: ? <3> [8]) [8]
P 2-fluoro-4-hydroxybenzoate + $NADP^+$ + H_2O <3> [8]
S 2-hydroxybenzoate + NADPH + O_2 <3> (Reversibility: ? <3> [8]) [8]
P 2,4-dihydroxybenzoate + $NADP^+$ + H_2O <3> [8]
S 2-methylbenzoate + NADPH + O_2 <3> (Reversibility: ? <3> [8]) [8]
P 2-methyl-4-hydroxybenzoate + $NADP^+$ + H_2O <3> [8]
S 3-chlorobenzoate + NADPH + O_2 <3> (Reversibility: ? <3> [8]) [8]
P 3-chloro-4-hydroxybenzoate + $NADP^+$ + H_2O <3> [8]
S 3-fluorobenzoate + NADPH + O_2 <3> (Reversibility: ? <3> [8]) [8]
P 3-fluoro-4-hydroxybenzoate + $NADP^+$ + H_2O <3> [8]
S 3-hydroxybenzoate + NADPH + O_2 <3> (Reversibility: ? <3> [8]) [8]
P 3,4-hydroxybenzoate + $NADP^+$ + H_2O <3> [8]
S 3-methoxybenzoate + NADPH + O_2 <3> (Reversibility: ? <3> [8]) [8]
P 3-methoxy-4-hydroxybenzoate + $NADP^+$ + H_2O <3> [8]
S 3-methylbenzoate + NADPH + O_2 <3> (Reversibility: ? <3> [8]) [8]
P 3-methyl-4-hydroxybenzoate + $NADP^+$ + H_2O <3> [8]
S 4-chlorobenzoate + NADPH + O_2 <2> (Reversibility: ? <2> [2]) [2]
P ?
S 4-methylbenzoate + NADPH + O_2 <2> (Reversibility: ? <2> [2]) [2]
P ?
S benzoate + NADPH + O_2 <1-4> (<3> specific for benzoate, NADPH is absolutely required, 30% activity with ascorbate as electron donor [3]) (Reversibility: ? <1-4> [1-3, 6]) [1-3, 6, 8]
P 4-hydroxybenzoate + $NADP^+$ + H_2O <1-4> [1-3, 6, 8]
S cinnamate + NADPH + O_2 <4> (<4> very low turnover rate [6]) (Reversibility: ? <4> [6]) [6]
P 4-coumarate + $NADP^+$ + H_2O <4> [6]
S nicotinate + NADPH + O_2 <3> (Reversibility: ? <3> [8]) [8]
P ?
S picolinate + NADPH + O_2 <3> (Reversibility: ? <3> [8]) [8]
P ?

Inhibitors

2,2'-dipyridyl <3> (<3> 0.05 mM; 50% inhbition [3]) [3]
4-chlorobenzoate <2> [2]
4-methylbenzoate <2> [2]
4-nitrobenzoate <2> (<2> 38% inhibition [2]) [2]
8-hydroxyquinoline <3> (<3> 5 mM, complete inhibition [3]) [3]
CO <1, 4> (<1> 44% inhibition [1]; <4> strong inhibition [6]) [1, 6]

Cu^{2+} <2, 3> (<2> 0.005 mM, 50% inhibition [2]; <3> 0.001-0.01 mM, complete inhibition [3]) [2, 3]
Hg^{2+} <2, 3> (<2> 0.005 mM, 70% inhibition [2]; <3> 0.001-0.01 mM, complete inhibition [3]) [2, 3]
KCN <1> (<1> 30 mM, 21% inhibition [1]) [1]
Mg^{2+} <3> (<3> 0.001-0.01 mM, complete inhibition [3]) [3]
Mn^{2+} <3> (<3> 0.001-0.01 mM, complete inhibition [3]) [3]
Mo^{2+} <3> (<3> 0.001-0.01 mM, complete inhibition [3]) [3]
N-ethylmaleimide <3> [3]
SKF-525A <1> (<1> P_{450} inhibitior, 1 mM, 51% inhibition [1]) [1]
Zn^{2+} <3> (<3> 0.001-0.01 mM, complete inhibition [3]) [3]
aminopterin <1> (<1> 0.5 mM, 10% inhibition [1]) [1]
benzaldehyde <2> [2]
benzoate methyl ester <2> (<2> slight inhibition [2]) [2]
benzyl acetate <1> (<1> 0.5 mM, 21% inhibition [1]) [1]
benzyl alcohol <1> (<1> 0.5 mM, 10% inhibition [1]) [1]
benzylformate <1> (<1> 0.5 mM, 10% inhibition [1]) [1]
diethyldithiocarbamate <1, 3> (<1> 0.5 mM, 21% inhibition [1]; <3> 5 mM, 95% inhibition [3]) [1, 3]
iodoacetamide <1> (<1> 5 mM, 43% inhibition [1]) [1]
iodoacetate <1> (<1> 5 mM, 23% inhibition [1]) [1]
m-hydroxybenzoate <2, 4> (<2> 12% inhibition [2]) [2, 6]
o-phenanthroline <3> (<3> 0.05 mM, 70% inhibition [3]) [3]
oxalate <3> (<3> 2 mM, 20% inhibition [3]) [3]
p-chloromercuribenzoate <3> [3]
p-hydroxymercuribenzoate <1> (<1> 0.052 mM, 39% inhibition [1]) [1]
phenylalanine <1> (<1> 0.5 mM, 21% inhibition [1]) [1]
quinacrine <1> (<1> 0.5 mM, 47% inhibition [1]) [1]
quinine-HCl <1> (<1> 0.5 mM, 42% inhibition [1]) [1]
salicylate <2, 4> (<2> 20% inhibition [2]) [2, 6]
trans-cinnamate <1> (<1> 0.5 mM, 55% inhibition [1]) [1]

Cofactors/prosthetic groups

FAD <1> (<1> 2fold stimulation of NADPH oxidation rate [1]) [1]
NADH <2> (<2> hydroxylation is not supported when NADH, NADPH, dithiothreitol, 2-mercaptoethanol and ascorbic acid are added alone to the reaction mixture, stimulation if added together with tetrahydropteridine [2]) [2]
NADPH <1-3> (<2> hydroxylation is not supported when NADH, NADPH, dithiothreitol, 2-mercaptoethanol and ascorbic acid are added alone to the reaction mixture, stimulation if added together with tetrahydropteridine [2]; <3> absolute requirement for NADPH [3]) [1-3]
tetrahydropteridine <2, 3> (<2> essential for activity, hydroxylation is not supported when NADH, NADPH, dithiothreitol, 2-mercaptoethanol and ascorbic acid are added alone to the reaction mixture, stimulation when added together with tetrahydropteridine [2]; <3> required as prosthetic group,

maximal activity with 6,7-dimethyl 5,6,7,8-tetrahydropteridine, 60% activity with biopterine and tetrahydrofolic acid [3]) [2, 3]
Additional information <1, 4> (<4> tetrahydropteridine has no effect on benzoate 4-hydroxylase activity [6]; <1> pteridine independent activity [1]) [1, 6]

Activating compounds

2-mercaptoethanol <2> (<2> hydroxylation is not supported when NADH, NADPH, dithiothreitol, 2-mercaptoethanol and ascorbic acid are added alone to the reaction mixture, stimulation when added together with tetrahydropteridine [2]) [2]
ascorbic acid <2, 3> (<2> hydroxylation is not supported when NADH, NADPH, dithiothreitol, 2-mercaptoethanol and ascorbic acid are added alone to the reaction mixture, stimulation when added together with tetrahydropteridine [2]; <3> 30% of activity with NADPH as electron donor [3]) [2, 3]
dithiothreitol <2> (<2> hydroxylation is not supported when NADH, NADPH, dithiothreitol, 2-mercaptoethanol and ascorbic acid are added alone to the reaction mixture, stimulation when added together with tetrahydropteridine [2]) [2]

Metals, ions

Fe^{2+} <2, 3> (<2> preincubation for 10 min in the presence of benzoate results in maximal activity [2]; <3> required for activity [3]) [2, 3]

Turnover number (min^{-1})

0.3 <4> (cinnamate, <4> reconstituted enzyme system consisting of benzoate 4-hydroxylase i.e. P_{450rm}, cytochrome P_{450} reductase, NADPH and dilauroylphosphatidylcholine [6]) [6]
2.4 <3> (benzoate, <3> reconstituted enzyme system consisting of benzoate 4-hydroxylase, cytochrome P_{450} reductase, NADPH and dilauroylphosphatidylcholine [7]) [7]
40 <4> (benzoate, <4> reconstituted enzyme system consisting of benzoate 4-hydroxylase i.e. P_{450rm}, cytochrome P_{450} reductase, NADPH and dilauroylphosphatidylcholine [6]) [6]
60 <4> (benzoate, <4> turnover rate in microsomes [6]) [6]
140 <3> (2-chlorobenzoate) [8]
220 <3> (2-methylbenzoate) [8]
230 <3> (picolinate) [8]
240 <3> (3-methoxybenzoate) [8]
240 <3> (benzoate, <3> turnover rate in microsomes [8]) [8]
240 <3> (nicotinate) [8]
250 <3> (2-hydroxybenzoate) [8]
260 <3> (2-fluorobenzoate) [8]
270 <3> (3-hydroxybenzoate) [8]
270 <3> (benzoate) [8]
290 <3> (3-fluorobenzoate) [8]
370 <3> (3-chlorobenzoate) [8]
370 <3> (3-methylbenzoate) [8]

Specific activity (U/mg)

0.041 <1> (<1> activity in cells grown on benzoate [4]) [4]
0.0888 <3> [3]
0.26 <3> (<3> activity in cells grown on benzoate, 2-chlorobenzoate or 3-chlorobenzoate [5]) [5]
33.92 <2> [2]

K_m-Value (mM)

0.019 <1> (NADPH) [1]
0.029 <1> (benzoate) [1]
0.045 <2> (6,7-dimethyltetrahydropterine) [2]
0.045 <2> (NADH) [2]
0.083 <3> (benzoate) [8]
0.086 <3> (3-fluorobenzoate) [8]
0.097 <3> (2-fluorobenzoate) [8]
0.127 <3> (3-chlorobenzoate) [8]
0.13 <3> (benzoate) [3]
0.16 <3> (NADPH) [3]
0.189 <3> (3-methoxybenzoate) [8]
0.28 <3> (2-hydroxybenzoate) [8]
0.3 <2> (benzoate) [2]
0.31 <3> (2-chlorobenzoate) [8]
0.673 <3> (3-methylbenzoate) [8]
1.6 <3> (2-methylbenzoate) [8]

K_i-Value (mM)

0.75 <2> (4-nitrobenzoate) [2]

pH-Optimum

6.2 <3> [3]
7.2 <2> [2]
7.6 <1> [1]

pH-Range

4.5-8 <3> (<3> approx. 35% of maximal activity at pH 4.5, approx. 15% of maximal activity at pH 8.0 [3]) [3]
5.5-8.2 <2> (<2> approx. 30% of maximal activity at pH 5.5, approx. 35% of maximal activity at pH 8.2 [2]) [2]

Temperature optimum (°C)

30 <3> [3]
34 <2> [2]

Temperature range (°C)

25-38 <3> (<3> rapid decrease of activity above 38°C and below 25°C [3]) [3]

4 Enzyme Structure

Molecular weight

120000 <2> (<2> gel filtration [2]) [2]

5 Isolation/Preparation/Mutation/Application

Source/tissue

mycelium <3> [3]

Localization

membrane <1> (<1> membrane associated [1]) [1]
microsome <4> (<4> more than 90% of activity [6]) [6]

Purification

<2> (protamine sulfate, heat treatment, calcium phosphate gel, ammonium sulfate, DEAE-Sephadex, Sephadex G-150 [2]) [2]
<3> (protamine sulfate, tricalcium phosphate gel, DEAE-cellulose, alumina C-γ gel [3]; DEAE-Sepharose, benzoate-agarose [8]) [3, 8]

Cloning

<1> [1]
<3> (coexpression of benzoate 4-hydroxylase and cytochrome P_{450} reductase in Aspergillus niger increases enzyme activity [7]) [7]

6 Stability

pH-Stability

5.5-7.5 <3> [3]

General stability information

<2>, freezing and thawing has no effect [2]
<2>, quite stable in presence of benzoate [2]
<3>, EDTA stabilizes [3]
<3>, benzoate stabilizes [3]
<3>, glutathione stabilizes [3]

Storage stability

<2>, 0-4°C, enzyme concentration 0.050 mg/ml, pH 6.0-7.5, 15 days, no loss of activity [2]
<2>, 0-4°C, pH 7.2-8.0, Tris-HCl buffer, 24 h, 50% loss of activity [2]
<3>, 4°C, 24 h, 40% loss of activity [3]

References

[1] McNamee, C.; Durham, D.R.: Properties of a membrane-associated benzoate-4-hydroxylase from Rhodotorula graminis. Biochem. Biophys. Res. Commun., **129**, 485-492 (1985)

[2] Reddy, C.C.; Vaidyanathan, C.S.: Purification and properties of benzoate-4-hydroxylase from a soil pseudomonad. Arch. Biochem. Biophys., **177**, 488-498 (1976)

[3] Reddy, C.C.; Vaidyanathan, C.S.: Purification, properties and induction of a specific benzoate-4-hydroxylase from Aspergillus niger (UBC 814). Biochim. Biophys. Acta, **384**, 46-57 (1975)

[4] Durham, D.R.: Initial reactions involved in the dissimilation of mandelate by Rhodotorula graminis. J. Bacteriol., **160**, 778-780 (1984)

[5] Shailubhai, K.; Sahasrabudhe, S.R.; Vora, K.A.; Modi, V.V.: Degradation of chlorinated derivatives of phenoxyacetic acid and benzoic acid by Aspergillus niger. FEMS Microbiol. Lett., **18**, 279-282 (1983)

[6] Fukuda, H.; Nakamura, K.; Sukita, E.; Ogawa, T.; Fujii, T.: Cytochrome P_{450rm} from Rhodotorula minuta catalyzes 4-hydroxylation of benzoate. J. Biochem., **119**, 314-318 (1996)

[7] Van Den Brink, J.M.; Van Den Hondel, C.A.M.J.J.; Van Gorcom, R.F.M.: Optimization of the benzoate-inducible benzoate *p*-hydroxylase cytochrome P_{450} enzyme system in Aspergillus niger. Appl. Microbiol. Biotechnol., **46**, 360-364 (1996)

[8] Faber, B.W.; van Gorcom, R.F.; Duine, J.A.: Purification and characterization of benzoate-*para*-hydroxylase, a cytochrome P_{450} (CYP53A1), from Aspergillus niger. Arch. Biochem. Biophys., **394**, 245-254 (2001)

Calcidiol 1-monooxygenase 1.14.13.13

1 Nomenclature

EC number
1.14.13.13

Systematic name
calcidiol,NADPH:oxygen oxidoreductase (1-hydroxylating)

Recommended name
calcidiol 1-monooxygenase

Synonyms
1-hydroxylase-25-hydroxyvitamin D3
25-hydroxy D3-1α-hydroxylase
25-hydroxycholecalciferol 1-hydroxylase
25-hydroxycholecalciferol 1-monooxygenase
25-hydroxycholecalciferol 1α-hydroxylase
25-hydroxycholecalciferol-1-hydroxylase
25-hydroxyvitamin D3 1α-hydroxylase
CYB27B1 <3, 4, 9, 10> [27, 28, 30, 31, 33]
hydroxylase, 25-hydroxcholecalciferol 1-
oxygenase, 25-hydroxycholecalciferol 1-mono-
vitamin D3 25- and 1α-hydroxylase <3> (<3> i.e. CYP2D25, microsomal, higher 25-hydroxylating than 1α-hydroxylating activity [33]) [33]

CAS registry number
9081-36-1

2 Source Organism

<1> *Gallus gallus* [1-5, 8-11, 13, 19, 21, 23]
<2> *Rattus norvegicus* [6, 7, 17, 22, 24]
<3> *Sus scrofa* [12, 14, 33]
<4> *Homo sapiens* [20, 26, 29-32]
<5> *Bos taurus* [15]
<6> *Cavia porcellus* [16]
<7> *mammalia* [18]
<8> *Mus musculus* [25]
<9> *Mus musculus* [27, 31]
<10> *Sus scrofa* [28]

3 Reaction and Specificity

Catalyzed reaction

calcidiol + NADPH + H^+ + O_2 = calcitriol + $NADP^+$ + H_2O (<4, 9> reconstitution of the enzyme system in E. coli cells, consisting of NADH-adrenodoxin reductase, adrenodoxin and CYP27B1 [31]; <4, 9, 10> cytochrome P-450 enzyme [27, 28, 30-32]; <8> amino acid sequence determination [25]; <1> amino acid composition and terminal sequence determination [21]; <1, 6-8> 25-hydroxyvitamin D3 1α-hydroxylase is a cytochrome P-450-dependent mixed-function oxidase [16, 18, 23, 25]; <1> kidney mitochondrial 25-hydroxyvitamin D_3-1α-hydroxylation system consists of a renal ferredoxin reductase (flavoprotein), renal ferredoxin and cytochrome P-450 [13]; <5> enzyme can be reconstituted with the 3 components: cytochrome P-$450_{D1\alpha}$, NADPH-renodoxin reductase and renodoxin [15]; <1-4, 8, 9> reconstituted with cytochrome P-450, adrenal ferredoxin, i.e. adrenodoxin, and adrenodoxin reductase [12, 13, 24-27, 31]; <1, 3, 7> ferredoxin, ferredoxin reductase and cytochrome P-450 [11, 13, 14, 18, 19]; <3> cytochrome P-450 type is specific for hydroxylation site of 25-hydroxyvitamin D3 [14])

Reaction type

oxidation
redox reaction
reduction

Natural substrates and products

S (24R),25-dihydroxycholecalciferol + NADPH + O_2 <4, 9> (Reversibility: ? <4, 9> [31]) [31]
P 1α,24,25-trihydroxycholecalciferol + $NADP^+$ + H_2O <4, 9> [31]
S 25-hydroxycholecalciferol + NADPH + O_2 <1-4, 6, 9> (<4, 9> recombinant in E. coli [27, 31]; <6> energy-dependent transhydrogenation is of importance [16]; <1, 2, 4, 9> central role in calcium regulation [1, 22, 31, 32]; <4> enzyme activity is correlated with phosphate content in serum [29]) (Reversibility: ? <1-4, 6, 9> [1, 8-11, 16, 19, 22, 26, 27, 29, 31-33]) [1, 8-11, 16, 19, 22, 26, 27, 29, 31-33]
P calcitriol + $NADP^+$ + H_2O <4, 9> [26, 31]

Substrates and products

S (24R),25-dihydroxycholecalciferol + NADPH + O_2 <4, 9> (<9> recombinant enzyme [27]; <4, 9> higher activity than with 25-hydroxycholecalciferol [26, 27, 31]; <4> reconstituted system [26]) (Reversibility: ? <4, 9> [26, 27, 31]) [26, 27, 31]
P 1α,24,25-trihydroxycholecalciferol + $NADP^+$ + H_2O <4, 9> [31]
S 23,25-dihydroxycholecalciferol + NADPH + O_2 <9> (<9> recombinant enzyme [27]; <9> low activity [27]) (Reversibility: ? <9> [27]) [27]
P ?
S 24-oxo-23,25-dihydroxycholecalciferol + NADPH + O_2 <9> (<9> recombinant enzyme [27]; <9> low activity [27]) (Reversibility: ? <9> [27]) [27]
P ?

S 24-oxo-25-hydroxycholecalciferol + NADPH + O_2 <4, 9> (<9> recombinant enzyme [27]; <4> low activity, reconstituted system [26]) (Reversibility: ? <4, 9> [26, 27]) [26, 27]

P ?

S calcidiol + NADPH + O_2 <1-10> (<3> enzyme shows vitamin D_3 25- and 1α-hydroxylase activity [33]; <9> recombinant enzyme [27]; <1-10> highly specific for calcidiol, calcidiol is identical with 25-hydroxycholecalciferol [1-33]) (Reversibility: ? <1-10> [1-33]) [1-33]

P calcitriol + $NADP^+$ + H_2O <1-10> (<1-10> calcitriol is identical with 1,25-dihydroxycholecalciferol [1-33]) [1-33]

S Additional information <1, 4, 9> (<4, 9> the 25-hydroxy group is essential for 1α-hydroxylating activity, the 24-hydroxy group enhances the activity, the 23-hydroxy group strongly reduces activity [27, 31]; <9> no activity with 24,25,26,27-tetranor-23-hydroxyvitamin D_3 and vitamin D_3 [27]; <4> not active with (23S),25-dihydroxycholecalciferol and 24-oxo-23,25-dihydroxycholecalciferol [26]; <1> not: cholecalciferol, dihydrotachysterol [1]) [1, 26, 27, 31]

P ?

Inhibitors

1,25-dihydroxycholecalciferol <1, 4, 6, 8> (<1, 4, 6, 8> calcitriol [1, 16, 20, 25, 30]; <4> decreases mRNA expression in less differentiated cells like Caco-2/AQ and COGA-1A and -1E [30]; <8> in vivo only inhibitory in vitamin D receptor containing cells, therefore: negative feedback inhibition is mediated by the enzyme through liganded vitamin D receptor [25]) [1, 16, 20, 25, 30]

22-oxacalcitriol <2> (<2> vitamin D_3 analogue, inhibits in vitro and in vivo [22]) [22]

25-hydroxy-3-deoxy-2-oxavitamin D_3 <1> (<1> vitamin D_3 analogue, competitive inhibition [23]) [23]

25-hydroxydihydrotachysterol3 <1> [1]

3-deoxy-2-oxa-9(11)-didehydro-25-hydroxyvitamin D3 <1> (<1> vitamin D_3 analogue, competitive inhibition [23]) [23]

A-homo-3-deoxy-2-oxa-25-hydroxyvitamin D_3 <1> (<1> vitamin D_3 analogue, competitive inhibition [23]) [23]

CN^- <7> [18]

CO <1, 6, 7, 9> (<2> no inhibition [17]; <1> 50% O_2/50% CO_2 incubation atmosphere [1]) [1, 6, 16, 18, 27]

Ca^{2+} <1> [4, 10]

Mn^{2+} <1> [10]

Sr^{2+} <1> [10]

Tris-chloride buffer <1> [1]

aminoglutethimide <1> (<1> enzyme of intact mitochondria [9]) [9]

antimycin A <1, 2, 7> (<6> no inhibition of hydroxylation [16]; <1> higher inhibition rate with succinate as electron donor than with malate [10]; <2> inhibition occurs with succinate reaction support, no inhibition together with malate [6]) [1, 6, 10, 18]

cyanide <1, 6, 7> [1, 16, 18]

dinitrophenol <1, 2, 7> (<1> higher inhibition rate with malate as electron donor than with succinate [10]) [1, 6, 10, 18]
diphenyl-*p*-phenylenediamine <1, 2> (<2> no inhibition [6]) [1, 17]
epidermal growth factor <4> (<4> i.e. EGF [30]; <4> decreases mRNA expression in less differentiated cells like Caco-2/AQ and COGA-1A and -1E [30]) [30]
ethylene glycol-bis(β-aminoethyl ether)-N,N'-tetraacetic acid <1> [4]
glutethimide <7> [18]
ketoconazole <9> [27]
metyrapone <1, 2, 7, 9> (<1> enzyme of intact mitochondria [9]) [1, 6, 9, 17, 18, 27]
natural inhibitors <1, 7> (<1> from rat and pig tissues [2]; <1> rat plasma 25-hydroxyvitamin D_3 binding protein [3]; <1> from rat serum [5]; <7> vitamin D binding protein [18]) [2, 3, 5, 18]
oligomycin <1> [1]
p-chloromercuribenzoate <1> [1]
p-trifluoromethoxyphenylhydrazone <6> [16]
phosphate <1> (<4> no inhibition [20]) [1, 4, 10]
rotenone <7> (<6> no inhibition of hydroxylation [16]) [18]
sucrose <1> (<1> hypertonic [1]) [1]
vitamin D_3 <4, 9> [27, 31]
Additional information <1> (<1> overview, inhibitory potency of several vitamin D_3 analogues [23]; <1> no inhibition with Ba^{2+}, acetate, sulfate [10]; <1> inhibition by renal mitochondrial protein kinase-catalyzed phosphorylation [8]) [8, 10, 23]

Cofactors/prosthetic groups

NADPH <1-9> [1-20, 24-27, 31]

Activating compounds

1,25-dihydroxycholecalciferol <4> (<4> increases mRNA expression in highly differentiated cells like Caco-2/15 [30]) [30]
8-bromo-cAMP <10> (<10> enhances mRNA expression level [28]) [28]
CN^- <1> (<1> activates with malate as electron donor [10]) [10]
epidermal growth factor <4> (<4> stimulates [32]; <4> i.e. EGF [30, 32]; <4> increases mRNA expression in highly differentiated cells like Caco-2/15 [30]) [30, 32]
forskolin <10> (<10> enhances mRNA expression level [28]) [28]
malate <1, 7> (<1, 7> supports 1α-hydroxylation as electron donor [1, 4, 10, 18]) [1, 4, 10, 18, 19, 23]
succinate <1, 2, 7> (<1, 2, 7> supports 1α-hydroxylation as electron donor [1, 6, 10, 18]) [1, 6, 10, 18]

Metals, ions

Mg^{2+} <1, 6> (<1> required [1, 19]; <6> 10 mM used in standard incubation [16]) [1, 16, 19]
Additional information <6, 7> (<6> highest conversion with a hypo-osmolar buffer [16]; <7> almost any other Krebs cycle substrate supports 1-α-hydroxylation [18]) [16, 18]

Turnover number (min^{-1})

4.4 <2> (25-hydroxycholecalciferol) [24]

Specific activity (U/mg)

0.0000001 <4> (<4> female human kidney [29]) [29]
0.0000002 <4, 9> (<4> male human kidney [29]; <9> substrate 23,25-dihydroxycholecalciferol [27,31]) [27, 29, 31]
0.00000042 <4> (<4> substrate 24-oxo-25-hydroxycholecalciferol, recombinant reconstituted system [26,31]) [26, 31]
0.00000058 <4> (<4> substrate 25-hydroxycholecalciferol, recombinant reconstituted system [26,31]) [26, 31]
0.0000011 <9> (<9> substrate 24-oxo-25-hydroxycholecalciferol [27, 31]) [27, 31]
0.0000015 <4, 9> (<4> substrate 24,25-dihydroxycholecalciferol [31]) [27, 31]
0.0000031 <1> [4]
0.0000043 <9> (<9> substrate 24,25-dihydroxycholecalciferol [27, 31]) [27, 31]
0.0000048 <1> (<1> with malate as electron source [4]) [4]
0.0000072 <3> (<3> purified cytochrome P-450 component of 1α-hydroxylase, reconstituted system [12]) [12]
0.0000082 <5> (<5> purified cytochrome P-450 component of 1α-hydroxylase, reconstituted system [15]) [15]
0.00125 <1> [1]
0.0048 <2> (<2> purified enzyme [24]) [24]
0.602 <1> (<1> reduction of cytochrome P-450 at non-saturating level of enzyme components [13]) [13]
0.687 <1> (<1> ferredoxin activity, reduction of cytochrome c [11]) [11]
6.758 <1> (<1> purified ferredoxin, reduction of cytochrome c [19]) [19]
11.38 <1> (<1> reconstituted system of cytochrome P-450, cytochrome b_5, adrenocortical ferredoxin reductase, and renal ferredoxin [19]) [19]
Additional information <2-4> (<4> higher activity in male kidney compared to female kidney [29]) [17, 22, 29, 33]

K_m-Value (mM)

0.000088 <4> (25-hydroxycholecalciferol) [20]
0.00097 <3> (25-hydroxycholecalciferol, <3> reconstituted system [12]) [12]
0.001 <6> (25-hydroxycholecalciferol) [16]
0.0011 <4> ((24R),25-dihydroxycholecalciferol, <4> reconstituted system [26]) [26, 31]
0.0013 <9> (24,25-dihydroxycholecalciferol) [27, 31]
0.0022 <1> (25-hydroxycholecalciferol) [1]
0.0027 <4, 9> (25-hydroxycholecalciferol, <4> reconstituted system [26]) [26, 27, 31]
0.014 <4> (25-hydroxycholecalciferol) [29]
0.89 <2> (25-hydroxycholecalciferol) [7]
Additional information <1> [10]

K_i-Value (mM)

0.001-0.002 <4,9> (vitamin D_3) [31]
0.16 <1> (metapyrone) [9]
1.1 <1> (aminoglutethimide) [9]

pH-Optimum

7.4 <1, 9> (<9> assay at [27]) [1, 27]
7.4-8.4 <6> [16]

pH-Range

7.1-7.7 <1> (<1> about 50% of activity maximum at pH 7.1 and pH 7.7 [1]) [1]

Temperature optimum (°C)

25 <2> (<2> assay at [7]) [7]
37 <1, 9> (<1,9> assay at [23, 27]) [23, 27]

4 Enzyme Structure

Molecular weight

11900 <1> (<1> ferredoxin component, SDS-PAGE [19]) [19]
12500 <1> (<1> ferredoxin component, gel filtration [11,13]) [11, 13]
49000 <5> (<5> cytochrome $P\text{-}450_{D1\alpha}$, SDS-PAGE [15]) [15]
50000 <9> (<9> recombinant enzyme [31]) [31]
53000 <1> (<1> MW of ferredoxin component, SDS-PAGE, gel filtration [11]) [11]
55000 <2, 8> (<8> SDS-PAGE after in vitro-translation [25]; <2> gel filtration [24]) [24, 25]
56370 <3> (<3> predicted from DNA sequence analysis [33]) [33]
Additional information <1, 3, 5, 7> (<1> kidney mitochondrial 25-hydroxyvitamin D_3-1α-hydroxylation system consists of a renal ferredoxin reductase (flavoprotein), renal ferredoxin and cytochrome P-450 [13]; <5> enzyme can be reconstituted with the 3 components: cytochrome $P\text{-}450_{D1\alpha}$, NADPH-renodoxin reductase and renodoxin [15]; <3,7> ferredoxin, ferredoxin reductase and cytochrome P-450 [12, 18]) [12, 13, 15, 18]

5 Isolation/Preparation/Mutation/Application

Source/tissue

CACO-2 <4> [30]
colonic carcinoma cell line <4> (<4> cell line COGA [30]) [30]
decidua <4> [20]
kidney <1-10> (<3, 10> proximal tubular cells [28, 33]; <2,4> renal cortex [22, 29]; <4> tumour tissue [29]) [1, 4-16, 18, 19, 21-25, 27-29, 31, 33]
liver <3> [33]

placenta <4> (<4> syncytiotrophoblast cells from preeclamptic pregnancies [32]) [32]
yolk sac <2> [17]

Localization

microsome <2, 3> (<3> vitamin D_3 25- and 1α-hydroxylase [33]) [17, 33]
mitochondrion <1-6, 9, 10> [1, 4-7, 9-16, 18, 19, 21-24, 28, 29, 31, 32]

Purification

<1> (partially [21]; purification of ferredoxin component [11, 13, 19]) [11, 13, 19, 21, 24]
<3> (purification of 1α-hydroxylating cytochrome P-450 [12,14]; vitamin D3 25- and 1α-hydroxylase [33]) [12, 14, 33]
<5> (purification of cytochrome P-450$_{D1\alpha}$ [15]) [15]

Cloning

<3> (<3> expression in COS cells, vitamin D_3 25- and 1α-hydroxylase, DNA sequence analysis [33]) [33]
<4> (<4> expression in Escherichia coli JM109, coexpression of adrenodoxin and NADPH-adrenodoxin reductase for reconstitution of the system [26]) [26]
<8> (<8> expression in COS cells, reconstituted with recombinant cytochrome P-450, recombinant adrenodoxin, and recombinant adrenodoxin reductase [25]) [25, 27]
<9> (<9> expression in Escherichia coli [27,31]) [27, 31]
<10> (<10> expression in LLC-PK1-cells [28]) [28]

Engineering

G125E <4> (<4> native mutant, no enzyme activity but normal expression level [26]) [26]
P382S <4> (<4> native mutant, no enzyme activity but normal expression level [26]) [26]
R107H <4> (<4> native mutant, no enzyme activity but normal expression level [26]) [26]
R335P <4> (<4> native mutant, no enzyme activity but normal expression level [26]) [26]
Additional information <8> (<8> construction and expression in COS cell line of a chimeric fusion protein: vitamin D receptor ligand-binding domain with the yeast GAL4 DNA-binding domain, also expression of lacZ containing reporter plasmid [25]) [25]

6 Stability

Storage stability

<2>, -30°C, stable without significant loss of activity for at least 6 months [24]
<2>, 0-4°C, rapid loss of activity [24]
<5>, -80°C, in the dark, stable for at least 3 months, cytochrome P-450$_{D1\alpha}$ [15]

References

[1] Gray, R.W.; Omdahl, J.L.; Ghazarian, J.G.; DeLuca, H. F.: 25-Hydroxycholecalciferol-1-hydroxylase. Subcellular location and properties. J. Biol. Chem., **247**, 7528-7532 (1972)
[2] Botham, K.M.; Tanaka, Y.; DeLuca, H.F.: 25-Hydroxyvitamin D3-1-hydroxylase. Inhibition in vitro by rat and pig tissues. Biochemistry, **13**, 4961-4966 (1974)
[3] Ghazarian, J.G.; Kream, B.; Botham, K.M.; Nickells, M. W.; DeLuca, H.F.: Rat plasma 25-hydroxyvitamin D3 binding protein: an inhibitor of the 25-hydroxyvitamin D_3-1α-hydroxylase. Arch. Biochem. Biophys., **189**, 212-220 (1978)
[4] Henry, H.L.; Norman, A.W.: Studies on calciferol metabolism. Regulation of 25-hydroxyvitamin D_3-1-hydroxylase in isolated renal mitochondria. Arch. Biochem. Biophys., **172**, 582-589 (1976)
[5] Botham, K.M.; Ghazarian, J.G.; Kream, B.E.; DeLuca, H. F.: Isolation of an inhibitor of 25-hydroxyvitamin D_3-1-hydroxylase from rat serum. Biochemistry, **15**, 2130-2135 (1976)
[6] Paulson, S.K.; DeLuca, H.F.: Subcellular location and properties of rat renal 25-hydroxyvitamin D3-1 α-hydroxylase. J. Biol. Chem., **260**, 11488-11492 (1985)
[7] Vieth, R.; Fraser, D.: Kinetic behavior of 25-hydroxyvitamin D-1-hydroxylase and -24-hydroxylase in rat kidney mitochondria. J. Biol. Chem., **254**, 12455-12460 (1979)
[8] Ghazarian, J.G.; Yanda, D.M.: Inhibition of 25-hydroxyvitamin D 1 α-hydroxylase by renal mitochondrial protein kinase-catalyzed phosphorylation. Biochem. Biophys. Res. Commun., **132**, 1095-1102 (1985)
[9] Ghazarian, J.G.; Jefcoate, C.R.; Knutson, J.C.; Orme-Johnson, W.H.; DeLuca, H.F.: Mitochondrial cytochrome P_{450}. A component of chick kidney 25-hydrocholecalciferol-1α-hydroxylase. J. Biol. Chem., **249**, 3026-3033 (1974)
[10] Henry, H.L.; Norman, A.W.: Studies on calciferol metabolism. IX. Renal 25-hydroxy-vitamin D_3-1 hydroxylase. Involvement of cytochrome P-450 and other properties. J. Biol. Chem., **249**, 7529-7535 (1974)
[11] Kulkoski, J.A.; Ghazarian, J.G.: Purification and characterization of the ferredoxin component of 25-hydroxycholecalciferol 1 α-hydroxylase. Biochem. J., **177**, 673-678 (1979)
[12] Gray, R.W.; Ghazarian, J.G.: Solubilization and reconstitution of kidney 25-hydroxyvitamin D_3 1 α- and 24-hydroxylases from vitamin D-replete pigs. Biochem. J., **259**, 561-568 (1989)
[13] Pedersen, J.I.; Ghazarian, J.G.; Orme-Johnson, N.R.; DeLuca, H.F.: Isolation of chick renal mitochondrial ferredoxin active in the 25-hydroxyvitamin D_3-1α-hydroxylase system. J. Biol. Chem., **251**, 3933-3941 (1976)
[14] Postlind, H.: Separation of the cytochromes P-450 in pig kidney mitochondria catalyzing 1 α-, 24- and 26-hydroxylations of 25-hydroxyvitamin D_3. Biochem. Biophys. Res. Commun., **168**, 261-266 (1990)

[15] Hiwatashi, A.; Nishii, Y.; Ichikawa, Y.: Purification of cytochrome P-450D1 α (25-hydroxyvitamin D3-1 α-hydroxylase) of bovine kidney mitochondria. Biochem. Biophys. Res. Commun., **105**, 320-327 (1982)

[16] Hagenfeldt, Y.; Pedersen, J.I.; Björkhem, I.: Properties of guinea-pig kidney 25-hydroxyvitamin D3 1 α-hydroxylase assayed by isotope dilution-mass spectrometry. Biochem. J., **250**, 521-526 (1988)

[17] Paulson, S.K.; Phelps, M.; DeLuca, H.F.: Assay and properties of rat yolk sac 25-hydroxyvitamin D3 1 α-hydroxylase. Biochemistry, **25**, 6821-6826 (1986)

[18] Lobaugh, B.; Almond, J.R.; Drezner, M.K.: Measurement of 25-hydroxyvitamin D 1-α-hydroxylase activity in mammalian kidney. Methods Enzymol., **123**, 159-167 (1986)

[19] Yoon, P.S.; DeLuca, H.F.: Purification and properties of vitamin D hydroxylases. Methods Enzymol., **67**, 430-440 (1980)

[20] Delvin, E.E.; Arabian, A.: Kinetics and regulation of 25-hydroxycholecalciferol 1 α-hydroxylase from cells isolated from human term decidua. Eur. J. Biochem., **163**, 659-662 (1987)

[21] Burgos-Trinidad, M.; Ismail, R.; Ettinger, R.A.; Prahl, J.M.; DeLuca, H.F.: Immunopurified 25-hydroxyvitamin D 1α-hydroxylase and 1,25-dihydroxyvitamin D 24-hydroxylase are closely related but distinct enzymes. J. Biol. Chem., **267**, 3498-3505 (1992)

[22] Grieff, M.; Dusso, A.; Mori, T.; Nishii, Y.; Slatopolsky, E.; Brown, A.J.: 22-Oxacalcitriol suppresses 25-hydroxycholecalciferol-1α-hydroxylase in rat kidney. Biochem. Biophys. Res. Commun., **185**, 191-196 (1992)

[23] Daniel, D.; Middleton, R.; Henry, H.L.; Okamura, W.H.: Inhibitors of 25-hydroxyvitamin D3 1a-hydroxylase: A-ring oxa analogs of 25-hydroxyvitamin D3. J. Org. Chem., **61**, 5617-5625 (1996)

[24] Nakamura, Y.; Eto, T.A.; Taniguchi, T.; Miyamoto, K.; Nagatomo, J.; Shiotsuki, H.; Sueta, H.; Higashi, S.; Okuda, K.I.; Setoguchi, T.: Purification and characterization of 25-hydroxyvitamin D_3 1α-hydroxylase from rat kidney mitochondria. FEBS Lett., **419**, 45-48 (1997)

[25] Takeyama, K.i.; Kitanaka, S.; Sato, T.; Kobori, M.; Yanagisawa, J.; Kato, S.: 25-Hydroxyvitamin D_3 1α-hydroxylase and vitamin D synthesis. Science, **277**, 1827-1830 (1997)

[26] Sawada, N.; Sakaki, T.; Kitanaka, S.; Takeyama, K.i.; Kato, S.; Inouye, K.: Enzymatic properties of human 25-hydroxyvitamin D_3 1α-hydroxylase: coexpression with adrenodoxin and NADPH-adrenodoxin reductase in Escherichia coli. Eur. J. Biochem., **265**, 950-956 (1999)

[27] Sakaki, T.; Sawada, N.; Takeyama, K.i.; Kato, S.; Inouye, K.: Enzymic properties of mouse 25-hydroxyvitamin D_3 1α-hydroxylase expressed in Escherichia coli. Eur. J. Biochem., **259**, 731-738 (1999)

[28] Yoshida, T.; Yoshida, N.; Nakamura, A.; Monkawa, T.; Hayashi, M.; Saruta, T.: Cloning of porcine 25-hydroxyvitamin D3 1α-hydroxylase and its regulation by cAMP in LLC-PK1 cells. J. Am. Soc. Nephrol., **10**, 963-970 (1999)

[29] Hagenfeldt, Y.; Berlin, T.: The human renal 25-hydroxyvitamin D_3-1α-hydroxylase: properties studied by isotope-dilution mass spectrometry. Eur. J. Clin. Invest., **22**, 223-228 (1992)

[30] Bareis, P.; Kallay, E.; Bischof, M.G.; Bises, G.; Hofer, H.; Potzi, C.; Manhardt, T.; Bland, R.; Cross, H.S.: Clonal differences in expression of 25-hydroxyvitamin D(3)-1α-hydroxylase, of 25-hydroxyvitamin D(3)-24-hydroxylase, and of the vitamin D receptor in human colon carcinoma cells: effects of epidermal growth factor and 1α,25-dihydroxyvitamin D(3). Exp. Cell Res., **276**, 320-327 (2002)
[31] Inouye, K.; Sakaki, T.: Enzymatic studies on the key enzymes of vitamin D metabolism; 1α-hydroxylase (CYP27B1) and 24-hydroxylase (CYP24). Biotechnol. Annu. Rev., **7**, 179-194 (2001)
[32] Diaz, L.; Arranz, C.; Avila, E.; Halhali, A.; Vilchis, F.; Larrea, F.: Expression and activity of 25-hydroxyvitamin D-1α-hydroxylase are restricted in cultures of human syncytiotrophoblast cells from preeclamptic pregnancies. J. Clin. Endocrinol. Metab., **87**, 3876-3882 (2002)
[33] Hosseinpour, F.; Norlin, M.; Wikvall, K.: Kidney microsomal 25- and 1α-hydroxylase in vitamin D metabolism: catalytic properties, molecular cloning, cellular localization and expression during development. Biochim. Biophys. Acta, **1580**, 133-144 (2002)

trans-Cinnamate 2-monooxygenase 1.14.13.14

1 Nomenclature

EC number
1.14.13.14

Systematic name
trans-cinnamate,NADPH:oxygen oxidoreductase (2-hydroxylating)

Recommended name
trans-cinnamate 2-monooxygenase

Synonyms
cinnamate 2-hydroxylase
cinnamate 2-monooxygenase
cinnamic 2-hydroxylase
cinnamic acid 2-hydroxylase
trans-cinnamic acid 2-hydroxylase

CAS registry number
53126-56-0

2 Source Organism

<1> *Cucumis sativus* (cucumber [1]) [1]
<2> *Melilotus alba* [2]

3 Reaction and Specificity

Catalyzed reaction
trans-cinnamate + NADPH + H^+ + O_2 = 2-hydroxycinnamate + $NADP^+$ + H_2O

Reaction type
oxidation
redox reaction
reduction

Natural substrates and products
S trans-cinnamate + NADPH + O_2 <1, 2> (Reversibility: ? <1, 2> [1, 2]) [1, 2]
P 2-hydroxycinnamate + $NADP^+$ + H_2O <1, 2> [1, 2]

Substrates and products

S trans-cinnamate + NADPH + O_2 <1> (Reversibility: ? <1> [1]) [1]

P 2-hydroxycinnamate + $NADP^+$ + H_2O <1> [1]

Inhibitors

KCN <1> (<1> 70% inhibition at 10 mM [1]) [1]

Triton X-100 <1> (<1> concentration higher than 0.5% deactivates the enzyme [1]) [1]

Additional information <1> (<1> UV-light treatment for 200 min leads to 30% inhibition of the enzyme [1]) [1]

Cofactors/prosthetic groups

NADH <2> (<2> less effective than NADPH [2]) [2]

NADPH <1, 2> [1, 2]

Activating compounds

Triton X-100 <1> (<1> 700% activation at 0.025% [1]) [1]

glucose 6-phosphate <2> (<2> 400% activation at 4 mM [2]) [2]

sulfonic acid <1> (<1> 700% activation at 0.003% [1]) [1]

Additional information <1> (<1> white light treatment for 200 min leads to 50% activation of the enzyme [1]) [1]

Specific activity (U/mg)

0.0004 <2> (<2> supernatant activity [2]) [2]

pH-Optimum

7 <2> [2]

5 Isolation/Preparation/Mutation/Application

Source/tissue

cotyledon <1> [1]

leaf <2> [2]

Localization

chloroplast <1> (<1> predominantly, membrane [1]) [1]

membrane <2> (<2> lamellar, 50% of the hydroxylase activity [2]) [2]

References

[1] Czichi, U.; Kindl, H.: Phenylalanine ammonia lyase and cinnamic acid hydroxylases as assembled consecutive enzymes on microsomal membranes of cucumber cotyledons: cooperation and subcellular distribution. Planta, **134**, 133-143 (1977)

[2] Gestetner, B.; Conn, E.E.: The 2-hydroxylation of trans-cinnamic acid by chloroplasts from Melilotus alba Desr. Arch. Biochem. Biophys., **163**, 617-624 (1974)

Cholestanetriol 26-monooxygenase 1.14.13.15

1 Nomenclature

EC number
1.14.13.15

Systematic name
5β-cholestane-3α,7α,12α-triol,NADPH:oxygen oxidoreductase (26-hydroxylating)

Recommended name
cholestanetriol 26-monooxygenase

Synonyms
5β-cholestane-3α,7α,12α-triol 26-hydroxylase
5β-cholestane-3α,7α,12α-triol 27-monooxygenase <1> (<1> enzyme was renamed into 27-monooxygenase because it hydroxylates the methyl group in position 27 [7]) [7]
5β-cholestane-3α,7α,12α-triol hydroxylase
cholestanetriol 26-hydroxylase
cholestanetriol 27-hydroxylase
cytochrome P-450A
hydroxylase, 5β-cholestane-3α,7α,12α-triol
oxygenase, cholestanetriol 26-mono-
vitamin D_3 25-hydroxylase

CAS registry number
52227-77-7

2 Source Organism

<1> *Rattus norvegicus* (male wistar rat [1]; female rat [8]; male Sprague Dawley, cholestryamine-treated rats [9]) [1-14]
<2> *Oryctolagus cuniculus* (similar enzyme catalyzing 25-hydroxylation of 5β-cholestane-3α,7α,12α-triol [10]) [10]
<3> *Mus musculus* (mouse [1]) [1]

3 Reaction and Specificity

Catalyzed reaction
5β-cholestane-3α,7α,12α-triol + NADPH + H^+ + O_2 = 5β-cholestane-3α,7α,12α,26-tetraol + $NADP^+$ + H_2O

Reaction type

oxidation
redox reaction
reduction

Natural substrates and products

S 5β-cholestane-3α,7α,12α-triol + NADPH + O_2 <1> (<1> conversion of cholesterol to cholic acid [1]; <1> mitochondrial cytochrome P_{450} forms a complex with the substrate [4]; <1> metabolism of cholesterol [9]) [1, 5, 9]
P ?

Substrates and products

S 1α-hydroxyvitamin D_3 + NADPH + O_2 <1> (<1> 25-hydroxylation [12, 13, 14]) (Reversibility: ? <1> [12, 13, 14]) [12, 13, 14]
P 1α,25-dihydroxyvitamin D_3 + $NADP^+$ + O_2 <1> [14]
S 5β-cholestane-3α,7α,12α-triol + NADPH + O_2 <1, 2, 3> (<1> mitochondrial enzyme is specific for 26-hydroxylation whereas the reaction with the microsomal enzyme hydroxylates at position 23, 24, 25 and 26 [2] <1, 2> microsomal enzyme also catalyzes 25-hydroxylation of 5β-cholestane-3α,7 α,12α-triol [10]; <1> isocitrate can act as electron donor [2, 3]) (Reversibility: ? <1, 2, 3> [1-10, 12, 13, 14]) [1-10, 12, 13, 14]
P 5β-cholestane-3α,7α,12α,27-tetrol + $NADP^+$ + H_2O <1> [14]
S 5β-cholestane-3α,7α-diol + NADPH + O_2 <1> (Reversibility: ? <1> [7, 8, 12]) [7, 8, 12]
P ?
S 5β-cholestane-3α-ol + NADPH + O_2 <1> (Reversibility: ? <1> [7]) [7]
P ?
S cholesterol + NADPH + O_2 <1> (Reversibility: ? <1> [8]) [8]
P ?

Inhibitors

1α-hydroxyvitamin D_3 <1> (<1> competes with 5β-cholestane-3α,7α,12α-triol [12]) [12]
5β-cholestane-3α,7α,12α-triol <1> (<1> inhibits 5β-cholestane-3α,7α-diol 27-monooxygenase activity [7]; <1> competes with 1α-hydroxyvitamin D_3 [12]) [7, 12]
5β-cholestane-3α,7α-diol <1> (<1> inhibits 5β-cholestane-3α,7α,12α-triol 27-monooxygenase activity only slightly [7]) [7]
7,8-benzoflavone <1> (<1> 68% inhibition at 0.04 mM [8]) [8]
CO <1> (<1> at a CO:O_2 ratio of 0.75 hydroxylation reaction is undetectable [2]; <1> inhibition maximally reversed by monochromatic light of 450 nm [3]; <1> 62% inhibition at 25%O_2/75%CO [6]) [3, 6, 8]
Ca^{2+} <1> (<1> complete inhibition at 1 mM [7]) [7]
N-bromosuccinimide <1> (<1> 50% loss of activity at 0.1 mM [12]) [12]
aminoglutethimide <1> (<1> slight inhibition [8]) [8]
dicoumarol <1> (<1> 10% inhibition at 0.05 mM [2]) [2]
metyrapone <1> (<1> slight inhibition of 27-monooxygenase [7]) [7]

p-chloromercuribenzoate <1> (<1> complete inhibition at 1 mM [7]; <1> complete inhibition at 0.08 mM [8]) [7, 8]
p-chloromercuriphenyl sulfonate <1> (<1> 80% inhibition at 0.1 mM [2]) [2]
phenyl isocyanide <1> (<1> 70% inhibition at 0.1 mM [2]; <1> complete inhibition at 1 mM [7]) [2, 7]

Cofactors/prosthetic groups

ATP <1> (<1> peroxisomal and mitochondrial enzyme require ATP [9]) [9]
NADPH <1> [1, 2, 6, 11]
cytochrome p450 <1> (<1> involvement of a cytochrome-P_{450}-dependent monooxygenase in the 26-hydroxylation of 5β-cholestane-3α,7α,12α-triol [3]; <1> possible function of a cytochrome-P_{450}-like entity in the intramitochondrial 26-hydroxylase system [2]; <1> mitochondrial cytochrome-P_{450} forms an enzyme-substrate complex with 5β-cholestane-3α,7α,12α-triol with K_m value very similar to the K_m value of 26-hydroxylation [4]; <1> inner mitochondrial membrane houses a species of cytochrome-P_{450} functional in 5β-cholestane-3α,7α,12α-triol 26-hydroxylation [6]) [2-4, 6, 8]

Metals, ions

Mg^{2+} <1> (<1> peroxisomal and mitochondrial enzyme requires Mg^{2+} [9]) [9]

Turnover number (min^{-1})

0.14 <1> (1α-hydroxyvitamin D_3) [13]
0.7 <1> (cholesterol) [8]
1.4 <1> (1α-hydroxyvitamin D_3) [12]
13 <1> (5β-cholestane-3α,7α-diol) [8, 12]
20 <1> (5β-cholestane-3α,7α,12α-triol) [13]
35.5 <1> (5β-cholestane-3α,7α,12α-triol) [8]
36 <1> (5β-cholestane-3α,7α,12α-triol) [12]

Specific activity (U/mg)

0.00233 <1> (<1> 1α-hydroxyvitamin D_3 as substrate [12]) [12]
0.114 <1> (<1> 5β-cholestane-3α,7α,12α-triol as substrate [12]) [12]
0.431 <1> [8]

K_m-Value (mM)

0.0045 <1> (5β-cholestane-3α,7α,12α-triol) [7]
0.0063 <1> (5β-cholestane-3α,7α,12α-triol) [8]
0.01 <1> (5β-cholestane-3α,7α-diol) [7]
0.01-0.02 <1> (O_2) [2]
0.05 <1> (5β-cholestane-3α,7α,12α-triol) [1]
0.06 <1> (5β-cholestane-3α,7α,12α-triol) [5]

K_i-Value (mM)

0.004 <1> (5β-cholestane-3α,7α,12α-triol) [7,12]

pH-Optimum

7 <1> [1]
7.7 <1> [7]

7.8 <1> [8]

pH-Range
6.5-8 <1> (<1> 27-hydroxylation [7]) [7]

4 Enzyme Structure

Molecular weight
52500 <1> (<1> SDS-PAGE [8, 14]) [8, 14]

5 Isolation/Preparation/Mutation/Application

Source/tissue
liver <1> [1-9, 10, 11-14]

Localization
microsome <1> [9, 10]
mitochondrial membrane <1> (<1> inner membrane-matrix [2]) [2, 6, 7]
mitochondrion <1> [1, 3, 4, 8, 9, 11-14]
peroxisome <1> [9]
Additional information <1> (<1> only the mitochondrial hydroxylase system is specific for C-26 position, the microsomal system is unspecific for C-26 position but rather more active for other adjacent positions [2]) [2]

Purification
<1> (partial [6]; partial purification of a ferredoxin-like iron-sulfur protein and a NADPH-ferredoxin reductase which are functional for 26-hydroxylation when reconstituted with partially purified liver mitochondrial cytochrome P-450 [11]) [6, 7, 8, 11, 12, 14]

Cloning
<1> [13, 14]

6 Stability

Temperature stability
45 <1> (<1> 90% loss of activity after 2 min [12]) [12]

Storage stability
<1>, -70°C, no loss of activity for 5 months [6]

References

[1] Okuda, K.; Hoshita, N.: Oxidation of 5-β-cholestane-3α,7α,12α-triol by rat-liver mitochondria. Biochim. Biophys. Acta, **164**, 381-388 (1968)

[2] Taniguchi, S.; Hoshita, N.; Okuda, K.: Enzymatic characteristics of CO-sensitive 26-hydroxylase system for 5β-cholestane-3α,7α,12α-triol in rat-liver mitochondria and its intramitochondrial localization. Eur. J. Biochem., **40**, 607-617 (1973)

[3] Okuda, K.; Weber, P.; Ullrich, V.: Photochemical action spectrum of the co-inhibited 5β-cholestane-3α,7α,12α-triol 26-hydroxylase system. Biochem. Biophys. Res. Commun., **74**, 1071-1076 (1977)

[4] Okuda, K.; Ruf, H.H.; Ullrich, V.: Spectral evidence for a liver mitochondrial cytochrome P_{450} involved in bile acid metabolism. Hoppe-Seyler's Z. Physiol. Chem., **358**, 689-694 (1977)

[5] Taniguchi, S.; Hoshita, N.; Okuda, K.: Enzymatic characteristics of CO-sensitive 26-hydroxylase system for 5β-cholestane-3α,7α,12α-triol in rat-liver mitochondria and its intramitochondrial localization. Eur. J. Biochem., **40**, 607-617 (1973)

[6] Sato, R.; Atsuta, Y.; Imai, Y.; Taniguchi, S.; Okuda, K.: Hepatic mitochondrial cytochrome P-450: isolation and functional characterization. Proc. Natl. Acad. Sci. USA, **74**, 5477-5481 (1977)

[7] Atsuta, Y.; Okuda, K.: Partial purification and characterization of 5β-cholestane-3α,7α,12α-triol and 5β-cholestane-3α,7α-diol 27-monooxygenase. J. Lipid Res., **23**, 345-351 (1982)

[8] Okuda, K.; Masumoto, O.; Ohyama, Y.: Purification and characterization of 5β-cholestane-3α,7α,12α-triol 27-hydroxylase from female rat liver mitochondria. J. Biol. Chem., **263**, 18138-18142 (1988)

[9] Thompson, S.L.; Krisans, S.K.: Evidence for peroxisomal hydroxylase activity in rat liver. Biochem. Biophys. Res. Commun., **130**, 708-716 (1985)

[10] Hansson, R.; Holmberg, I.; Wikvall, K.: 25-Hydroxylation vitamin D_3 and side chain hydroxylations of 5β-cholestane-3α,7α,12α-triol by purified rabbit and rat liver microsomal cytochromes P-450. J. Biol. Chem., **256**, 4345-4349 (1981)

[11] Atsuta, Y.; Okuda, K.: Isolation of rat liver mitochondrial ferredoxin and its reductase active in the 5β-cholestane-3α,7α,12α-triol 26 hydroxylase. J. Biol. Chem., **253**, 4653-4658 (1978)

[12] Ohyama, Y.; Masumoto, O.; Usui, E.; Okuda, K.: Multi-functional property of rat liver mitochondrial cytochrome P-450. J. Biochem., **109**, 389-393 (1991)

[13] Akiyoshi-Shibata, M.; Usui, E.; Sakaki, T.; Yabusaki, Y.; Noshiro, M.; Okuda, K.; Ohkawa, H.: Expression of rat liver vitamin D_3 25-hydroxylase cDNA in Saccharomyces cerevisiae. FEBS Lett., **280**, 367-370 (1991)

[14] Usui, E.; Noshiro, M.; Ohyama, Y.; Okuda, K.: Unique property of liver mitochondrial P_{450} to catalyze the two physiologically important reactions involved in both cholesterol catabolism and vitamin D activation. FEBS Lett., **274**, 175-177 (1990)

Cyclopentanone monooxygenase 1.14.13.16

1 Nomenclature

EC number
1.14.13.16

Systematic name
cyclopentanone,NADPH:oxygen oxidoreductase (5-hydroxylating, lactonizing)

Recommended name
cyclopentanone monooxygenase

Synonyms
cyclopentanone oxygenase

CAS registry number
37364-15-1

2 Source Organism

<1> *Pseudomonas sp.* (NCIB 9872 [1-5]) [1-5]
<2> *Comamonas sp.* (reclassified, formerly named Pseudomonas sp. NCIB 9872 [6]) [6]

3 Reaction and Specificity

Catalyzed reaction
cyclopentanone + NADPH + H^+ + O_2 = 5-valerolactone + $NADP^+$ + H_2O

Reaction type
oxidation
redox reaction
reduction

Natural substrates and products
S cyclopentanone + NADPH + O_2 <1, 2> (Reversibility: ? <1, 2> [1-6]) [1-6]
P 5-valerolactone + $NADP^+$ + H_2O

Substrates and products
S 2-methylcyclohexanone + NADPH + O_2 <1> (Reversibility: ? <1> [1, 2]) [1, 2]
P 1-oxa-2-oxo-3-methylcycloheptane + $NADP^+$ + H_2O

S butan-2-one + NADPH + O_2 <1> (Reversibility: ? <1> [1, 2]) [1, 2]
P ?
S cyclobutanone + NADPH + O_2 <1> (Reversibility: ? <1> [1, 2]) [1, 2]
P butyrolactone + $NADP^+$ + H_2O
S cycloheptanone + NADPH + O_2 <1> (<1> poor substrate [1,2]) (Reversibility: ? <1> [1, 2]) [1, 2]
P 1-oxa-2-oxocyclooctane + $NADP^+$ + H_2O
S cyclohexanone + NADPH + O_2 <1> (Reversibility: ? <1> [1, 2]) [1, 2]
P 1-oxa-2-oxocycloheptane + $NADP^+$ + H_2O
S cyclooctanone + NADPH + O_2 <1> (<1> poor substrate [1,2]) (Reversibility: ? <1> [1, 2]) [1, 2]
P 1-oxa-2-oxocyclononane + $NADP^+$ + H_2O
S cyclopentanone + NADPH + O_2 <1, 2> (Reversibility: ? <1, 2> [1-6]) [1-6]
P 5-valerolactone + $NADP^+$ + H_2O
S norbornanone + NADPH + O_2 <1> (Reversibility: ? <1> [1, 2]) [1, 2]
P ?

Inhibitors
5,5'-dithiobis(2-nitrobenzoate) <1> [1, 2]
arsenite <1> [1]
bathocuproine <1> [1, 2]
bathophenanthroline <1> [1]
diethyldithiocarbamate <1> [1, 2]
iodoacetamide <1> [1]
p-hydroxymercuribenzoate <1> [1, 2]

Cofactors/prosthetic groups
FAD <1, 2> (<1> FAD : enzyme ratio is 2.2 to 3.5 [1, 4]) [1, 2, 4-6]
NADPH <1, 2> [1-3, 5, 6]

Specific activity (U/mg)
4.3-5 <1> [1, 2, 4]

K_m-Value (mM)
0.001 <1> (O_2) [1, 4]
0.15 <1> (cyclopentanone) [4]

pH-Optimum
7-7.5 <1> [2]
7.7 <1> [1]

4 Enzyme Structure

Molecular weight
194000 <1> (<1> ultracentrifugation [4]) [4]
200000 <1> (<1> gel filtration, sedimentation velocity [1,2]) [1, 2, 4]

Subunits

homotetramer <1> (<1> α_4, 4 * 50000, SDS-Page [5]) [5]
trimer or tetramer <1> (<1> 3 or 4 * 54000-58000, SDS-PAGE [1, 2, 4]) [1, 2, 4]

5 Isolation/Preparation/Mutation/Application

Purification

<1> [1, 2, 4]

Cloning

<2> (overexpression in Escherichia coli [6]) [6]

6 Stability

Oxidation stability

<1>, reduction of enzyme by visible light in presence of EDTA [1]

References

[1] Griffin, M.; Trudgill, P.W.: Purification and properties of cyclopentanone oxygenase of Pseudomonas NCIB 9872. Eur. J. Biochem., **63**, 199-209 (1976)
[2] Griffin, M.; Trudgill, P.W.: The purification of cyclopentanone oxygenase from Pseudomonas N.C.I.B.9872. Biochem. Soc. Trans., **1**, 1255-1258 (1973)
[3] Griffin, M.; Trudgill, P.W.: The metabolism of cyclopentanol by Pseudomonas N.C.I.B. 9872. Biochem. J., **129**, 595-603 (1972)
[4] Trudgill, P.W.: Cyclopentanone 1,2-monooxygenase from Pseudomonas N.C.I.B. 9872. Methods Enzymol., **188**, 77-81 (1990)
[5] Bes, M.T.; Roberts, S.M.; Wan, P.W.H.: Oxidative biotransformations by microorganisms: production of chiral synthons by cyclopentanone monooxygenase from Pseudomonas N.C.I.B. 9872. J. Mol. Catal., B Enzym., **1**, 127-134 (1996)
[6] Iwaki, H.; Hasegawa, Y.; Wang, S.; Kayser, M.M.; Lau, P.C.K.: Cloning and characterization of a gene cluster involved in cyclopentanol metabolism in Comamonas sp. strain N.C.I.M.B. 9872 and biotransformations effected by Eschericha coli-expressed cyclopentanone 1,2-monooxygenase. Appl. Environ. Microbiol., **68**, 5671-5684 (2002)

Cholesterol 7α-monooxygenase 1.14.13.17

1 Nomenclature

EC number
1.14.13.17

Systematic name
cholesterol,NADPH:oxygen oxidoreductase (7α-hydroxylating)

Recommended name
cholesterol 7α-monooxygenase

Synonyms
CYPVII
Cholesterol 7-α-hydroxylase
Cholesterol 7-α-monooxygenase
cholesterol 7α-hydroxylase
oxygenase, cholesterol 7α-mono-

CAS registry number
9037-53-0

2 Source Organism

<1> *Rattus norvegicus* (<1> enzyme complex reconstituted from cytochrome P-450 and NADPH-cytochrome P-450 reductase [14, 15]; <1> enzyme expression and activity is regulated by circadian rhythm [19]) [1-5, 7-15, 17, 19]
<2> *Oryctolagus cuniculus* (<2> enzyme complex reconstituted from cytochrome P-450, NADPH-cytochrome P-450 reductase [15, 16] and cytochrome b_5 [16]) [8, 15, 16]
<3> *Homo sapiens* [6, 7, 17]
<4> *Mesocricetus auratus* (golden syrian hamster, 3fold higher activity in males compared to females [18]; <4> sucrose-rich diet reduces enzyme activity to 25% [18]) [18]

3 Reaction and Specificity

Catalyzed reaction
cholesterol + NADPH + H^+ + O_2 = 7α-hydroxycholesterol + $NADP^+$ + H_2O

Reaction type

oxidation
redox reaction
reduction

Natural substrates and products

S cholesterol + NADPH + O_2 <1, 3> (<1, 3> rate-limiting step in bile acid formation [2, 3, 6, 9, 12]) (Reversibility: ? <1, 3> [2, 3, 6, 9, 12]) [2, 3, 6, 9, 12]

P 7α-hydroxycholesterol + $NADP^+$ + H_2O <1, 3> [2, 3, 6, 9, 12]

Substrates and products

S 5α-cholestan-3β-ol + NADPH + O_2 <1> (Reversibility: ?<1> [12]) [12]

P 5α-cholestan-3β,7α-diol + $NADP^+$ + H_2O <1> [12]

S cholesterol + NADPH + O_2 <1-4> (<1> hydroxylation of cholesterol with strict regio- and stereoselectivity, inert towards other sterols and intermediates in conversion of cholesterol to bile acids [12]) (Reversibility: ? <1-4> [1-19]) [1-19]

P 7α-hydroxycholesterol + $NADP^+$ + H_2O <1-4> [1-19]

S Additional information <1, 2> (<1, 2> specificity of reconstituted enzyme system [14-16]) [14-16]

P ?

Cofactors/prosthetic groups

NADPH <1-3> [1-16]
cytochrome b_5 <2> (<2> essential component of the cholesterol 7α-hydroxylase system [16]) [16]
heme <1, 3> (<1, 3> cytochrome P-450 dependent enzyme [3, 5-7, 9, 10]; <1> 8 nmol of heme per mg of protein [10]; <1> 9 nmol of cytochrome P-450 per mg of protein [12]) [3, 5-7, 9, 10, 12]

Activating compounds

Triton X-100 <1> (<1> stimulates [10]) [10]
thiol-containing substances <1> (<1> e.g. mercaptoethanol, dithiothreitol or cysteamine enhance activity [4]) [4]
Additional information <1> (<1> a protein purified from rat liver cytosol with minumum molecular weight of 25000 stimulats cholesterol 7α-hydroxylase activity in presence of glutathione or thioredoxin [13]) [13]

Turnover number (min^{-1})

2.5 <1> (cholesterol, <1> isoenzyme I [10]) [10]
4.67 <1> (cholesterol, <1> isoenzyme II [10]) [10]
50 <1> (cholesterol) [12]

Specific activity (U/mg)

0.00015 <3> [7]
0.00015 <3> [7]
0.0006 <1> (<1> enzyme activity depends on circadian rhythm, minimum activity at 2 p.m. [19]) [19]

0.00077 <3> (<3> after incubation with 1 unit alkaline phosphatase for 30 min [17]) [17]
0.00092 <1> (<1> after incubation with 1 unit alkaline phosphatase for 30 min [17]) [17]
0.001 <1> (<1> enzyme activity depends on circadian rhythm, maximum activity at 10 p.m. [19]) [19]
0.00196 <3> [17]
0.00213 <1> [17]
0.00294 <3> (<3> after incubation with 5 units cAMP-dependent protein kinase for 30 min [17]) [17]
0.00315 <1> (<1> after incubation with 5 units cAMP-dependent protein kinase for 30 min [17]) [17]
0.01 <1> [7]
0.01 <1> [7]
36.7 <1> [10]
36.7 <1> [10]
156 <1> [12]
156 <1> [12]

K_m-Value (mM)
0.015 <1> (cholesterol, <1> reconstituted enzyme complex [14]) [14]
0.02 <1> (O_2) [2]
0.031 <1> (cholesterol, <1> isoenzyme II [10]) [10]
0.052 <1> (cholesterol, <1> isoenzyme I [10]) [10]
Additional information <1> [12]

pH-Optimum
7.4 <1> [12]

4 Enzyme Structure

Molecular weight
Additional information <1, 2> (<2> MW of cytochrome P-450: 48000, SDS-PAGE [16]; <1> 51000, SDS-PAGE [5]) [5, 16]

Subunits
? <1> (<1> x * 53000, SDS-PAGE [10]; <1> x * 52000, SDS-PAGE [12]) [10, 12]
? <1, 3> (<1,3> x * 51000, SDS-PAGE, expressed in E. coli [17]) [17]

Posttranslational modification
side-chain modification <1, 3> (<1,3> enzyme activity decreases after dephosphorylation with alkaline phosphatase and increases after phosphorylation with cAMP-dependent protein kinase [17]) [17]
Additional information <1> (<1> enzyme has a partial requirement for phospholipid [3]) [3]

5 Isolation/Preparation/Mutation/Application

Source/tissue
liver <1-4> [1-3, 5-14, 16, 18, 19]

Localization
microsome <1-4> [2, 3, 6-12, 14, 16, 18]

Purification
<1> (purification of P-450 7α [5]; 2 isoenzymes: I and II [10]) [3, 5, 7, 10, 12]
<2> (purification of cytochrome P-450 [16]) [16]
<3> [7]

Cloning
<1, 3> (expression in Escherichia coli [17]) [17]

References

[1] Mitton, J.R.; Scholan, N.A.; Boyd, G.S.: The oxidation of cholesterol in rat liver sub-cellular particles. The cholesterol-7-α-Hydroxylase enzyme system. Eur. J. Biochem., **20**, 569-579 (1971)
[2] Boyd, G.S.; Grimwade, A.M.; Lawson, M.E.: Studies on rat-liver microsomal cholesterol 7α-hydroxylase. Eur. J. Biochem., **37**, 334-340 (1973)
[3] Rush, W.R.; Gibson, G.G.; Parke, D.V.: The purification and reconstitution of cytochrome P-450-dependent cholesterol 7 α-hydroxylase activity from rat liver microsomal fractions. Biochem. Soc. Trans., **8**, 102-103 (1980)
[4] Van Cantfort, J.; Gielen, J.: Cholesterol 7α-hydroxylase. 2. Biochemical properties and participation of endogenous cholesterol in the assay in vitro. Eur. J. Biochem., **55**, 33-40 (1975)
[5] Andersson, S.; Boström, H.; Danielsson, H.; Wikvall, K.: Purification from rabbit and rat liver of cytochromes P-450 involved in bile acid biosynthesis. Methods Enzymol., **111**, 364-377 (1985)
[6] Erickson, S.K.; Bösterling, B.: Cholesterol 7 α-hydroxylase from human liver: partial purification and reconstruction into defined phospholipid-cholesterol vesicles. J. Lipid Res., **22**, 872-876 (1981)
[7] Nguyen, L.B.; Shefer, S.; Salen, G.; Ness, G.; Tanaka, R.D.; Packin, V.; Thomas, P.; Shore, V.; Batta, A.: Purification of cholesterol 7 α-hydroxylase from human and rat liver and production of inhibiting polyclonal antibodies. J. Biol. Chem., **265**, 4541-4546 (1990)
[8] Kalles, I.; Wikvall, K.: Role of sulfhydryl groups in catalytic activity of purified cholesterol 7α-hydroxylase system from rabbit and rat liver microsomes. Biochem. Biophys. Res. Commun., **100**, 1361-1369 (1981)
[9] Waxman, D.J.: Rat hepatic cholesterol 7 α-hydroxylase: biochemical properties and comparison to constitutive and xenobiotic-inducible cytochrome P-450 enzymes. Arch. Biochem. Biophys., **247**, 335-345 (1986)
[10] Chiang, J.Y.L.; Miller, W.F.; Lin, G.M.: Regulation of cholesterol 7 α-hydroxylase in the liver. Purification of cholesterol 7 α-hydroxylase and the im-

munochemical evidence for the induction of cholesterol 7 α-hydroxylase by cholestyramine and circadian rhythm. J. Biol. Chem., **265**, 3889-3897 (1990)
[11] Suckling, K.E.; Jackson, B.; Suffolk, R.A.; Houghton, J.D.; Suckling, C.J.: Effects of 6,6-difluorocholestanol and 7,7-difluorocholestanol on hepatic enzymes of cholesterol metabolism. Biochim. Biophys. Acta, **1002**, 401-404 (1989)
[12] Ogishima, T.; Deguchi, S.; Okuda, K.: Purification and characterization of cholesterol 7 α-hydroxylase from rat liver microsomes. J. Biol. Chem., **262**, 7646-7650 (1987)
[13] Danielsson, H.; Kalles, I.; Wikvall, K.: Regulation of hydroxylations in biosynthesis of bile acids. Isolation of a protein from rat liver cytosol stimulating reconstituted cholesterol 7 α-hydroxylase activity. J. Biol. Chem., **259**, 4258-4262 (1984)
[14] Ozasa, S.; Boyd, G.S.: Cholesterol 7 α-hydroxylase of rat liver. Studies on the solubilisation, resolution and reconstitution of the enzyme complex. Eur. J. Biochem., **119**, 263-272 (1981)
[15] Hansson, R.; Wikvall, K.: Properties of reconstituted cholesterol 7 α-hydroxylase system from rat and rabbit liver microsomes. Eur. J. Biochem., **93**, 419-426 (1979)
[16] Miki, N.; Miura, R.; Miyake, Y.: Purification and characterization of cholesterol 7 α-hydroxylase cytochrome P-450 of untreated rabbit liver microsomes. J. Biochem., **101**, 1087-1094 (1987)
[17] Nguyen, L.B.; Shefer, S.; Salen, G.; Chiang, J.Y.L.; Patel, M.: Cholesterol 7α-hydroxylase activities from human and rat liver are modulated in vitro posttranslationally by phosphorylation/dephosphorylation. Hepatology, **24**, 1468-1474 (1996)
[18] Souidi, M.; Parquet, M.; Ferezou, J.; Lutton, C.: Modulation of cholesterol 7α-hydroxylase and sterol 27-hydroxylase activities by steroids and physiological conditions in hamster. Life Sci., **17**, 1585-1593 (1999)
[19] Yamada, M.; Nagatomo, J.; Setoguchi, Y.; Kuroki, N.; Higashi, s.; Setoguchi, T.: Circadian Rhytms of sterol 12α-hydroxylase, cholesterol 7α-hydroxylase and DBP involved in rat cholesterol catabolism. Biol. Chem., **381**, 1149-1153 (2000)

4-Hydroxyphenylacetate 1-monooxygenase 1.14.13.18

1 Nomenclature

EC number
1.14.13.18

Systematic name
4-hydroxyphenylacetate,NAD(P)H:oxygen oxidoreductase (1-hydroxylating)

Recommended name
4-hydroxyphenylacetate 1-monooxygenase

Synonyms
4-HPA 1-hydroxylase
4-hydroxyphenylacetate 1-hydroxylase
4-hydroxyphenylacetic 1-hydroxylase
oxygenase, 4-hydroxyphenylacetate 1-mono-

CAS registry number
55326-44-8

2 Source Organism

<1> *Bacillus sp.* (PHPXAa-B) [5]
<2> *Pseudomonas acidovorans* [1]
<3> *Flavobacterium sp.* [2]
<4> *Xanthomonas sp.* (124X) [3]
<5> *Pseudomonas sp.* (CBS3) [4]

3 Reaction and Specificity

Catalyzed reaction
4-hydroxyphenylacetate + NAD(P)H + H^+ + O_2 = homogentisate + $NAD(P)^+$ + H_2O

Reaction type
oxidation
redox reaction
reduction

Substrates and products

S 4-hydroxy-2-methylphenylacetate + NAD(P)H + O_2 <2> (Reversibility: ? <2> [1]) [1, 1]

P ?

S 4-hydroxyhydratropate + NAD(P)H + O_2 <2> (Reversibility: ? <2> [1]) [1]

P 2-methylhomogentisate + $NAD(P)^+$ + H_2O <2> [1]

S 4-hydroxyphenoxyacetate + ? <2> (Reversibility: ? <2> [1]) [1, 1]

P hydroquinone + glycolate + ? <2> [1]

S 4-hydroxyphenylacetate + NAD(P)H + O_2 <1-5> (Reversibility: ? <1-5> [1-5]) [1-5]

P homogentisate + $NAD(P)^+$ + H_2O <1-5> [1-5]

S Additional information <2> (low activity with: 4-hydroxyphenylpropionate, 4-hydroxy-3-methylphenylacetate, 3,4-dihydroxyphenylacetate) [1]

P ?

Inhibitors

3,4-dihydroxyphenylacetic acid <2> [1]
4-hydroxy-3-methylphenylacetic acid <2> [1]
4-hydroxyphenylpropionic acid <2> [1]
DL-4-hydroxymandelic acid <2> [1]
KCl <2> [1]

Cofactors/prosthetic groups

FAD <2> (FAD required, <2> [1]; no effect, <4> [3]) [1]
NADH <2> (same maximal velocity with NADH and NADPH, <2> [1]; no activity with NADH, <4> [3]) [1]
NADPH <2, 4> (same maximal velocity with NADH and NADPH, <2> [1]) [1, 3]

Metals, ions

Mg^{2+} <2> (<2> required [1]; <4> no effect [3]) [1]

Specific activity (U/mg)

2.2 <2> [1]

K_m-Value (mM)

0.031 <2> (4-hydroxyphenylacetate) [1]
0.067 <2> (O_2) [1]
0.095 <2> (NADH) [1]
0.25 <2> (NADPH) [1]

pH-Optimum

7.3 <2> [1]

Temperature optimum (°C)

25 <2> (<2> assay at [1]) [1]

5 Isolation/Preparation/Mutation/Application

Purification

<2> [1]

6 Stability

Storage stability

<2>, 4°C, in the presence of dithioerythritol [1]

References

[1] Hareland, W.A.; Crawford, R.L.; Chapman, P.J.; Dagley, S.: Metabolic function and properties of 4-hydroxyphenylacetic acid 1-hydroxylase from Pseudomonas acidovorans. J. Bacteriol., **121**, 272-285 (1975)
[2] Van den Tweel, W.J.J.; Smits, J.P.; de Bont, J.A.M.: Catabolism of DL-α-phenylhydracrylic, phenylacetic and 3- and 4-hydroxyphenylacetic acid via homogentisic acid in a Flavobacterium sp.. Arch. Microbiol., **149**, 207-213 (1988)
[3] Van den Tweel, W.J.J.; Janssens, R.J.J.; De Bont, J.A. M.: Degradation of 4-hydroxyphenylacetate by Xanthobacter 124X. Antonie Leeuwenhoek, **52**, 309-318 (1986)
[4] Klages, U.; Markus, A.; Lingens, F.: Degradation of 4-chlorophenylacetic acid by a Pseudomonas species. J. Bacteriol., **146**, 64-68 (1981)
[5] Crawford, R.L.: Hydroxylation of 4-hydroxyphenoxyacetate by a Bacillus sp.. FEMS Microbiol. Lett., **4**, 233-234 (1978)

Taxifolin 8-monooxygenase 1.14.13.19

1 Nomenclature

EC number
1.14.13.19

Systematic name
taxifolin,NAD(P)H:oxygen oxidoreductase (8-hydroxylating)

Recommended name
taxifolin 8-monooxygenase

Synonyms
taxifolin hydroxylase

CAS registry number
39307-19-2

2 Source Organism

<1> *Pseudomonas sp.* [1]

3 Reaction and Specificity

Catalyzed reaction
taxifolin + NAD(P)H + H^+ + O_2 = 2,3-dihydrogossypetin + $NAD(P)^+$ + H_2O (A flavoprotein, converting a flavanol into a flavanone. Also acts on fustin, but not on catechin, quercitin or mollisacidin)

Reaction type
oxidation
redox reaction
reduction

Natural substrates and products
S taxifolin + NADH + O_2 <1> (Reversibility: ? <1> [1]) [1]
P 2,3-dihydrogossypetin + NAD^+ + H_2O <1> [1]
S taxifolin + NADPH + O_2 <1> (Reversibility: ? <1> [1]) [1]
P 2,3-dihydrogossypetin + $NADP^+$ + H_2O <1> [1]

Substrates and products
S (-)-fustin + NADH + O_2 <1> (Reversibility: ? <1> [1]) [1]
P ? + NAD^+ + H_2O <1> (<1> product not identified [1]) [1]

S taxifolin + NADH + O_2 <1> (Reversibility: ? <1> [1]) [1]
P 2,3-dihydrogossypetin + NAD^+ + H_2O <1> [1]
S taxifolin + NADPH + O_2 <1> (Reversibility: ? <1> [1]) [1]
P 2,3-dihydrogossypetin + $NADP^+$ + H_2O <1> [1]

Cofactors/prosthetic groups

FAD <1> [1]
NADH <1> (<1> twice oxidized compared to NADPH [1]) [1]
NADPH <1> [1]

pH-Optimum

8 <1> [1]

5 Isolation/Preparation/Mutation/Application

Source/tissue

cell culture <1, 2> [1, 20]

Purification

<1> (partial [1]) [1]

6 Stability

pH-Stability

6 <1> (<1> 20% of maximal activity [1]) [1]

Temperature stability

40 <1> (<1> 25% loss of activity in 10 min [1]) [1]
50 <1> (<1> 95% loss of activity in 10 min [1]) [1]

References

[1] Jeffrey, A.M.; Knight, M.; Evans, W.C.: The bacterial degradation of flavonoids. Hydroxylation of the A-ring of taxifolin by a soil pseudomonad. Biochem. J., **130**, 373-381 (1972)

2,4-Dichlorophenol 6-monooxygenase 1.14.13.20

1 Nomenclature

EC number
1.14.13.20

Systematic name
2,4-dichlorophenol,NADPH:oxygen oxidoreductase (6-hydroxylating)

Recommended name
2,4-dichlorophenol 6-monooxygenase

Synonyms
2,4-dichlorophenol hydroxylase
2,4-dichlorophenol monooxygenase
oxygenase, 2,4-dichlorophenol 6-mono
oxygenase, 2,4-dichlorophenol mono-

CAS registry number
82047-82-3
92767-55-0

2 Source Organism

<1> *Pseudomonas cepacia* [3]
<2> *Acinetobacter sp.* [1]
<3> *Ralstonia eutropha* (basonym Alcaligenes eutrophus [2,5]; JMP134 [2,4,5]; TFD41 [4]) [2, 4, 5]
<4> *Pseudomonas putida* (PaW85 [4]) [4]
<5> *Alcaligenes paradoxus* (2811P [4]) [4]
<6> *Sphingomonas paucimobilis* (1443 [4]) [4]
<7> *Burkholderia cepacia* (DBO131 [4]) [4]
<8> *Rhodoferax fermentans* (TFD23 [4]; B6-9 [4]) [4]
<9> *Burkholderia mallei* (TFD6 [4]) [4]
<10> *Burkholderia sp.* (TFD2 [4]; RASC [4]) [4]
<11> *Rhodopseudomonas palustris* (M1 [4]) [4]
<12> *Sphingomonas sp.* (TFD44 [4]; B6-10 [4]; B6-5 [4]; EML146 [4]) [4]
<13> *Protobacteria* (S1 [6]) [6]

3 Reaction and Specificity

Catalyzed reaction

2,4-dichlorophenol + NADPH + H^+ + O_2 = 3,5-dichlorocatechol + $NADP^+$ + H_2O

Reaction type

oxidation
redox reaction
reduction

Natural substrates and products

S 2,4-dichlorophenol + NADPH + O_2 <1-13> (Reversibility: ? <1-13> [1-6]) [1-6]
P 3,5-dichlorocatechol + $NADP^+$ + H_2O

Substrates and products

S 2,3-dichlorophenol + NADPH + O_2 <13> (Reversibility: ? <13> [6]) [6]
P ?
S 2,4,5-trichlorophenol + NADPH + O_2 <3> (Reversibility: ? <3> [2]) [2]
P ?
S 2,4,5-trichlorophenol + NADPH + O_2 <13> (Reversibility: ? <13> [6]) [6]
P ?
S 2,4,6-trichlorophenol + NADPH + O_2 <13> (Reversibility: ? <13> [6]) [6]
P ?
S 2,4-dibromophenol + NADPH + O_2 <3, 13> (Reversibility: ? <3, 13> [2, 6]) [2, 6]
P ?
S 2,4-dichlorophenol + NADPH + O_2 <1-13> (Reversibility: ? <1-13> [1-6]) [1-6]
P 3,5-dichlorocatechol + $NADP^+$ + H_2O
S 2,5-dichlorophenol + NADPH + O_2 <13> (Reversibility: ? <13> [6]) [6]
P ?
S 2,6-dichlorophenol + NADPH + O_2 <13> (Reversibility: ? <13> [6]) [6]
P ?
S 2-chloro-4-nitrophenol + NADPH + O_2 <3> (Reversibility: ? <3> [2]) [2]
P ?
S 2-chlorophenol + NADPH + O_2 <2, 3, 13> (<2,3> incompletely metabolized [1,2]) (Reversibility: ? <2, 3, 13> [1, 2, 6]) [1, 2, 6]
P ?
S 2-cresol + NADPH + O_2 <2> (<2> incompletely metabolized [1]) (Reversibility: ? <2> [1]) [1]
P ?
S 2-methylphenol + NADPH + O_2 <13> (Reversibility: ? <13> [6]) [6]
P ?
S 3,4-dichlorophenol + NADPH + O_2 <13> (Reversibility: ? <13> [6]) [6]
P ?
S 3,5-dichlorophenol + NADPH + O_2 <13> (Reversibility: ? <13> [6]) [6]

P ?

S 3-chlorophenol + NADPH + O_2 <13> (Reversibility: ? <13> [6]) [6]

P ?

S 4-bromophenol + NADPH + O_2 <2, 13> (<2> incompletely metabolized [1]) (Reversibility: ? <2, 13> [1, 6]) [1, 6]

P ?

S 4-chloro-2-methylphenol + NADPH + O_2 <2, 3, 13> (Reversibility: ? <2, 3, 13> [1, 2, 6]) [1, 2, 6]

P ?

S 4-chloro-2-nitrophenol + NADPH + O_2 <3> (Reversibility: ? <3> [2]) [2]

P ?

S 4-chlorophenol + NADPH + O_2 <2, 3, 13> (Reversibility: ? <2, 3, 13> [1, 2, 6]) [1, 2, 6]

P ?

S Additional information <2> (<2> pseudosubstrates evoke oxidation of NAD(P)H and oxygen consumption without themselves undergoing hydroxylation, the product is hydrogen peroxide: 2,5-dichlorophenol, 2,6-dichlorophenol, 3,4-dichlorophenol, 2, 4,5-trichlorophenol [1]) [1]

P ?

Inhibitors

2,4,5-trichlorophenol <1> [3]
2,4,6-trichlorophenol <1> [3]
2,4-dichlorophenol <2, 3> (<2,3> above 1 mM [1,2]) [1, 2]
$AgNO_3$ <2, 3> [1, 2]
$CuSO_4$ <2, 3> [1, 2]
$FeCl_3$ <2> [1]
$FeSO_4$ <2, 3> [1, 2]
Hg^{2+} <1-3> (<2,3> $HgCl_2$ [1,2]) [1-3]
KCN <2> (no effect, <3> [2]) [1]
KCl <3> (<3> high salt concentration inhibits, 0.05 M: 40%, 0.1 M: 50%, 0.2 M: 70% [2]) [2]
SDS <1> [3]
Zn^{2+} <1> [3]
diethyldithiocarbamate <2, 3> [1, 2]
dithionitrobenzoate <3> [2]
dithiothreitol <2, 3> (<2,3> slight [1,2]) [1, 2]
guanidine hydrochloride <1> [3]
o-phenanthroline <3> [2]
p-chloromercuribenzoate <1, 2> [1, 3]
urea <1> [3]

Cofactors/prosthetic groups

FAD <1-3, 13> (<1-3> flavoprotein [1-3]; <2> FAD cannot be replaced by riboflavin or FMN [1]; <2> 1 mol of enzyme contains: 1.9 mol FAD [1]; <3> 1 mol of enzyme contains: 3.5 mol FAD [2]; <1,13> 1 mol of enzyme contains: 4 mol FAD [3,6]; <1> requires exogenous addition of FAD [3]) [1-3, 6]

NADH <1-3, 13> (<2> 5% of the activity with NADPH [1]; <3> 50% of the activity with NADPH [2]; <1> NADPH preferred to NADH [3]) [1-3, 6]
NADPH <1-3, 13> [1-3, 6]

Turnover number (min^{-1})
410 <2> (2,4-dichlorophenol) [1]

Specific activity (U/mg)
1.14 <3> [5]
1.46 <2> [1]
1.48 <3> [2]
1.8 <13> (<13> without FAD [6]) [6]
1.9 <1> [3]
2.4 <13> [6]

K_m-Value (mM)
0.002 <3> (2,4-dichlorophenol, <3> + NADPH [2]) [2]
0.003 <13> (2,4-dichlorophenol) [6]
0.004 <2, 3> (2,4-dichlorophenol, <2> + NADPH [1]; <3> + NADH [2]) [1, 2]
0.014 <1> (2,4-dichlorophenol) [3]
0.024 <13> (NADPH) [6]
0.042 <13> (NADH) [6]
0.048 <3> (NADPH, <3> + 2,4-dichlorophenol [2]) [2]
0.057 <2> (O_2) [1]
0.083 <2> (NADPH) [1]
0.1 <1> (NADPH) [3]
0.3 <3> (NADH, <3> + 2,4-dichlorophenol [2]) [2]

pH-Optimum
7.3-8 <3> (<3> phosphate buffer [2]) [2]
7.6 <2> [1]
8.1-8.5 <3> (<3> 100 mM Tris-HCl buffer [2]) [2]
8.5 <1> [3]

Temperature optimum (°C)
29 <2> [1]
37 <1> [3]

4 Enzyme Structure

Molecular weight
224000 <3> (<3> ultracentrifugation [2]) [2]
240000 <2> (<2> gel filtration [1]) [1]
245000 <3> (<3> gel filtration [5]) [5]
256000 <13> (<13> gel filtration, gradient PAGE [6]) [6]
275000 <1> (<1> gel filtration and ultracentrifugation [3]) [3]

Subunits

tetramer <1-3> (<1> α_4, 4 * 69000, SDS-PAGE [3]; <2> α_4, 4 * 63000, SDS-PAGE [1]; <3> $\alpha_2\beta_2$, 2 * 67000 + 2 * 45000, SDS-PAGE [2]; <3> α_4, 4 * 66000, SDS-PAGE [5]; <13> α_4, 4 * 64000, SDS-PAGE [6]) [1-3, 5, 6]

5 Isolation/Preparation/Mutation/Application

Purification

<1> (<1> single step affinity chromatography [3]) [3]
<2> [1]
<3> (<3> plasmid pJP4-encoded enzyme from Alcaligenes eutrophus JMP134 transferred to Pseudomonas putida 3CB5 and Alcaligenes eutrophus 335 [2]) [2, 4, 5]
<13> [6]

Cloning

<3> (expression in Pseudomonas putida [2]) [2]

6 Stability

General stability information

<2>, 2,4-dichlorophenol, 0.1 mM, stabilizes during purification [1]
<2>, EDTA, 0.1 mM, stabilizes during purification [1]
<2>, FAD, 0.01 mM, stabilizes during purification [1]
<2>, dithiothreitol, 1 mM, stabilizes during purification [1]

Storage stability

<2>, -20°C, 3 months stable [1]

References

[1] Beadle, C.A.; Smith, A.R.W.: The purification and properties of 2,4-dichlorophenol hydroxylase from a strain of Acinetobacter species. Eur. J. Biochem., **123**, 323-332 (1982)

[2] Liu, T.; Chapman, P.J.: Purification and properties of a plasmid-encoded 2,4-dichlorophenol hydroxylase. FEBS Lett., **173**, 314-318 (1984)

[3] Radjendirane, V.; Bhat, M.A.; Vaidyanathan, C.S.: Affinity purification and characterization of 2,4-dichlorophenol hydroxylase from Pseudomonas cepacia [published erratum appears in Arch Biochem Biophys 1992 Jul;296(1):354]. Arch. Biochem. Biophys., **288**, 169-176 (1991)

[4] Farhana, L.; Fulthorpe, R.R.; Harbour, C.; New, P.B.: Monoclonal antibodies to 2,4-dichlorophenol hydroxylase as probes for the 2,4-degradative phenotype. Can. J. Microbiol., **44**, 920-928 (1998)

[5] Farhana, L.; New, P.B.: The 2,4-dichlorophenol hydroxylase of Alcaligenes eutrophus JMP134 is a homotetramer. Can. J. Microbiol., **43**, 202-205 (1997)
[6] Makdessi, K.; Lechner, U.: Purification and characterization of 2,4-dichlorophenol hydroxylase isolated from a bacterium of the α-2 subgroup of the Proteobacteria. FEMS Microbiol. Lett., **157**, 95-101 (1997)

Flavonoid 3'-monooxygenase 1.14.13.21

1 Nomenclature

EC number
1.14.13.21

Systematic name
flavonoid,NADPH:oxygen oxidoreductase (3'-hydroxylating)

Recommended name
flavonoid 3'-monooxygenase

Synonyms
NADPH:flavonoid-3'-hydroxylase
flavonoid 3'-hydroxylase
flavonoid 3-hydroxylase (erroneous)
oxygenase, flavonoid 3'-mono-
oxygenase, flavonoid 3-mono- (erroneous)

CAS registry number
75991-44-5
85340-98-3 (oxygenase, flavonoid 3-mono-)

2 Source Organism

<1> *Dianthus caryophyllus* (carnation [8]) [8]
<2> *Matthiola incana* (enzyme present only in lines of organism with wild-type allele b+ [9]) [1, 9]
<3> *Columnea hybrida* [2]
<4> *Sinningia cardinalis* (syn. Rechsteineria [3]) [3]
<5> *Antirrhinum majus* (snapdragon [4]) [4]
<6> *Zea mays* [5]
<7> *Verbena hybrida* [6]
<8> *Petroselinum hortense* (parsley [7]) [7]
<9> *Petunia hybrida* [10]
<10> *Perilla frutescens* [11]
<11> *Torenia sp.* [12]
<12> *Citrus sinensis* (sweet orange [13]) [13]

3 Reaction and Specificity

Catalyzed reaction

a flavonoid + NADPH + H^+ + O_2 = a 3'-hydroxy flavonoid + $NADP^+$ + H_2O

Reaction type

oxidation
redox reaction
reduction

Natural substrates and products

S apigenin + NADPH + O_2 <3, 4> (Reversibility: ? <3, 4> [2, 3]) [2, 3]
P luteolin + $NADP^+$ + H_2O
S dihydrokaempferol + NADPH + O_2 <2> (Reversibility: ? <2> [1]) [1]
P dihydroquercetin + $NADP^+$ + H_2O
S naringenin + NADPH + O_2 <2, 4> (Reversibility: ? <2, 4> [1, 3]) [1, 3]
P eriodictyol + $NADP^+$ + H_2O

Substrates and products

S apigenin + NADPH + O_2 <3, 4, 6, 8, 10, 11> (Reversibility: ? <3, 4, 6, 8, 10, 11> [2, 3, 5, 7, 11, 12]) [2, 3, 5, 7, 11, 12]
P luteolin + $NADP^+$ + H_2O <3, 4> [2, 3]
S dihydrokaempferol + NADPH + O_2 <1, 2, 8-12> (<6> no activity [5]) (Reversibility: ? <1, 2, 8-12> [1, 7-13]) [1, 7-13]
P dihydroquercetin + $NADP^+$ + H_2O
S kaempferol + NADPH + O_2 <6, 8, 11, 12> (Reversibility: ? <6, 8, 11, 12> [5, 7, 12, 13]) [5, 7, 12, 13]
P quercetin + $NADP^+$ + H_2O
S naringenin + NADPH + O_2 <1, 2, 4, 6, 8-12> (Reversibility: ? <1, 2, 4, 6, 8-12> [1, 3, 5, 7-13]) [1, 3, 5, 7-13]
P eriodictyol + $NADP^+$ + H_2O <2, 4, 6, 8, 12> [1, 3, 5, 7, 13]
S Additional information <2, 8> (<8> not: prunin [7]; <2> not: 4-coumarate, 4-coumaroyl-CoA [9]) [7, 9]
P ?

Inhibitors

CO <6, 8, 12> (<6, 12> inhibition reduced by irradiation with 450 nm light during incubation [5, 13]; <8> partial [7]) [5, 7, 13]
EDTA <2, 8> (<8> partial [7]) [7, 9]
$FeCl_3$ <6> [5]
KCN <2, 4, 8> (<4,8> partial [3, 7]) [3, 7, 9]
N-ethylmaleimide <4, 6> (<8> 1 mM: activation [7]; <4> 2 mM: partial inhibition [3]) [3, 5]
$NADP^+$ <6, 8> [5, 7]
NaN_3 <6, 8> (<8> partial [7]) [5, 7]
α,α'-dipyridyl <6> (<6> partial [5]) [5]
cytochrome c <6, 8, 12> [5, 7, 13]
diethyldicarbonate <2, 4, 8> (<8> partial [7]) [3, 7, 9]

ketoconazole <4, 12> (<4> cytochrome P_{450} inhibitor [3]) [3, 13]
p-chloromercuribenzoate <2, 4, 6, 8> (<2, 4, 6> partial [3, 5, 9]) [3, 5, 7, 9]
tetcyclacis <4> [3]

Cofactors/prosthetic groups
NADH <2, 5> (<5> much lower product yield than with NADPH [4]; <6> no activity [5]; <2> about 12% of the activity with NADPH [9]) [4, 9]
NADPH <1-8, 12> [1-9, 13]
cytochrome P_{450} <6, 8, 9, 12> (<6> 0.21 nmol per mg of protein [5]; <8> cytochrome P-450-dependent enzyme [7]) [5, 7, 10, 13]

Activating compounds
N-ethylmaleimide <8> (<8> 1 mM: activation [7]) [7]

Specific activity (U/mg)
Additional information <3> [2]

K_m-Value (mM)
0.0008 <11> (naringenin) [12]
0.003 <11> (kaempferol) [12]
0.004 <11> (dihydrokaempferol) [12]
0.0058 <3> (NADPH, <3> with kaempferol [2]) [2]
0.00715 <3> (kaempferol) [2]
0.019 <10> (apigenin) [11]
0.02 <10> (dihydrokaempferol) [11]
0.02 <10> (naringenin) [11]
0.021 <11> (apigenin) [12]
24 <12> (naringenin) [13]

pH-Optimum
7.2 <2> (<2> naringenin [9]) [9]
7.4-7.6 <12> [13]
7.5 <4, 5, 8> (<4> apigenin [3]) [3, 4, 7]
8.5 <6> [5]

pH-Range
6.8-8.4 <8> (<8> half-maximal activity at pH 6.8 and 8.4 [7]) [7]
Additional information <6> (<6> sharp drop in activity on either side of the optimum of pH 8.5 [5]) [5]

Temperature optimum (°C)
25 <8> [7]
30 <6> [5]
30-37 <12> [13]

Temperature range (°C)
10-25 <8> (<8> 10°C: 30% of maximal activity, 25°C: optimum [7]) [7]

4 Enzyme Structure

Subunits

? <10> (x * 57500, deduced from gene sequence [11]) [11]

5 Isolation/Preparation/Mutation/Application

Source/tissue

bud <2> (<2> most active in developmental stages 2-5 of bud and flower formation [1]) [1]
cell culture <8, 12> [7, 13]
flower <5> [4]
petal <9> (<9> of flowers at early stage of development [10]) [10]
seedling <6> [5]

Localization

microsome <1-8, 12> [2-9, 13]

Cloning

<9> [10]
<10> (predominantly expressed in red form of plant [11]) [11]
<11> [12]

6 Stability

Temperature stability

0 <8> (<8> addition of 14 mM mercaptoethanol, 15% sucrose, half-life 250 min [7]) [7]
10 <8> (<8> addition of 14 mM mercaptoethanol, 15% sucrose, half-life 180 min [7]) [7]
20 <8> (<8> addition of 14 mM mercaptoethanol, 15% sucrose, half-life 60 min [7]) [7]
30 <8> (<8> addition of 14 mM mercaptoethanol, 15% sucrose, half-life 24 min [7]) [7]
100 <6> (<6> 10 min, activity destroyed [5]) [5]

General stability information

<6>, sucrose or glycerol plus dithiothreitol stabilize during purification and storage at -70°C [5]
<8>, 14 mM mercaptoethanol plus 15% sucrose stabilize, mercaptoethanol can be replaced by 1.4 mM dithiothreitol [7]

Storage stability

<4>, -80°C, microsomal preparations containing 10% sucrose, frozen in liquid nitrogen, stable [3]
<6>, -70°C, sucrose or glycerol plus dithiothreitol stabilize during purification and storage [5]
<8>, -70°C, storage of microsomes in presence of 14 mM mercaptoethanol and 15% sucrose, stable for several weeks [7]

References

[1] Dangelmayr, B.; Stotz, G.; Spribille, R.; Forkmann, G.: Relationship between flower development, anthocyanin accumulation and activity of enzymes involved in flavonoid biosynthesis in Matthiola incana R.Br.. Z. Naturforsch. C, **38**, 551-555 (1983)
[2] Stich, K.; Forkmann, G.: Studies on columnidin biosynthesis with flower extracts from Columnea hybrida. Z. Naturforsch. C, **43**, 311-314 (1988)
[3] Stich, K.; Forkmann, G.: Enzymatic synthesis of 4'- and 3',4'-hydroxylated flavanones and flavones with flower extracts of Sinningia cardinalis. Z. Naturforsch. C, **42**, 1193-1199 (1987)
[4] Forkmann, G.; Stotz, G.: Genetic control of flavanone 3-hydroxylase activity and flavonoid 3'-hydroxylase activity in Antirrhinum majus (Snapdragon). Z. Naturforsch. C, **36**, 411-416 (1981)
[5] Larson, R.L.; Bussard, J.B.: Microsomal flavonoid 3'-monooxygenase from maize seedlings. Plant Physiol., **80**, 483-486 (1986)
[6] Stotz, G.; Spribille, R.; Forkmann, G.: Flavonoid biosynthesis in flowers of Verbena hybrida. J. Plant Physiol., **116**, 173-183 (1984)
[7] Hagmann, M.L.; Heller, W.; Grisebach, H.: Induction and characterization of a microsomal flavonoid 3-hydroxylase from parsley cell cultures. Eur. J. Biochem., **134**, 547-554 (1983)
[8] Spribille, R.; Forkmann, G.: Chalcone synthesis and hydroxylation of flavonoids in 3'-position with enzyme preparations from flowers of Dianthus caryophyllus L. (carnation). Planta, **155**, 176-182 (1982)
[9] Forkmann, G.; Heller, W.; Grisebach, H.: Anthocyanin biosynthesis in flowers of Matthiola incana flavanone 3- and flavonoid 3'-hydroxylases. Z. Naturforsch. C, **35**, 691-695 (1980)
[10] Brugliera, F.; Barri-Rewell, G.; Holton, T.A.; Mason, J.G.: Isolation and characterization of a flavonoid 3'-hydroxylase cDNA clone corresponding to the Ht1 locus of Petunia hybrida. Plant J., **19**, 441-451 (1999)
[11] Kitada, C.; Gong, Z.; Tanaka, Y.; Yamazaki, M.; Saito, K.: Differential expression of two cytochrome P_{450}s involved in the biosynthesis of flavones and anthocyanins in chemo-varietal forms of Perilla frutescens. Plant Cell Physiol., **42**, 1338-1344 (2001)
[12] Ueyama, Y.; Suzuki, K.i.; Fukuchi-Mizutani, M.; Fukui, Y.; Miyazaki, K.; Ohkawa, H.; Kusumi, T.; Tanaka, Y.: Molecular and biochemical characterization of torenia flavonoid 3'-hydroxylase and flavone synthase II and modification of flower color by modulating the expression of these genes. Plant Sci., **163**, 253-263 (2002)
[13] Doostdar, H.; Shapiro, J.P.; Niedz, R.; Burke, M.D.; McCollum, T.G.; McDonald, R.E.; Mayer, R.T.: A cytochrome P_{450} mediated naringenin 3'-hydroxylase from sweet orange cell cultures. Plant Cell Physiol., **36**, 69-77 (1995)

Cyclohexanone monooxygenase 1.14.13.22

1 Nomenclature

EC number
1.14.13.22

Systematic name
cyclohexanone:NADPH:oxygen oxidoreductase (6-hydroxylating, 1,2-lactonizing)

Recommended name
cyclohexanone monooxygenase

Synonyms
CHMO <1, 10, 11> [22-24, 26]
CMO <1, 9> [20, 25]
cyclohexanone oxygenase
cyclohexanone 1, 2-mono-oxygenase
cyclohexanone mono-oxygenase
oxygenase, cyclohexanone mono-

CAS registry number
52037-90-8
59088-27-6

2 Source Organism

<1> *Acinetobacter sp.* (NCIB 9871 [1, 2, 4-6, 8, 9, 12, 13, 15-17, 19-21, 23, 27-29]) [1, 2, 4-6, 8, 9, 12, 13, 15-17, 19-21, 23, 27-29]
<2> *Xanthobacter sp.* [3, 7, 10, 11]
<3> *Pseudomonas aeruginosa* (strain PA 09501 [19]) [19]
<4> *Pseudomonas sp.* [11]
<5> *Arthrobacter sp.* (strain CA1, low activity [14]) [14]
<6> *Nocardia globerula* (enzyme exists in 2 electrophoretically distinct isoforms: type 1 and type 2 [18]) [17, 18]
<7> *Nocardia sp.* [19]
<8> *Xanthobacter autotrophicus* [7]
<9> *Acinetobacter sp.* (NCIB 9871 [25, 30]) [25, 30]
<10> *Acinetobacter calcoaceticus* (NCIB 9871 [24]; inducible by cyclohexanone [24]) [22, 24]
<11> *Exophiala jeanselmei* (strain KUFI-6N [26]; inducible by cyclohexanone [26]) [26]

3 Reaction and Specificity

Catalyzed reaction

cyclohexanone + NADPH + H^+ + O_2 = 6-hexanolide + $NADP^+$ + H_2O (a flavoprotein (FAD): acts on a number of other cyclic ketones; <1> stereochemistry and enantioselectivity of the oxidation reaction depends strongly on the structure and stereochemistry of the substrates [27, 28]; <1> ter-ter mechanism [1]; <1> a stable 4a-hydroperoxy-flavin carries out a nucleophilic attack on the ketone [1]; <1> re-side specificity of cofactor binding, stereochemistry [5]; <1> kinetic studies [23]; <1> stereochemistry [13]; <1,10> enantioselectivity [13, 21, 22]; <1> regioselectivity [2, 13]; <1, 3, 7> mechanistic study [13, 15, 19]; <1> general mechanistic scheme [2, 12]; <1> reaction mechanism [15, 23])

Reaction type

Baeyer-Villiger reaction <1> [1, 2, 6, 8, 13, 15]
oxidation
oxygen ring insertion reaction <1, 2> [3, 6]
redox reaction
reduction

Natural substrates and products

S (S)-dithiane sulfoxide + NADPH + O_2 <1> (<1> recombinant enzyme in E. coli or Saccharomyces cerevisiae [21]) (Reversibility: ? <1> [21]) [21]
P ?
S 1-oxa-2-oxocycloheptane + NADPH + O_2 <2> (<2> oxidation [10]) (Reversibility: ? <2> [10]) [10]
P ?
S bicyclo[3.2.0]hept-2-en-6-one + NADPH + O_2 <10> (<10> racemic substrate [22]; <10> recombinant in E. coli [22]) (Reversibility: ? <10> [22]) [22]
P ?
S cyclohexan-1,2-dione + NADPH + O_2 <2> (Reversibility: ? <2> [10]) [10]
P 1-oxa-2,3-dioxo-cycloheptane + $NADP^+$ + H_2O
S cyclohexan-1,4-dione + NADPH + O_2 <2> (Reversibility: ? <2> [10]) [10]
P 1-oxa-2,5-dioxo-cycloheptane + $NADP^+$ + H_2O
S cyclohexanone + NADPH + O_2 <1, 2, 5, 8, 10> (<5> pathway [14]; <2,8> depending on inducing growth substrate [7]; <1> enzyme plays an important role in degradation of keto-compounds by microorganisms [1]) (Reversibility: ? <1, 2, 5, 8, 10> [1, 7, 10, 14, 24]) [1, 7, 10, 14, 24]
P 1-oxa-2-oxocycloheptane + $NADP^+$ + H_2O
S dithiane + NADPH + O_2 <1> (<1> recombinant enzyme in E. coli or Saccharomyces cerevisiae [21]) (Reversibility: ? <1> [21]) [21]
P (R)-dithiane sulfoxide + $NADP^+$ + H_2O <1> [21]
S methyl pentyl sulfide + NADPH + O_2 <1, 10> (<10> recombinant in E. coli [22]; <1> recombinant enzyme in E. coli or Saccharomyces cerevisiae [21]) (Reversibility: ? <1, 10> [21, 22]) [21, 22]

P (R)-methyl phenyl sulfoxide + $NADP^+$ + H_2O
S n-butyl methyl sulfide + NADPH + O_2 <1> (<1> recombinant enzyme in E. coli or Saccharomyces cerevisiae [21]) (Reversibility: ? <1> [21]) [21]
P ?
S tert-butyl methyl sulfide + NADPH + O_2 <1> (<1> recombinant enzyme in E. coli or Saccharomyces cerevisiae [21]) (Reversibility: ? <1> [21]) [21]
P ?
S Additional information <1> (<10> inductive growth on cyclohexanol, addition of cyclohexanone [24]; <1> production of optically pure sulfoxides by biotransformation in whole cell systems of several sulfides, dithianes and dithiolanes [21]) [21, 24]
P ?

Substrates and products
S (+)-camphor + NADPH + O_2 <1, 6> (Reversibility: ? <1, 6> [1, 2, 17]) [1, 2, 17]
P ?
S (+)-dihydrocarvone + NADPH + O_2 <1, 6> (Reversibility: ? <1, 6> [2, 17]) [2, 17]
P ?
S 1-oxa-2-oxocycloheptane + NADPH + O_2 <2> (<2> oxidation [7]) (Reversibility: ? <2> [7]) [7]
P ?
S 1-phenyl-2-propanone + NADPH + O_2 <2, 4> (Reversibility: ? <2, 4> [11]) [11]
P ?
S 2-hexyl-cyclopentanone + NADPH + O_2 <10> (Reversibility: ? <10> [24]) [24]
P ?
S 2-hydroxycyclohexanone + NADPH + O_2 <1-3, 6, 7> (<2,8> no activity [7]) (Reversibility: ? <1-3, 6, 7> [3, 10, 17-19]) [3, 10, 17-19]
P 1-oxa-2-oxo-3-hydroxycycloheptane + $NADP^+$ + H_2O
S 2-methylcyclohexanone + NADPH + O_2 <1-3, 6, 7> (Reversibility: ? <1-3, 6, 7> [1, 3, 13, 17-19]) [1, 3, 13, 17-19]
P 1-oxa-2-oxo-3-methylcycloheptane + $NADP^+$ + H_2O
S 2-methylcyclohexyl boronic acid + NADPH + O_2 <1> (<1> racemic substrate [8]) (Reversibility: ? <1> [8]) [8]
P 2-methylcyclohexanol + BO3- + $NADP^+$ H_2O <1> [8]
S 2-norbornanone + NADPH + O_2 <1> (Reversibility: ? <1> [2]) [2]
P ?
S 2-phenyl-1-ethanal + NADPH + O_2 <2, 4> (Reversibility: ? <2, 4> [11]) [11]
P ?
S 2-phenylcyclohexanone + NADPH + O_2 <1> (<1> slight [2]) (Reversibility: ? <1> [2]) [2]
P 1-oxa-2-oxo-3-phenylcycloheptane + $NADP^+$ + H_2O

S 2-phenylethyl methyl sulfide + NADPH + O_2 <1> (Reversibility: ? <1> [27]) [27]
P 2-phenylethyl methyl sulfoxide <1> [27]
S 2-phenylpropyl methyl sulfide + NADPH + O_2 <1> (Reversibility: ? <1> [27]) [27]
P 2-phenylpropyl methyl sulfoxide <1> [27]
S 2-thiacyclohexanone + NADPH + O_2 <1> (<1> substrate inactivates enzyme after a few turnovers [6]) (Reversibility: ? <1> [6]) [6]
P 1-oxa-2-oxo-3-thiacycloheptane + $NADP^+$ + H_2O
S 4-hydroxycyclohexanone + NADPH + O_2 <1, 6> (Reversibility: ? <1, 6> [17]) [17]
P 1-oxa-2-oxo-5-hydroxycycloheptane + $NADP^+$ + H_2O
S 4-hydroxyethyl methyl sulfide + NADPH + O_2 <1> (Reversibility: ? <1> [20]) [20]
P ?
S 4-methylcyclohexanone + NADPH + O_2 <1, 6, 10> (Reversibility: ? <1, 6, 10> [1, 17, 24]) [1, 17, 24]
P 1-oxa-2-oxo-5-methylcycloheptane + $NADP^+$ + H_2O
S 4-tert-butylcyclohexanone + NADPH + O_2 <1> (<1> slight [2]) (Reversibility: ? <1> [2]) [2]
P 1-oxa-2-oxo-5-tert-butylcycloheptane + $NADP^+$ + H_2O
S D-fenchone + NADPH + O_2 <1> (Reversibility: ? <1> [2]) [2]
P ?
S S-γ-thiobutyrolactone + NADPH + O_2 <1> (<1> substrate irreversibly inactivates enzyme after a few turnovers [6]) (Reversibility: ? <1> [6]) [6]
P ?
S allyl methyl sulfide + NADPH + O_2 <1> (Reversibility: ? <1> [20]) [20]
P ?
S bicyclo[3.2.0]hept-2-en-6-one + NADPH + O_2 <10> (<10> racemic substrate [22]; <10> recombinant from E. coli [22]) (Reversibility: ? <10> [22]) [22]
P ?
S butanal + NADPH + O_2 <1> (Reversibility: ? <1> [12]) [12]
P ?
S cis-hex-2-enyl phenyl selenide + NADPH + O_2 <1> (Reversibility: ? <1> [9]) [9]
P ?
S cuprizone + NADPH + O_2 <1, 6> (Reversibility: ? <1, 6> [17]) [17]
P ?
S cyclobutanone + NADPH + O_2 <1-3, 6, 7, 11> (Reversibility: ? <1-3, 6, 7, 11> [1, 3, 17-19, 26]) [1, 3, 17-19, 26]
P 1-oxa-2-oxo-cyclopentane + $NADP^+$ + H_2O
S cyclodecanone + NADPH + O_2 <11> (<11> low activity [26]) (Reversibility: ? <11> [26]) [26]
P 1-oxa-2-oxocycloendecane + $NADP^+$ + H_2O
S cycloheptanone + NADPH + O_2 <1-3, 6, 7, 11> (Reversibility: ? <1-3, 6, 7, 11> [1, 3, 17-19, 26]) [1, 3, 17-19, 26]

P 1-oxa-2-oxo-cyclooctane + $NADP^+$ + H_2O

S cyclohexan-1,2-dione + NADPH + O_2 <1, 2, 6> (Reversibility: ? <1, 2, 6> [3, 10, 17]) [3, 10, 17]

P 1-oxa-2,3-dioxo-cycloheptane + $NADP^+$ + H_2O

S cyclohexan-1,4-dione + NADPH + O_2 <1, 2> (Reversibility: ? <1, 2> [2, 3, 10]) [2, 3, 10]

P 1-oxa-2,5-dioxo-cycloheptane + $NADP^+$ + H_2O

S cyclohexanone + NADPH + O_2 <1, 2, 5, 6, 8-11> (<1> activity limited to cyclic ketones [23]) (Reversibility: ? <1, 2, 5, 6, 8-11> [1-3, 6, 7, 10-18, 23-28]) [1-3, 6, 7, 10-18, 23-28]

P 6-hexanolide + $NADP^+$ + H_2O <1, 2, 8-11> (<1,2,8-11> product: ε-caprolactone i.e. 1-oxa-2-oxocycloheptane [1,7,10,15,17,23-26]) [1, 7, 10, 15, 17, 23-26]

S cyclohexyl methyl sulfide + NADPH + O_2 <1> (Reversibility: ? <1> [20]) [20]

P cyclohexyl methyl sulfoxide + $NADP^+$ + H_2O <1> [20]

S cyclooctanone + NADPH + O_2 <1-3, 6, 7, 11> (<2> low activity [3]) (Reversibility: ? <1-3, 6, 7, 11> [1, 3, 17-19, 26]) [1, 3, 17-19, 26]

P 1-oxa-2-oxo-cyclononane + $NADP^+$ + H_2O

S cyclopentanone + NADPH + O_2 <1-3, 6, 7, 11> (Reversibility: ? <1-3, 6, 7, 11> [1, 3, 17-19, 26]) [1, 3, 17-19, 26]

P 1-oxa-2-oxo-cyclohexane + $NADP^+$ + H_2O

S cyclopentyl methyl sulfide + NADPH + O_2 <1> (Reversibility: ? <1> [20]) [20]

P cyclopentyl methyl sulfoxide + $NADP^+$ + H_2O <1> [20]

S δ-thiovalerolactone + NADPH + O_2 <1> (<1> substrate inactivates enzyme after a few turnovers [6]) (Reversibility: ir <1> [6]) [6]

P ?

S dihydrocarvone + NADPH + O_2 <1> (Reversibility: ? <1> [1]) [1]

P ?

S ε-thiocaprolactone + NADPH + O_2 <1> (<1> substrate inactivates enzyme after a few turnovers [6]) (Reversibility: ir <1> [6]) [6]

P ?

S ethyl *p*-tolyl sulfide + NADPH + O_2 <1> (Reversibility: ? <1> [12, 16, 28]) [12, 16, 28]

P (S)-ethyl *p*-tolyl sulfoxide + $NADP^+$ + H_2O <1> [28]

S ethylene monothiocarbonate + NADPH + O_2 <1> (<1> substrate inactivates enzyme after a few turnovers [6]) (Reversibility: ? <1> [6]) [6]

P ?

S ethylene monothiocarbonate + NADPH + O_2 <1> (<1> substrate inactivates enzyme after a few turnovers [6]) (Reversibility: ir <1> [6]) [6]

P ?

S γ-thiobutyrolactone + NADPH + O_2 <1> (<1> substrate inactivates enzyme after a few turnovers [6]) (Reversibility: ir <1> [6]) [6]

P ?

S iodide + NADPH + O_2 <1> (Reversibility: ? <1> [12]) [12]

P ?

S isopropyl methyl sulfide + NADPH + O_2 <1> (Reversibility: ? <1> [20]) [20]
P ?
S methyl pentyl sulfide + NADPH + O_2 <1, 10> (<10> recombinant from E. coli [22]) (Reversibility: ? <1, 10> [20, 22]) [20, 22]
P (R)-methyl phenyl sulfoxide + $NADP^+$ + H_2O <10> [22]
S n-butylboronic acid + NADPH + O_2 <1> (Reversibility: ? <1> [12]) [12]
P n-butanol + BO_3^- + $NADP^+$ + H_2O
S n-octylboronic acid + NADPH + O_2 <1> (Reversibility: ? <1> [12]) [12]
P n-octanol + BO_3^- + $NADP^+$ + H_2O
S norcamphor + NADPH + O_2 <1, 6> (Reversibility: ? <1, 6> [1, 17]) [1, 17]
P ?
S octyl methyl sulfide + NADPH + O_2 <1> (Reversibility: ? <1> [20]) [20]
P ?
S phenyl allyl selenide + NADPH + O_2 <1> (Reversibility: ? <1> [12]) [12]
P ?
S phenyl allyl sulfide + NADPH + O_2 <1> (Reversibility: ? <1> [12]) [12]
P ?
S phenyl methyl selenide + NADPH + O_2 <1> (Reversibility: ? <1> [12]) [12]
P ?
S phenyl propargyl selenide + NADPH + O_2 <1> (Reversibility: ? <1> [9]) [9]
P ?
S phenylboronic acid + NADPH + O_2 <1> (Reversibility: ? <1> [12]) [12]
P phenol + BO_3^- + $NADP^+$ + H_2O
S syn-7-benzyloxymethyl-2-norbonen-5-one + NADPH + O_2 <1> (Reversibility: ? <1> [2]) [2]
P ?
S tert-butyl ethyl sulfide + NADPH + O_2 <1> (<1> low activity [20]) (Reversibility: ? <1> [20]) [20]
P ?
S tert-butyl methyl sulfide + NADPH + O_2 <1> (Reversibility: ? <1> [20]) [20]
P ?
S tert-butyl vinyl sulfide + NADPH + O_2 <1> (Reversibility: ? <1> [20]) [20]
P ?
S thiane + NADPH + O_2 <1> (Reversibility: ? <1> [12, 15]) [12, 15]
P ?
S thiane sulfoxide + NADPH + O_2 <1> (Reversibility: ? <1> [12]) [12]
P ?
S trans-hex-2-enyl phenyl selenide + NADPH + O_2 <1> (Reversibility: ? <1> [9]) [9]
P ?
S triethylphosphite + NADPH + O_2 <1> (Reversibility: ? <1> [12]) [12]
P ?

S Additional information <1, 2, 4> (<1> active with diverse alkyl aryl sulfides, dialkyl sulfides and dialkyldisulfides [28]; <1> active with benzyl alkyl sulfides with alkyl groups from methyl to hexyl and *para*-alkylbenzyl groups from methyl to butyl [27]; <1> high enantioselectivity in oxidation of sulfides [20, 27, 28]; <1> peroxide-like oxidation catalyzed [8]; <1> ability to convert aryl and alkyl boronic acids into phenols and alcohols, racemic 2-methylcyclohexyl boronic acid is processed to 2-methylcyclohexanol [8]; <1> oxidation of propargylic and allylic selenides [9]; <1,2> absolutely specific for NADPH as electron donor [1, 10]; <2, 4> no activity towards straight-chain alkanones [3,11]) [1, 3, 8-11, 20, 27, 28]

P ?

Inhibitors

2-thiacyclohexanone <1> (<1> substrate inactivates enzyme after a few turnovers [6]) [6]
5,5'-dithiobis(2-nitrobenzoate) <1, 2, 6, 11> (<11> completely [26]) [1, 3, 17, 26]
5-deaza-FAD <1> (<1> competitive [1,15]) [1, 15]
Cu^{2+} <11> (<11> completely [26]) [26]
$CuCl_2$ <2> [3]
EDTA <1, 6> [17]
$FeCl_3$ <2> [3]
$NADP^+$ <1> (<1> competitive against NADPH [15, 23]) [15, 23]
S-γ-thiobutyrolactone <1> (<1> substrate inactivates enzyme after a few turnovers [6]) [6]
arsenate <2> [3]
bathocuproine <1, 6> (<1,6> slight inhibition [17]) [17]
δ-thiovalerolactone <1> (<1> substrate inactivates enzyme after a few turnovers [6]) [6]
ethylene monothiocarbonate <1> (<1> substrate inactivates enzyme after a few turnovers [6]) [6]
iodoacetamide <1, 6> [17]
p-hydroxymercuribenzoate <1, 2, 6> [1, 3, 17]
quinacrine <11> [26]
quinine <11> [26]
Additional information <1, 6> (<1> inactivation mechanism [6]; <1, 6> NADPH protects against sulfhydryl active agents [17]) [17]

Cofactors/prosthetic groups

1-deaza-FAD <1> (<1> reactivates apoenzyme [15]) [15]
6-methyl-FAD <1> (<1> reactivates apoenzyme [1]) [1]
9-aza-FAD <1> (<1> reactivates apoenzyme [1, 15]) [1, 15]
FAD <1-3, 6, 7, 9, 11> (<11> binding motif, comparison of N-terminal amino acid sequences of several species [26]; <1> reactivates apoenzyme [15, 17]; <6> 0.62 mol of FAD per mol of holoenzyme type 1, 0.75 mol of FAD per mol of holoenzyme type 2 [18]; <1> enzyme-bound FAD-4a-OOH is the actual oxygenation reagent [8]) [1-3, 8, 12, 15-19, 23, 25, 26]

FMN <2> (<1> no activity with [17]; <2> involved in catalytic mechanism [3]; <2> 1 FMN molecule per molecule of protein [3,11]) [3, 11]
NADH <8> (<8> NADH or NADPH as electron donor [7]; <2> no activity [7,10]) [7]
NADPH <1-11> (<1,2> electron donor [1-3]; <2> NADPH [7]; <8> NADH or NADPH [7]; <1,2,4,6> absolutely specific for NADPH as electron donor [1,10,11,18]) [1-23, 25, 26]
Additional information <1> (<1> enzyme forms stable complex with 8-hydroxy-5-deazaflavin which reacts with oxidized 3-acetylpyridine adenine dinucleotide phosphate [5]) [5]

Activating compounds

7-chloro-8-demethyl-FAD <1> (<1> reactivates apoenzyme [1,15]) [1, 15]

Metals, ions

iron <1> (<1> iron-hem dependent enzyme [20]) [20]

Turnover number (min^{-1})

13.6 <1> (ε-thiocaprolactone, <1> substrate inactivates enzyme after a few turnovers [6]) [6]
14.4 <1> (δ-thiovalerolactone, <1> substrate inactivates enzyme after a few turnovers [6]) [6]
41 <1> (cis-hex-2-enyl phenyl selenide) [9]
41 <1> (γ-thiobutyrolactone, <1> substrate inactivates enzyme after a few turnovers [6]) [6]
55 <1> (trans-hex-2-enyl phenyl selenide) [9]
143 <1> (ethylene monothiocarbonate, <1> substrate inactivates enzyme after a few turnovers [6]) [6]
504 <1> (2-methylcyclohexyl boronic acid) [8]
585 <1> (phenyl propargyl selenide) [9]
1818 <1> (cyclohexanone) [15]
Additional information <1> [8, 15]

Specific activity (U/mg)

2.1 <1, 2> [1, 3, 11]
8-10 <1> (partially purified enzyme, substrate 4-methylcyclohexanone [6]) [6]
9.9 <6> (<6> purified enzyme [17]) [17]
14 <9> (<9> purified recombinant His-tagged enzyme expressed in yeast [25]) [25]
15 <9> (<9> purified native enzyme [25]) [25]
19.9 <9> (<9> purified recombinant His-tagged enzyme expressed in E. coli [25]) [25]
21 <1> (<1> purified enzyme [17]) [17]
678 <11> (<11> purified enzyme [26]) [26]
Additional information <10> (<10> activity is dependent on substrate concentration and buffer conditions [22]) [22, 24]

K_m-Value (mM)

0.00048 <11> (cyclohexanone) [26]
0.0005 <2> (cyclohexanone, <2> below [3,10]) [3, 10]
0.0025 <1> (n-octylboronic acid) [12]
0.0035 <1> (cuprizone) [17]
0.004 <1> (cyclohexanone) [1]
0.006 <1> (cyclohexanone) [6, 12, 17]
0.0068 <1> (trans-hex-2-enyl phenyl selenide) [9]
0.009 <1> (4-methylcyclohexanone) [17]
0.009 <1> (n-butylboronic acid) [12]
0.0093 <2> (2-hydroxycyclohexanone) [10]
0.0096 <1> (cis-hex-2-enyl phenyl selenide) [9]
0.01 <1> (3-thiacyclohexanone, <1> substrate inactivates enzyme after a few turnovers [6]) [6]
0.0112 <1> (2-hydroxycyclohexanone) [17]
0.012 <1> (2-methylcyclohexanone) [17]
0.0125 <2> (NADPH) [3]
0.0142 <11> (NADPH) [26]
0.016 <1> (4-methylcyclohexanone, <1> substrate inactivates enzyme after a few turnovers [6]) [6]
0.0185 <6> (NADPH, <6> isozyme type 1 [18]) [18]
0.02 <1, 6> (NADPH, <6> isozyme type 2 [18]) [1, 15, 18]
0.021 <1> (4-thiacyclohexanone) [6]
0.024 <1> (thiane, <1> substrate inactivates enzyme after a few turnovers [6]) [6, 12, 15]
0.025 <1> (butanal) [12]
0.03 <1> (phenyl allyl selenide) [12]
0.043 <1> (phenylboronic acid) [12]
0.044 <1> (phenyl methyl selenide) [12]
0.07 <1> (2-methylcyclohexyl boronic acid) [8]
0.1 <1> (O_2, <1> below [15]) [15]
0.11 <1> (phenyl allyl sulfide) [12]
0.128 <1> (phenyl propagyl selenide) [9]
0.14 <1> (norcamphor) [17]
0.143 <2> (cyclohexan-1,2-dione) [10]
0.17 <1> (cyclohexan-1,2-dione) [17]
0.178 <1> (thiane sulfoxide) [12]
0.19 <2> (cyclohexan-1,4-dione) [10]
0.2 <1> (cycloheptanone) [17]
0.22 <1> (ethyl *p*-tolyl sulfide) [12]
0.33 <1> (1-phenyl-2-propanone) [12]
0.35 <1> (2-phenyl-1-ethanal) [12]
0.36 <1> (triethyl phosphite) [12]
0.39 <1> (4-hydroxycyclohexanone) [17]
0.64 <1> (dihydrocarvone) [17]
1 <1> ((+)-camphor) [17]
1.8 <1> (cyclooctanone) [17]

2.5 <1> (iodide) [12]
3.6 <1> (cyclopentanone) [1, 17]
Additional information <1, 6> [8, 17, 18]

K_i-Value (mM)
0.038 <1> ($NADP^+$, <1> pH 9.0, 25°C [23]) [15,23]
0.062 <1> (5-deaza-FAD) [1,15]
2 <1> (2-thiocyclohexanone) [6]
2 <1> (S-γ-thiobutyrolactone) [6]
2 <1> (ethylene monothiocarbonate) [6]
2.9 <1> (δ-thiovalerolactone) [6]

pH-Optimum
6.5 <2> [10]
7.2-7.7 <8> [7]
7.5-10.1 <6> (<6> broad, isozyme type 2 [18]) [18]
8 <11> [26]
8 <6> (<6> sharp, isozyme type 1 [18]) [18]
8.4 <6> [17]
8.5-9 <2> [7, 10]
8.8 <2> [3, 11]
9 <1> [1, 17]

pH-Range
5.5-10.5 <11> (<11> no activity below pH 5.5 and above pH 10.5 at 30°C [26]) [26]
6-10.5 <6> (<6> pH 6.0: about 40% of activity maximum with isozymes type 1 and type 2, pH 10.5: about 10% of activity maximum with isozyme type 1, about 60% of activity maximum with isozyme type 2 [18]) [18]
6-11 <2> (<2> at pH 6.0 and 11.0: about 10% of activity maximum [3]) [3]

Temperature optimum (°C)
25 <1, 10> (<1, 10> assay at [22, 23]) [22, 23]
30 <1, 2, 4, 6, 11> (<1, 2, 4, 6, 11> assay at [3, 11, 15, 17, 26]) [3, 11, 15, 17, 26]

4 Enzyme Structure

Molecular weight
50000 <2> (<2> gel filtration [3]) [3]
53000 <6> (<6> low speed sedimentation without reaching equilibrium [17]) [17]
56000 <1> (<1> native PAGE [15]) [15]
59000 <1> (<1> low speed sedimentation without reaching equilibrium [17]) [17]
60800 <9> (<9> wild-type, electrospray mass spectrometry [25]) [25]
61620 <9> (<9> recombinant His-tagged enzyme expressed in E. coli [25]) [25]

61670 <9> (<9> recombinant His-tagged enzyme expressed in yeast [25]) [25]
74000 <11> (<11> gel filtration [26]) [26]

Subunits
monomer <1, 2, 6, 11> (<11> 1 * 74000, SDS-PAGE [26]; <1> 1 * 56000, SDS-PAGE [15]; <2> 1 * 51000, SDS-PAGE [3,11]; <6> 1 * 58000-59000, SDS-PAGE [17,18]) [3, 11, 15, 17, 18, 26]

5 Isolation/Preparation/Mutation/Application

Localization
cytoplasm <2, 4, 6> [3, 10, 11, 18]

Purification
<1> (recombinant from Escherichia coli [23]; partially [6]; recombinant from Escherichia coli and Saccharomyces cerevisiae [21]) [1, 6, 13, 15, 17, 21, 23]
<2> [3, 7, 11]
<6> [17]
<8> [7]
<9> (native enzyme and recombinant His-tagged enzyme from Escherichia coli and Saccharomyces cerevisiae [25]; recombinant from bacterial expression system [30]) [25, 30]
<10> (industrial scale production [24]; partially, recombinant from Escherichia coli [22]) [22, 24]
<11> [26]

Crystallization
<6> (microdialysis method, saturated ammonium sulfate solution [17]) [17]

Cloning
<1> (expression in Escherichia coli [23]; overexpression in Escherichia coli and Saccharomyces cerevisiae [21]; expression in Escherichia coli, amino acid and DNA sequence analysis, identification of potential flavin- and nicotinamide-binding sites [29]; overexpression in Escherichia coli, amino acid sequence determination [4]) [4, 21, 23]
<9> (expression of His-tagged enzyme in Escherichia coli and Saccharomyces cerevisiae [25]; expression in bacterial expression system, DNA and amino acid sequence analysis [30]) [25, 30]
<10> (overexpression in Escherichia coli [22]) [22]

Engineering
Additional information <9> (<9> His-tagged recombinant enzyme expressed in E. coli is, despite the tag, similar to the wild-type enzyme, whereas the recombinant His-tagged enzyme expressed in yeast is posttranslationally modified, Escherichia coli is the preferred expression system [25]) [25]

Application

synthesis <1, 10> (<10> chiral catalysis for the laboratory-scale transformation of racemic and prochiral ketones to chiral lactones and organic sulfur compounds to optically active sulfoxides, as a whole cell preparation and as an isolated immobilized enzyme [22]; <1> production of optically pure sulfoxides by biotransformation in whole cell systems of several sulfides, dithianes and dithiolanes [21]) [21, 22]

6 Stability

pH-Stability

7-8 <11> (<11> stable [26]) [26]

Temperature stability

40 <6> (<6> 2 min, 40% loss of activity for isozyme type 1, 7% loss of activity for isozyme type 2 [18]) [18]
45 <6> (<6> 1 min, 40% loss of activity for isozyme type 1, 18% loss of activity for isozyme type 2 [18]) [18]

Oxidation stability

<6>, enzyme is reduced by visible light in presence of EDTA under anaerobic conditions [17]

General stability information

<10>, immobilization stabilizes [22]

Storage stability

<1>, -20°C, more than 1 year [1]
<1>, apoenzyme and enzyme complexed with 8-hydroxy-5-deazaflavin is stable on ice for several weeks [5]
<2>, -20°C, stable [3, 11]
<2>, 4°C, in presence of cyclohexane or 0.1 mM NADPH [11]
<10>, -18°C, crude homogenates, negligible loss of activity, 2 months [24]
<10>, 25°C, 1 M sodium sulfate, half-life 1 week [22]
<10>, 25°C, half-life 1 day [22]
<1, 6>, -25°C, pH 7.1 [17]

References

[1] Trudgill, P.W.: Cyclohexanone 1,2-monooxygenase from Acinetobacter NCIMB 9871. Methods Enzymol., **188**, 70-77 (1990)
[2] Abril, O.; Ryerson, C.C.; Walsh, C.; Whitesides, G.M.: Enzymatic Baeyer-Villiger type oxidations of ketones catalyzed by cyclohexanone oxygenase. Bioorg. Chem., **17**, 41-52 (1989)
[3] Trower, M.K.; Buckland, R.M.; Griffin, M.: Characterization of an FMN-containing cyclohexanone monooxygenase from a cyclohexane-grown Xanthobacter sp. Eur. J. Biochem., **181**, 199-206 (1989)

[4] Chen, Y.C.; Peoples, O.P.; Walsh, C.T.: Acinetobacter cyclohexanone monooxygenase: gene cloning and sequence determination. J. Bacteriol., **170**, 781-789 (1988)
[5] Manstein, D.J.; Massey, V.; Ghisla, S.; Pai, E.F.: Stereochemistry and accessibility of prosthetic groups in flavoproteins. Biochemistry, **27**, 2300-2305 (1988)
[6] Latham, J.A.; Walsh, C.: Mechanism-based in activation of the flavoenzyme cyclohexanone oxygenase during oxygenation of cyclic thiol ester substrates. J. Am. Chem. Soc., **109**, 3421-3427 (1987)
[7] Magor, A.M.; Warburton, J.; Trower, M.K.; Griffin, M.: Comparative study of the ability of three Xanthobacter species to metabolize cycloalkanes. Appl. Environ. Microbiol., **52**, 665-671 (1986)
[8] Latham, J.A.; Walsh, C.: Retention of configuration in oxidation of a chiral boronic acid by the flavoenzyme cyclihexanone oxygenase. J. Chem. Soc. Chem. Commun., **1986**, 527-528 (1986)
[9] Latham, J.A.; Branchaud, B.P.; Chen, Y.C.J.; Walsh, C.: Allylic and propargylic phenyl selenide oxygenation by cyclohexanone oxygenase: [2,3]-sigmatropic rearrangement of the enzyme-generated selenoxide. J. Chem. Soc. Chem. Commun., **1986**, 528-530 (1986)
[10] Trower, M.K.; Buckland, R.M.; Higgins, R.; Griffin, M.: Isolation and characterization of a cyclohexane-metabolizing Xanthobacter sp.. Appl. Environ. Microbiol., **49**, 1282-1289 (1985)
[11] Trower, M.K.; Buckland, R.M.; Griffin, M.: Characterization of a cyclohexane hydrolase and a cyclohexanone mono-oxygenase from a cyclohexane-degrading Xanthobacter sp.. Biochem. Soc. Trans., **13**, 463-464 (1985)
[12] Branchaud, B.P.; Walsh, C.T.: Functional group diversity in enzymatic oxygenation reactions catalyzed by bacterial flavin-containing cyclohexanone oxygenase. J. Am. Chem. Soc., **107**, 2153-2161 (1985)
[13] Schwab, J.M.; Li, W.; Thomas, L.P.: Cyclohexanone oxygenase: stereochemistry, enantioselectivity, and regioselectivity of an enzyme-catalyzed Baeyer-Villiger reaction. J. Am. Chem. Soc., **105**, 4800-4808 (1983)
[14] Ougham, H.J.; Trudgill, P.W.: Metabolism of cyclohexaneacetic acid and cyclohexanebutyric acid by Arthrobacter sp. strain CA1. J. Bacteriol., **150**, 1172-1182 (1982)
[15] Ryerson, C.C.; Ballou, D.P.; Walsh, C.: Mechanistic studies on cyclohexanone oxygenase. Biochemistry, **21**, 2644-2655 (1982)
[16] Light, D.R.; Waxman, D.J.; Walsh, C.: Studies on the chirality of sulfoxidation catalyzed by bacterial flavoenzyme cyclohexanone monooxygenase and hog liver flavin adenine dinucleotide containing monooxygenase. Biochemistry, **21**, 2490-2498 (1982)
[17] Donoghue, N.A.; Norris, D.B.; Trudgill, P.W.: The purification and properties of cyclohexanone oxygenase from Nocardia globerula CL1 and Acinetobacter NCIB 9871. Eur. J. Biochem., **63**, 175-192 (1976)
[18] Norris, D.B.; Trudgill, P.W.: Multiple forms of cyclohexanone oxygenase from Nocardia globerula CL1. Eur. J. Biochem., **63**, 193-198 (1976)

[19] Walsh, C.; Branchaud, B.; Fox, B.; Latham, J.: Mechanistic studies on mercuric ion reductase and cyclohexanone oxygenase: pharmacologic and toxicologic aspects. Colloq. Ges. Biol. Chem. Mosbach, **34**, 140-148 (1983)
[20] Colonna, S.; Gaggero, N.; Carrea, G.; Pasta, P.: A new enzymic enantioselective synthesis of dialkyl sulfoxides catalyzed by monooxygenases. Chem. Commun., **1997**, 439-440 (1997)
[21] Chen, G.; Kayser, M.M.; Mihovilovic, M.D.; Mrstik, M.E.; Martinez, C.A.; Stewart, J.D.: Asymmetric oxidations at sulfur catalyzed by engineered strains that overexpress cyclohexane monooxygenase. New J. Chem., **23**, 827-832 (1999)
[22] Zambianchi, F.; Pasta, P.; Carrea, G.; Colonna, S.; Gaggero, N.; Woodley, J.M.: Use of isolated cyclohexanone monooxygenase from recombinant Escherichia coli as a biocatalyst for Baeyer-Villiger and sulfide oxidations. Biotechnol. Bioeng., **78**, 489-496 (2002)
[23] Sheng, D.; Ballou, D.P.; Massey, V.: Mechanistic studies of cyclohexanone monooxygenase: chemical properties of intermediates involved in catalysis. Biochemistry, **40**, 11156-11167 (2001)
[24] Barclay, S.S.; Woodley, J.M.; Lilly, M.D.; Spargo, P.L.; Pettman, A.J.: Production of cyclohexanone monooxygenase from Acinetobacter calcoaceticus for large scale Baeyer-Villiger monooxygenase reactions. Biotechnol. Lett., **23**, 385-388 (2001)
[25] Cheesman, M.J.; Kneller, M.B.; Kelly, E.J.; Thompson, S.J.; Yeung, C.K.; Eaton, D.L.; Rettie, A.E.: Purification and characterization of hexahistidine-tagged cyclohexanone monooxygenase expressed in Saccharomyces cerevisiae and Escherichia coli. Protein Expr. Purif., **21**, 81-86 (2001)
[26] Hasegawa, Y.; Nakai, Y.; Tokuyama, T.; Iwaki, H.: Purification and characterization of cyclohexanone 1,2-monooxygenase from Exophiala jeanselmei strain KUFI-6N. Biosci. Biotechnol. Biochem., **64**, 2696-2698 (2000)
[27] Pasta, P.; Carrea, G.; Holland, H.L.; Dallavalle, S.: Synthesis of chiral benzyl alkyl sulfoxides by cyclohexanone monooxygenase from Acinetobacter NCIB 9871. Tetrahedron, **6**, 933-936 (1995)
[28] Carrea, G.; Redigolo, B.; Riva, S.; Colonna, S.; Gaggero, N.; Battistel, E.; Bianchi, D.: Effects of substrate structure on the enantioselectivity and stereochemical course of sulfoxidation catalyzed by cyclohexanone monooxygenase. Tetrahedron, **3**, 1063-1068 (1992)
[29] Chen, Y.C.; Peoples, O.P.; Walsh, C.T.: Acinetobacter cyclohexanone monooxygenase: gene cloning and sequence determination. J. Bacteriol., **170**, 781-789 (1988)
[30] Kneller, M.B.; Cheesman, M.J.; Rettie, A.E.: ESI- and MALDI-MS analysis of cyclohexanone monooxygenase from Acinetobacter NCIB 9871. Biochem. Biophys. Res. Commun., **282**, 899-903 (2001)

3-Hydroxybenzoate 4-monooxygenase 1.14.13.23

1 Nomenclature

EC number
1.14.13.23

Systematic name
3-hydroxybenzoate,NADPH:oxygen oxidoreductase (4-hydroxylating)

Recommended name
3-hydroxybenzoate 4-monooxygenase

Synonyms
3-hydroxybenzoate 4-hydroxylase
EC 1.14.99.13 (formerly)
oxygenase, 3-hydroxybenzoate 4-mono-

CAS registry number
37256-76-1

2 Source Organism

<1> *Pseudomonas testosteroni* [1]
<2> *Aspergillus niger* [2]
<3> *Bacillus sp.* [3]
<4> *Comamonas testosteroni* [4, 5]

3 Reaction and Specificity

Catalyzed reaction
3-hydroxybenzoate + NADPH + H^+ + O_2 = 3,4-dihydroxybenzoate + $NADP^+$ + H_2O

Reaction type
oxidation
redox reaction
reduction

Natural substrates and products
S 3-hydroxybenzoate + NADPH + O_2 <1-4> (Reversibility: ? <1-4> [1-5]) [1-5]
P 3,4-dihydroxybenzoate + $NADP^+$ + H_2O

Substrates and products

S 2,3-dihydroxybenzoate + NADPH + O_2 <1, 4> (Reversibility: ? <1, 4> [1, 5]) [1, 5]
P ?
S 2,5-dihydroxybenzoate + NADPH + O_2 <1> (Reversibility: ? <1> [1]) [1]
P ?
S 2-fluoro-5-hydroxybenzoate + NADPH + O_2 <1> (Reversibility: ? <1> [1]) [1]
P ?
S 3,5-dihydroxybenzoate + NADPH + O_2 <1> (Reversibility: ? <1> [1]) [1]
P ?
S 3-hydroxyanthranilate + NADPH + O_2 <1> (Reversibility: ? <1> [1]) [1]
P ?
S 3-hydroxybenzoate + NADPH + O_2 <1-4> (Reversibility: ? <1-4> [1-5]) [1-5]
P 3,4-dihydroxybenzoate + $NADP^+$ + H_2O
S 4-fluoro-3-hydroxybenzoate + NADPH + O_2 <1> (Reversibility: ? <1> [1]) [1]
P ?
S gentisate + NADPH + O_2 <4> (Reversibility: ? <4> [4, 5]) [4, 5]
P ?

Inhibitors

4-hydroxy-3-iodomethylbenzoate <4> (<4> inhibition not reversed in presence of dithiotreitol [4]) [4]
N-ethylmaleimide <2> [2]
N-iodosuccinimide <4> (<4> inhibition reversed in presence of dithiotreitol [4]) [4]
diethyldithiocarbamate <2> [2]
heavy metal ions <2> [2]
iodoacetamide <4> (<4> inhibition reversed in presence of dithiotreitol [4]) [4]
m-aminobenzoate <2> [2]
o-iodosobenzoate <4> (<4> inhibition reversed in presence of dithiotreitol [4]) [4]
o-phenanthroline <2> [2]
p-hydroxymercuribenzoate <2> [2]
salicylaldoxime <2> [2]

Cofactors/prosthetic groups

FAD <1-4> (<1-4> flavoprotein [1-5]) [1-5]
NADH <1> (<1> poor substitute for NADPH [1]) [1]
NADPH <1, 2, 4> (<4> specific requirement for NADPH [4,5]) [1, 2, 4, 5]

Specific activity (U/mg)

0.01 <3> [3]
5.17 <4> [5]
6.03 <4> [4]
11.8 <1> [1]

K_m-Value (mM)

0.02 <1> (NADPH, <1> + 2-fluoro-5-hydroxybenzoate [1]) [1]
0.03 <1> (3-hydroxybenzoate, <1> + NADPH [1]) [1]
0.03 <4> (NADPH) [5]
0.03 <4> (*m*-hydroxybenzoate) [5]
0.04 <1> (3,5-dihydroxybenzoate, <1> + NADPH [1]) [1]
0.05 <1> (NADPH, <1> + 3-hydroxyanthranilate or 3,5-dihydroxybenzoate [1]) [1]
0.06 <1> (2,3-dihydroxybenzoate, <1> + NADPH [1]) [1]
0.07 <1> (NADPH, <1> + 3-hydroxybenzoate or 2,3-dihydroxybenzoate [1]) [1]
0.12 <1> (4-fluoro-3-hydroxybenzoate, <1> + NADPH [1]) [1]
0.14 <1> (NADPH, <1> + 2,5-dihydroxybenzoate [1]) [1]
0.15 <1> (NADPH, <1> + 4-fluoro-3-hydroxybenzoate [1]) [1]
0.19 <2> (3-hydroxybenzoate) [2]
0.2 <2> (NADPH) [2]
0.5 <1> (2,5-dihydroxybenzoate, <1> + NADPH [1]) [1]
3 <1> (NADH, <1> + 3-hydroxybenzoate [1]) [1]

pH-Optimum

6.2 <4> (<4> in potassium phosphate buffer [4]) [4]
7.2 <2> [2]
7.3 <4> (<4> in Tris-HCl buffer [4]) [4]

Temperature optimum (°C)

30 <1, 2> (<1,2> assay at [1,2]) [1, 2]

4 Enzyme Structure

Molecular weight

145000 <1> (<1> sedimentation equilibrium centrifugation [1]) [1]

Subunits

? <4> (<4> x * 71000, SDS-PAGE [4,5]) [4, 5]

5 Isolation/Preparation/Mutation/Application

Source/tissue

mycelium <2> [2]

Purification

<1> [1]
<2> (<2> partial [2]) [2]
<4> [5]

Crystallization

<4> (<4> 1.4-1.6 M ammonium sulfate, 4-8% dioxane, pH 6.5 [4]) [4]

6 Stability

General stability information

<2>, 2-mercaptoethanol stabilizes [2]

References

[1] Michalover, J.M.; Ribbons, D.W.: 3-Hydroxybenzoate 4-hydroxylase from Pseudomonas testosteroni. Biochem. Biophys. Res. Commun., **55**, 888-896 (1973)

[2] Kumar, R.P.; Rao, P.V.S.; Vaidyanathan, C.S.: *m*-Hydroxybenzoate 4-hydroxylase from Aspergillus niger: purification and properties. Indian J. Biochem. Biophys., **10**, 184-190 (1973)

[3] Mashetty, S.B.; Manohar, S.; Karegoudar, T.B.: Degradation of 3-hydroxybenzoic acid by a Bacillus species. Indian J. Biochem. Biophys., **33**, 145-148 (1996)

[4] Chen, R.; Oki, H.; Scott, R.P.; Yamaguchi, H.; Kusunoki, M.; Matsuura, Y.; Chaen, H.; Tsugita, A.; Hosokawa, K.: Crystallization and further characterization of *meta*-hydroxybenzoate 4-hydroxylase from Comamonas testosteroni. Res. Commun. Biochem. Cell Mol. Biol., **2**, 253-274 (1998)

[5] Chen, R.; Oki, H.; Chaen, H.; Hosokawa, K.: Studies on *m*-hydroxybenzoate 4-hydroxylase from Comamonas testosteroni I. Purification and characterization. Res. Commun. Biochem. Cell Mol. Biol., **1**, 304-322 (1997)

3-Hydroxybenzoate 6-monooxygenase 1.14.13.24

1 Nomenclature

EC number
1.14.13.24

Systematic name
3-hydroxybenzoate,NADH:oxygen oxidoreductase (6-hydroxylating)

Recommended name
3-hydroxybenzoate 6-monooxygenase

Synonyms
3-HBA-6-hydroxylase
3-hydroxybenzoate 6-hydroxylase
3-hydroxybenzoic acid-6-hydroxylase
m-hydroxybenzoate 6-hydroxylase
oxygenase, 3-hydroxybenzoate 6-mono-

CAS registry number
51570-26-4

2 Source Organism

<1> *Pseudomonas aeruginosa* (strain T1 [1,2]) [1, 2]
<2> *Pseudomonas cepacia* [3, 4]
<3> *Micrococcus sp.* [5, 8-10]
<4> *Klebsiella pneumoniae* (strain M5a1 [6]) [6]
<5> *Rhodococcus erythropolis* (strain S1 [7]) [7]

3 Reaction and Specificity

Catalyzed reaction
3-hydroxybenzoate + NADH + H^+ + O_2 = 2,5-dihydroxybenzoate + NAD^+ + H_2O (<1> A-stereospecificity [2]; <2> mechanism [4])

Reaction type
oxidation
redox reaction
reduction

Natural substrates and products

S 3-hydroxybenzoate + NADH + O_2 <3> (<3> key role in metabolism of aromatic compounds [5]) (Reversibility: ? <1> [5]) [5]
P 2,5-dihydroxybenzoate + NAD^+ + H_2O

Substrates and products

S 2,3-dihydroxybenzoate + NADH + O_2 <1> (<1> 10.5% of the reaction with 3-hydroxybenzoate [1]; <3> not [5]) (Reversibility: ? <1> [1, 5]) [1, 5]
P 2,3,5-trihydroxybenzoate + NAD^+ + H_2O <1> [1]
S 3,4-dihydroxybenzoate + NADH + O_2 <1> (<3> not [5]; <1> 9% of the reaction with 3-hydroxybenzoate [1]) (Reversibility: ? <1> [1, 5]) [1, 5]
P 2,4,5-trihydroxybenzoate + NAD^+ + H_2O <1> [1]
S 3,5-dihydroxybenzoate + NADH + O_2 <1> (<1> 7% of the reaction with 3-hydroxybenzoate [1]) (Reversibility: ? <1> [1]) [1]
P 2,3,5-trihydroxybenzoate + NAD^+ + H_2O <1> [1]
S 3-hydroxy-5-methylbenzoate + NADH + O_2 <1> (<1> 31% of the reaction with 3-hydroxybenzoate [1]) (Reversibility: ? <1> [1]) [1]
P 2,5-dihydroxy-3-methylbenzoate + NAD^+ + H_2O <1> [1]
S 3-hydroxybenzoate + NADH + O_2 <1-4> (<3> strictly specific for 3-hydroxybenzoate [5]) (Reversibility: ? <1-4> [1-7, 9, 10]) [1-7, 9, 10]
P 2,5-dihydroxybenzoate + NAD^+ + H_2O <2> [3, 4, 6]
S 4-fluoro-3-hydroxybenzoate + NADH + O_2 <1> (<1> 36% of the reaction with 3-hydroxybenzoate [1]) (Reversibility: ? <1> [1]) [1]
P 4-fluoro-2,5-dihydroxybenzoate + NAD^+ + H_2O <1> [1]
S Additional information <3> (<3> not: salicylate [5]; <3> not: *p*-hydroxybenzoate [5]; <3> acts also on a number of analogs of 3-hydroxybenzoate substituted in the 2, 3, 5 and 6 positions [5]) [5]
P ?

Inhibitors

Cibacron blue <3> (<3> 50% inhibition at 0.04 mM [5]) [5]
Cu^{2+} <3, 4> (<3> 85% inhibition at 0.01 mM, inhibition can be reversed by thiol reagents [5]; <4> inhibition at 0.02 mM, partially reversed by 2-mercaptoethanol [6]) [5, 6]
Fe^{2+} <4> (<4> inhibition at 0.02 mM, partially reversed by 2-mercaptoethanol [6]) [6]
Hg^{2+} <3, 4> (<3> 85% inhibition at 0.01 mM, inhibition can be reversed by thiol reagents [5]; <4> inhibition at 0.02 mM, partially reversed by 2-mercaptoethanol [6]) [5, 6]
N-bromosuccinimide <3> (<3> 85-90% loss of activity at 0.04 mM, substrate protects, 10 mM N-acetylimidazole restores 85-90% of enzyme activity [9]; <3> 0.04 mM, reduced binding affinity of substrate [10]) [9, 10]
diethylpyrocarbonate <3> (<3> 80% loss of activity after 12 min at 0.1 mM, pseudo-first-order kinetics, substrate protects, 0.1 M hydroxylamine restores 75-85% of enzyme activity [9]; <3> 0.1 mM, reduced binding affinity of substrate [10]) [9, 10]
p-chloromercuribenzoate <3> (<3> at low concentrations [5]) [5]

phenylglyoxal <3> (<3> 70% loss of activity after 30 min at 0.5 mM, dialysis results in recovery of the enzyme, preincubation with the substrate causes significant protection [8]; <3> 1 mM, reduced binding affinity of substrate [10]) [8, 10]
urea <3> (<3> 70% loss of activity at 1 M [10]) [10]

Cofactors/prosthetic groups
FAD <1-5> (<1-3> flavoprotein [1,3,5]; <1> 0.9 mol of FAD per mol of enzyme [1]; <2> 1 mol of FAD per mol of enzyme [3]; <3> 1 mol FAD per mol enzyme, exogenous addition of FAD required to regenerate oxidized FAD during catalysis [5]) [1, 3, 5, 6, 7]
NADH <1-3> [1-5, 6, 10]
NADPH <1-3> (<1> can act instead of NADH, more slowly [1]; <2> NADH and NADPH utilized with similar efficiencies [3]; <3> strict preference for NADH over NADPH [5]; <4> NADH and NADPH utilized with similar efficiencies [6]) [1, 3, 5, 6]

Metals, ions
Additional information <3> (no requirement for metal ions) [5]

Specific activity (U/mg)
8.42 <4> [6]
21.4 <3> [5]
23.7 <2> [3]
24 <1> [1]

K_m-Value (mM)
0.04 <2> (NADH) [3]
0.046 <2> (NADH) [4]
0.063 <3> (3-hydroxybenzoate) [5]
0.064 <2> (NADPH) [3]
0.07 <4> (NADH) [6]
0.07 <4> (NADPH) [6]
0.095 <3> (NADH) [5]
0.13 <2> (O_2) [4]
0.16 <1> (NADH) [1]
0.162 <2> (3-hydroxybenzoate) [3]
0.168 <2> (3-hydroxybenzoate) [3]
0.19 <2> (3-hydroxybenzoate) [4]
0.8 <1> (NADPH) [1]
Additional information <3> (<3> K_m at various pH values [10]) [10]

pH-Optimum
8 <2, 3, 4> [3, 5, 6, 10]

pH-Range
6.5-11 <4> (<4> no activity below pH 6.5 and above pH 11 [6]) [6]

Temperature optimum (°C)

23 <2> (<2> assay at [4]) [4]
25-30 <4> [6]
30 <3> (<3> assay at [8]) [8]

Temperature range (°C)

15-40 <4> (<4> no activity below 15°C and above 40°C [6]) [6]

4 Enzyme Structure

Molecular weight

40000-45000 <2> (<2> gel filtration, SDS-PAGE [3]) [3]
41500 <4> (<4> gel filtration [6]) [6]
67000 <3> (<3> SDS-PAGE [5]) [5]
70000 <3> (<3> gel filtration [5]) [5]
85000 <1> (<1> sedimentation equilibrium measurement [1]) [1]

Subunits

monomer <2, 3, 4> (<2> 1 * 44000, SDS-PAGE [3]; <3> 1 * 67000, SDS-PAGE [5]; <4> 1 * 42000, SDS-PAGE [6]) [3, 5, 6]

5 Isolation/Preparation/Mutation/Application

Purification

<1> (homogeneity [1]) [1]
<2> (homogeneity [3]) [3]
<3> (single step purification using substrate-mediated interaction of the enzyme with blue-Sepharose, homogeneity [5]; <3> 98% homogeneity [9]) [5, 8, 9, 10]
<4> (homogeneity [6]) [6]

6 Stability

pH-Stability

7-8.5 <4> (<4> in phosphate and Tris-HCl buffer [6]) [6]

General stability information

<4>, complete inactivation at room temperature without glycerol after 18 h [6]

Storage stability

<4>, 4°C, optimal storage in 50 mM phosphate or Tris buffer pH 8 with 10% glycerol [6]

References

[1] Groseclose, E.E.; Ribbons, D.W.: 3-Hydroxybenzoate 6-hydroxylase from Pseudomonas aeruginosa. Biochem. Biophys. Res. Commun., **55**, 897-903 (1973)
[2] You, K.S.; Arnold, L.J.; Kaplan, N.O.: The stereospecificity of bacterial external flavoprotein monooxygenases for nicotinamide adenine dinucleotide. Arch. Biochem. Biophys., **180**, 550-554 (1977)
[3] Wang, L.H.; Hamzah, R.Y.; Yu, Y.; Tu, S.C.: Pseudomonas cepacia 3-hydroxybenzoate 6-hydroxylase: induction, purification, and characterization. Biochemistry, **26**, 1099-1104 (1987)
[4] Yu, Y.; Wang, L.H.; Tu, S.C.: Pseudomonas cepacia 3-hydroxybenzoate 6-hydroxylase: stereochemistry, isotope effects, and kinetic mechanism. Biochemistry, **26**, 1105-1110 (1987)
[5] Rajasekharan, S.; Rajasekharan, R.; Vaidyanathan, C.S.: Substrate-mediated purification and characterization of a 3-hydroxybenzoic acid-6-hydroxylase from Micrococcus. Arch. Biochem. Biophys., **278**, 21-25 (1990)
[6] Suarez, M.; Ferrer, E.; Garrido-Pertierra, A.; Martin, M.: Purification and characterization of the 3-hydroxybenzoate-6-hydroxylase from Klebsiella pneumoniae. FEMS Microbiol. Lett., **126**, 283-290 (1995)
[7] Suemori, A.; Nakajima, K.; Kurane, R.; Nakamura, Y.: Physicochemical and immunochemical characterization of salicylate 5-hydroxylase, *m*-hydroxybenzoate 6-hydroxylase and *p*-hydroxybenzoate 3-hydroxylase from Rhodococcus erythropolis. Biotechnol. Lett., **17**, 1063-1068 (1995)
[8] Sumathi, S.; Dasgupta, D.; Vaidyanathan, C.S.: Chemical modification of 3-HBA-6-hydroxylase by phenylglyoxal: kinetic and physicochemical studies on the modified enzyme. Indian J. Biochem. Biophys., **35**, 266-272 (1998)
[9] Sumathi, S.; Dasgupta, D.: Reactivity of 3-HBA-6-hydroxylase with diethylpyrocarbonate and N-bromosuccinimide: Effect of chemical modifications on kinetic and spectral properties of the enzyme. Biotechnol. Prog., **16**, 577-582 (2000)
[10] Suresh, S.; Dasgupta, D.: Interaction of 3-hydroxybenzoate with 3-hydroxybenzoate-6-hydroxylase. Biotechnol. Prog., **17**, 1026-1031 (2001)

Methane monooxygenase 1.14.13.25

1 Nomenclature

EC number
1.14.13.25

Systematic name
methane,NAD(P)H:oxygen oxidoreductase (hydroxylating)

Recommended name
methane monooxygenase

Synonyms
MMO <2-4> [4, 6, 8, 10, 13, 23-32]
methane hydroxylase
methane mono-oxygenase
oxygenase, methane mono-
pMMO <2, 3, 5-10> (<2, 3, 5-10> particulate, membrane-bound enzyme form [23, 24-32]) [23, 24-32]
sMMO <2, 3, 5, 7, 10> (<2, 3, 5, 7, 10> soluble, cytoplasmic enzyme form [23, 24-31]; <10> multicomponent enzyme [31]) [23, 24-31]

CAS registry number
51961-97-8

2 Source Organism

<1> *Methylobacterium sp.* [7, 13]
<2> *Methylococcus capsulatus* (enzyme form sMMO [26, 30]; enzyme form pMMO is copper-inducible [25, 33]; methanotroph type I [23]; Bath [1, 3, 4, 9-13, 17-20, 23-25, 30, 33]) [1, 3, 4, 9-13, 17-20, 23-26, 30, 33]
<3> *Methylosinus trichosporium* (enzyme form sMMO [29]; strain IMV 3011 [32]; methanotroph type II [23]; strain OB3b [2, 3, 5, 8, 13-16, 21, 22, 27, 29]) [2, 3, 5, 8, 13-16, 21-23, 27, 29, 32]
<4> *Methylosinus sporium* (strain 5 [6]) [6]
<5> *Methylomonas sp.* (methanotroph type I [23]) [23]
<6> *Methylobacter sp.* (methanotroph type I [23]) [23]
<7> *Methylomicrobium sp.* (methanotroph type I [23]) [23]
<8> *Methylocaldum sp.* (methanotroph type I [23]) [23]
<9> *Methylosphaera sp.* (methanotroph type I [23]) [23]
<10> *Methylocystis sp.* (enzyme form sMMO [28]; strain WI 14 [28]; methanotroph type II [23, 28]) [23, 28, 31]
<11> *Beijerinckia indica* (only sMMO [23]) [23]

3 Reaction and Specificity

Catalyzed reaction

methane + NAD(P)H + H^+ + O_2 = methanol + $NAD(P)^+$ + H_2O (<2> structural model for component protein B of sMMO [30]; <2> mechanism of pMMO [25]; <2> modeling of interaction between the different protein components of sMMO [24]; <1-3> mechanism [4, 13, 15, 18, 21, 23]; <3> kinetic model of protein component interaction [14])

Reaction type

oxidation
redox reaction
reduction

Natural substrates and products

S formate + NAD(P)H + O_2 <2, 3> (<2,3> assay with whole cells [25,27]) (Reversibility: ? <2, 3> [25, 27]) [25, 27]
P ?
S methane + NAD(P)H + O_2 <2, 3, 5-10> (<2,3,5-10> key enzyme in oxidation of methane [23]; <2> initial step in the assimilation of methane in bacteria that grow with methane as sole carbon and energy source [4]) (Reversibility: ? <2, 3, 5-10> [4, 5, 23]) [4, 5, 23]
P methanol + $NAD(P)^+$ + H_2O <2, 3, 5-10> [23]

Substrates and products

S 1-butene + NAD(P)H + O_2 <2, 3> (Reversibility: ? <2, 3> [1, 15]) [1, 15]
P 1,2-epoxybutane + $NAD(P)^+$ + H_2O <2, 3> [1, 15]
S 2,3-dimethylpentane + NAD(P)H + O_2 <2> (Reversibility: ? <2> [18]) [18]
P 3,4-dimethylpentan-2-ol + $NAD(P)^+$ + H_2O <2> [18]
S 2-methylpropane + NAD(P)H + O_2 <2> (Reversibility: ? <2> [18]) [18]
P 2-methylpropan-2-ol + 2-methylpropan-1-ol + $NAD(P)^+$ + H_2O <2> [18]
S CO + NAD(P)H + O_2 <2, 3> (Reversibility: ? <2, 3> [1, 2, 4]) [1, 2, 4]
P CO_2 + $NAD(P)^+$ + H_2O
S adamantane + NAD(P)H + O_2 <2> (Reversibility: ? <2> [18]) [18]
P 1-adamantanol + 2-adamantanol + $NAD(P)^+$ + H_2O <2> [18]
S ammonia + NAD(P)H + O_2 <2> (Reversibility: ? <2> [4]) [4]
P hydroxylamine + $NAD(P)^+$ + H_2O
S benzene + NAD(P)H + O_2 <2> (Reversibility: ? <2> [1, 4]) [1, 4]
P cyclohexanol + phenol + hydroquinone + $NAD(P)^+$ + H_2O <2> [1, 4]
S benzene + NAD(P)H + O_2 <3, 10> (Reversibility: ? <3, 10> [15, 28]) [15, 28]
P phenol + $NAD(P)^+$ + H_2O <3, 10> [15, 28]
S β-pinene + NAD(P)H + O_2 <2> (Reversibility: ? <2> [18]) [18]
P 6,6-dimethylbicyclo[3.1.1]hept-2-ene-2-methanol + β-pinene oxide + $NAD(P)^+$ + H_2O <2> [18]
S bromobenzene + NAD(P)H + O_2 <10> (<10> sMMO [28]) (Reversibility: ? <10> [28]) [28]

P bromophenol + NAD(P)$^+$ + H_2O <10> [28]
S bromomethane + NAD(P)H + O_2 <2> (Reversibility: ? <2> [17]) [17]
P ?
S butane + NAD(P)H + O_2 <2, 3, 10> (Reversibility: ? <2, 3, 10> [1, 2, 28]) [1, 2, 28]
P 1-butanol + 2-butanol + NAD(P)$^+$ + H_2O <2, 10> (<10> only 2-butanol, sMMO [28]) [1, 28]
S butylene + NAD(P)H + O_2 <10> (<10> sMMO [28]) (Reversibility: ? <10> [28]) [28]
P butylene oxide + NAD(P)$^+$ + H_2O <10> [28]
S chlorobenzene + NAD(P)H + O_2 <10> (<10> sMMO [28]) (Reversibility: ? <10> [28]) [28]
P chlorophenol + NAD(P)$^+$ + H_2O <10> [28]
S chloromethane + NAD(P)H + O_2 <2> (Reversibility: ? <2> [4]) [4]
P formaldehyde + NAD(P)$^+$ + H_2O + ? <2> [4]
S chloronaphthalene + NAD(P)H + O_2 <10> (<10> sMMO [28]) (Reversibility: ? <10> [28]) [28]
P chloronaphthol + NAD(P)$^+$ + H_2O <10> [28]
S chloropentane + NAD(P)H + O_2 <10> (<10> sMMO [28]) (Reversibility: ? <10> [28]) [28]
P chloropentanol + NAD(P)$^+$ + H_2O <10> [28]
S cis-1,3-dimethylcyclohexane + NAD(P)H + O_2 <2> (Reversibility: ? <2> [18]) [18]
P 3,5-dimethylcyclohexanol + 1-cis-3-dimethylcyclohexanol + NAD(P)$^+$ + H_2O + 1-trans-3-dimethylcyclohexanol <2> (<2> 1-trans-3-dimethylcyclohexanol is produced in a low concentration [18]) [18]
S cis-1,4-dimethylcyclohexane + NAD(P)H + O_2 <2> (Reversibility: ? <2> [18]) [18]
P 1-cis-4-dimethylcyclohexanol + NAD(P)$^+$ + H_2O + trans-2,5-dimethylcyclohexanol <2> (<2> trans-2,5-dimethylcyclohexanol is produced in a low concentration [18]) [18]
S cis-2-butene + NAD(P)H + O_2 <2, 3> (Reversibility: ? <2, 3> [1, 15, 18]) [1, 15, 18]
P cis-2,3-epoxybutane + cis-2-buten-1-ol + 2-butanone + NAD(P)$^+$ + H_2O <2, 3> [1, 15, 18]
S cyclohexane + NAD(P)H + O_2 <2, 3, 10> (<10> sMMO [28]) (Reversibility: ? <2, 3, 10> [1, 4, 15, 28]) [1, 4, 15, 28]
P cyclohexanol + NAD(P)$^+$ + H_2O <2, 3, 10> [1, 4, 15, 28]
S cyclohexene + NAD(P)H + O_2 <3> (Reversibility: ? <3> [15]) [15]
P epoxycyclohexane + 2-cyclohexen-1-ol + NAD(P)$^+$ + H_2O <3> [15]
S cytochrome c + NAD(P)H + O_2 <10> (<10> sMMO [28]) (Reversibility: ? <10> [28]) [28]
P reduced cytochrome c + NAD(P)$^+$ + H_2O
S dichloromethane + NAD(P)H + O_2 <2> (Reversibility: ? <2> [4]) [4]
P CO + Cl^- + NAD(P)$^+$ + H_2O <2> [4]
S diethyl ether + NAD(P)H + O_2 <2, 10> (<10> sMMO [28]) (Reversibility: ? <2, 10> [1, 28]) [1, 28]

P ethanol + ethanal + $NAD(P)^+$ + H_2O <2, 10> [1, 4, 28]

S dimethyl ether + NAD(P)H + O_2 <2> (<3> no activity [2]) (Reversibility: ? <2> [1, 4]) [1, 4]

P methanol + formaldehyde + $NAD(P)^+$ + H_2O <2> [4]

S duroquinone + NAD(P)H + O_2 <2> (<2> enzyme form pMMO [25]) (Reversibility: ? <2> [25]) [25]

P duroquinol + $NAD(P)^+$ + H_2O <2> [25]

S ethane + NAD(P)H + O_2 <2, 3> (Reversibility: ? <2, 3> [1, 2, 4]) [1, 2, 4]

P ethanol + $NAD(P)^+$ + H_2O <2> [1]

S ethene + NAD(P)H + O_2 <2, 10> (<10> sMMO [28]) (Reversibility: ? <2, 10> [1, 4, 19, 20, 28]) [1, 4, 19, 20, 28]

P epoxyethane + $NAD(P)^+$ + H_2O <2, 10> [1, 4, 19, 20, 28]

S fluorobenzene + NAD(P)H + O_2 <10> (<10> sMMO [28]) (Reversibility: ? <10> [28]) [28]

P fluorophenol + $NAD(P)^+$ + H_2O <10> [28]

S formate + NAD(P)H + O_2 <2, 3> (<2,3> assay with whole cells [25,27]) (Reversibility: ? <2, 3> [25, 27]) [25, 27]

P ?

S furan + NAD(P)H + O_2 <3> (Reversibility: ? <3> [5]) [5]

P ?

S heptane + NAD(P)H + O_2 <2, 10> (<10> sMMO [28]) (Reversibility: ? <2, 10> [1, 18, 28]) [1, 18, 28]

P 1-heptanol + 2-heptanol + $NAD(P)^+$ + H_2O <2, 10> (<10> position of hydroxylation cannot be determined exactly [28]) [1, 28]

S hexane + NAD(P)H + O_2 <2, 10> (<10> sMMO [28]) (Reversibility: ? <2, 10> [1, 4, 18, 28]) [1, 4, 18, 28]

P 1-hexanol + 2-hexanol + $NAD(P)^+$ + H_2O <2, 10> (<10> position of hydroxylation cannot be determined exactly [28]) [1, 28]

S isobutan + NAD(P)H + O_2 <3> (Reversibility: ? <3> [15]) [15]

P 2-methyl-1-propanol + 2-methyl-2-propanol + $NADP^+$ + H_2O <3> [15]

S isopentane + NAD(P)H + O_2 <3> (Reversibility: ? <3> [15]) [15]

P 2-methylbutan-1-ol + 3-methylbutan-1-ol + 2-methylbutan-2-ol + 3-methylbutan-2-ol + $NADP^+$ + H_2O <3> [15]

S methane + NAD(P)H + O_2 <1-10> (Reversibility: ? <1-10> [1-30]) [1-30]

P methanol + $NAD(P)^+$ + H_2O <1-10> [1-30]

S methylene cyclohexane + NAD(P)H + O_2 <2> (Reversibility: ? <2> [18]) [18]

P 1-cyclohexane-1-methanol + methylene cyclohexane oxide + 4-hydroxymethylene cyclohexane + $NAD(P)^+$ + H_2O <2> [18]

S naphthalene + NAD(P)H + O_2 <2, 10> (<10> sMMO [28]) (Reversibility: ? <2, 10> [4, 28]) [4, 28]

P α-naphthol + β-naphthol + $NAD(P)^+$ + H_2O <2, 10> [4, 28]

S octane + NAD(P)H + O_2 <2> (Reversibility: ? <2> [1]) [1]

P 1-octanol + 2-octanol + $NAD(P)^+$ + H_2O <2> [1]

S pentane + NAD(P)H + O_2 <2, 10> (<10> sMMO [28]) (Reversibility: ? <2, 10> [1, 18, 28]) [1, 18, 28]

P 1-pentanol + 2-pentanol + NAD(P)$^+$ + H_2O <2, 10> (<10> position of hydroxylation cannot be determined exactly [28]) [1, 18, 28]

S phenylalanine + NAD(P)H + O_2 <2> (Reversibility: ? <2> [1]) [1]

P tyrosine + NAD(P)$^+$ + H_2O <2> [1]

S propane + NAD(P)H + O_2 <2, 3, 10> (Reversibility: ? <2, 3, 10> [1, 2, 15, 28]) [1, 2, 15, 28]

P 1-propanol + 2-propanol + NAD(P)$^+$ + H_2O <2, 3, 10> (<10> only 2-propanol, sMMO [28]) [1, 15, 28]

S propene + NAD(P)H + O_2 <2, 3> (Reversibility: ? <2, 3> [1, 3, 5, 15]) [1, 3, 5, 15]

P 1,2-epoxypropane + NAD(P)$^+$ + H_2O <2, 3> [1, 4, 5, 15]

S propylene + NAD(P)H + O_2 <2, 10> (<2,10> enzyme form sMMO [24,25,28]) (Reversibility: ? <2, 10> [24, 25, 28]) [24, 25, 28]

P propylene oxide + NADP$^+$ + H_2O <2, 10> [24, 25, 28]

S pyridine + NAD(P)H + O_2 <2> (Reversibility: ? <2> [1]) [1]

P pyridine N-oxide + NAD(P)$^+$ + H_2O <2> [1, 4]

S styrene + NAD(P)H + O_2 <2> (Reversibility: ? <2> [1, 4, 18]) [1, 4, 18]

P styrene epoxide + NAD(P)$^+$ + H_2O <2> [1, 4]

S toluene + NAD(P)H + O_2 <2> (Reversibility: ? <2> [1]) [1]

P benzyl alcohol + cresol + NAD(P)$^+$ + H_2O <2> [1]

S toluene + NAD(P)H + O_2 <3> (Reversibility: ? <3> [15]) [15]

P benzyl alcohol + NAD(P)$^+$ + H_2O <3> [15]

S toluene + NAD(P)H + O_2 <10> (<10> sMMO [28]) (Reversibility: ? <10> [28]) [28]

P cresol + NAD(P)$^+$ + H_2O <10> [28]

S trans-2-butene + NAD(P)H + O_2 <2, 3> (Reversibility: ? <2, 3> [1, 4, 15, 18]) [1, 4, 15, 18]

P trans-2,3-epoxybutane + trans-2-buten-1-ol + NAD(P)$^+$ + H_2O <2, 3> [1, 4, 15, 18]

S trichloromethane + NAD(P)H + O_2 <2> (Reversibility: ? <2> [4]) [4]

P CO_2 + Cl^- + NAD(P)$^+$ + H_2O <2> [4]

S xylene + NAD(P)H + O_2 <10> (<10> sMMO [28]) (Reversibility: ? <10> [28]) [28]

P xylenol + NAD(P)$^+$ + H_2O <10> [28]

S Additional information <2, 3, 5, 7, 10> (<10> sMMO expressed at low copper concentration shows low substrate specificity, while pMMO expressed at high copper concentration shows high substrate specificity [28]; <3> oxidation of norborneols [15]; <2> very non-specific oxygenase [1]; <3> oxidation of deuterated compounds [15]; <2,3,5,7,10> broad specificity of soluble, cytoplasmic enzyme form [23]; <2> broad specificity [18]) [1, 15, 18, 23, 28]

P ?

Inhibitors

2,3-dimercaptopropan-1-ol <2> [17]

2,4-dichloro-(6-phenylphenoxy)ethylamine hydrochloride <3> (<3> no inhibition [3]) [2]

2,4-dichloro-(6-phenylphenoxy)ethyldiethylamine <3> [2]
2-heptyl-4-hydroxyquinoline-N-oxide <2> (<2> pMMO, at 0.05 mM [25]) [25]
2-mercaptoethanol <3> [2]
3-amino-1,2,4-triazole <3> [2]
8-hydroxyquinoline <3> (<2> compound C not inhibited [19]) [3, 4]
Ag^{+} <2> [10]
Cd^{2+} <2> (<2> soluble enzyme form more than the membrane-bound form [10]) [10]
Co^{2+} <10> (<10> slightly, sMMO [28]) [28]
Cu^{+} <2, 3, 5-10> (<2> pMMO [25]; <2,3,5-10> soluble enzyme form sMMO [10,23,29]) [10, 23, 25, 29]
Cu^{2+} <2, 3, 10> (<10> causes protein aggregation [28]; <2,3,10> soluble enzyme form, sMMO [10,23,29]; <2> irreversible in vivo and in vitro by inactivation of reductase component [23]) [10, 23, 28, 29]
Fe^{2+} <2> (<2> slightly, membrane-bound enzyme form [10,25]) [10, 25]
Fe^{3+} <2, 3> (<2,3> slightly, soluble enzyme form sMMO [10,29]) [10, 29]
Hg^{2+} <10> (<10> complete inhibition at 0.01 mM [28]) [28]
KCN <2, 3> (<2> pMMO [25]) [2, 25]
Ni^{2+} <3, 10> (<3,10> sMMO [28,29]; <10> causes protein aggregation [28]) [28, 29]
SKF 525A <3> [2]
Zn^{2+} <2, 3, 10> (<3,10> sMMO [28,29]; <10> causes protein aggregation [28]) [10, 28, 29]
acetylene <2, 3> (<2,3> pMMO [25,27]; <2,3> suicide substrate [23]; <2> component C not inhibited [19]) [3, 4, 23, 25, 27]
acriflavin <2> [17]
allylthiourea <3> [2]
ammonium chloride <3> (<3> slightly [29]) [29]
choramphenicol <3> (<3> sMMO [29]) [29]
cysteine <2> (<2> pMMO [25]) [25]
dichloromethane <3> (<3> competitively in presence of formate [27]) [27]
dimercaptopropanol <3> [2]
dithiothreitol <3> [2]
ethyl carbamate <2> [17]
methylamine <3> (<3> slightly [29]) [29]
myxothiazol <2> (<2> pMMO, suicide substrate [25]) [25]
o-phenanthroline <3> [2]
thioacetamide <3> [2]
thiosemicarbazide <3> [2]
thiourea <3> [2]
trichlorethylene <3> (<3> non-competitively [27]; <3> pMMO [27]) [27]
Additional information <2, 3> (<3> different chlorinated hydrocarbons cause different inhibition patterns [27]; <2> sMMO is completely inhibited by monomeric component D, but not by dimeric, monomeric MMOD is interfering with the catalytically active complex between component A hydroxylase and component protein B [24]) [24, 27]

Cofactors/prosthetic groups

FAD <2-4> (<2-4> protein C, reductase component: contains 1 mol FAD per mol protein [4-6,13,19,20,22,23,24]; <2> characterization of FAD redox centre of component C [9]) [4-6, 9, 13, 19, 20, 22, 23, 24]
NADH <1-10> (<10> in vivo only NADH can be the electron donor [28]; <2> preferred [4]; <3> not: pMMO [32]) [1-30, 33]
NADPH <2, 3, 10> (<10> only in vitro [28]) [1, 2, 4, 14, 17, 1, 28, 29]
ascorbate <3> [21]
cytochrome c <3> (<3> one of three protein components is a soluble CO-binding cytochrome c [2]) [2]
succinate <3> (<3> electron donor, membrane-bound enzyme [8]) [8]
Additional information <2, 3, 10> (<2> component D contains no metal ions or organic cofactors [24]; <2,10> component B has no prosthetic group [13,31]; <3> succinate, particulate enzyme functions in vitro with either succinate or NADH as electron donor, soluble enzyme functions only with NADH [8]) [8, 13, 21, 24, 31]

Activating compounds

Cu^{2+} <2> (<2> pMMO, optimal at 0.3 mM [25]) [25]
Fe^{3+} <2> (<2> pMMO, optimal at 5.0 mM [25]) [25]

Metals, ions

Cu^{+} <2> (<2> pMMO, requirement for, contains 12-15 Cu^{+} ions per molecule of enzyme [33]) [33]
Cu^{2+} <2, 3, 5-10> (<2> 14.5 atoms per molecule of enzyme pMMO, type II copper centre [25]; <3> pMMO contains tightly bound copper, EDTA has no effect [32]; <2,3> pMMO: 15 molecules of copper per mol of enzyme [23]; <2,3,5-10> copper genetically regulates the enzyme activity of the soluble and membrane-bound form [8-10,23]; <2> component C contains no copper [20]; <3> cytochrome component contains 0.3-0.8 atoms copper per molecule [2]; <3> copper-containing protein component contains one copper atom per molecule [2]) [2, 8-10, 23, 25, 32]
Mn^{2+} <2> (<2> pMMO, low content [25]) [25]
Mo^{2+} <2> (<2> pMMO, low content [25]) [25]
Ni^{2+} <2> (<2> protein B contains 0.04 mol Ni^{2+} per mol protein [11]) [11]
Zn^{2+} <1, 2> (<2> pMMO [25]; <1> component A, hydroxylase component: contains 0.5 mol zinc per mol protein [7]; <2> 0.2-0.5 mol zinc per mol protein [12]) [7, 12, 25]
iron <1-4, 10> (<10> 3.6 mol of iron per mol of hydroxylase component A [28]; <2> 2.5 atoms per enzyme molecule of pMMO [25]; <10> component B of sMMO [31]; <2,3> pMMO: 2 iron molecules per mol of enzyme [23]; <2,3> sMMO: the α-subunit of component A contains a non-haem bis-micro-hydroxo-bridged binuclear iron centre [23]; <3> cytochrome component contains 1 atom iron per molecule [2]; <2> protein A, hydroxylase component: contains a binuclear iron center [4]; <3> 4.3 mol Fe per mol enzyme [5]; <4> 2.1 mol Fe per mol enzyme [6]; <1> 2.8 mol Fe per mol protein [7]; <2> 2.3 mol Fe per mol protein [12]; <3> contains oxo-bridged binuclear

iron clusters [16,22]; <3> contains hydroxo-bridged binuclear iron clusters [22]; <2> protein C, reductase component: contains 1 [Fe2-S2] [4,13,23]; <3> 2 mol Fe per mol enzyme [5]; <4> 1 mol Fe per mol enzyme [6]; <3> a [2Fe-2S]cluster [22]; <2> characterization of [Fe2-S2] redox centre of component C [9,19]; <2> 2 g atom iron [19]; <2> 1 mol of [2Fe-2S(S-Cys)4]centre per mol protein [19]; <2> 1.3-1.5 atoms iron per molecule [20]) [2, 4-7, 9, 12, 13, 16, 19, 20, 22, 23, 25, 28, 31]
Additional information <2, 10> (<2> copper-induced iron-uptake [25]; <10> sMMO contains no metal ions [31]) [25, 31]

Turnover number (min^{-1})
222 <3> (methane) [5]
264 <3> (propene) [5]
456 <3> (furan) [5]
Additional information <2, 3> (<3> turnover numbers of component A [16]) [16, 24]

Specific activity (U/mg)
0.0008 <10> (<10> sMMO, substrate chloronaphthalene [28]) [28]
0.0012 <10> (<10> sMMO, substrate naphthalene [28]) [28]
0.0039 <10> (<10> sMMO, substrate chloropentane [28]) [28]
0.011 <2> (<2> purified enzyme form pMMO [25]) [25]
0.0122 <10> (<10> sMMO, low copper growth concentration, growth substrate nitrate [28]) [28]
0.0191 <10> (<10> sMMO, substrate butylene [28]) [28]
0.0254 <10> (<10> sMMO, substrate propylene [28]) [28]
0.0336 <10> (<10> sMMO, substrate ethylene [28]) [28]
0.0605 <2> (<2> whole cells, substrate formate [25]) [25]
0.072 <2> (<2> component A [12]) [12]
0.1 <2> (<2> about, all 3 components individually [10]) [10]
0.185 <4> (<4> component A [6]) [6]
0.205 <10> (<10> sMMO, substrate diethylic ether [28]) [28]
0.208 <1> (<1> purified component A hydroxylase [7]) [7]
0.265 <2> (<2> protein A after reconstitution of iron [4]) [4]
0.334 <10> (<10> sMMO, substrate propane [28]) [28]
0.764 <3> (<3> all components, substrate propene [16]) [16]
1.66 <2> (<2> sMMO protein B triple mutant G10A/G13Q/G16A [30]) [30]
2.284 <2> (<2> His-tagged sMMO protein B triple mutant G10A/G13Q/G16A [30]) [30]
3.99 <2> (<2> sMMO protein B mutant G13Q [30]) [30]
5.09 <2> (<2> sMMO, wild-type enzyme [30]) [30]
5.71 <2> (<2> purified component C [20]) [20]
6 <2, 3> (<2> purified protein C [4,19]) [2, 4, 19]
7.3 <2> (<2> protein B [11]) [11]
11.37 <2> (<2> purified protein B [4]) [4]
17.5 <4> (<4> component C [6]) [6]
26.1 <3> (<3> purified enzyme, substrate propene [5,22]) [5, 22]

Additional information <2, 3, 10> (<10> component protein B is involved in enzyme regulation and enhances the activity 10fold [28]; <2> recombinant shows higher activity than the wild-type [26]; <2> overview [1]) [1, 2, 4, 26, 28]

K_m-Value (mM)

0.003 <2> (methane) [4]
0.0064 <10> (NADH, <10> sMMO component C reductase [28]) [28]
0.0168 <2> (O_2) [4]
0.025 <3> (methane) [5]
0.035 <3> (furan) [5]
0.05 <3> (NADH) [5]
0.0558 <2> (NADH, <2> with methane [4]) [4]
0.066 <3> (methane) [2]
5.2 <10> (NADPH, <10> sMMO component C reductase [28]) [28]
Additional information <2, 3> [4, 11, 27]

pH-Optimum

6.5-7 <2> [17]
6.9-7 <3> [2]
7 <2> (<2> assay at [24]; <2> enzyme form sMMO [24]) [24]
7.2 <10> (<10> assay at [28]) [28]
7.5 <3> (<3> furan, propene [5]) [5]
8.5-9 <2> (<2> component C [19]) [19]

pH-Range

6.4-7.4 <3> (<3> pH 6.4: about 20% of activity maximum, pH 7.4: about 25% of activity maximum [2]) [2]

Temperature optimum (°C)

25 <2> (<2> assay at [24]; <2> enzyme form sMMO [24]) [24]
30 <3> (<3> assay at, hydroxylase component [16]) [16]
45 <2> (<2> assay at [4,19]; <2> assay at, protein B [11]) [4, 11, 17, 19]

4 Enzyme Structure

Molecular weight

9400 <3> (<3> component: small protein, native PAGE [2]) [2]
13000 <3> (<3> component: CO-binding cytochrome c, native PAGE [2]) [2]
15100 <3> (<3> component B, gel filtration [5]) [5]
15800 <3> (<3> component B containing FAD and $[Fe_2\text{-}S_2]$-cluster, gel filtration [22]) [22]
15850 <2> (<2> component protein B of sMMO, mass spectrometry [26,30]) [26, 30]
17000 <2> (<2> component: protein B, gel filtration [4]) [4]
18000 <10> (<10> component protein B of sMMO [28]) [28]
24000 <2> (<2> component D of sMMO, forms homodimers, gel filtration [24]) [24]

32000 <10> (<10> component B [31]) [31]
38000-38550 <2> (<2> component: protein C, gel filtration [4]) [4]
38300-38400 <3> (<3> component C: reductase, gel filtration [5]) [5]
39700 <3> (<3> component C NADH-reductase, gel filtration [22]) [22]
40000 <4> (<4> component C acceptor reductase [6]) [6]
41000 <10> (<10> component C reductase of sMMO [28]) [28]
42000 <2> (<2> component C [20]) [20]
47000 <3> (<3> component: copper-containing protein, native PAGE [2]) [2]
94000 <2> (<2> pMMO [33]) [33]
99000 <2> (<2> pMMO, mass spectrometry [25]) [25]
210000 <2> (<2> component A, analytical ultracentrifugation [12]) [12]
220000 <1, 2, 4> (<1,4> component A: hydroxylase, gel filtration [6,7]; <2> component: protein A, gel filtration [4]) [4, 6, 7]
229000 <10> (<10> component A hydroxylase of sMMO [28]) [28]
240000 <3> (<3> component A hydroxylase [16]) [16]
241000-246000 <3> (<3> protein A hydroxylase, gel filtration [5]) [5]
245000 <3> (<3> protein A hydroxylase, gel filtration [22]) [22]
Additional information <1-4, 10> (<2> protein B shows unusual behaviour in gel filtration [30]; <10> sMMO is a multicomponent enzyme consisting of a hydroxylase, a protein B and a reductase [28,31]; <1-4> enzyme system consists of 3 protein components A, B, C [2-7,13-16,19,20,23]; <3> 3 components: 1 soluble CO-binding cytochrome c, 1 copper-containing protein, and 1 small protein, SDS-PAGE [2]; <3> complex formation of protein components [14]; <2> structure, review [13]; <1-4> see under subunits: molecular weights of the subunits of components [2-7,12]) [2-7, 12-16, 19, 20, 23, 30, 31]

Subunits

? <1-4, 10> (<10> component A of sMMO: 2 * 57000 + 2 * 43000 + 2 * 23000, $\alpha_2\beta_2\gamma_2$, SDS-PAGE [28]; <2> component A: 2 * 54000 α + 2 * 42000 β + 2 * 17000 γ, SDS-PAGE and analytical ultracentrifugation [12]; <2> protein A: 2 * 54000-60630, α+ 2 * 42000-44720, β + 2 * 17000-19840, γ [4]; <3> component A: 2 * 54400 α, 2 * 43000 β + 2 * 22700 γ, sedimentation velocity, SDS-PAGE, amino acid analysis [5]; <4> component A: 2 * 56000 α + 2 * 40000 β + 2 * 20000 γ, SDS-PAGE [6]; <1> component A: 2 * 55000 α + 2 * 40000 β + 2 * 20000 γ, SDS-PAGE [7]) [2-7, 12, 24, 28, 31]
dimer <2, 10> (<2> component D of sMMO: 2 * 12000, SDS-PAGE [24]; <10> component B: 2 * 15100, SDS-PAGE [31]) [24, 31]
trimer <2> (<2> 1 * 45000 + 1 * 26000 + 1 * 23000, pMMO, SDS-PAGE [33]; <2> 1 * 47000 + 1 * 27000 + 1 * 25000, pMMO, mass spectrometry and SDS-PAGE [25]) [25, 33]
Additional information <1-4, 10> (<2> pMMO subunit A has acetylene binding ability [25]; <2> sMMO consists of 4 components: a hydroxylase, a reductase, a protein B and a proteinD [24]; <10> sMMO is a multicomponent enzyme consiting of a hydroxylase, a protein B and a reductase [31]; <2,3> pMMO: 3 subunits [23]; <2,3> sMMO: component B has a regulatory role, component A is a hydroxylase [23]; <2,3> sMMO: 3 protein components, component A has $\alpha_2\beta_2\gamma_2$-subunit structure [23]; <2,3> component A is a hy-

droxylase [2-7,13-16,19,20,23]; <1-4> enzyme system consists of 3 protein components A, B, C [2-7,13-16,19,20]; <3> 3 components: 1 soluble CO-binding cytochrome c, 1 copper-containing protein, and 1 small protein, SDS-PAGE [2]; <1-4> see under molecular weight for the size of the protein components [2-7]; <3> complex formation of protein components [14]; <2> structure, review [13]) [2-7, 13-16, 19, 20, 23-25, 31]

5 Isolation/Preparation/Mutation/Application

Localization

cytoplasm <1-5, 7, 10> (<2, 3, 5, 7, 10> cytoplasmatic, soluble enzyme form termed sMMO [23-26, 28, 29, 31]; <2, 3, 5, 7, 10> enzyme from cells grown under conditions of low copper availability [4, 8, 23, 28, 29]) [1, 3, 4, 6, 8, 10, 13, 17, 18, 20, 23-26, 28, 29, 31]
membrane <2, 3, 5-10> (<2, 3, 5-10> membrane-bound particulate enzyme form termed pMMO [23, 25, 27, 28, 32, 33]; <2, 3, 5-10> enzyme from cells grown under conditions of high copper availability is membrane-bound [4, 8, 23, 28, 33]) [4, 8, 21, 23, 25, 27, 28, 32, 33]

Purification

<1> (hydroxylase component [7]) [7]
<2> (sMMO protein B wild-type and mutants recombinant from Escherichia coli by affinity chromatography, high salt concentration increases the binding stability between protein B and hydroxylase of sMMO [30]; pMMO after induction with copper, kinetic type I with respect to O_2-sensitivity [33]; pMMO [25]; recombinant component protein B of sMMO as glutathione-S-transferase fusion protein from Escherichia coli [26]; protein B [11]; component A [12]; component C [19,20]; component D of sMMO recombinant from Escherichia coli as His-tagged thioredoxin-fusion protein, the thioredoxin is cleaved off during purification by factor Xa [24]) [4, 11, 12, 19, 20, 24-26, 30, 33]
<3> (sMMO [29]; pMMO [32]; 3 components: a soluble CO-binding cytochrome c, a copper-containing protein, another small protein [2]; improved purification protocol using stabilizing agents [5]; hydroxylase component A [16]; partial [21]) [2, 5, 15, 16, 21, 22, 29, 32]
<4> (protein A and C [6]) [6]
<10> (sMMO with all components [28]; component B of sMMO [31]) [28, 31]

Renaturation

<3> (copper ions increase the stability of exfoliated pMMO [32]) [32]

Cloning

<2> (genetic analysis of pMMO genes [33]; expression of His-tagged triple mutant G10A/G13Q/G16A and G13Q mutant in Escherichia coli [30]; expression of component protein B of sMMO as glutathione-S-transferase fusion protein in Escherichia coli [26]; overexpression of an additional protein component D of sMMO encoded by orfY as thioredoxin-fusion protein with His-

Tag in Escherichia coli, protein component D is termed MMOD [24]; <2> DNA sequence analysis [24]) [24, 30, 33]
<3> (transcription of sMMO is repressed at Cu^{2+} concentration above 0.00086 mM per g dry cell weight [29]) [29]
<10> (expression in Escherichia coli, determination of complete sMMO DNA gene sequence, phylogenetic analysis [28]) [28]
<2, 3> (both enzyme forms: genetic structure and expression systems, review [23]; regulation model, copper plays regulatory role on cis level [23]) [23]

Engineering

G10A/G13Q/G16A <2> (<2> reduced activity [30]; <2> His-tagged protein B of sMMO, triple mutant is resistant to degradation in contrast to the wild-type, N-terminus is responsible for unusual mobility in size exclusion chromatography and proteolytic sensitivity of protein B [30]) [30]
G13Q <2> (<2> reduced activity [30]; <2> sMMO, alteration of a cleavage site in component protein B [26,30]; <2> enhanced temperature stability compared to wild-type, site-directed mutagenesis [26]) [26, 30]
Additional information <2> (<2> native occurence of a truncated form of pMMO with different molecular weight [33]; <2> native parallel occurence of full length and 2 N-terminal truncated forms of regulatory component protein B of sMMO, truncated forms are inactive [26]) [26, 33]

6 Stability

pH-Stability

6-7 <2> (<2> protein B, stable, rapid loss of activity above and below [11]) [11]
6.8 <2> (<2> or below: pMMO is irreversibly inactivated [33]) [33]

Temperature stability

45 <2> (<2> protein B, half-life: 10.2 min [11]) [11]
55 <2> (<2> protein A, 10 min [20]) [20]
Additional information <10> (<10> N-terminal truncated sMMO component B increases the heat stability of the sMMO hydroxylase [31]) [31]

Oxidation stability

<2>, sensitive to O_2, 3 kinetically distinct froms of pMMO with respect to O_2 tension, type I is stable with moderate activity, type II is highly unstable to oxygen, type III is an intermediate form [33]

General stability information

<2>, 5 mM thioglycollate, 5 mM dithiothreitol, 5 mM NADH stabilize component C at 0°C [19]
<2>, pMMO is very unstable in vitro [33]
<2>, protein A unstable to successive freezing and thawing [4]
<2>, protein B stable to successive freezing and thawing [11]
<2>, protein B, in crude form requires addition of protease inhibitor phenylmethylsulfonyl fluoride [4]

<2>, protein C requires presence of thiol protective agent, e.g. sodium thioglycolate throughout purification [4]
<2>, sMMO is more stable and easier to purify than pMMO [24]
<3>, copper ions increase the stability of exfoliated pMMO [32]
<3>, no component of the enzyme is stable to freezing [2]
<3>, succinate stabilizes the membrane-bound enzyme [8]
<10>, Fe^{2+} stabilizes component C of sMMO [28]
<2, 3>, instability of enzyme in crude extract [3, 8]

Storage stability

<2>, -20°C, 50% glycerol, component A, stable for several weeks [12]
<2>, -80°C, pMMO kinetically type I with respect to O_2-sensitivity, repeated freeze-thaw-cycles, stable [33]
<2>, 0°C, protein C, loss of 60-90% of activity within 20 h [20]
<2>, 0°C, proteins A and B, stable for at least 24 h [20]
<2>, 4°C, reducing argon or nitrogen atmosphere, no loss of activity after 1 week [25]
<3>, -80°C, more than 1 year [5]
<3>, 4°C, membrane-bound enzyme, activity is lost after 24 h, can be stabilized by succinate [8]
<10>, 4°C, sMMO component C reductase, 90% loss of activity after 120 h, can be restored by Fe^{2+} [28]
<10>, sMMO component A hydroxylase is very unstable, splits into inactive subunits when stored frozen even for short periods [28]
<10>, sMMO component protein B is stable when stored frozen [28]

References

[1] Colby, J.; Stirling, D.I.; Dalton, H.: The soluble methane mono-oxygenase of Methylococcus capsulatus (Bath). Its ability to oxygenate n-alkanes, n-alkenes, ethers, and alicyclic, aromatic and heterocyclic compounds. Biochem. J., **165**, 395-402 (1977)

[2] Tonge, G.M.; Harrison, D.E.F.; Higgins, I.J.: Purification and properties of the methane mono-oxygenase enzyme system from Methylosinus trichosporium OB3b. Biochem. J., **161**, 333-344 (1977)

[3] Stirling, D.I.; Dalton, H.: Properties of the methane mono-oxygenase from extracts of Methylosinus trichosporium OB3b and evidence for its similarity to the enzyme from Methylococcus capsulatus (Bath). Eur. J. Biochem., **96**, 205-212 (1979)

[4] Pilkington, S.J.; Dalton, H.: Soluble methane monooxygenase from Methylococcus capsulatus Bath. Methods Enzymol., **188**, 181-190 (1990)

[5] Fox, B.G.; Froland, W.A.; Jollie, D.R.; Lipscomb, J.D.: Methane monooxygenase from Methylosinus trichosporium OB3b. Methods Enzymol., **188**, 191-202 (1990)

[6] Pilkington, S.J.; Dalton, H.: Purification and characterization of the soluble methane monooxygenase from Methylosinus sporium 5 demonstrates the

highly conserved nature of this enzyme in methanotrophs. FEMS Microbiol. Lett., **78**, 103-108 (1991)
[7] Patel, R.N.; Savas, J.C.: Purification and properties of the hydroxylase component of methane monooxygenase. J. Bacteriol., **169**, 2313-2317 (1987)
[8] Cornish, A.; MacDonald, J.; Burrows, K.J.; King, T.S.; Scott, D.; Higgins, I.J.: Succinate as an in vitro electron donor for the particulate methane monooxygenase of Methylosinus trichosporium. Biotechnol. Lett., **7**, 319-324 (1985)
[9] Lund, J.; Dalton, H.: Further characterisation of the FAD and Fe2S2 redox centres of component C, the NADH:acceptor reductase of the soluble methane monooxygenase of Methylococcus capsulatus (Bath). Eur. J. Biochem., **147**, 291-296 (1985)
[10] Green, J.; Prior, S.D.; Dalton, H.: Copper ions as inhibitors of protein C of soluble methane monooxygenase of Methylococcus capsulatus (Bath). Eur. J. Biochem., **153**, 137-144 (1985)
[11] Green, J.; Dalton, H.: Protein B of soluble methane monooxygenase from Methylococcus capsulatus (Bath). A novel regulatory protein of enzyme activity. J. Biol. Chem., **260**, 15795-15801 (1985)
[12] Woodland, M.P.; Dalton, H.: Purification and characterization of component A of the methane monooxygenase from Methylococcus capsulatus (Bath). J. Biol. Chem., **259**, 53-59 (1984)
[13] Dalton, H.; Smith, D.D.S.; Pilkington, S.J.: Towards a unified mechanism of biological methane oxidation. FEMS Microbiol. Lett., **87**, 201-208 (1990)
[14] Fox, B.G.; Liu, Y.; Dege, J.E.; Lipscomb, J.D.: Complex formation between the protein components of methane monooxygenase from Methylosinus trichosporium OB3b. Identification of sites of component interaction. J. Biol. Chem., **266**, 540-550 (1991)
[15] Rataj, M.J.; Kauth, J.E.; Donnelly, M.I.: Oxidation of deuterated compounds by high specific activity methane monooxygenase from Methylosinus trichosporium. Mechanistic implications. J. Biol. Chem., **266**, 18684-18690 (1991)
[16] Fox, B.G.; Lipscomb, J.D.: Purification of a high specific activity methane monooxygenase hydroxylase component from a type II methanotroph. Biochem. Biophys. Res. Commun., **154**, 165-170 (1988)
[17] Colby, J.; Dalton, H.: Some properties of a soluble methane mono-oxygenase from Methylococcus capsulatus strain Bath. Biochem. J., **157**, 495-497 (1976)
[18] Green, J.; Dalton, H.: Substrate specificity of soluble methane monooxygenase. Mechanistic implications. J. Biol. Chem., **264**, 17698-17703 (1989)
[19] Colby, J.; Dalton, H.: Characterization of the second prosthetic group of the flavoenzyme NADH-acceptor reductase (component C) of the methane mono-oxygenase from Methylococcus capsulatus (Bath). Biochem. J., **177**, 903-908 (1979)
[20] Colby, J.; Dalton, H.: Resolution of the methane mono-oxygenase of Methylococcus capsulatus (Bath) into three components. Purification and properties of component C, a flavoprotein. Biochem. J., **171**, 461-468 (1978)

[21] Tonge, G.M.; Harrison, D.E.F.; Knowles, C.J.; Higgins, I.J.: Properties and partial purification of the methane-oxidising enzyme system from Methylosinus trichosporium. FEBS Lett., **58**, 293-299 (1975)
[22] Fox, B.G.; Froland, W.A.; Dege, J.E.; Lipscomb, J.D.: Methane monooxygenase from Methylosinus trichosporium OB3b. Purification and properties of a three-component system with high specific activity from a type II methanotroph. J. Biol. Chem., **264**, 10023-10033 (1989)
[23] Murrell, J.C.; Gilbert, B.; McDonald, I.R.: Molecular biology and regulation of methane monooxygenase. Arch. Microbiol., **173**, 325-332 (2000)
[24] Merkx, M.; Lippard, S.J.: Why orfY? Characterization of MMOD, a long overlooked component of the soluble methane monooxygenase from Methylococcus capsulatus (Bath). J. Biol. Chem., **277**, 5858-5865 (2002)
[25] Zahn, J.A.; DiSpirito, A.A.: Membrane-associated methane monooxygenase from Methylococcus capsulatus (Bath). J. Bacteriol., **178**, 1018-1029 (1996)
[26] Lloyd, J.S.; Bhambra, A.; Murrell, J.C.; Dalton, H.: Inactivation of the regulatory protein B of soluble methane monooxygenase from Methylococcus capsulatus (Bath) by proteolysis can be overcome by a Gly to Gln modification. Eur. J. Biochem., **248**, 72-79 (1997)
[27] Lontoh, S.; DiSpirito, A.A.; Semrau, J.D.: Dichloromethane and trichloroethylene inhibition of methane oxidation by the membrane-associated methane monooxygenase of Methylosinus trichosporium OB3b. Arch. Microbiol., **171**, 301-308 (1999)
[28] Grosse, S.; Laramee, L.; Wendlandt, K.D.; McDonald, I.R.; Miguez, C.B.; Kleber, H.P.: Purification and characterization of the soluble methane monooxygenase of the type II methanotrophic bacterium Methylocystis sp. strain WI 14. Appl. Environ. Microbiol., **65**, 3929-3935 (1999)
[29] Jahng, D.; Wood, T.K.: Metal ions and chloramphenicol inhibition of soluble methane monooxygenase from Methylosinus trichosporium OB3b. Appl. Microbiol. Biotechnol., **45**, 744-749 (1996)
[30] Brandstetter, H.; Whittington, D.A.; Lippard, S.J.; Frederick, C.A.: Mutational and structural analyses of the regulatory protein B of soluble methane monooxygenase from Methylococcus capsulatus (Bath). Chem. Biol., **6**, 441-449 (1999)
[31] Shinohara, Y.; Uchiyama, H.; Yagi, O.; Kusakabe, I.: Purification and characterization of component B of a soluble methane monooxygenase from Methylocystis sp. M. J. Ferment. Bioeng., **85**, 37-42 (1998)
[32] Xin, J.Y.; Cui, J.R.; Hu, X.X.; Li, S.B.; Xia, C.G.; Zhu, L.M.; Wang, Y.Q.: Particulate methane monooxygenase from Methylosinus trichosporium is a copper-containing enzyme. Biochem. Biophys. Res. Commun., **295**, 182-186 (2002)
[33] Nguyen, H.H.T.; Elliott, S.J.; Yip, J.H.K.; Chan, S.I.: The particulate methane monooxygenase from Methylococcus capsulatus (Bath) is a novel copper-containing three-subunit enzyme. Isolation and characterization. J. Biol. Chem., **273**, 7957-7966 (1998)

Phosphatidylcholine 12-monooxygenase 1.14.13.26

1 Nomenclature

EC number
1.14.13.26

Systematic name
1-acyl-2-oleoyl-sn-glycero-3-phosphocholine,NADH:oxygen oxidoreductase (12-hydroxylating)

Recommended name
phosphatidylcholine 12-monooxygenase

Synonyms
oleate Δ^{12}-hydroxylase
oxygenase, oleate Δ^{12}-mono
ricinoleic acid synthase

CAS registry number
77950-95-9

2 Source Organism

<1> *Ricinus communis* (castor bean [1]) [1, 2, 4, 5]
<2> *Lesquerella fendleri* [3]

3 Reaction and Specificity

Catalyzed reaction
1-acyl-2-oleoyl-sn-glycero-3-phosphocholine + NADH + H^+ + O_2 = 1-acyl-2-[(S)-12-hydroxyoleoyl]-sn-glycero-3-phosphocholine + NAD^+ + H_2O

Reaction type
oxidation
redox reaction
reduction

Natural substrates and products
S 1-acyl-2-oleoyl-sn-glycero-3-phosphocholine + NADH + O_2 <1> (Reversibility: ? <1> [4]) [4]
P 1-acyl-2-[(S)-12-hydroxyoleoyl]-sn-glycero-3-phosphocholine + NAD^+ + H_2O

Substrates and products

S 1-acyl-2-oleoyl-sn-glycero-3-phosphocholine + NADH + O_2 <1> (Reversibility: ? <1> [1]) [1]

P 1-acyl-2-[(S)-12-hydroxyoleoyl]-sn-glycero-3-phosphocholine + NAD^+ + H_2O

S cis-linoleic acid + NADH + O_2 <2> (Reversibility: ? <2> [3]) [3]

P densipolic acid + NAD + H_2O

S cis-oleic acid + NADH + O_2 <2> (Reversibility: ? <2> [3]) [3]

P ricinoleic acid + NAD + H_2O

S eicosenoic acid + NADH + O_2 <2> (Reversibility: ? <2> [3]) [3]

P lesquerolic acid + NAD + H_2O

Inhibitors

$FeSO_4$ <1> [1]

Additional information <1> (<1> no inhibition by KCN, CO and metapyrone [1]) [1]

Cofactors/prosthetic groups

NADH <1> [1]

Activating compounds

EDTA <1> (<1> slight stimulation [1]) [1]

K_m-Value (mM)

0.004 <1> (O_2) [1]

pH-Optimum

6.3 <1> [4]

7 <1> (<1> assay at [1]) [1]

Temperature optimum (°C)

23 <1> [4]

30 <1> (<1> assay at [1]) [1]

4 Enzyme Structure

Subunits

? <1> (<1> x * 44000, deduced from gene sequence [2]) [2]

5 Isolation/Preparation/Mutation/Application

Source/tissue

seed <1, 2> [1, 3, 5]

Localization

endoplasmic reticulum <1> [5]

microsome <1> [1, 4]

Cloning

<1> (fatty acyl desaturase homolog [2]) [2]

<2> (encoded enzyme has both hydroxylase and desaturase activities [3]) [3]

References

[1] Moreau, R.A.; Stumpf, P.K.: Recent studies of the enzymic synthesis of ricinoleic acid by developing castor beans. Plant Physiol., **67**, 672-676 (1981)

[2] Van de Loo, F.J.; Broun, P.; Turner, S.; Somerville, C.: An oleate 12-hydroxylase from Ricinus communis L. is a fatty acyl desaturase homolog. Proc. Natl. Acad. Sci. USA, **92**, 6743-6747 (1995)

[3] Broun, P.; Boddupalli, S.; Somerville, C.: A bifunctional oleate 12-hydroxylase: desaturase from Lesquerella fendleri. Plant J., **13**, 201-210 (1998)

[4] Lin, J.T.; McKeon, T.A.; Goodrich-Tanrikulu, M.; Stafford, A.E.: Characterization of oleoyl-12-hydroxylase in castor microsomes using the putative substrate, 1-acyl-2-oleoyl-sn-glycero-3-phosphocholine. Lipids, **31**, 571-577 (1996)

[5] Richards, D.E.; Taylor, R.D.; Murphy, D.J.: Localization and possible substrate requirement of the oleate 12-hydroxylase of developing Ricinus communis seeds. Plant Physiol. Biochem., **31**, 89-94 (1993)

4-Aminobenzoate 1-monooxygenase 1.14.13.27

1 Nomenclature

EC number
1.14.13.27

Systematic name
4-aminobenzoate,NAD(P)H:oxygen oxidoreductase (1-hydroxylating, decarboxylating)

Recommended name
4-aminobenzoate 1-monooxygenase

Synonyms
4-aminobenzoate hydroxylase
oxygenase, 4-aminobenzoate mono-
Additional information (cf. EC 1.14.13.1)

CAS registry number
98668-55-4

2 Source Organism

<1> *Agaricus bisporus* (mushroom [1,3]) [1-6]

3 Reaction and Specificity

Catalyzed reaction
4-aminobenzoate + NAD(P)H + H^+ + O_2 = 4-hydroxyaniline + $NAD(P)^+$ + H_2O + CO_2 (<1> A-stereospecificity [2])

Reaction type
decarboxylation
oxidation
redox reaction
reduction

Substrates and products
S 2-amino-5-chlorobenzoate + NAD(P)H + O_2 <1> (Reversibility: ? <1> [1]) [1]
P 4-chloro-2-hydroxyaniline + $NAD(P)^+$ + H_2O_2 + CO_2 <1> (<1> H_2O_2 formed simultaneously with hydroxylation [1]) [1]

S 3,4-diaminobenzoate + NAD(P)H + O_2 <1> (<1> 17.3% of the reaction with 4-aminobenzoate [1]) (Reversibility: ? <1> [1]) [1]

P 2-amino-4-hydroxyaniline + $NAD(P)^+$ + H_2O_2 + CO_2 <1> (<1> H_2O_2 formed simultaneously with hydroxylation [1]) [1]

S 3-chloro-4-aminobenzoate + NAD(P)H + O_2 <1> (Reversibility: ? <1> [4]) [4]

P 2-chloro-4-hydroxyaniline + NAD(P)H + O_2 <1> [4]

S 4-amino-2-chlorobenzoate + NAD(P)H + O_2 <1> (Reversibility: ? <1> [1]) [1]

P 3-chloro-4-hydroxyaniline + $NAD(P)^+$ + H_2O_2 + CO_2 <1> (<1> H_2O_2 formed simultaneously with hydroxylation [1]) [1]

S 4-aminobenzoate + NAD(P)H + O_2 <1> (Reversibility: ? <1> [1-4]) [1-4]

P 4-hydroxyaniline + $NAD(P)^+$ + H_2O + CO_2 <1> [1, 3]

S 4-aminosalicylate + NAD(P)H + O_2 <1> (Reversibility: ? <1> [1, 3]) [1, 3]

P 3,4-dihydroxyaniline + $NAD(P)^+$ + H_2O_2 + CO_2 <1> (<1> H_2O_2 formed simultaneously with hydroxylation [1]) [1]

S 4-hydroxybenzoate + NAD(P)H + O_2 <1> (<1> 8.4% of the reaction with 4-aminobenzoate [1]) (Reversibility: ? <1> [1]) [1]

P quinol + $NAD(P)^+$ + H_2O_2 + CO_2 <1> (<1> H_2O_2 formed simultaneously with hydroxylation [1]) [1]

S anthranilate + NAD(P)H + O_2 <1> (Reversibility: ? <1> [1, 3]) [1, 3]

P 2-hydroxyaniline + $NAD(P)^+$ + H_2O_2 + CO_2 <1> (<1> H_2O_2 formed simultaneously with hydroxylation [1]) [1]

S Additional information <1> (<1> not: salicylate [3]; <1> not: 4-aminobenzyl-alcohol [3]; <1> not: aniline [3]; <1> not: benzoic acid [3]; <1> several aromatic compounds stimulate NADH oxidation, but no hydroxylated product is formed, all of the O_2 consumed is converted to H_2O_2 [1]) [1, 3]

P ?

Inhibitors

$AgNO_3$ <1> (<1> 93% inactivation at 0.1 mM [1]) [1]

Br^- <1> (<1> 37% inactivation at 0.1 M [1]) [1]

Cl^- <1> (<1> 30% inactivation at 0.1 M [1]) [1]

Cu^{2+} <1> (<1> severe inhibition at 1 mM [3]) [3]

$CuSO_4$ <1> (<1> 91% inactivation at 0.1 mM [1]) [1]

F^- <1> (<1> 20% inactivation at 0.1 M [1]) [1]

Fe^{2+} <1> (<1> slight inhibition at 1 mM [3]) [3]

Hg^{2+} <1> (<1> complete inhibition at 0.1 mM [3]) [3]

$HgCl_2$ <1> (<1> 92% inactivation at 0.1 mM [1]) [1]

I^- <1> (<1> 62% inactivation at 0.1 M [1]) [1]

NO_3^- <1> (<1> 45% inactivation at 0.1 M [1]) [1]

SCN^- <1> (<1> 60% inactivation at 0.1 M [1]) [1]

p-chloromercuribenzoate <1> (<1> 94% inactivation at 0.1 mM [1]; <1> complete inhibition at 0.1 mM [3]) [1, 3]

Additional information <1> (<1> amphoteric and anionic detergents cause inactivation [4]) [4]

Cofactors/prosthetic groups

FAD <1> (<1> flavoprotein [1,3]; <1> 0.91 mol of FAD per mol of enzyme [1]) [1, 3, 5]
NADH <1> [1, 3]
NADPH <1> (<1> 62% of the activity with NADH [1]; <1> 68% of the activity with NADH [1,3]) [1, 3]

Specific activity (U/mg)

0.5 <1> [3]
0.91 <1> [4]
28 <1> [2]

K_m-Value (mM)

0.0136 <1> (NADH) [1]
0.0204 <1> (4-aminobenzoate) [1]
0.1117 <1> (4-aminobenzoate, <1> at pH 7 [4]) [4]
0.133 <1> (NADPH) [1]
0.2 <1> (O_2) [1]
0.4739 <1> (4-aminobenzoate, <1> at pH 8 [4]) [4]

pH-Optimum

6-7 <1> (<1> NADPH [1]) [1]
6.5-8 <1> (<1> NADH [1]) [1]
7 <1> [3, 4]

Temperature optimum (°C)

40 <1> [1]

4 Enzyme Structure

Molecular weight

49000 <1> (<1> gel filtration [1]) [1]
50000 <1> (<1> SDS-PAGE [1]) [1]
50950 <1> (<1> calculated from amino acid residues [6]) [6]

Subunits

monomer <1> (<1> 1 * 49000, SDS-PAGE [1]) [1]

5 Isolation/Preparation/Mutation/Application

Source/tissue

fruitbody <1> [3]

Purification

<1> (homogeneity [1]) [1, 2, 3, 4]

Cloning

<1> (fusion protein with glutathione-S-transferase [4]; DNA fragments encoding various partial amino acid sequences of the enzyme [5]) [4, 5, 6]

6 Stability

pH-Stability

6-8 <1> (<1> stable [1]) [1]

Temperature stability

30 <1> (<1> 60 min, little loss of activity in presence of 0.1 mM FAD and 0.02% bovine serum albumin [1]) [1]
35 <1> (<1> 10 min, 50 mM potassium phosphate buffer, pH 7.0, 0.1 mM FAD, 0.02% bovine serum albumin, stable [1]) [1]
40 <1> (<1> 10 min, 50 mM potassium phosphate buffer, pH 7.0, 0.1 mM FAD, 0.02% bovine serum albumin, about 70% loss of activity [1]) [1]
45 <1> (<1> 10 min, 50 mM potassium phosphate buffer, pH 7.0, 0.1 mM FAD, 0.02% bovine serum albumin, 97% loss of activity [1]) [1]
50 <1> (<1> 10 min, 50 mM potassium phosphate buffer, pH 7.0, 0.1 mM FAD, 0.02% bovine serum albumin, 98% loss of activity [1]) [1]

Storage stability

<1>, -20°C, 10 mM potassium phosphate buffer, pH 7.0, 20% glycerol, 6 months, stable [3]
<1>, -20°C, 50 mM potassium phosphate buffer, pH 7.0, 20% glycerol, 1 mM 2-mercaptoethanol, 0.01 mM FAD, stable for 1 month [1]

References

[1] Tsuji, H.; Ogawa, T.; Bando, N.; Sasaoka, K.: Purification and properties of 4-aminobenzoate hydroxylase, a new monooxygenase from Agaricus bisporus. J. Biol. Chem., **261**, 13203-13209 (1986)

[2] Tsuji, H.; Ogawa, T.; Bando, N.; Sasaoka, K.: Stereospecificity of NAD(P)H oxidation catalyzed by 4-aminobenzoate hydroxylase from Agaricus bisporus. Biochim. Biophys. Acta, **991**, 380-382 (1989)

[3] Tsuji, H.; Ogawa, T.; Bando, N.; Sasaoka, K.: A unique enzyme catalyzing the formation of 4-hydroxyaniline from 4-amino-benzoic acid in Agaricus bisporus. Biochem. Biophys. Res. Commun., **130**, 633-639 (1985)

[4] Mizutani, Y.; Narikawa, T.; Satoh, T.; Sakurai, N.; Kaji, H.; Yamada, S.; Samejima, T.: A new UV method for serum γ-glutamyltransferase assay using recombinant 4-aminobenzoate hydroxylase as a coupling enzyme. J. Biochem., **126**, 347-353 (1999)

[5] Tsuji, H.; Kimoto, M.; Watanabe, H.; Sasagawa, T.; Oka, T.; Yamashita, H.; Okita, M.: Epitope mapping of monoclonal antibodies against 4-aminobenzoate hydroxylase from Agaricus bisporus. Biochim. Biophys. Acta, **1425**, 628-631 (1998)

[6] Tsuji, H.; Oka, T.; Kimoto, M.; Hong, Y.M.; Natori, Y.; Ogawa, T.: Cloning and sequencing of cDNA encoding 4-aminobenzoate hydroxylase from Agaricus bisporus. Biochim. Biophys. Acta, **1309**, 31-36 (1996)

3,9-Dihydroxypterocarpan 6a-monooxygenase 1.14.13.28

1 Nomenclature

EC number
1.14.13.28

Systematic name
(6aR,11aR)-3,9-dihydroxypterocarpan,NADPH:oxygen oxidoreductase (6a-hydroxylating)

Recommended name
3,9-dihydroxypterocarpan 6a-monooxygenase

Synonyms
3,9-dihydroxypterocarpan 6a-hydroxylase
oxygenase, 3,9-dihydroxypterocarpan 6α-mono-

CAS registry number
92584-16-2

2 Source Organism

<1> *Glycine max* [1-6]

3 Reaction and Specificity

Catalyzed reaction
(6aR,11aR)-3,9-dihydroxypterocarpan + NADPH + H^+ + O_2 = (6aS,11aS)-3,6a,9-trihydroxypterocarpan + $NADP^+$ + H_2O

Reaction type
oxidation
redox reaction
reduction

Natural substrates and products
S (6aR,11aR)-3,9-dihydroxypterocarpan + NADPH + O_2 <1> (<1>, the product of the reaction is the biosynthetic precursor of the phytoalexin glyceollin in soybean [1,6]) (Reversibility: ? <1> [1, 6]) [1, 6]
P (6aS,11aS)-3,6a,9-trihydroxypterocarpan + $NADP^+$ + H_2O <1> [1, 6]

Substrates and products

S (6aR,11aR)-3,9-dihydroxypterocarpan + NADPH + O_2 <1> (Reversibility: ? <1> [1, 4, 6]) [1, 4, 6]

P (6aS,11aS)-3,6a,9-trihydroxypterocarpan + $NADP^+$ + H_2O <1> [1, 4, 6]

Inhibitors

cytochrome c <1> [1]

Cofactors/prosthetic groups

FAD <1> (<1>, at 0.05 mM with 0.05 mM FMN and 0.01 mM NADPH: 132% relative activity to NADPH alone [1]) [1]
FMN <1> (<1>, at 0.05 mM with 0.05 mM FAD and 0.01 mM NADPH: 132% relative activity to NADPH alone [1]) [1]
NADPH <1> (<1>, absolute requirement for, with 0.05 mM FAD: 110% relative activity to only NADPH, with 0.05 mM FMN: 126% relative activity to NADPH alone [1]) [1-3, 6]
cytochrome P_{450} <1> (<1>, possibly a heme-thiolate protein [1-3]) [1-3]
Additional information <1> (<1>, with NADH, ascorbic acid, FAD, FMN or 6,7-dimethyl-5,6,7,8-tetrahydropterine alone no activity can be detected [1]) [1]

Metals, ions

Mg^{2+} <1> (<1>, increases activity [1]) [1]

Specific activity (U/mg)

0.00006 <1> (<1>, at 2 h after inoculation with zoospores from incompatible, heat resistant Phytophtora megasperma f. sp. glycinea [4]) [4]
0.00054 <1> (<1>, at 8 h after inoculation with zoospores from incompatible, heat resistant Phytophtora megasperma f. sp. glycinea [4]) [4]
Additional information <1> [2, 3]

K_m-Value (mM)

0.0001 <1> ((6aR,11aR)-3,9-dihydroxypterocarpan, <1>, recombinant enzyme expressed in yeast [6]) [6]
0.00016 <1> ((6aR,11aR)-3,9-dihydroxypterocarpan, <1>, enzyme from elicitor-treated soybean cells [6]) [6]

pH-Optimum

7.2-8.2 <1> (<1>, in Tris/HCl and potassium phosphate [1]) [1]
7.4 <1> (<1>, recombinant enzyme and soybean microsomes [6]) [6]

Temperature optimum (°C)

30 <1> (<1>, recombinant enzyme and soybean microsomes [6]) [6]
40 <1> [1]

Temperature range (°C)

30 <1> (<1>, 75% of activity maximum at 30°C [1]) [1]

4 Enzyme Structure

Subunits

? <1> (<1>, x * 55000, SDS-PAGE [2,3]) [2, 3]

Posttranslational modification

glycoprotein <1> [3]

5 Isolation/Preparation/Mutation/Application

Source/tissue

cell culture <1> [1, 2, 6]
seedling <1> [1, 4]

Localization

microsome <1> [1, 2, 6]

Purification

<1> [2, 3]

Renaturation

<1> (<1>, three artichoke NADPH-cytochrome c reductase isoforms are able to reconstitute the enzyme with cytochrome P-450 purified from elicitor-challenged soybean cell cultures [5]) [5]

Cloning

<1> (<1>, CYP93A1 cDNA isolated from elicitor-induced soybean cells encodes the enzyme, CYP93A1 expressed in Saccharomyces cerevisiae [6]) [6]

6 Stability

Temperature stability

30 <1> (<1>, half-life: 9 min, in 0.08 M potassium phosphate buffer, pH 7.5, 14 mM 2-mercaptoethanol, 20% sucrose, crude enzyme extract [1]) [1]

Storage stability

<1>, -70°C, storage of microsomes, 50% loss of activity in 3-4 weeks [1]

References

[1] Hagmann, M.L.; Heller, W.; Grisebach, H.: Induction of phytoalexin synthesis in soybean. Stereospecific 3,9-dihydroxypterocarpan 6a-hydroxylase from elicitor-induced soybean cell cultures. Eur. J. Biochem., **142**, 127-131 (1984)

[2] Kochs, G.; Grisebach, H.: Phytoalexin synthesis in soybean: purification and reconstitution of cytochrome P_{450} 3,9-dihydroxypterocarpan 6a-hydroxylase

and separation from cytochrome P_{450} cinnamate 4-hydroxylase. Arch. Biochem. Biophys., **273**, 543-553 (1989)
[3] Kochs, G.; Werck-Reichhart, D.; Grisebach, H.: Further characterization of cytochrome P_{450} involved in phytoalexin synthesis in soybean: cytochrome P_{450} cinnamate 4-hydroxylase and 3,9-dihydroxypterocarpan 6a-hydroxylase. Arch. Biochem. Biophys., **293**, 187-194 (1992)
[4] Bonhoff, A.; Loyal, R.; Ebel, J.; Grisebach, H.: Race: Cultivar-specific induction of enzymes related to Phytoalexin biosynthesis in soybean roots following infection with Phytophtora megasperma f. sp. glycinea. Arch. Biochem. Biophys., **246**, 149-154 (1986)
[5] Benveniste, I.; Lesot, A.; Hasenfratz, M.; Kochs, G.; Durst, F.: Multiple forms of NADPH-cytochrome P_{450} reductase in higher plants. Biochem. Biophys. Res. Commun., **177**, 105-112 (1991)
[6] Schopfer, C.R.; Kochs, G.; Lottspeich, F.; Ebel, J.: Molecular characterization and functional expression of dihydroxypterocarpan 6a-hydroxylase, an enzyme specific for pterocarpanoid phytoalexin biosynthesis in soybean (Glycine max L.). FEBS Lett., **432**, 182-186 (1998)

4-Nitrophenol 2-monooxygenase 1.14.13.29

1 Nomenclature

EC number
1.14.13.29

Systematic name
4-nitrophenol,NADH:oxygen oxidoreductase (2-hydroxylating)

Recommended name
4-nitrophenol 2-monooxygenase

Synonyms
4-nitrophenol hydroxylase
4-nitrophenol-2-hydroxylase
CYP 2E1
cytochrome P-450 2E1
cytochrome P-450 isozyme 3a
oxygenase, 4-nitrophenol 2-mono-

CAS registry number
91116-87-9

2 Source Organism

<1> *Nocardia sp.* [1]
<2> *Oryctolagus cuniculus* (New Zealand white male rabbit [2]) [2]
<3> *Ovis aries* (sheep [3]) [3]
<4> *Rattus norvegicus* (EtOH-treated rats [4]; male Sprague-Dawley rats, 7 weeks old [6]) [4, 5, 6]
<5> *Homo sapiens* (human [7]) [7]

3 Reaction and Specificity

Catalyzed reaction
4-nitrophenol + NADH + H^+ + O_2 = 4-nitrocatechol + NAD^+ + H_2O

Reaction type
oxidation
redox reaction
reduction

Substrates and products

S 3-nitrophenol + NADH + O_2 <1> (Reversibility: ? <1> [1]) [1]
P ?
S 4-nitrophenol + NADH + O_2 <1-5> (<1> NADPH: 50% of the activity with NADH [1]) (Reversibility: ? <1-5> [1-5, 7]) [1-5, 7]
P 4-nitrocatechol + NAD^+ + H_2O <1-4> [1-5]
S chlorzoxazone + NADH + O_2 <1> (<5> 6-hydroxylation [7]) (Reversibility: ? <1> [1]) [1]
P ?
S Additional information <1> (<1> no substrates: 2-nitrophenol, 2,4-dinitrophenol [1]) [1]
P ?

Inhibitors

4-nitrophenol <2, 5> (<2> substrate inhibition above 0.1 mM [2]; <5> mutual competitive inhibition with chlorzoxazone [7]) [2, 7]
Co^{2+} <1> (<1> slight effect, crude enzyme extract [1]) [1]
Cu^{2+} <1> (<1> 94% inhibition at 1 mM,crude enzyme extract [1]) [1]
Fe^{2+} <1> (<1> slight effect, crude enzyme extract [1]) [1]
Fe^{3+} <1> (<1> slight effect, crude enzyme extract [1]) [1]
Hg^{2+} <1> (<1> 63% inhibition at 1 mM, crude enzyme extract [1]) [1]
N-methylmaleimide <1> (<1> 84% inhibition at 5 mM [1]) [1]
Ni^{2+} <1> (<1> slight effect, crude enzyme extract [1]) [1]
Sn^{2+} <1> (<1> 57% inhibition at 1 mM,crude enzyme extract [1]) [1]
α-naphthoflavone <5> (<5> slight [7]) [7]
catalase <1> (<1> 30% inhibition at 1000 units [1]) [1]
chlorzoxazone <5> (<5> mutual competitive inhibition with 4-nitrophenol [7]) [7]
coumarin <5> (<5> slight [7]) [7]
diethyldithiocarbamate <5> (<5> 50% inhibition at 0.002-0.003 mM [7]) [7]
ethanol <4> [5]
horseradish peroxidase <1> (<1> complete inhibition at 25 units [1]) [1]
mephenytoin <5> (<5> slight [7]) [7]
p-chloromercuribenzoate <1> (<1> 81% inhibition at 1 mM [1]) [1]
quinidine <5> (<5> slight [7]) [7]
sulfaphenazole <5> (<5> slight [7]) [7]
troleandomycin <5> (<5> slight [7]) [7]
Additional information <1> (<1> inhibition at high ionic strength of all common buffers and salts e.g. phosphate, Tris, KCl, $(NH_4)_2SO_4$, 60% inhibition above 300 mM, crude enzyme extract [1]) [1]

Cofactors/prosthetic groups

FAD <1> (<1> flavoprotein, FMN cannot replace FAD [1]) [1]
NADH <1-4> [1-4]
NADPH <1> (<1> 50% of the activity with NADH [1]) [1]

Activating compounds

α-naphthoflavone <5> (<5> 60% activiation at 25 mM [7]) [7]

Turnover number (min^{-1})

7.6 <2> (4-nitrophenol, <2> reconstituted system of purified isozyme 3a, dilaroylglyceryl-3-phosphorylcholine and NADPH-cytochrome P-450 reductase at pH 7.6 [2]) [2]

Specific activity (U/mg)

0.000037 <4> (<4> microsomal fraction [6]) [6]
0.000494 <3> [3]

K_m-Value (mM)

0.0083 <1> (4-nitrocatechol) [1]
0.021 <5> (4-nitrocatechol, <5> cDNA expressed enzyme [7]) [7]
0.03 <5> (4-nitrocatechol) [7]
Additional information <3> [3]

K_i-Value (mM)

0.042 <5> (4-nitrophenol, <5> inhibition of chlorzoxazone 6-hydroxylation [7]) [7]
0.047 <5> (chlorzoxazone, <5> inhibition of 4-nitrophenol hydroxylation [7]) [7]

pH-Optimum

6.8 <2, 3, 4> (<2,3> crude enzyme extract [2,3]; <4> assay at [5]) [2, 3, 5]
7.3 <1> (<1> crude enzyme extract [1]) [1]

Temperature optimum (°C)

37 <4> (<4> assay at [5]) [5]
40 <1> (<1> crude enzyme extract [1]) [1]

4 Enzyme Structure

Molecular weight

40000 <5> (<5> SDS-PAGE [7]) [7]

5 Isolation/Preparation/Mutation/Application

Source/tissue

liver <2, 4, 5> [2, 4, 5, 6, 7]
lung <3> [3]

Localization

cytosol <1> [1]
microsome <2, 3, 4, 5> [2, 3, 5, 6, 7]

Purification

<2> [2]
<5> [7]

Cloning

<5> [7]

6 Stability

General stability information

<1>, dialysis of crude enzyme extract, complete loss of activity [1]

Storage stability

<1>, -20°C, 85% loss of activity after 10 days [1]

<1>, 0-4°C, 30% loss of activity after 7 days, 70% loss of activity after 14 days, crude enzyme extract [1]

References

[1] Mitra, D.; Vaidyanathan, C.S.: A new 4-nitrophenol 2-hydroxylase from a Nocardia sp. Biochem. Int., **8**, 609-615 (1984)

[2] Koop, D.R.: Hydroxylation of *p*-nitrophenol by rabbit ethanol-inducible cytochrome P-450 isozyme 3a. Mol. Pharmacol., **29**, 399-404 (1986)

[3] Arinc, E.; Aydan, A.: Lung microsomal *p*-nitrophenol hydroxylase - characterization and reconstitution of its activity. Comp. Biochem. Physiol. B, **97**, 455-460 (1990)

[4] Reinke, L.A.; Moyer, M.J.: *p*-Nitrophenol hydroxylation. A microsomal oxidation which is highly inducible by ethanol. Drug Metab. Dispos., **13**, 548-552 (1985)

[5] Allis, J.W.; Robinson, B.L.: A kinetic assay for *p*-nitrophenyl hydroxylase in rat liver microsomes. Anal. Biochem., **219**, 49-52 (1994)

[6] Hanioka, N.; Watanabe, K.; Yoda, R.; Ando, M.: Effect of alachlor on hepatic cytochrome P 450 enzymes in rats. Drug Chem. Toxicol., **25**, 25-37 (2002)

[7] Tassaneeyakul, W.; Veronese, M.E.; Birkett, D.J.; Gonzalez, F.J.; Miners, J.O.: Validation of 4-nitrophenol as an in vitro substrate probe for human liver CYP2E1 using cDNA expression and microsomal kinetic techniques. Biochem. Pharmacol., **46**, 1975-1981 (1993)

Leukotriene-B_4 20-monooxygenase 1.14.13.30

1 Nomenclature

EC number
1.14.13.30

Systematic name
(6Z,8E,10E,14Z)-(5S,12R)-5,12-dihydroxyicosa-6,8,10,14-tetraenoate,NAD-PH:oxygen oxidoreductase (20-hydroxylating)

Recommended name
leukotriene-B_4 20-monooxygenase

Synonyms
CYP4F14 <4> [14]
CYP4F2 <2> [12, 13]
CYP4F3 <2> [9, 11, 12]
CYPIVF2
CYPIVF3
Cytochrome P_{450}-LTB-ω
LTB_4 20-hydroxylase
LTB_4 ω-hydroxylase
leukotriene-B_4 20-monooxygenase
cytochrome P_{450} 4F3 <2> [9, 11]
leukotriene B_4 20-hydroxylase
leukotriene-B_4 ω-hydroxylase
oxygenase, leukotriene B_4 20-mono-

CAS registry number
90119-11-2

2 Source Organism

<1> *Rattus norvegicus* [1, 3, 8]
<2> *Homo sapiens* [2-7, 9, 10-13]
<3> *Oryctolagus cuniculus* [3]
<4> *Mus musculus* [14]

3 Reaction and Specificity

Catalyzed reaction

(6Z,8E,10E,14Z)-(5S,12R)-5,12-dihydroxyicosa-6,8,10,14-tetraenoate + NADPH + H^+ + O_2 = (6Z,8E,10E,14Z)-(5S,12R)-5,12,20-trihydroxyicosa-6,8,10,14-tetraenoate + $NADP^+$ + H_2O (a heme-thiolate protein P_{450}; not identical with EC 1.14.13.34)

Reaction type

ω-oxidation
redox reaction
reduction

Natural substrates and products

S (6Z,8E,10E,14Z)-(5S,12R)-5,12-dihydroxyicosa-6,8,10,14-tetraenoate + NADPH + O_2 <1-3> (<1-3> inactivation of leukotriene B_4, a mediator of inflammation [1-13]) [1-13]
P (6Z,8E,10E,14Z)-(5S,12R)-5,12,20-trihydroxyicosa-6,8,10,14-tetraenoate + $NADP^+$ + H_2O

Substrates and products

S 5-deoxy-leukotriene B_4 + NADPH + O_2 <2> (Reversibility: ? <2> [5]) [5]
P 5-deoxy-20-hydroxy-leukotriene B_4
S 5-epi-leukotriene B_4 + NADPH + O_2 <2> (Reversibility: ? <2> [5]) [5]
P 5-epi-20-hydroxy-leukotriene B_4
S 6-trans-12-epi-leukotriene B_4 + NADPH + O_2 <2> (Reversibility: ? <2> [5]) [5]
P 6-trans-12-epi-20-hydroxy-leukotriene B_4
S leukotriene B_4 + NADPH + O_2 <1-4> (<1-4> leukotriene B_4 is identical with (6Z,8E,10E,14Z)-(5S,12R)-5,12-dihydroxyicosa-6,8,10,14-tetraenoate [1-14]) (Reversibility: ? <1-4> [1-14]) [1-14]
P 20-hydroxy-leukotriene B_4 + $NADP^+$ + H_2O
S prostaglandin A_1 + NADPH + O_2 <1, 4> (<4> recombinant CYP4F14 [14]) (Reversibility: ? <1, 4> [8, 14]) [8, 14]
P 20-hydroxy-prostaglandin A_1 + $NADP^+$ + H_2O
S Additional information <2> (<2> overview, stereochemical requirement for substrate specificity [5]; <2> recombinant protein CYP4F3, wide substrate specificity [11] ; <2> recombinant protein CYP4F2 [13]; <4> recombinant protein CYP4F14 [14]) [5, 11, 13, 14]
P ?

Inhibitors

5-aminosalicylic acid <2> (<2> 50% inhibition at 0.05 mM [10]) [10]
CO <1, 2> (<2> recombinant protein, expressed in yeast [11]) [7, 8, 11]
acetylenic fatty acids <2> (<2> suicidal inactivation [4]) [4]
α-naphthoflavone <1> [1]
ferricytochrome c <2> [7]
leukotriene B_4 <1> (<1> substrate prostaglandin A_1 [8]) [8]

leukotriene B_5 <2> [5]
metyrapone <1, 2> (<2> recombinant protein, expressed in yeast, 35% inhibition at 2 mM [11]) [1, 11]
prostaglandin A_1 <1> [8]
Additional information <2> (<2> overview: inhibition by alternative substrates [5]) [5]

Cofactors/prosthetic groups
NADH <1> [1]
NADPH <1-3> [1-8]
cytochrome P_{450} <1, 2> (<1,2> enzyme contains cytochrome P-450 [1,2,6-8]) [1, 2, 6-8]

Activating compounds
Additional information <1> (<1> hepatic enzyme is not induced by phenobarbital or 3-methylcholanthrene [1]) [1]

Specific activity (U/mg)
0.154 <2> (<2> microsomal preparation [6]) [6]
30 <2> (<2> recombinant protein, expressed in yeast [11]) [11]

K_m-Value (mM)
0.00022 <2> (leukotriene B_4) [2]
0.0003 <2> (leukotriene B_4) [7]
0.0006 <2> (leukotriene B_4, <2> recombinant protein CYP4F3, expressed in yeast [11]) [11]
0.0008 <2> (NADPH) [7]
0.001 <2> (NADPH) [2]
0.002 <2> (leukotriene B_4) [5]
0.0038 <2> (6-trans-12-epi-leukotriene B_4) [5]
0.004 <2> (leukotriene B_4, <2> recombinant isoform CYP4F3A, expressed in COS-7 cells [9]) [9]
0.0056 <2> (5-deoxy-leukotriene B_4) [5]
0.0066 <2> (5-epi-leukotriene B_4) [5]
0.04 <1> (leukotriene B_4) [1]
0.04 <1> (prostaglandin A_1) [8]
0.042 <1> (leukotriene B_4) [8]
0.06 <2> (leukotriene B_4, <2> recombinant protein CYP4F2, expressed in yeast [11]) [11]
0.0661 <1> (leukotriene B_4, <1> cosubstrate NADPH [1]) [1]
0.105 <2> (leukotriene B_4, <2> recombinant isoform CYP4F3B, expressed in COS-7 cells [9]) [9]
0.868 <1> (leukotriene B_4, <1> cosubstrate NADH [1]) [1]
Additional information <2> (<2> recombinant protein CYP4F3, kinetic parameters of various substrates [11]; <2> recombinant protein CYP4F2, kinetic parameters of various substrates [13]) [11, 13]

K_i-Value (mM)
0.001 <2> (leukotriene B_5) [5]
0.038 <1> (prostaglandin A_1) [8]
0.043 <1> (leukotriene B_4, <1> substrate prostaglandin A_1 [8]) [8]

pH-Optimum
7.4 <1> [8]
7.5 <2> (<2> recombinant protein, expressed in yeast [11]) [2, 11]
7.5-8 <2> [6]

pH-Range
6.8-9.5 <2> (<2> pH 6.8: about 50% of activity maximum, pH 9.5: about 70% of activity maximum [2]) [2]

Temperature optimum (°C)
37 <2> (<2> assay at [2,6]) [2, 6]

4 Enzyme Structure

Molecular weight
55000 <2> (<2> recombinant CYP4F3, expressed in yeast, SDS-PAGE [11]) [11]
56300 <2> (<2> recombinant CYP4F2, expressed in yeast, SDS-PAGE [13]) [13]
59800 <4> (<4> CYP4F14, predicted from gene sequence [14]) [14]
61000 <2> (<2> recombinant isoforms CYP4F3A and CYP4F3B expressed in COS-7 cells, western blot analysis [9]) [9]

5 Isolation/Preparation/Mutation/Application

Source/tissue
Hep-G2 <2> (<2> constitutively expressed in hepatoma cell line Hep-G2 [12]) [12]
kidney <2> (<2> CYP4F2 [12]) [12]
leukocyte <1-3> (<1-3> polymorphonuclear [2-6,10]) [2-6, 10]
liver <1, 2, 4> (<2> isoform CYP4F3B [9]; <2> CYP4F2 [12]; <4> CYP4F14 [14]) [1, 8, 9, 12, 14]
neutrophil <2> (<2> isoform CYP4F3A [9]; <2> CYP4F2 not expressed in neutrophils [12]) [7, 9, 11, 12]

Localization
cytosol <2> [2]
microsome <1, 2, 4> [1, 5-8, 9, 11, 13, 14]
Additional information <1> (<1> not cytosol [1]) [1]

Purification

<2> (recombinant protein CYP4F3, expressed in yeast [11]; recombinant protein CYP4F2, expressed in yeast [13]) [11, 13]

Cloning

<2> (two isoforms of CYP4F3, CYP4F3A expressed in neutrophils and CYP4F3B expressed in fetal and adult liver [9]; CYP4F3 gene contains 14 exons and 13 introns [9]; tissue-specific expression is regulated by alternative promoter usage and mutually exclusive exon splicing, exon 3 and 4 [9]; expression in transfected COS-7 cells [9]; chromosomal localization 19p [9]; expression in yeast cells [11,13]; CYP4F2 [12,13]) [9, 11-13]
<4> (CYP4F14, expression in yeast cells [14]) [14]

6 Stability

Temperature stability

50 <1> (<1> completely inactivated after 1 h [1]) [1]

Storage stability

<2>, -70°C, 20% glycerol,1 mM dithithreitol, 3 days and 10 days, 50% resp. 100% [13]

References

[1] Romano, M.C.; Eckardt, R.D.; Bender, P.E.; Leonard, T. B.; Straub, K.M.; Newton, J.F.: Biochemical characterization of hepatic microsomal leukotriene B_4 hydroxylases. J. Biol. Chem., **262**, 1590-1595 (1987)
[2] Soberman, R.J.; Harper, T.W.; Murphy, R.C.; Austen, K. F.: Identification and functional characterization of leukotriene B_4 20-hydroxylase of human polymorphonuclear leukocytes. Proc. Natl. Acad. Sci. USA, **82**, 2292-2295 (1985)
[3] Powell, W.S.: Properties of leukotriene B_4 20-hydroxylase from polymorphonuclear leukocytes. J. Biol. Chem., **259**, 3082-3089 (1984)
[4] Shak, S.; Reich, N.O.; Goldstein, I.M.; Ortiz de Montellano, P.R.: Leukotriene B_4 ω-hydroxylase in human polymorphonuclear leukocytes. Suicidal inactivation by acetylenic fatty acids. J. Biol. Chem., **260**, 13023-13028 (1985)
[5] Soberman, R.J.; Okita, R.T.; Fitzsimmons, B.; Rokach, J.; Spur, B.; Austen, K.F.: Stereochemical requirements for substrate specificity of LTB_4 20-hydroxylase. J. Biol. Chem., **262**, 12421-12427 (1987)
[6] Soberman, R.J.; Okita, R.T.: Leukotriene B_4 20-hydroxylase of human polymorphonuclear leukocytes. Methods Enzymol., **163**, 349-353 (1988)
[7] Sumimoto, H.; Takeshige, K.; Minakami, S.: Characterization of human neutrophil leukotriene B_4 ω-hydroxylase as a system involving a unique cytochrome P-450 and NADPH-cytochrome P-450 reductase. Eur. J. Biochem., **172**, 315-324 (1988)

[8] Sumimoto, H.; Kusunose, E.; Gotoh, Y.; Kusunose, M.; Minakami, S.: Leukotriene B_4 ω-hydroxylase in rat liver microsomes: identification as a cytochrome P-450 that catalyzes prostaglandin A_1 ω-hydroxylation, and participation of cytochrome b_5. J. Biochem., **108**, 215-221 (1990)
[9] Christmas, P.; Ursino, S.R.; Fox, J.W.; Soberman, R.J.: Expression of the CYP4F3 gene. Tissue-specific splicing and alternative promoters generate high and low K_m forms of leukotriene B_4 ω-hydroxylase. J. Biol. Chem., **274**, 21191-21199 (1999)
[10] Ikehata, A.; Hiwatashi, N.; Kinouchi, Y.; Ito, K.; Yamazaki, H.; Toyota, T.: 5-Aminosalicylic acid inhibits leukotriene B_4 ω-hydroxylase activity in human polymorphonuclear leukocytes. Prostaglandins Leukot. Essent. Fatty Acids, **50**, 317-320 (1994)
[11] Kikuta, Y.; Kusunose, E.; Sumimoto, H.; Mizukami, Y.; Takeshige, K.; Sakaki, T.; Yabusaki, Y.; Kusunose, M.: Purification and characterization of recombinant human neutrophil leukotriene B_4 ω-hydroxylase (cytochrome P450 4F3). Arch. Biochem. Biophys., **355**, 201-205 (1998)
[12] Kikuta, Y.; Miyauchi, Y.; Kusunose, E.; Kusunose, M.: Expression and molecular cloning of human liver leukotriene B_4 ω-hydroxylase (CYP4F2) gene. DNA Cell Biol., **18**, 723-730 (1999)
[13] Kikuta, Y.; Kusunose, E.; Kusunose, M.: Characterization of human liver leukotriene B_4 ω-hydroxylase P_{450} (CYP4F2). J. Biochem., **127**, 1047-1052 (2000)
[14] Kikuta, Y.; Kasyu, H.; Kusunose, E.; Kusunose, M.: Expression and catalytic activity of mouse leukotriene B_4 ω-hydroxylase, CYP4F14. Arch. Biochem. Biophys., **383**, 225-232 (2000)

2-Nitrophenol 2-monooxygenase 1.14.13.31

1 Nomenclature

EC number
1.14.13.31

Systematic name
2-nitrophenol,NADPH:oxygen 2-oxidoreductase (2-hydroxylating, nitrite-forming)

Recommended name
2-nitrophenol 2-monooxygenase

Synonyms
nitrophenol oxygenase
oxygenase, 2-nitrophenol

CAS registry number
104520-84-5

2 Source Organism

<1> *Pseudomonas putida* (strain B2 [1, 2, 3]) [1, 2, 3]
<2> *Bacillus sphaericus* (strain JS905 [4]) [4]

3 Reaction and Specificity

Catalyzed reaction
2-nitrophenol + NADPH + H^+ + O_2 = catechol + nitrite + $NADP^+$ + H_2O

Reaction type
oxidation
redox reaction
reduction

Natural substrates and products
S 2-nitrophenol + NADPH + O_2 <1, 2> (<1> involved in metabolism of nitro-aromatic compounds by a strain of Pseudomonas putida [1]) (Reversibility: ? <1, 2> [1, 4]) [1, 4]
P ?

Substrates and products
S 2-nitrophenol + NADPH + O_2 <1> (Reversibility: ? <1> [1-3]) [1, 2, 3]
P catechol + nitrite + $NADP^+$ + H_2O <1> [1, 2]

S 2-nitrophenol + NADPH + O_2 <2> (Reversibility: ? <2> [4]) [4]
P 3-nitrocatechol + $NADP^+$ + H_2O <2> [4]
S 3-methyl-2-nitrophenol + NADPH + O_2 <1> (Reversibility: ? <1> [2]) [2]
P 3-methylcatechol + $NADP^+$ + H_2O
S 4-chloro-2-nitrophenol + NADPH + O_2 <1> (Reversibility: ? <1> [1]) [1]
P 4-chlorocatechol + $NADP^+$ + H_2O
S 4-formyl-2-nitrophenol + NADPH + O_2 <1> (<1> weak activity [1]) (Reversibility: ? <1> [1]) [1]
P 4-formyl-1,2-benzenediol + $NADP^+$ + H_2O
S 4-methyl-2-nitrophenol + NADPH + O_2 <1> (Reversibility: ? <1> [1, 2]) [1, 2]
P 4-methylcatechol + $NADP^+$ + H_2O
S 4-nitroresorcinol + NADPH + O_2 <2> (Reversibility: ? <2> [4]) [4]
P ?
S 5-methyl-2-nitrophenol + NADPH + O_2 <1> (Reversibility: ? <1> [2]) [2]
P 5-methylcatechol + $NADP^+$ + H_2O
S Additional information <1> (<1> broad specificity towards alkylated and halogenated *o*-nitrophenols, overview [2]; <1> not: 4-carboxy-2-nitrophenol, 4-nitro-2-nitrophenol [1]; <2> 2-component enzyme system comprising of a flavoprotein reductase and an oxygenase [4]) [1, 2, 4]
P ?

Inhibitors

2,4-dinitrophenol <1> (<1> competitive inhibition [2]) [2]
2-nitrophenol <1> (<1> substrate inhibition above 0.02 mM [2]) [2]
Cu^{2+} <1> [1]
N-ethylmaleimide <1> (<1> 80-90% inhibition at 0.008 mmol per g protein, irreversible by dilution, inhibition does not affect enzyme activity, but retards translocation of the substrate into the cytoplasm [3]) [3]

Cofactors/prosthetic groups

FAD <2> [4]
NADH <1, 2> (<1> very low affinity [2]) [2, 4]
NADPH <1, 2> [1, 2, 4]

Metals, ions

Ca^{2+} <1> (<1> 30% stimulation at 4 mM [1]) [1]
Mg^{2+} <1> (<1> 50% stimulation at 4 mM, EDTA reverses stimulation [1]; <1> stimulates [2]) [1, 2]
Mn^{2+} <1> (<1> 50% stimulation at 4 mM [1]; <1> stimulates [2]) [1, 2]

Specific activity (U/mg)

0.04 <2> [4]
2 <1> [2]
12 <1> (<1> with 4-chloro-2-nitrophenol as substrate [1]) [1]
50 <1> (<1> with 4-methyl-2-nitrophenol as substrate [1]) [1]
60 <1> (<1> with 2-nitrophenol as substrate [1]) [1]

K_m-Value (mM)
0.008 <1> (2-nitrophenol) [2]
0.14 <1> (NADPH) [2]
2 <1> (NADH) [2]

K_i-Value (mM)
0.0005 <1> (2,4-dinitrophenol) [2]

pH-Optimum
7.5 <1> (<1> assay at [1]) [1]
7.5-8 <1> (<1> *o*-nitrophenol, 4-chloro-2-nitrophenol [2]) [2]

pH-Range
6.5-9 <1> (<1> pH 6.5: about 60% of activity maximum, pH 9.0: about 55% of activity maximum [2]) [2]

4 Enzyme Structure

Molecular weight
58000 <1> (<1> gel filtration [2]) [2]
65000 <1> (<1> SDS-PAGE [2]) [2]

Subunits
monomer <1> (<1> 1 * 65000, SDS-PAGE [2]) [2]

5 Isolation/Preparation/Mutation/Application

Localization
soluble <1> [1]

Purification
<1> (strain B2 [2]) [2]
<2> (partial [4]) [4]

6 Stability

pH-Stability
7.5-8 <1> (<1> highest stability [2]) [2]

Temperature stability
4 <1> (<1> stable for several hours in 20 mM phosphate buffer or 50 mM Tris buffer [2]) [2]
40 <1> (<1> 2 min, complete inactivation in absence of *o*-nitrophenol, 2-nitrophenol above 0.2 mM prevents heat inactivation [2]) [2]

General stability information
<1>, 2-nitrophenol stabilizes against heat inactivation [2]
<1>, glycerol stabilizes during storage at -20°C [2]

Storage stability

<1>, -20°C, 50% glycerol, less than 25% loss of activity after 5 days [2]

<1>, 4°C, 20 mM phosphate buffer, 50 mM Tris buffer, pH 7.5, 4 mM DTT, less than 35% loss of activity after 5 days [2]

References

[1] Zeyer, J.; Kocher, H.P.; Timmis, K.N.: Influence of *para*-substituents on the oxidative metabolism of *o*-nitrophenols by Pseudomonas putida B2. Appl. Environ. Microbiol., **52**, 334-339 (1986)

[2] Zeyer, J.; Kocher, H.P.: Purification and characterization of a bacterial nitrophenol oxygenase which converts *ortho*-nitrophenol to catechol and nitrite. J. Bacteriol., **170**, 1789-1794 (1988)

[3] Folsom, B.R.: Characterization of 2-nitrophenol uptake system of Pseudomonas putida B2. J. Ind. Microbiol. Biotechnol., **19**, 123-129 (1997)

[4] Kadiyala, V.; Spain, J.C.: A two-component monooxygenase catalyzes both the hydroxylation of *p*-nitrophenol and the oxidative release of nitrite from 4-nitrocatechol in Bacillus sphaericus JS905. Appl. Environ. Microbiol., **64**, 2479-2484 (1998)

Albendazole monooxygenase 1.14.13.32

1 Nomenclature

EC number
1.14.13.32

Systematic name
albendazole,NADPH:oxygen oxidoreductase (sulfoxide-forming)

Recommended name
albendazole monooxygenase

Synonyms
albendazole oxidase
albendazole sulfoxidase

CAS registry number
101299-59-6

2 Source Organism

<1> *Rattus norvegicus* [1, 2]
<2> *Sus scrofa* [1]

3 Reaction and Specificity

Catalyzed reaction
albendazole + NADPH + H^+ + O_2 = albendazole S-oxide + $NADP^+$ + H_2O

Reaction type
oxidation
redox reaction
reduction
sulfoxidation

Natural substrates and products
S albendazole + NADPH + O_2 <1, 2> (Reversibility: ? <1, 2> [1]) [1, 2]
P albendazole S-oxide + $NADP^+$ + H_2O

Substrates and products
S albendazole + NADPH + O_2 <1, 2> (Reversibility: ? <1, 2> [1]) [1, 2]
P albendazole S-oxide + $NADP^+$ + H_2O

Inhibitors

methimazole <1> (<1> competitive inhibitor, thermal treatment of the microsomes inhibited the methimazole oxidation [2]) [2]
thiobenzamide <1> [1]
tranylcypromine <1> (<1> competitive inhibitor [1]) [1]

Cofactors/prosthetic groups

FAD <1, 2> (<1,2> flavoprotein, flavin-containing monooxygenase [1]) [1]
NADPH <1, 2> [1]

Specific activity (U/mg)

0.00059 <1> [1]
0.013 <1> [2]

K_m-Value (mM)

0.0536 <1> (albendazole) [1]

pH-Optimum

7.4 <1> (<1> assay conditions [1,2]) [1, 2]

4 Enzyme Structure

Molecular weight

59000 <1> (<1> SDS-PAGE [2]) [2]

5 Isolation/Preparation/Mutation/Application

Source/tissue

liver <1, 2> [1, 2]

Localization

endoplasmic reticulum <1, 2> [1, 2]

6 Stability

Temperature stability

Additional information <1> (<1> 2 min. 50°C in assay buffer in the absence of NADPH inactivates the enzyme [2]) [2]

References

[1] Fargetton, X.; Galtier, P.; Delatour, P.: Sulfoxidation of albendazole by a cytochrome P_{450}-independent monooxygenase from rat liver microsomes. Vet. Res. Commun., **10**, 317-324 (1986)

[2] Moroni, P.; Buronfosse, T.; Longin-Sauvageon, C.; Delatour, P.; Benoit, E.: Chiral sulfoxidation of albendazole by the flavin adenine dinucleotide-containing and cytochrome P_{450}-dependent monooxygenases from rat liver microsomes. Drug Metab. Dispos., **23**, 160-165. (1995)

4-Hydroxybenzoate 3-monooxygenase [NAD(P)H] 1.14.13.33

1 Nomenclature

EC number
1.14.13.33

Systematic name
4-hydroxybenzoate,NAD(P)H:oxygen oxidoreductase (3-hydroxylating)

Recommended name
4-hydroxybenzoate 3-monooxygenase [NAD(P)H]

Synonyms
4-hydroxybenzoate 3-hydroxylase
4-hydroxybenzoate 3-monooxygenase (NAD(P)H)
4-hydroxybenzoate 3-monooxygenase (NAD(P)H_2)
4-hydroxybenzoate-3-hydroxylase
oxygenase, 4-hydroxybenzoate 3-mono- (reduced nicotinamide adenine dinucleotide (phosphate))
Additional information (cf. EC 1.14.13.2)

CAS registry number
95471-33-3

2 Source Organism

<-1> no activity in *Ralstonia paucula* [2]
<1> *Corynebacterium cyclohexanicum* [1]
<2> *Burkholderia cepacia* (strain PB4 [2]) [2]

3 Reaction and Specificity

Catalyzed reaction
4-hydroxybenzoate + NAD(P)H + H^+ + O_2 = 3,4-dihydroxybenzoate + $NAD(P)^+$ + H_2O

Reaction type
oxidation
redox reaction
reduction

Substrates and products

S 4-hydroxybenzoate + NAD(P)H + O_2 <1, 2> (<1> enzyme is highly specific for 4-hydroxybenzoate, but uses NADH and NADPH at approximately equal rates [1]; <2> after growth on 4-aminobenzoate [2]) (Reversibility: ? <1, 2> [1, 2]) [1, 2]

P 3,4-dihydroxybenzoate + $NAD(P)^+$ + H_2O

Inhibitors

2,4-dihydroxybenzoate <1> (<1> over 70% inhibition at 10 mM [1]) [1]
2,5-dihydroxybenzoate <1> (<1> over 70% inhibition at 10 mM [1]) [1]
2-hydroxybenzoate <1> (<1> over 70% inhibition at 10 mM [1]) [1]
3-hydroxybenzoate <1> [1]
4-aminobenzoate <1> (<1> over 70% inhibition at 10 mM [1]) [1]
4-fluorobenzoate <1> [1]
Cl^- <1> (<1> non-competitive [1]) [1]

Cofactors/prosthetic groups

FAD <1> (<1> flavoprotein, 0.8 FAD per enzyme molecule [1]) [1]
NADH <1> [1]
NADPH <1> [1]

Metals, ions

Mg^{2+} <1> (<1> required for full activity in low concentrations of phosphate buffer, most effective at 5 mM [1]) [1]

Specific activity (U/mg)

0.12 <2> (<2> crude cell extract [2]) [2]
30.5 <1> (<1> NADPH [1]) [1]

K_m-Value (mM)

0.035 <1> (4-hydroxybenzoate) [1]
0.045 <1> (NADH, <1> pH 7.0-8.4 [1]) [1]
0.063-0.17 <1> (NADPH, <1> K_m increases as pH rises from 7.0 to 8.4 [1]) [1]

K_i-Value (mM)

0.029-0.045 <1> (Cl^-) [1]

pH-Optimum

7.1 <1> (<1> NADPH [1]) [1]
7.8 <1> (<1> NADH [1]) [1]

pH-Range

6.2-8.8 <1> (<1> pH 6.2: about 50% of activity maximum, pH 8.8: about 70% of activity maximum, NADPH [1]) [1]
6.5-8.8 <1> (<1> 50% of activity maximum at pH 6.5 and pH 8.8, NADH [1]) [1]

Temperature optimum (°C)

50 <1> [1]

4 Enzyme Structure

Molecular weight

47000 <1> (<1> PAGE of native and denatured enzyme [1]) [1]

Subunits

monomer <1> (<1> 1 * 47000, PAGE of native and denatured enzyme [1]) [1]

5 Isolation/Preparation/Mutation/Application

Localization

soluble <1> [1]

Purification

<1> [1]

6 Stability

pH-Stability

4.3-8.5 <1> (<1> 4°C, 24 h, stable [1]) [1]

Temperature stability

60 <1> (<1> 10 min, stable [1]) [1]

65 <1> (<1> rapid inactivation above, *p*-hydroxybenzoate protects against heat inactivation [1]) [1]

Storage stability

<1>, -20°C, several months [1]

References

[1] Fujii, T.; Kaneda, T.: Purification and properties of NADH/NADPH-dependent *p*-hydroxybenzoate hydroxylase from Corynebacterium cyclohexanicum. Eur. J. Biochem., **147**, 97-104 (1985)

[2] Peres, C.M.; Russ, R.; Lenke, H.; Agathos, S.N.: Biodegradation of 4-nitrobenzoate, 4-aminobenzoate and their mixtures: new strains, unusual metabolites and insights into pathway regulation. FEMS Microbiol. Ecol., **37**, 151-159 (2001)

Leukotriene-E_4 20-monooxygenase 1.14.13.34

1 Nomenclature

EC number
1.14.13.34

Systematic name
(7E,9E,11Z,14Z)-(5S,6R)-6-(cystein-S-yl)-5-hydroxyicosa-7,9,11,14-tetraenoate,NADPH:oxygen oxidoreductase (20-hydroxylating)

Recommended name
leukotriene-E_4 20-monooxygenase

Synonyms
leukotriene-E_4 ω-hydroxylase
oxygenase, leukotriene E4 20-mono

CAS registry number
111940-51-3

2 Source Organism

<1> *Rattus norvegicus* (male Sprague Dawley rats [1, 2]; phenobarbital-treated [2]) [1, 2]

3 Reaction and Specificity

Catalyzed reaction
(7E,9E,11Z,14Z)-(5S,6R)-6-(cystein-S-yl)-5-hydroxyicosa-7,9,11,14-tetraenoate + NADPH + H^+ + O_2 = 20-hydroxy-leukotriene E_4 + $NADP^+$ + H_2O

Reaction type
oxidation
redox reaction
reduction

Natural substrates and products
S (7E,9E,11Z,14Z)-(5S,6R)-6-(cystein-S-yl)-5-hydroxyicosa-7,9,11,14-tetraenoate + NADPH + O_2 <1> (<1> metabolism of leukotriene E_4 and N-acetyl-leukotriene E_4 in rat liver microsomes [1]) (Reversibility: ? <1> [1]) [1]
P 20-hydroxy-leukotriene E_4 + $NADP^+$ + H_2O <1> [1]

Substrates and products

S (7E,9E,11Z,14Z)-(5S,6R)-6-(cystein-S-yl)-5-hydroxyicosa-7,9,11,14-tetraenoate + NADPH + O_2 <1> (<1> leukotriene E_4 [1,2]) (Reversibility: ? <1> [1, 2]) [1, 2]

P 20-hydroxy-leukotriene E_4 + NADP + H_2O <1> (<1> ω-hydroxy-leukotriene E_4 [1, 2]) [1, 2]

S N-acetyl-leukotriene E_4 + NADPH + O_2 <1> (<1> at a rate six times lower than leukotriene E_4 [1]) (Reversibility: ? <1> [1, 2]) [1, 2]

P N-acetyl-ω-hydroxy-leukotriene E_4 + NADP + H_2O <1> [1, 2]

Inhibitors

(6Z,8E,10E,14Z)-(5S,12R)-5,12-dihydroxyeicosa-6,8,10,14-tetraenoate <1> (<1> competitively inhibits leukotriene E_4 ω-oxidation [2]) [2]

halothane <1> (<1> induces a transient decrease in leukotriene ω-oxidation both in vivo and in vitro, halothane metabolism-dependent trifluoracetylation of hepatic target proteins might play a role in the inactivation of the enzyme [2]) [2]

Cofactors/prosthetic groups

NADPH <1> [1, 2]

O_2 <1> [1, 2]

Specific activity (U/mg)

Additional information <1> [1]

K_i-Value (mM)

0.08 <1> ((6Z,8E,10E,14Z)-(5S,12R)-5,12-dihydroxyeicosa-6,8,10,14-tetraenoate) [2]

Temperature optimum (°C)

37 <1> (<1> assay at [1]) [1]

5 Isolation/Preparation/Mutation/Application

Source/tissue

kidney <1> (<1> the specific activity is about 12 times lower than for liver [1]) [1]

liver <1> (<1> highest activity [1]) [1, 2]

lung <1> (<1> the specific activity is about 12 times lower than for liver [1]) [1]

Localization

microsome <1> [1, 2]

References

[1] Írning, L.: ω-Oxidation of cysteine-containing leukotrienes by rat-liver microsomes. Eur. J. Biochem., **170**, 77-85 (1987)
[2] Huwyler, J.; Jedlitschky, G.; Keppler, D.; Gut, J.: Halothane metabolism. Impairment of hepatic ω-oxidation of leukotrienes in vivo and in vitro. Eur. J. Biochem., **206**, 869-897 (1992)

Anthranilate 3-monooxygenase (deaminating) 1.14.13.35

1 Nomenclature

EC number
1.14.13.35

Systematic name
anthranilate,NADPH:oxygen oxidoreductase (3-hydroxylating, deaminating)

Recommended name
anthranilate 3-monooxygenase (deaminating)

Synonyms
EC 1.14.12.2 (formerly)
anthranilate 2,3-dioxygenase (deaminating)
anthranilate 2,3-hydroxylase (deaminating)
anthranilate hydroxylase
anthranilate hydroxylase (deaminating)
anthranilic hydroxylase

CAS registry number
37256-68-1

2 Source Organism

<1> *Trichosporon cutaneum* (yeast [1,3,5-7]) [1, 3, 5-7]
<2> *Aspergillus niger* (UBC 814 [8,11]; inducible enzyme [11]) [2, 4, 8-12]

3 Reaction and Specificity

Catalyzed reaction
anthranilate + NADPH + H^+ + O_2 = 2,3-dihydroxybenzoate + $NADP^+$ + NH_3 (<1> mechanism proposed involving imine formation and hydrolysis during the reaction with the flavin peroxide formed from reduced enzyme flavin and molecular oxygen [1,3]; <1> mechanism [5]; <2> possible involvement of superoxide anion (O_2^-) in the reaction [9])

Reaction type
deamination
hydroxylation
oxidation
redox reaction
reduction

Natural substrates and products

S anthranilate + NADPH + O_2 <1, 2> (<2> metabolism of indole [4]; <1> enzyme in degradation of L-tryptophan [7]) (Reversibility: ? <1, 2> [4, 7]) [4, 7]

P 2,3-dihydroxybenzoate + $NADP^+$ + NH_3 <1, 2> [4, 7]

Substrates and products

S 2-hydrazinebenzoate + NADH + O_2 <1> (<2> not [12]) (Reversibility: ? <1> [1]) [1]

P ?

S 2-thiobenzoate + NADPH + O_2 <1> (Reversibility: ? <1> [3]) [3]

P ?

S 3-hydroxyanthranilate + NADPH + O_2 <1> (<2> no activity [10]) (Reversibility: ? <1> [1]) [1]

P ?

S 3-methylanthranilate + NADPH + O_2 <1> (<2> no activity [10]) (Reversibility: ? <1> [1, 3]) [1, 3]

P ? + $NADP^+$ + NH_3 <1> (<1> the proposed structure of the product contains a ketone function at the position 2 [3]) [3]

S 4-fluoroanthranilate + NADPH + O_2 <1> (Reversibility: ? <1> [3]) [3]

P 4-fluoro-2,3-dihydroxybenzoate + $NADP^+$ + NH_3 <1> [3]

S 5-fluoroanthranilate + NADPH + O_2 <1> (Reversibility: ? <1> [3]) [3]

P 5-fluoro-2,3-dihydroxybenzoate + $NADP^+$ + NH_3 <1> [3]

S N,N-dimethylanthranilate + NADPH + O_2 <1> (Reversibility: ? <1> [1]) [1]

P ?

S N-methylanthranilate + NADPH + O_2 <1> (<2> not [12]) (Reversibility: ? <1> [1]) [1]

P 2,3-dihydroxybenzoate + $NADP^+$ + NH_3 <1> [1]

S anthranilate + NADPH + O_2 <1, 2> (Reversibility: ? <1, 2> [1-4, 6-12]) [1-4, 6-12]

P 2,3-dihydroxybenzoate + $NADP^+$ + NH_3 <1, 2> (<1> 2,3-dihydroxybenzoate i.e. *o*-pyrocatechuate, oxygen atom at the 3-position of the product 2,3-dihydroxybenzoate originates from O_2, that at the 2-position is derived from H_2O [1]) [1-4, 6-12]

S Additional information <2> (<2> no activity with: 3-hydroxyanthranilic acid, benzoic acid, salicylic acid, *m*-hydroxybenzoic acid, *p*-hydroxybenzoic acid, *m*-aminobenzoic acid, *p*-aminobenzoic acid, methylanthranilate or ethylanthranilate [10,12]) [10, 12]

P ?

Inhibitors

1,10-phenanthroline <2> (<2> 50% inhibition at 0.25 mM, 80% inhibition at 0.5 mM and 95% inhibition at 1 mM [2]; <2> 46% inhibition at 0.5 mM , 60% inhibition at 1 mM, 80% inhibition at 3 mM, anthranilic acid protects the enzyme from inhibition [8]; 46% inhibition at 0.5 mM, 60% inhibition at 1 mM [12]) [2, 8, 10, 12]

2,2'-dipyridyl <2> (<2> 50% inhibition at 0.5 mM, 80% inhibition at 1 mM [2]; <2> 27% inhibition at 0.5 mM, 41% inhibition at 1 mM [8]; <2> 40% inhibition at 1 mM [12]) [2, 8, 10, 12]
3-hydroxyanthranilic acid <2> [12]
3-hydroxybenzoic acid <2> [12]
8-hydroxyquinoline <2> (<2> 30% inhibition at 0.5 mM, 40% inhibition at 1 mM [2]; <2> 23% inhibition at 0.5 mM, 37% inhibition at 1 mM [8]; <2> 37% inhibition at 1 mM [12]) [2, 8, 12]
Cd^{2+} <2> [12]
Cu^{2+} <2> [12]
EDTA <2> (<2> 7% inhibition at 1 mM [8]) [8]
Hg^{2+} <2> [12]
N-ethylmaleimide <2> (<2> 95% inhibition at 0.5 mM [12]) [10, 12]
NaN_3 <2> (<2> 42% inhibition at 0.5 mM [12]) [10, 12]
cycloheximide <2> (<2> inhibits the enzyme induction [11]) [11]
diethyldithiocarbamate <2> [10, 12]
heavy metal ions <2> [10, 12]
p-chloromercuribenzoate <2> [10]
p-hydroxymercuribenzoate <2> (<2> 96% inhibition at 0.5 mM [12]) [12]
salicylaldoxime <2> (<2> 40% inhibition at 0.5 mM, 52% inhibition at 1 mM [8]; <2> 42% inhibition at 0.5 mM, 58% inhibition at 1 mM [12]) [8, 10, 12]
Additional information <2> (<2> not: atebrin, aminopterin [2,12]; <2> reversal of 1,10-phenanthroline inhibition by ferric-EDTA, ferrous-EDTA, ferric citrate and cytochrome c [8]; <2> complete inhibition when superoxide dismutase is included in the reaction mixture for the assay of the enzyme [9]) [2, 8, 9, 12]

Cofactors/prosthetic groups

FAD <1, 2> (<1,2> flavoprotein [1-3,5]; <1> 2 mol FAD per mol enzyme [1]) [1-3, 5]
NADPH <1, 2> (<1> enzyme uses the re-face of the flavin ring [6]; <1,2> absolute requirement for NADPH [7,10,12]) [1-12]
Additional information <2> (<2> 2% of the activity when NADPH is replaced by NADH, no effect: FAD, FMN [10,12]) [10, 12]

Activating compounds

2-aminonicotinate <1> (<1> activates [1]) [1]
3-hydroxyanthranilate <2> (<2> induces [11]) [11]
L-tryptophan <1, 2> (<1,2> induces [7,11]) [7, 11]
N-formyl anthranilate <1> (<1> activates [1]) [1]
anthranilate <2> (<2> induces, the enzyme activity increases with an increase in the concentration of anthranilate in the growth medium, optimal amounts of the enzyme are synthesized when the concentration of anthranilate is 1 mg/ml, further increase in the concentration results in a considerable decrease in the growth of the organism as well in the enzyme activity [11]) [11]
indole <2> (<2> 0.02% induces the enzyme 516-fold [4]) [4]

kynurenine <2> (<2> induces [11]) [11]
salicylate <1> (<1> activates [1,5]) [1, 5]
salicylate <1> (<1> induces [7]) [7]

Metals, ions

iron <2> (<2> Fe^{2+} probably required for activity [12]; <2> contains 2 gatom of non-heme iron per mol [2]; <2> enzyme-bound iron, participation of Fe in reaction, omission of Fe in the growth medium yields inactive preparation [8]; <2> the partially purified enzyme is not activated by any metal ion but a considerable decrease in anthranilate hydroxylase activity occurs when the organism is grown on a medium deprived of iron [8]) [2, 8]

Turnover number (min^{-1})

Additional information <1> [5]

Specific activity (U/mg)

0.0026 <2> (<2> organism grown on the standard medium with kynurenine [11]) [11]
0.0032 <2> (<2> organism grown on the standard medium with tryptophan [11]) [11]
0.0056 <2> (<2> organism grown on the standard medium with 3-hydroxyanthranilate [11]) [11]
0.0058 <2> (<2> organism grown on the standard medium with anthranilate [11]) [11]
0.041 <1> (<1> extracts of tryptophan-grown cells [7]) [7]
0.05 <2> (<2> in cell extracts grown on glucose [4]) [4]
0.0507 <2> [12]
0.0507 <2> (<2> partially purified enzyme [12]) [12]
0.193 <2> [2]
0.28 <2> (<2> organism grown on iron-deficient medium, no addition of other constituents [8]) [8]
1.02 <1> (<1> partially purified enzyme [7]) [7]
2.2 <2> (<2> organism grown on iron-deficient medium, addition of 1 mM ferric citrate, preincubation time: 5 min [8]) [8]
4.2 <2> (<2> organism grown on iron-deficient medium, addition of 1 mM ferric citrate, preincubation time: 10 min [8]) [8]
4.4 <1> [1]
4.6 <2> (<2> organism grown on iron-deficient medium, addition of 1 mM ferric-EDTA [8]) [8]
4.9 <2> (<2> organism grown on iron-deficient medium, addition of 1 mM ferric citrate, preincubation time: 15, 20 or 30 min [8]) [8]
5.6 <2> (<2> organism grown on standard iron-sufficient medium [8]) [8]
25.8 <2> (<2> in cell extracts grown on glucose plus indole [4]) [4]
Additional information <2> [10]

K_m-Value (mM)

0.15 <2> (anthranilic acid) [10, 12]
0.16 <2> (NADPH) [10, 12]
Additional information <1> [5]

pH-Optimum

6 <2> (<2> for optimal induction [11]) [11]
7.7 <1> [1]
8 <2> (<2> assay at [4]) [4]
8.2 <2> (<2> assay at [8,11]) [8, 10-12]

pH-Range

5.5-9.8 <1> (<1> less than 1% of maximal activity at pH 5.5 and pH 9.8 [1]) [1]
7.5-9.5 <2> (<2> at pH 7.5 or pH 9.0 about 60% of activity maximum [10]; <2> pH 7.5: about 80% of activity maximum, pH 9.5: about 55% of activity maximum, [12]) [10, 12]

Temperature optimum (°C)

25 <1> (<1> assay at [1]) [1]

4 Enzyme Structure

Molecular weight

89000 <2> (<2> gel filtration [2]) [2]
95000 <1> (<1> gel filtration [1]) [1]

Subunits

dimer <1, 2> (<1> 2 * 50000, SDS-PAGE [1]; <2> 2 * 42000, SDS-PAGE [2]) [1, 2]

5 Isolation/Preparation/Mutation/Application

Source/tissue

mycelium <2> [10, 12]

Purification

<1> (using ammonium sulfate precipitation, column chromatography on DE23-cellulose, phenyl-Sepharose and S-300 Sephacryl [1]; partial, using heat treatment, DEAE-cellulose column chromatography and elution with a linear gradient of 0 to 0.1 M KCl in phosphate buffer [7]) [1, 7]
<2> (using protamine sulfate treatment, DEAE-cellulose treatment, ammonium sulfate precipitation, fractionation on Biogel P-100 column, successive negative adsorption on alumina-gel, tricalcium phosphate gel and DEAE-cellulose column, and positive adsorption on a DEAE-Sephadex A-50 column [2]; partial, using centrifugation, protamine sulfate treatment, treatment with diethylaminoethyl-cellulose and filtration through a Buchner funnel [8]; using protamine sulfate treatment, DEAE-cellulose treatment and alumina C-γ treatment [10]; partial, using protamine sulfate treatment, DEAE-cellulose treatment, alumina C-γ treatment and hydroxylapatite treatment [12]) [2, 8, 10, 12]

6 Stability

pH-Stability

5.9-9.1 <1> (<1> 4°C, stable for at least 3 days [1]) [1]

Temperature stability

4 <2> (<2> half-life: 14 h [2]) [2]

Organic solvent stability

acetone <2> (inactivates the enzyme [10,12]) [10, 12]

General stability information

<2>, glutathione stabilizes [10, 12]

<2>, the enzyme is inactivated by treatment with ammonium sulfate or by adsorption on DEAE-cellulose or CM-cellulose, filtration through Sephadex G-25 or dialysis against 0.025 M sodium phosphate buffer, pH 7, containing 1 mM GSH irreversibly inactivates the enzyme [10, 12]

Storage stability

<1>, -20°C, in the dark, Na^+/K^+ phosphate buffer, pH 7.4, 0.1 mM EDTA, stable for at least 6 months [1]

<2>, -20°C, 50% inactivation of the purified enzyme after 2 days [10, 12]

<2>, -20°C, ammonium sulfate precipitate, stable for at least 3 months [2]

<2>, frozen, mycelium, no appreciable loss of activity after 1 week [10]

References

[1] Powlowski, J.B.; Dagley, S.; Massey, V.; Ballou, D.P.: Properties of anthranilate hydroxylase (deaminating), a flavoprotein from Trichosporon cutaneum. J. Biol. Chem., **262**, 69-74 (1987)

[2] Subramanian, V.; Vaidyanathan, C.S.: Anthranilate hydroxylase from Aspergillus niger: new type of NADPH-linked nonheme iron monooxygenase. J. Bacteriol., **160**, 651-655 (1984)

[3] Powlowski, J.; Ballou, D.P.; Massey, V.: Studies of the oxidative half-reaction of anthranilate hydroxylase (deaminating) with native and modified substrates. J. Biol. Chem., **265**, 4969-4975 (1990)

[4] Kamath, A.V.; Vaidyanathan, C.S.: New pathway for the biodegradation of indole in Aspergillus niger. Appl. Environ. Microbiol., **56**, 275-280 (1990)

[5] Powlowski, J.; Massey, V.; Ballou, D.P.: Reactions of anthranilate hydroxylase with salicylate, a nonhydroxylated substrate analogue. Steady state and rapid reaction kinetics. J. Biol. Chem., **264**, 5606-5612 (1989)

[6] Manstein, D.J.; Pai, E.F.: Absolute stereochemistry of flavins in enzyme-catalyzed reactions. Biochemistry, **25**, 6807-6816 (1986)

[7] Anderson, J.J.; Dagley, S.: Catabolism of tryptophan, anthranilate, and 2,3-dihydroxybenzoate in Trichosporon cutaneum. J. Bacteriol., **146**, 291-297 (1981)

[8] Kumar, R.P.; Sreeleela, N.S.; Rao, P.V.S.; Vaidyanathan, C.S.: Anthranilate hydroxylase from Aspergillus niger: evidence for the participation of iron in the double hydroxylation reaction. J. Bacteriol., **113**, 1213-1216 (1973)
[9] Kumar, R.P.; Ravindranath, S.D.; Vaidyanathan, C.S.; Rao, N.A.: Mechanism of hydroxylation of aromatic compounds. II. Evidence for the involvement of superoxide anions in enzymatic hydroxylations. Biochem. Biophys. Res. Commun., **49**, 1422-1426 (1972)
[10] Rao, P.V.S.; Sreeleela, N.S.; Kumar, R.P.; Vaidyanathan, C.S.: Anthranilic acid hydroxylase (Aspergillus niger). Methods Enzymol., **17A**, 510-513 (1970)
[11] Rao, P.V.S.; Sreeleela, N.S.; Premakumar, R.; Vaidyanathan, C.S.: Regulation of the pathway for the degradation of anthranilate in Aspergillus niger. J. Bacteriol., **107**, 100-105 (1971)
[12] Sreeleela, N.S.; SubraRao, P.V.; Premkumar, R.; Vaidyanathan, C.S.: A new anthranilic acid hydroxylase from Aspergillus niger. Purification and properties. J. Biol. Chem., **244**, 2293-2298 (1969)

5-O-(4-Coumaroyl)-D-quinate 3'-monooxygenase 1.14.13.36

1 Nomenclature

EC number
1.14.13.36

Systematic name
trans-5-O-(4-coumaroyl)-D-quinate,NADPH:oxygen oxidoreductase (3'-hydroxylating)

Recommended name
5-O-(4-coumaroyl)-D-quinate 3'-monooxygenase

Synonyms
5-O-(4-coumaroyl)-D-quinate/shikimate 3'-hydroxylase
coumaroylquinate (coumaroylshikimate) 3'-monooxygenase

CAS registry number
112131-08-5

2 Source Organism

<1> *Daucus carota* [1]

3 Reaction and Specificity

Catalyzed reaction
trans-5-O-(4-coumaroyl)-D-quinate + NADPH + H^+ + O_2 = trans-5-O-caffeoyl-D-quinate + $NADP^+$ + H_2O (Also acts on trans-5-O-(4-coumaroyl)shikimate)

Reaction type
oxidation
redox reaction
reduction

Natural substrates and products
S trans-5-O-(4-coumaroyl)-D-quinate + NADPH + O_2 <1> (<1> specific for the trans isomer, final step in the chlorogenic acid pathway, reaction is greatly enhanced by irradiation with blue/uv light [1]) (Reversibility: ? <1> [1]) [1]
P trans-5-O-caffeoyl-D-quinate + $NADP^+$ + H_2O <1> [1]

S trans-5-O-(4-coumaroyl)shikimate + NADPH + O_2 <1> (<1> 70% of the activity compared to trans-5-O-(4-coumaroyl)-D-quinate, specific for the trans isomer [1]) (Reversibility: ? <1> [1]) [1]
P trans-5-O-caffeoylshikimate + $NADP^+$ + H_2O <1> [1]

Substrates and products
S trans-5-O-(4-coumaroyl)-D-quinate + NADPH + O_2 <1> (<1> specific for the trans isomer, final step in the chlorogenic acid pathway, reaction is greatly enhanced by irradiation with blue/uv light [1]) (Reversibility: ? <1> [1]) [1]
P trans-5-O-caffeoyl-D-quinate + $NADP^+$ + H_2O <1> [1]
S trans-5-O-(4-coumaroyl)shikimate + NADPH + O_2 <1> (<1> 70% of the activity compared to trans-5-O-(4-coumaroyl)-D-quinate, specific for the trans isomer [1]) (Reversibility: ? <1> [1]) [1]
P trans-5-O-caffeoylshikimate + $NADP^+$ + H_2O <1> [1]

Inhibitors
cytochrome c <1> (<1> 76% inhibition at 0.1 mM [1]) [1]
diethyldicarbonate <1> (<1> 31% inhibition at 2 mM [1]) [1]
p-chloromercuribenzoate <1> (<1> 32% inhibition at 0.5 mM [1]) [1]
tetcyclacis <1> (<1> 38% inhibition at 0.01 mM [1]) [1]

Cofactors/prosthetic groups
NADPH <1> (<1> strictly specific [1]) [1]

Activating compounds
KCN <1> (<1> 2fold activation at 1-10 mM [1]) [1]
diethyldithiocarbamate <1> (<1> 2fold activation at 0.2-2 mM [1]) [1]
light exposure <1> (<1> 4fold activation after 20 h irradiation at 350 nm [1]) [1]

pH-Optimum
7.5 <1> [1]

5 Isolation/Preparation/Mutation/Application

Source/tissue
cell culture <1> [1]

Localization
microsome <1> [1]

6 Stability

pH-Stability
6.3 <1> (<1> half-maximal activity [1]) [1]
8.6 <1> (<1> half-maximal activity [1]) [1]

Temperature stability

30 <1> (<1> half-life: 10 min [1]) [1]

Storage stability

<1>, -70°C, stable for several months [1]

References

[1] Kuhnl, T.; Koch, U.; Heller, W.; Wellmann, E.: Chlorogenic acid biosynthesis: Characterization of a light-induced microsomal 5-O-(4-coumaroyl)-D-quinate/shikimate 3'-hydroxylase from carrot (Daucus carota L.) cell suspension cultures. Arch. Biochem. Biophys., **258**, 226-232 (1987)

Methyltetrahydroprotoberberine 14-monooxygenase 1.14.13.37

1 Nomenclature

EC number
1.14.13.37

Systematic name
(S)-N-methylcanadine,NADPH:oxygen oxidoreductase (14-hydroxylating)

Recommended name
methyltetrahydroprotoberberine 14-monooxygenase

Synonyms
(S)-cis-N-methyltetrahydroprotoberberine-14-hydroxylase
methyltetrahydroprotoberberine 14-hydroxylase
oxygenase, (S)-cis-N-methyltetrahydroberberine 14-mono-

CAS registry number
113478-42-5

2 Source Organism

<1> *Corydalis vaginans* (Papaveraceae [1]) [1]

3 Reaction and Specificity

Catalyzed reaction
(S)-N-methylcanadine + NADPH + H^+ + O_2 = allocryptopine + $NADP^+$ + H_2O (<1> stereo- and regiospecificity [1])

Reaction type
oxidation
redox reaction
reduction

Natural substrates and products
S (S)-cis-N-methylcanadine + NADPH + O_2 <1> (<1> (R)-cis/trans-N-methyl-canadine is no substrate [1]; <1> hydroxylation in benzylisoquinoline metabolism [1]) (Reversibility: ? <1> [1]) [1]
P allocryptopine + $NADP^+$ + H_2O <1> [1]

Substrates and products

S (S)-cis-N-methylcanadine + NADPH + O_2 <1> (<1> (R)-cis/trans-N-methyl-canadine is no substrate [1]; <1> no activity without molecular oxygen [1]) (Reversibility: ? <1> [1]) [1]
P allocryptopine + $NADP^+$ + H_2O <1> [1]
S (S)-cis-N-methylcorydalmine + NADPH + O_2 <1> (Reversibility: ? <1> [1]) [1]
P muramine + $NADP^+$ + H_2O
S (S)-cis-N-methylstylopine + NADPH + O_2 <1> (Reversibility: ? <1> [1]) [1]
P protopine + $NADP^+$ + H_2O <1> [1]
S (S)-cis-N-methyltetrahydropalmitine + NADPH + O_2 <1> (Reversibility: ? <1> [1]) [1]
P muramine + $NADP^+$ + H_2O
S (S)-cis-N-methyltetrahydrothalifendine + NADPH + O_2 <1> (Reversibility: ? <1> [1]) [1]
P thalictrisine + $NADP^+$ + H_2O

Inhibitors

CO <1> (<1> reversible by light [1]) [1]
ancymidol <1> [1]
cytochrome c <1> [1]
ketoconazole <1> [1]
metyrapone <1> [1]
prochloraz <1> [1]

Cofactors/prosthetic groups

NADH <1> (<1> 33% of the activity with NADPH, NAD^+ or $NADP^+$ are ineffective [1]) [1]
NADPH <1> [1]
cytochrome P_{450} <1> [1]

Specific activity (U/mg)

0.0255 <1> [1]

K_m-Value (mM)

0.0125 <1> (cis-N-methylcanadine) [1]
0.0625 <1> (NADPH) [1]

pH-Optimum

8.5 <1> [1]

pH-Range

6.5-9.7 <1> (<1> half-maximal activity at pH 6.5 and pH 9.7 [1]) [1]

Temperature optimum (°C)

30 <1> [1]

Temperature range (°C)

10-45 <1> (<1> half-maximal activity at 10°C and 45°C [1]) [1]

5 Isolation/Preparation/Mutation/Application

Source/tissue
cell suspension culture <1> [1]

Localization
microsome <1> (<1> membrane bound [1]) [1]

References

[1] Rueffer, M.; Zenk, M.H.: Enzymatic formation of protopines by a microsomal cytochrome P-450 system of Corydalis vaginans. Tetrahedron Lett., **28**, 5307-5310 (1987)

Anhydrotetracycline monooxygenase 1.14.13.38

1 Nomenclature

EC number
1.14.13.38

Systematic name
anhydrotetracycline,NADPH:oxygen oxidoreductase (6-hydroxylating)

Recommended name
anhydrotetracycline monooxygenase

Synonyms
ATC oxygenase
anhydrotetracycline oxygenase
oxygenase, anhydrotetracycline

CAS registry number
70766-62-0

2 Source Organism

<1> *Streptomyces aureofaciens* (50/137, UV-light induced mutant of strain 84/25 [1-4]; high- and low-production strain [5]) [1-7]

3 Reaction and Specificity

Catalyzed reaction
anhydrotetracycline + NADPH + H^+ + O_2 = 12-dehydrotetracycline + $NADP^+$ + H_2O

Reaction type
oxidation
redox reaction
reduction

Natural substrates and products
S anhydrotetracycline + NADPH + O_2 <1> (<1> tetracycline and chlortetracycline biosynthetic pathways [1]; <1> involved in the biosynthesis of tetracyclines [3]; <1> chlortetracycline biosynthetic pathway [6]) (Reversibility: ? <1> [1, 3, 6]) [1, 3, 6]
P 12-dehydrotetracycline + $NADP^+$ + H_2O

Substrates and products

S anhydrooxytetracycline + NADPH + O_2 <1> (Reversibility: ? <1> [2]) [2]
P ?
S anhydrotetracycline + NADPH + O_2 <1> (Reversibility: ? <1> [1-4]) [1-4]
P 12-dehydrotetracycline + $NADP^+$ + H_2O <1> [2-4]

Inhibitors

2,5-dithiobis(2-nitrobenzoic acid) <1> (<1> 40% inhibition at 1 mM, irreversible [1]) [1]
Ca^{2+} <1> (<1> above 2 mM [2]) [2]
Co^{2+} <1> [2]
Cu^{2+} <1> [2]
Fe^{3+} <1> [2]
Mg^{2+} <1> (<1> above 1 mM [2]) [2]
Mn^{2+} <1> [2]
N-ethylmaleimide <1> (<1> complete inhibition at 0.1 M, irreversible [1]) [1]
Ni^{2+} <1> [2]
ammonium sulfate <1> [2]
chlortetracycline <1> [2]
iodoacetamide <1> (<1> 14% inhibition at 1 mM, irreversible [1]) [1]
oxytetracycline <1> [2]
tetracycline <1> [2]

Cofactors/prosthetic groups

NADPH <1> [1-4]
ascorbate <1> (<1> increase of activity [1]) [1]

Activating compounds

Additional information <1> (<1> addition of supernatant of Streptomyces aureofaciens or Streptomyces rimosus crude extract after boiling and centrifugation increases activity 7-fold [1]) [1]

Metals, ions

Ca^{2+} <1> (<1> increase of activity at 0.1-2 mM [2]) [2]
Co^{2+} <1> (<1> increase of activity [2]) [2]
Cu^{2+} <1> (<1> increase of activity [2]) [2]
Fe^{2+} <1> (<1> increase of activity [1,2]) [1, 2]
Fe^{3+} <1> (<1> increase of activity [2]) [2]
Mg^{2+} <1> (<1> increase of activity at 0.1-1 mM [2]) [2]
Mn^{2+} <1> (<1> increase of activity [2]) [2]
Ni^{2+} <1> (<1> increase of activity [2]) [2]

Specific activity (U/mg)

0.00264 <1> (<1> low production strain, membrane fraction [5]) [5]
0.0078 <1> (<1> high production strain, cytoplasm [5]) [5]
0.44 <1> [3]
Additional information <1> (<1> overview: different activities in various subcellular fractions [5]) [5]

K_m-Value (mM)
0.022 <1> (anhydrotetracycline) [2]

pH-Optimum
7.3-7.4 <1> (<1> assay at [1,3,4]) [1, 3, 4]

Temperature optimum (°C)
28 <1> (<1> assay at [3]) [3]
29-30 <1> (<1> assay at [4]) [4]

4 Enzyme Structure

Molecular weight
115000 <1> (<1> gel filtration [1]) [1]

Subunits
dimer <1> (<1> 2 * 57500, SDS-PAGE [1,3]) [1, 3]

5 Isolation/Preparation/Mutation/Application

Localization
cytoplasm <1> (<1> predominantly, high-production strain, distribution in the cell is influenced by benzylthiocyanate [5]) [1-3, 5]
membrane <1> (<1> predominantly, low-production strain [5]) [5]

Purification
<1> (partial [2]; homogeneity [3]) [1-3]

Application
pharmacology <1> (<1> synthesis of chlortetracycline [6]) [6]

6 Stability

General stability information
<1>, loss of activity during purification [1]

Storage stability
<1>, -25°C, crude extract stable for several weeks [1]
<1>, 4°C, crude extract stable for several days [1]

References

[1] Vancurova, I.; Flieger, M.; Volc, J.; Benes, M.J.; Novotna, J.; Neuzil, J.; Behal, V.: Partial purification and characterization of anhydrotetracycline oxygenase of Streptomyces aureofaciens. J. Basic Microbiol., **27**, 529-533 (1987)

[2] Behal, V.; Neuzil, J.; Hostalek, Z.: Effect of tetracycline derivates and some cations on the activity of anhydrotetracycline oxygenase. Biotechnol. Lett., **5**, 537-542 (1983)
[3] Vancurova, I.; Volc, J.; Flieger, M.; Neuzil, J.; Novotna, J.; Vlach, J.; Behal, V.: Isolation of pure anhydrotetracycline oxygenase from Streptomyces aureofaciens. Biochem. J., **253**, 263-267 (1988)
[4] Neuzil, J.; Novotna, J.; Vancurova, I.; Behal, V.; Hostalek, Z.: A direct-injection reversed-phase liquid chromatographic micromethod for studying the kinetics of terminal reactions of tetracycline biosynthesis. Anal. Biochem., **181**, 125-129 (1989)
[5] Erban, V.; Trilisenko, L.V.; Novotna, J.; Behal, V.; Kulaev, I.S.; Hostalek, Z.: Subcellular localization if enzymes in Streptomyces areofaciens and its alteration by benzyl thiocyanate. Folia Microbiol., **32**, 411-416 (1987)
[6] Behal, V.; Gregrova-Prusakova, J.; Hostalek, Z.: Effect of inorganic phosphate and benzyl thiocyanate of the activity of anhydrotetracycline oxygenase in Streptomyces aureofaciens. Folia Microbiol., **27**, 102-106 (1982)
[7] Li, X.M.; Novotna, J.; Vohradsky, J.; Weiser, J.: Major proteins related to chlortetracycline biosynthesis in a Streptomyces aureofaciens production strain studied by quantitative proteomics. Appl. Microbiol. Biotechnol., **57**, 717-724 (2001)

Nitric-oxide synthase 1.14.13.39

1 Nomenclature

EC number
1.14.13.39

Systematic name
L-arginine,NADPH:oxygen oxidoreductase (nitric-oxide-forming)

Recommended name
nitric-oxide synthase

Synonyms
EC-NOS <2-4> (<2-4> isoform III, in endothelial cells [37]) [37]
NADPH-diaphorase
NO synthase
cb-NOS <1, 3, 5> (<1,3,5> isoform I, constitutive, from brain [37]) [37]
e-NOS <3, 4> (<3,4> endothelial isoform III [47,51]) [47, 51]
endothelium-derived relaxation factor-forming enzyme
endothelium-derived relaxing factor synthase
i-NOS <1-3> (<1-3> isoform II, inducible [37,47,49]) [37, 47, 49]
n-NOS <1, 3, 5> (<1,3,5> isoform I, neuronal enzyme [37,47]) [37, 47]
nitric oxide synthase
nitric oxide synthetase
synthetase, nitric oxide

CAS registry number
125978-95-2

2 Source Organism

<1> *Rattus norvegicus* (Wistar strain [1, 13, 43]; Sprague-Dawley [7, 8, 14, 39, 41, 47]; 2 isoforms [1, 8, 13]) [1-3, 7-9, 13, 14, 18, 19, 21, 23, 24, 31, 37, 39, 41, 43, 47]
<2> *Mus musculus* (strain C57BL/6 [50]; cytokine-inducible in macrophage [11, 17, 30, 32, 34, 49]; 3 isoforms exist in mouse macrophages differentiated by substrate specificity and Ca^{2+}-dependence [15]) [3, 4, 6, 9, 11, 12, 15, 17, 20, 28, 30, 32-34, 36, 37, 45, 47, 49, 50]
<3> *Homo sapiens* (inducible [30, 48]) [5, 16, 26, 27, 29, 30, 37, 44, 47, 48]
<4> *Bos taurus* (endothelial isoform [51]; 3 isoforms exist in bovine aortic endothelial cells differentiated by substrate specificity and Ca^{2+}-dependence [15]) [14, 15, 25, 37, 38, 40, 46, 51]

<5> *Sus scrofa* [10, 22, 37]
<6> *Felis catus* (mongrel cat [35]) [35]
<7> *Ovis aries* [42]

3 Reaction and Specificity

Catalyzed reaction

L-arginine + n NADPH + n H^+ + m O_2 = citrulline + nitric oxide + n $NADP^+$ (The enzyme in brain, but not that induced in lung or liver by endotoxin, requires Ca^{2+}. The stoichiometry is not clear, but may involve a two-electron and a one-electron oxidation step; <1,2> possible mechanism [9]; <1,2> cytochrome P-450 enzyme [3,9]; <2> alignment of heme-binding domain amino acid sequences of NOS [28]; <2> dimerization is required, activation of NO-synthesis by enabling electron transfer between the reductase and the oxygenase domains, isolated monomers are inactive [32]; <2> structure-function study of macrophage enzyme [34]; <2> subunits align in head-to-head manner with oxygenase domains interacting to form a dimer and reductase domains existing as independent extensions [36]; <1-5> isoform I and II are regulated by Ca^{2+}/calmodulin, isoform II is Ca^{2+}-independent, but requires calmodulin, inducible by cytokines [37]; <2> relation between structure, function and binding of prosthetic groups during dissociation, unfolding and renaturation [49])

Reaction type

oxidation
redox reaction
reduction

Natural substrates and products

S L-arginine + NADPH + O_2 <1-5> (<1-5> physiological functions and pathophysiology of the isoforms [37]; <3> a cytokine-inducible, calcium independent and a constitutive, calcium dependent form [5]; <1,2> acts as signal molecule for neurotransmission, vasorelaxation, and cytotoxity [3, 5, 11]; <2> enzyme of mammalian immune, cardiovascular and neural systems, synthesizing the free radical nitric oxide or a NO-releasing product [12]; <1, 2> soluble cytochrome P-450 enzyme in eukaryotes [9]) (Reversibility: ? <1-5> [1, 3, 5, 9, 11, 12, 30, 37, 40, 50]) [1, 3, 5, 9, 11, 12, 30, 37, 40, 50]

P citrulline + NO + $NADP^+$ <3, 4> [5, 40, 50]

S Additional information <1, 3-5> (<4> enzyme shows also superoxide formation activity [40]; <1, 3, 5> NO represents the endogenous activator of soluble guanylyl cyclase [7, 10, 21, 26]) [7, 10, 21, 26, 40]

P ?

Substrates and products

S L-Ala-L-Arg + NADPH + O_2 <2, 4> (<2, 4> endothelial microsomes, macrophage [15]) (Reversibility: ? <2, 4> [15]) [15]

P ?
S L-Arg-L-Arg + NADPH + O_2 <2, 4> (<2,4> endothelial, microsomes, macrophage [15]) (Reversibility: ? <2, 4> [15]) [15]
P ?
S L-Arg-L-Arg-L-Arg + NADPH + O_2 <2, 4> (<2,4> endothelial microsomes [15]) (Reversibility: ? <2, 4> [15]) [15]
P ?
S L-Arg-L-Phe + NADPH + O_2 <2, 4> (Reversibility: ? <2, 4> [15]) [15]
P ?
S L-arginine + NADPH + O_2 <1-7> (<2> capacity to synthesize NO only through dimerization and binding of heme and tetrahydrobiopterin [34]; <2> tetrahydrobiopterin is absolutely required for partial reaction 1 [33]; <2> dimeric structure is required for enzyme activity [30,32]; <1,2> the overall reaction proceeds via 2 partial reactions: reaction 1 converts L-arginine into L-N^{γ}-hydroxyarginine, reaction 2 converts L-N^{γ}-hydroxyarginine into citrulline and nitric oxide [31,32]; <1,2> specific for NADPH, 5-electron oxidation of L-arginine [9]) (Reversibility: ir <1> [31]; ? <1-7> [1-22, 24-30, 32-51]) [1-22, 24-51]
P citrulline + NO + $NADP^+$ <1-7> (<1> guanidino-nitrogen of L-arginine is oxidized to form NO and citrulline [19]; <1,2> probably via N^{ω}-hydroxy-L-arginine [3]; <1,3,5> the product is a guanylyl-cyclase-relaxing factor, that is identical with nitric oxide or a NO-releasing compound [7,10,21,26]) [1-22, 24-51]
S L-homoarginine + NADPH + O_2 <1, 2, 4> (<1> poor substrate [1]; <2,4> constitutive endothelial membrane-bound and inducible soluble macrophage enzyme [15]) (Reversibility: ? <1, 2, 4> [1, 15]) [1, 15]
P ?
S N^{γ}-hydroxy-L-arginine + H_2O_2 <2> (<2> tetrahydrobiopterin-free [33]) (Reversibility: ? <2> [33]) [33]
P citrulline + N^{δ}-cyanoornithine + NO_2^- + NO_3- <2> (<2> NO_2^-/NO_3^- as aerobic decomposition products from NO^- [33]) [33]
S $N\gamma$-hydroxy-L-arginine + NADPH + O_2 <1, 2, 4> (<2> reaction is possible without tetrahydrobiopterin, can also use H_2O_2 instead of NADPH and O_2 [33]; <1,2> substrate is intermediate between reaction 1 and 2 to form citrulline and NO from L-arginine [31,33]; <2,4> best substrate [15]) (Reversibility: ir <1> [31]; ? <2, 4> [4, 15, 33, 34, 51]) [4, 15, 31, 33, 34, 51]
P citrulline + $NADP^+$ + NO <2> [4, 33]
S ferricyanide + NADPH + O_2 <4> (Reversibility: ? <4> [51]) [51]
P ferrocyanide + $NADP^+$ + H_2O
S nitroblue tetrazolium + NADPH <1> (<1> NADPH-diaphorase reaction [7]) (Reversibility: ? <1> [7]) [7]
P nitroblue tetrazolium-flavazone + $NADP^+$ <1> [7]
S oxidized cytochrome c + NADPH + O_2 <2-5> (<2> wild-type and mutants [45]; <5> reaction is enhanced by addition of calmodulin at 0.0002 mM [10]) (Reversibility: ? <2-5> [10, 27, 30, 36, 45, 51]) [10, 27, 30, 36, 45, 51]
P reduced cytochrome c + $NADP^+$ + H_2O

S Additional information <1-5> (<4> enzyme shows also superoxide formation activity, uneffected by L-arginine, inhibited by tetrahydrobiopterin and diphenyleneiodonium [40]; <2> dimeric enzyme and subunits are equivalent in catalyzing electron transfer from NADPH to cytochrome c, dichlorophenolindiphenol, and ferricyanide [34]; <3> enzyme can also Ca^{2+}/calmodulin-dependently produce superoxide in absence of tetrahydropterin and in depletion of L-arginine, which is inhibited by tetrahydropterin, cyanide and imidazole [29]; <2> N^{γ}-hydroxylation is the first step of the reaction, N^{γ}-hydroxy-L-arginine being an intermediate in the L-arginine to NO pathway [4]; <1> the enzyme exhibits NADPH-diaphorase activity, uncoupled from nitric oxide synthase activity [7]; <1,2> D-arginine is no substrate [1,34]; <5> the reductase domain has a broad substrate specificity, catalyzes a moderate Ca^{2+}/calmodulin independent hydroxylation when the enzyme is reconstituted with purified P-450 [10]) [1, 4, 7, 10, 29, 34, 40]

P ?

Inhibitors

1-phenylimidazole <4> (<4> reversible inhibition of endothelial enzyme, competitive versus L-arginine and tetrahydrobiopterin, no inhibition of cytochrome c reduction [46]) [46]

2',3'-dialdehyde of NADPH <1> (<1> at concentrations of 40times the apparent K_m-value or after prolonged incubation, independent of Ca^{2+}/calmodulin, L-arginine or tetrahydrobiopterin, NADPH prevents inhibition, the NADPH-diaphorase activity of the enzyme is less sensitive than the nitric oxide synthase activity [7]) [7]

6(R,S)-methyl-5-deazatetrahydropterin <2> [6]

7-nitroindazole <4, 7> (<4> reversible inhibition of endothelial enzyme, competitive versus tetrahydrobiopterin, no inhibition of cytochrome c reduction [46]; <7> weak inhibition [42]) [42, 46]

CO <1, 2> (<1> partially purified rat cerebellum enzyme [9]) [3, 9, 33]

CO/O_2 <2> (<2> 80%:20%, mixture [9]) [9]

Ca^{2+} <1> (<1> preincubation at 37°C leads to time-dependent inhibition of the enzyme [39]) [39]

EDTA <1, 4, 6> (<4> inhibits at concentrations above 0.01 mM [51]; <1,6> brain enzyme [1,35]; <2> no inhibition [12]) [1, 35, 51]

H_2O_2 <3> (<3> alters heme group, decrease in activity [44]) [44]

L-canavanine <1> (<7> not inhibitory [42]; <1> liver enzyme, slight inhibition of brain enzyme [1]) [1]

N-(4-aminobutyl)-5-chloro-2-naphthalene sulfonamide <1> [2]

N-(6-aminohexyl)-1-naphthalene sulfonamide <1> [2]

N-(6-aminohexyl)-5-chloro-1-naphthalene sulfonamide <6> (<6> i.e. W-7 [35]; <6> calmodulin antagonist above 0.01 mM [35]) [35]

N-[(1,3-benzodioxol-5-yl)methyl]-1-[2-(1H-imidazol-1-yl)pyrimidin-4-yl]-4-(methoxycarbonyl)-piperazine-2-acetamide <2> (<2> inhibition of dimer formation in vivo and in vitro, efficiency is dependent on enzyme source [30]) [30]

NO <7> (<7> feedback inhibition [42]) [42]
N^{γ},N^{γ}-dimethyl-L-arginine <4> [38]
N^{γ}-amino-L-arginine <1> [8]
N^{γ}-hydroxy-N^{γ}-methyl-L-arginine <2> (<2> preincubation at 37°C leads to irreversible inactivation, substrates protect [4]) [4]
N^{γ}-iminoethyl-L-ornithine <1> (<1> competitive inhibitor [21]) [21]
N^{γ}-monomethyl-L-arginine <1-4, 6, 7> (<1-3> L-arginine protects against enzyme inactivation, thus inactivation occurs at or near active site [47]; <1> in presence of tetrahydrobiopterin 0.004 mM the neuronal isoform is inactivated [47]; <1,3> endothelial and neuronal isoforms: reversible inhibition [47]; <2> inducible isoform: after preincubation irreversible, time- and concentration-dependent inactivation, without preincubation reversible inhibition [47]; <1,7> slightly [41,42]; <1> not D-isomer, strong, competitive [21]; <4> inhibits citrulline formation, not cytochrome c reduction [25]) [1, 8, 14-16, 21, 25, 26, 35, 38, 41, 42, 47]
N^{γ}-nitro-L-arginine <1-4, 6, 7> (<1-3> L-arginine protects against enzyme inactivation, thus inactivation occurs at or near active site [47]; <1,3> irreversible inactivation of neuronal and endothelial isoform after preincubation, unaffected by tetrahydrobiopterin [47]; <2> reversible inhibitor of inducible isoform from macrophage [47]; <1> competitive inhibitor [21]; <4> inhibits citrulline formation, not cytochrome c reduction [25]) [14-16, 21, 25, 35, 38, 42, 47]
N^{γ}-nitro-L-arginine methyl ester <1, 3, 4, 7> (<4> complete inhibition [51]; <1> very slightly, only L-isomer and in presence of tetrahydrobiopterin and NADPH [41]; <3> only L-isomer, inhibits NO and citrulline production from L-arginine as well as superoxide formation in absence of tetrahydropterin [29]) [29, 38, 41, 42, 51]
agmatine <3> (<3> causes an increase in NADPH oxidase activity of the enzyme [44]; <3> at lower concentration than the Ki value agmatine leads to time-, concentration-, NADPH- and calmodulin-dependent inhibition of the neuronal enzyme in presence of calmodulin [44]) [44]
calcineurin <1> [39]
calmidazolium <1-3> (<2> complete inhibition [50]; <1> in absence of calmodulin [43]; <1,2> calmodulin antagonist [8,50]) [8, 16, 43, 50]
cyanide <1, 3> (<1> pretreatment [31]; <3> heme-blocker inhibits superoxide formation after pretreatment of the enzyme [29]) [29, 31]
di-2-thienyliodonium <2> (<2> competitive, irreversible, complete, time and temperature dependent inhibition [20]) [20]
diphenyleneiodonium <2, 4> (<4> inhibition of superoxide production of recombinant isoform III [40]; <2> competitive, irreversible, complete, time and temperature dependent inhibition [20]) [20, 40]
ethylene glycol bis(β-amino-ethylether)-N,N,N',N'-tetraacetic acid <1, 2, 4, 6> (<1,2,4> i.e. EGTA, complete inhibition of cytosolic enzyme, partial inhibition of particulate enzyme [12,15,18]) [12, 15, 18, 35]
imidazole <3, 4> (<4> inhibition of the endothelial enzyme, competitive versus L-arginine, no inhibition of cytochrome c reduction [46]; <3> heme-

blocker inhibits superoxide formation after pretreatment of the enzyme [29]) [29, 46]
iodoniumdiphenyl <2> (<2> competitive, irreversible, complete, time and temperature dependent inhibition [20]) [20]
nitroblue tetrazolium <1, 4, 5> (<1> potent non-competitive inhibitor, partially reversible by tetrahydrobiopterin [7]) [7, 10, 25]
trifluoperazine <1, 3, 4> (<1> in absence of calmodulin [43]; <3> inhibition in the presence of Ca^{2+}, reversible by calmodulin [16]; <4> inhibits cytochrome c reductase activity [25]; <1,2> no inhibitor of macrophage enzyme [6,18]) [2, 16, 25, 39, 43]
Additional information <1, 2, 4> (<4> N^{γ},$N^{\gamma'}$-dimethyl-L-arginine has no inhibitory effect [38]; <1> no inhibitor of NADPH-diaphorase activity: methotrexate [7]; <1,2> the macrophage enzyme is not inhibited by calmodulin antagonists (N-4-aminobutyl-), (N-6-aminohexyl)-5-chloro-2-naphthalene sulfonamide [12,18,19]; <1> the macrophage enzyme is not inhibited by calmodulin antagonists (N-6-aminohexyl)-1-naphthalene sulfonamide [18,19]) [7, 12, 18-19, 38]

Cofactors/prosthetic groups

2',3'-dialdehyde analogue of NADPH <1> (<1> activation, can substitute for NADPH at low concentrations, inhibitory at concentrations of 40times the apparent K_m-value or after prolonged incubation [7]) [7]
2,6-dichlorophenolindophenol <5> (<5> activation [10]) [10]
5,6,7,8-tetrahydro-L-biopterin <1-7> (<2> 0.04 mol per mol of subunit [49]; <2> only wild-type [45]; <4> stimulates [40,46]; <4> not required for activity [38]; <2> 0.19 mol bound per mol of dimer [34]; <2> required for the first partial reaction, formation of N^{γ}-hydroxy-L-arginine [33]; <3> absolute requirement, recombinant from Pichia pastoris [27]; <4> stimulates 9fold [25]; <1-3,6> required [18,19,26,31-35]; <3> stimulates 4fold at 0.001 mM [5]; <1> i.e. (6R)-2-amino-4-hydroxy-6-(L-erythro-1,2-dihydroxypropyl)-5,6,7,8-tetrahydropteridine, 6R-isomer, requirement, biopteroflavoprotein, 1 mol tetrahydrobiopterin per mol enzyme dimer [7]; <2> presumably tightly enzyme-bound [17]; <2> enzyme purified in absence of biopterin contains substoichiometric concentration, if purified in presence of biopterin it contains 1 mol biopterin per mol MW 130000 subunit [6]; <2> activity is correlated directly to bound biopterin concentration [6]; <2> enhances initial rate of NO-formation [4]) [4-7, 9, 10, 12, 14, 16-19, 22, 24-27, 31-35, 40-42, 45-47, 49-51]
FAD <1-4, 6> (<3> no flavin [48]; <2> wild-type and mutant C415H contain1 mol per mol of subunit [45]; <4> major source of superoxide production in absence of tetrahydrobiopterin [40]; <2> 0.49 mol per mol of dimer [34]; <3> 0.56 mol per mol of recombinant enzyme [27]; <1-4> required [19,26,32,38]; <2> no activation by the addition of exogenous FAD [4]; <1> absolute requirement for FAD [13]; <2> 1 mol per mol of enzyme subunit [17]; <2> 2.2 mol FAD per mol of enzyme dimer [12]; <1> 1 mol FAD per mol enzyme dimer [7]; <2> non-covalently bound FAD [12]; <1,2> FAD containing flavoprotein [3,6,17]; <2> slight activation by exogeneous FAD [17];

<2> tightly enzyme-bound [4,6]; <5> the enzyme bears Ca^{2+}/calmodulin dependent FAD and FMN containing reductase domain which transfers electrons from NADPH to a variety of acceptors [10]) [3, 4, 6, 7, 10, 12, 13, 17-20, 26, 27, 32-36, 38, 40, 41, 45, 49-51]
FMN <1-4, 6> (<3> no flavin [48]; <2> wild-type and mutant C415H contain 0.8 and 0.9 mol per mol of subunit, respectively [45]; <2> 0.71 mol per mol of dimer [34]; <3> 0.79 mol per mol of recombinant enzyme [27]; <2-4> required [26,32,34,38]; <2> no activation by the addition of exogenous FMN [4]; <2> 1 mol per mol of enzyme subunit [17]; <2> 1.1 mol FMN per mol enzyme dimer [12]; <1> 1 mol FMN per mol enzyme dimer [7]; <2> tightly enzyme-bound [4,6]; <1,2> FMN containing flavoprotein [3,6,17]; <5> the enzyme bears Ca^{2+}/calmodulin dependent FAD and FMN containing reductase domain which transfers electrons from NADPH to a variety of acceptors [10]) [3, 4, 6, 7, 10, 12, 17, 20, 26, 27, 32-36, 38, 45, 49-51]
NADPH <1-7> (<2> at high concentration inhibits dimer reconstitution from subunits [34]; <1-3,6> dependent on [8,13,18,21,26,34,35]; <1> requirement, specific for, NADPH-diaphorase activity requires higher NADPH concentrations than nitric oxide formation [7]; <2,4> NADPH-dependent dioxygenase [15]; <1> crude preparation requires only NADPH as cofactor [21]) [1-22, 24-36, 38-42, 45-47, 49, 50]
calmodulin <1-7> (<4> enzyme-bound is required, supplemented stimulates [51]; <3> no stimulation with exogenous calmodulin, inducible isoform from liver [48]; <2> 15fold stimulation of cytochrome c reduction of wild-type and mutants C415A and C415H [45]; <1,2,4,6,7> required [32,34-36,38,39,42]; <3> Ca^{2+}/calmodulin is required for superoxide formation in absence of tetrahydropterin [29]; <3,4> dependent on, endothelial enzyme [27,46]; <1> activation, potent stimulator of purified, not crude, enzyme preparation [2]; <5> the enzyme bears a Ca^{2+}/calmodulin dependent FAD and FMN containing reductase domain which transfers electrons from NADPH to a variety of acceptors [10]; <2> murine macrophage enzyme is Ca^{2+}/calmodulin independent [11,12]; <1> rat neutrophil enzyme is calmodulin independent [19,24]; <2,4> Ca^{2+}/calmodulin stimulates cytochrome c reductase activity [25,45]; <1> NADPH-diaphorase activity of the enzyme is Ca^{2+}/calmodulin independent [7]) [2, 7, 10, 13, 14, 16, 25-27, 32, 34-36, 38, 39, 41, 42, 45-48, 51]
cytochrome c <5> (<5> activation [10]) [10]
nitroblue tetrazolium <5> (<5> activation [10]) [10]

Activating compounds

(6R,S)-methyl-tetrahydropterin <2> (<2> activation in the absence of biopterin, not as effective as tetrahydrobiopterin [6]) [6]
dithiothreitol <1-4> (<1> requirement [18]) [18, 45-47, 49, 51]
interferon γ <2> (<2> activates [12,49]) [12, 49]
lipopolysaccharide <2> (<2> from E. coli, activates [49]) [49]
Additional information <3> (<3> neopterin derivatives are completely inactive and do not bind to the enzyme [5,26]) [5, 26]

Metals, ions

Ca^{2+} <1-7> (<3> no stimulation with exogenous Ca^{2+}, inducible isoform from liver [48]; <1> retina enzyme is dependent on Ca^{2+} [43]; <1,4,7> required [39,41,42,46,51]; <3> Ca^{2+}/calmodulin is required for citrulline/NO formation and for superoxide formation in absence of tetrahydropterin [29]; <3,4> absolutely dependent on [26,38]; <1> cerebellum enzyme is Ca^{2+}-dependent [13,43]; <1> requirement, brain, not liver or lung [1]; <2,4> insoluble enzyme [15]; <2,4> macrophage enzyme [15,18]; <2> slight activation [12]; <1,6> required, calmodulin independent [19,35]; <4,5> activation only in the presence of calmodulin [22,25]; <2,4> constitutive endothelial and inducible membrane bound macrophage enzyme are strictly Ca^{2+}-dependent [15]; <5> the enzyme bears Ca^{2+}/calmodulin dependent cytochrome P-450 reductase activity which catalyzes cytochrome c reduction [10]) [1, 2, 5, 10, 12-16, 19, 21, 22, 25, 26, 29, 35, 38, 39, 41-43, 45-48, 51]

iron <1-3> (<2> 0.83 mol per mol of subunit [49]; <2> heme ligand is bound via C415 [45]; <3> 0.88 mol per mol of recombinant enzyme monomer [44]; <2> required [34]; <2> 0.9-1.2 mol heme per mol of dimer [34]; <1,2> protoporphyrin IX heme [9,33,34,36,49]; <1,2> 2 mol iron-protoporphyrin IX per mol enzyme dimer, the heme-iron is ferric, EPR-and light absorbance spectroscopy [3]; <2> mouse macrophage enzyme: cytochrome P-450 type hemoprotein [9]; <2> naturally occuring neuronal mutant with a 105-amino acid deletion in the heme-binding domain as a result of in-frame mutation by specific alternative splicing, contains heme, but shows no L-arginine and NADPH-dependent citrulline-forming activity in presence of Ca^{2+}-promoted calmodulin, the heme coordination geometry is highly abnormal [28]; <3> 0.8 mol per mol of subunit [27]; <2-4> heme-iron [27,32,34,51]) [3, 9, 27-29, 31-34, 36, 44, 45, 49, 51]

zinc <3> (<3> 0.43 mol per mol of subunit [27]) [27]

Turnover number (min^{-1})

39 <1> (nitroblue tetrazolium, <1> NADPH-diaphorase activity [7]) [7]
65 <1> (L-arginine) [7]

Specific activity (U/mg)

0.0000000335 <1> (<1> crude extract [39]) [39]
0.000000305 <1> (<1> crude extract [41]) [41]
0.000009 <2> (<2> liver mitochondria [50]) [50]
0.000017 <2> (<2> liver [50]) [50]
0.000033 <2> (<2> brain mitochondria [50]) [50]
0.000043 <1> (<1> crude extract [21]) [21]
0.000073 <2> (<2> purified enzyme, without tetrahydropterin [6]) [6]
0.00035 <2> (<2> purified enzyme, with tetrahydropterin [6]) [6]
0.00054 <2> (<2> brain [50]) [50]
0.00074 <1> (<1> crude extract [8]) [8]
0.0098 <1> (<1> purified enzyme [13]) [13]
0.01 <5> (<5> reductase activity [10]) [10]

0.031 <2, 4> (<4> purified pancreatic enzyme [38]; <2> purified recombinant enzyme, NADPH/O_2-supported reduction of N^{γ}-hydroxy-L-arginine, tetrahydrobiopterin-free, 25°C [33]) [33, 38]
0.074 <3> (<3> purified enzyme [16]) [16]
0.12 <1, 2> (<1> NADPH-diaphorase activity [7]; <1> purified enzyme [19]; <2> purified recombinant enzyme, H_2O_2-supported reduction of N^{γ}-hydroxy-L-arginine, tetrahydrobiopterin-free, 25°C [33]) [7, 19, 33]
0.143 <4> (<4> purified recombinant enzyme [51]) [51]
0.17 <1, 2> (<2> purified recombinant enzyme, NADPH/O_2-supported reduction of N^{γ}-hydroxy-L-arginine, with tetrahydrobiopterin, 25°C [33]) [7, 33]
0.181 <3> (<3> partially purified enzyme [5]) [5]
0.19 <3> (<3> purified recombinant enzyme [27]) [27]
0.34-0.35 <3, 4> (<3> purified enzyme, recombinant, determined as NADPH-oxidase activity [27]) [14, 27]
0.41 <4> (<4> substrate N^{γ}-hydroxy-L-arginine, purified enzyme [51]) [51]
0.73 <5> (<5> purified enzyme [22]) [22]
0.815 <2> (<2> purified recombinant enzyme, H_2O_2-supported reduction of N^{γ}-hydroxy-L-arginine, with tetrahydrobiopterin, 25°C [33]) [33]
0.94-0.96 <1> (<1> purified enzyme [18]) [2, 18]
1 <3> (<3> purified enzyme [44]) [44]
1.06 <2> (<2> purified enzyme [12]) [12]
1.1 <2> (<2> purified enzyme [34,36]) [34, 36]
1.62 <2> (<2> purified enzyme [17]) [17]
1.9 <1> (<1> purified enzyme [24]) [24]
38 <3> (<3> purified enzyme, inducible isoform, cytochrome c reductase activity [48]) [48]
Additional information <1-3, 7> (<2> wild-type and mutants [45]; <3> assay method [44]; <7> activity in urinary tract tissues [42]; <1> activity in lung and liver only after induction by endotoxin [1]) [1, 9, 32, 42, 44, 45]

K_m-Value (mM)

0.000004 <4> (calmodulin) [25]
0.00002 <1> (tetrahydrobiopterin) [7]
0.0003 <2> (NADPH) [12]
0.0004 <4> (NADPH) [14, 25]
0.0008 <1> (2',3'-dialdehyde NADPH) [7]
0.002-0.0028 <1, 2, 4> (L-arginine, <1> NADPH [7]) [2, 7, 12, 14]
0.0021 <4> (L-arginine) [51]
0.0023 <4> (L-arginine) [14]
0.0028 <2> (L-arginine) [12]
0.0029 <2> (L-arginine, <2> isoform III [37]) [37]
0.0033-0.0043 <1, 4> (L-arginine, <4> NADPH, bovine [14]) [7, 14]
0.0039 <3> (L-arginine, <3> recombinant from Pichia pastoris [27]) [27]
0.008-0.0084 <1, 5> (dichlorophenolindophenol, <1> with L-arginine [21]) [10, 21]
0.0084 <1> (L-arginine) [21]

0.0091 <7> (L-arginine, <7> detrusor [42]) [42]
0.0099 <7> (L-arginine, <7> urethra [42]) [42]
0.011 <6> (L-arginine, <6> with FAD and tetrahydropterin [35]) [35]
0.014-0.018 <1, 2, 4> (L-arginine) [8, 13, 17, 38]
0.016 <5> (nitroblue tetrazolium) [10]
0.019 <4> (N^{γ}-hydroxy-L-arginine) [51]
0.022 <1, 3> (L-arginine, <3> in presence of tetrahydrobiopterin and L-arginine [48]) [19, 48]
0.028 <2> (N^{γ}-hydroxy-L-arginine, <2> nonlinear regression analysis [4]) [4]
0.0314 <2> (N^{γ}-hydroxy-L-arginine, <2> Eadie-Hofstee graph [4]) [4]
0.0323-0.035 <1, 5> (L-arginine, <1> NADPH, diaphorase activity [7]; <5> cytochrome c reduction [10]) [7, 10, 18]
0.036 <2> (N^{γ}-hydroxy-L-arginine, <2> 25°C, with tetrahydrobiopterin [33]) [33]
0.041 <6> (L-arginine, <6> without tetrahydropterin and FAD [35]) [35]
0.097 <1> (L-arginine, <1> with Ca^{2+}, 0.01 mM [41]) [41]
0.11 <1> (L-arginine, <1> without Ca^{2+} [41]) [41]
0.129 <2> (N^{γ}-hydroxy-L-arginine, <2> 25°C, tetrahydrobiopterin-free [33]) [33]
Additional information <1, 4> (<4> K_m of Ca^{2+}: 0.0003 mM [25]) [8, 25]

K_i-Value (mM)

0.0000022 <2> (N-[(1,3-benzodioxol-5-yl)methyl]-1-[2-(1H-imidazol-1-yl)-pyrimidin-4-yl]-4-(methoxycarbonyl)-piperazine-2-acetamide) [30]
0.00002 <3> (N^{γ}-nitro-L-arginine, <3> irreversible inactivation of endothelial isoform after preincubation [47]) [47]
0.000041 <6> (N^{γ}-nitro-L-arginine) [35]
0.00009 <1> (N^{γ}-nitro-L-arginine, <1> irreversible inactivation of neuronal isoform after preincubation [47]) [47]
0.0002 <4> (N^{γ}-nitro-L-arginine) [25]
0.00031 <6> (N^{γ}-monomethyl-L-arginine) [35]
0.0004 <1> (N^{γ}-nitro-L-arginine) [21]
0.0007 <1,3> (N^{γ}-monomethyl-L-arginine, <3> endothelial isoform [47]) [21,47]
0.0008 <4> (7-nitroindazole) [46]
0.0012 <1> (N^{γ}-iminoethyl-L-ornithine) [21]
0.0015 <4> (N^{γ}-nitro-L-arginine) [14,38]
0.0018 <4> (N^{γ}-monomethyl-L-arginine) [25]
0.002 <1> (N^{γ}-monomethyl-L-arginine, <1> neuronal isoform in presence of 0.004 mM tetrahydrobiopterin [47]) [47]
0.0025 <4> (N^{γ}-monomethyl-L-arginine) [14]
0.0026 <2> (N^{γ}-monomethyl-L-arginine, <2> inducible isoform from macrophage after preincubation [47]) [47]
0.0034 <4> (N^{γ},N^{γ}-dimethyl-L-arginine) [38]
0.0039 <2> (N^{γ}-monomethyl-L-arginine, <2> inducible isoform from macrophage without preincubation [47]) [47]

0.0054 <4> (N^{γ}-monomethyl-L-arginine) [38]
0.0065 <1> (N^{γ}-monomethyl-L-arginine, <1> neuronal isoform [47]) [47]
0.007 <4> (nitroblue tetrazolium) [25]
0.0081 <2> (N^{γ}-nitro-L-arginine, <2> reversible inhibition of inducible macrophage isoform [47]) [47]
0.0155 <4> (N^{γ}-nitro-L-arginine methyl ester) [38]
0.0265 <2> (N^{γ}-hydroxy-N^{γ}-methyl-L-arginine) [4]
0.05 <4> (1-phenylimidazole) [46]
0.05 <4> (imidazole) [46]
0.66 <3> (agmatine, <3> at lower concentration than the Ki value agmatine leads to time-, concentration-, NADPH- and calmodulin-dependent inhibition of the neuronal enzyme in presence of calmodulin [44]) [44]

pH-Optimum
7 <3> (<3> assay at [5]) [5]
7-8 <1> [18, 19]
7.2 <1> (<1> assay at [1,21]) [1, 21]
7.4 <1-4> (<1-4> assay at [8,29,47,50,51]) [8, 29, 38, 47, 50, 51]
7.5 <2, 6> (<2> assay at [17]) [17, 35]
7.6 <1> (<1> assay at [39]) [39]
7.8 <2> (<2> assay at [34,49]) [34, 49]
8 <3> (<3> assay at [26]) [26]
Additional information <1> (<1> pI: 5.6 [19]) [19]

pH-Range
6.6-8 <6> [35]

Temperature optimum (°C)
23 <3> (<3> about, assay at [29]) [29]
25 <2, 4> (<2,4> assay at [33,51]) [33, 51]
32 <2> (<2> assay at [50]) [50]
37 <1-7> (<1-7> assay at [1-22, 24, 26, 34, 35, 39, 42, 43, 47-49]) [1-22, 24, 26, 34, 35, 38, 39, 42, 43, 47-49]

4 Enzyme Structure

Molecular weight
55000 <2> (<2> oxygenase subunit domain, gel filtration [32]) [32]
74000 <2> (<2> reductase subunit domain, gel filtration [36]) [36]
112000 <2> (<2> oxygenase subunit domain, gel filtration [36]) [36]
150000 <1> (<1> gel filtration [19]) [19]
200000 <1, 5> (<1> gel filtration [2]; <5> above, gel filtration [22]) [2, 22]
250000 <2> (<2> gel filtration [12]) [12]
260000 <2> (<2> gel filtration [17]) [17]
300000 <1, 4> (<1,4> gel filtration [14,18]) [14, 18]
320000 <4> (<4> gel filtration [38]) [38]
Additional information <2> (<2> dissociates irreversibly into subunits in the absence of L-arginine, FAD and tetrahydrobiopterin [12]) [12]

Subunits

? <1-5> (<1,3> x * 160000, isoform I, amino acid sequence determination [37]; <1-4> 130000-133000, isoforms II and III, amino acid sequence determination [37]; <3> x * 135000, SDS-PAGE [29]; <4> x * 152000, SDS-PAGE, Western Blot [25]; <5> x * 160000, SDS-PAGE [22]) [22, 25, 29, 37]

dimer <1-4> (<4> 2 * 130000-150000, SDS-PAGE [51]; <2> 2 * 56000, oxygenase subunit domain, SDS-PAGE [36]; <3> 2 * 131000-133000, recombinant, SDS-PAGE [27]; <2> 2 * 130000, SDS-PAGE [12,17,32,34,49]; <4> 2 * 150000, SDS-PAGE, gel filtration at high salt concentration [14]; <1> 2 * 150000, SDS-PAGE [18]; <3,4> 2 * 160000, SDS-PAGE [16,38]) [12, 14, 16-18, 27, 32, 34, 36, 38, 49, 51]

monomer <1-4> (<2> 1 * 74000, reductase domain, SDS-PAGE [36]; <1,4> 1 * 150000, SDS-PAGE [2, 13, 14, 19]; <3> 1 * 160000, SDS-PAGE [5]) [2, 5, 13, 14, 19, 36]

Additional information <1-3> (<2> D-arginine inhibits reconstitution of dimer from subunits [34]; <2> dissociation of dimer into subunits at pH 6.8 [34]; <2> dimeric structure is required for enzyme activity, interaction between subunits via oxygenase domains [32,36]; <2> dimer formation, each subunit consists of: one oxygenase domain containing heme, tetrahydrobiopterin, substrate binding site and one reductase domain containing FAD, FMN, calmodulin, NADPH binding site [32, 36, 49]; <3> gel filtration: native protein is a dimer, especially in presence of tetrahydrobiopterin and L-arginine, but dissociates at low temperature to monomers during gel electrophoresis [27]; <1,2> significant amino acid sequence homology to NADPH-cytochrome P-450-type hemoprotein [9]; <2> subunit composition of dimeric enzyme [34]) [9, 27, 32, 34, 36, 49]

5 Isolation/Preparation/Mutation/Application

Source/tissue

DLD-1 <3> [37]

adrenal gland <1, 3-5> (<1,3,5> isoform I [37]) [14, 37]

brain <1-6> (<1,3,5> isoform I [37]; <1,3> Ca^{2+}-dependent isoform [5,13]) [1-3, 5, 7-10, 13, 14, 16, 21-24, 31, 35, 37, 39, 43, 47, 50]

brain stem <1, 4> (<1,4> cerebrum shows higher activity than cerebellum [14]) [14]

cerebellum <1, 3-5> [5, 7-10, 13, 14, 22, 31, 43, 47]

cerebral cortex <6> [35]

cerebrum <1, 4> [14, 39]

cervical carcinoma cell <3> (<3> cell line ME-180, constitutive expression [26]) [26]

chondrocyte <3> [37]

detrusor <7> [42]

endothelium <1-5> (<3> umbilical vein cells [47]; <4> constitutive, from pulmonary artery [46]; <1-3> liver, lung, adrenal gland, colon, isoform II

[37]; <1-5> isoform III [37,40]; <1,3,5> lung, uterus, stomach [37]; <2,4> aorta, cell culture [15]) [15, 27, 29, 37, 40, 46, 47, 51]
forebrain <1> (<1> synaptosomal fraction [21]) [21]
glioblastoma cell <3> (<3> cell line A-172, american type [30]) [30]
hepatocyte <1-3> (<1-3> isoform II [37]) [37]
hippocampus <6> [35]
hypothalamus <6> [35]
kidney <1, 3, 5> (<1> renal cortex [41]; <1,3,5> macula densa cells, isoform I [37]) [37, 41]
liver <1-3> (<3> inducible isoform [48]) [1, 37, 48, 50]
lung <1> [1]
macrophage <1-4> (<1-3> in liver, lung, kidney, isoform II [37]; <2> RAW 264.7 cells [30,32-34,36,47,49]; <2> Ca^{2+}-independent form [11,12]; <1,2> cytokine-activated [3,11,30]) [3, 4, 6, 9, 11, 12, 15, 17-20, 30, 32-34, 36, 37, 45, 47, 49]
mesencephalon <6> [35]
neostriatum <6> [35]
neuron <1-3, 5> (<1,3,5> peripheral nitrergic nerves, isoform I [37] <2> central nervous system, natural variant with 105-amino acid deletion in the heme-binding domain [28]) [28, 37, 44]
neutrophil <1> (<1> peritoneal polymorphonuclear [19,24]) [19, 24]
neutrophil <1> (<1> polymorphonuclear [19,24]) [19, 24]
nucleus <6> [35]
pancreas <1, 3-5> (<1,3,5> islet cells, isoform I [37]) [37, 38]
peritoneum <1> [19, 24]
pituitary gland <1, 4> [14, 25]
pons <6> [35]
retina <1> (<1> constitutive, soluble form [43]) [43]
smooth muscle <1> [37]
spinal cord <1, 3, 5> (<1,3,5> isoform I [37]) [37]
spleen <1-3> (<1-3> red pulp, eosinophils and neutrophils, isoform II [37]) [37]
sympathetic ganglion <1, 3, 5> (<1,3,5> isoform I [37]) [37]
thalamus <6> [35]
ureter <7> [42]
urethra <7> [42]
Additional information <1-5> (<1-5> 3 isoforms: 1. neuronal, soluble isoform I is constitutively expressed in brain and other tissues and Ca^{2+}-regulated, 2. soluble isoform II is usually not constitutively expressed, but inducible in macrophages and other cells, 3. isoform III is membrane-bound and expressed in endothelial cells [37,43]; <2> 3 distinct isoforms: 1. a membrane-associated, constitutive enzyme from the vascular endothelium, 2. a soluble, constitutive enzyme from neuronal cells, 3. an endotoxin- and cytokine-inducible enzyme exemplified by that from murine macrophages [33]; <2> activity of constitutive enzymes is regulated by binding of calmodulin and Ca^{2+}, the inducible enzyme is regulated by binding of calmodulin, not by Ca^{2+} [33]) [33, 37, 43]

Localization

cytosol <1-4, 7> [6, 7, 9, 13, 15, 16, 19, 21, 24, 26, 41-44, 47]
membrane <1, 2, 4> (<2,4> constitutive endothelial enzyme: predominantly membrane-bound, inducible macrophage enzyme: equally distributed between cytosol and membrane, small constitutive membrane-bound portion in murine macrophages [15]) [13, 15, 41]
mitochondrion <2> [50]
Additional information <1-5> (<1-5> 3 isoforms: 1. neuronal, soluble isoform I is constitutively expressed in brain and other tissues and Ca^{2+}-regulated, 2. soluble isoform II is usually not constitutively expressed, but inducible in macrophages and other cells, 3. isoform III is membrane-bound and expressed in endothelial cells [37,43]; <2> 3 distinct isoforms: 1. a membrane-associated, constitutive enzyme from the vascular endothelium, 2. a soluble, constitutive enzyme from neuronal cells, 3. an endotoxin- and cytokine-inducible enzyme exemplified by that from murine macrophages [33]; <2> activity of constitutive enzymes is regulated by binding of calmodulin and Ca^{2+}, the inducible enzyme is regulated by binding of calmodulin, not by Ca^{2+} [33]) [33, 37, 43]

Purification

<1> (cerebellum wild-type and recombinant brain enzyme [31]; partial [9]; 2',5'-ADP-affinity and anion exchange chromatography [2,19,24]; sequential affinity chromatography on 2',5'-ADP-agarose and calmodulin Sepharose 4B [13]; 2',5'-ADP-agarose affinity chromatography [3,9]) [2, 3, 7, 9, 13, 18, 19, 24, 31, 32]
<2> (from interferon-γ- and lipopolysaccharide-activated macrophage [49]; wild-type and mutants from insect cells [45]; dimeric enzyme and subunits [34,36,49]; recombinant from Escherichia coli [33]; heterodimer [32]; wild-type and natural mutant, recombinant from Escherichia coli and insect cells [28]; inducible form from macrophage [17]; 2',5'-ADP-agarose affinity chromatography [4,6,9]; sequential anion-exchange-, affinity chromatogaphy [17]; gel filtration chromatography [12]; partial [20]) [4, 6, 9, 12, 17, 20, 28, 32-34, 36, 45, 49]
<3> (recombinant wild-type of inducible liver isoform from Escherichia coli with and without His-tag, requires inclusion of tetrahydrobiopterin in purification buffer [48]; wild-type recombinant from insect cells [29,44]; recombinant from Pichia pastoris [27]; partial, 2',5'-ADP-agarose affinity chromatography [5]; affinity and size exclusion chromatography [16]) [5, 16, 27, 29, 44, 48, 51]
<4> (recombinant endothelial isoform from human embryonic kidney cells [51]; recombinant endothelial isoform III from insect cells [40]; affinity chromatography [14]) [14, 25, 38, 40]
<5> (ammonium sulfate precipitation (30% saturation), 2',5'-ADP-affinity chromatography [22]) [10, 22]

Renaturation

<2> (refolding after treatment/equilibration with 5 M urea in presence of L-arginine and tetrahydrobiopterin [49]) [49]

Crystallization

<2> (orthorhombic crystals from NOS oxygenase domain lacking the N-terminal 114 residues, preparation in presence of imidazole, structure analysis via x-ray diffraction, also with bound inhibitor N-[(1,3-benzodioxol-5-yl)-methyl]-1-[2-(1H-imidazol-1-yl)pyrimidin-4-yl]-4-(methoxycarbonyl)-piperazine-2-acetamide [30]; x-ray crystal structure of the heme-binding domains of neuronal wild-type and mutant with deletion in the heme-binding domain [28]) [28, 30]

Cloning

<1> (rat brain enzyme expressed in human 293 kidney cells transfected with a vector encoding rat brain enzyme [3,31]) [3, 31]
<2> (expression of neuronal isoform wild type and mutants C415H, C415A in Spodoptera frugiperda cells via baculovirus infection [45]; expression in Escherichia coli, coexpression of calmodulin is necessary [33]; expression of His-tagged NOS oxygenase domain lacking the N-terminal 114 residues in Escherichia coli [30]; neuronal wild-type and natural mutant, specifically expressed in the central nervous system, expression in Escherichia coli and in insect cells of Spodoptera frugiperda via baculovirus infection [28]) [28, 30, 45]
<3> (inducible liver isoform is expressed in Escherichia coli with and without His-tag, requires coexpression of calmodulin [48]; neuronal enzyme expressed in Spodoptera frugiperda cells via baculovirus infection [44]; wild-type endothelial enzyme is expressed in insect cells via baculovirus infection [29]; endothelial enzyme expressed in Pichia pastoris using a highly inducible alcohol oxidase promoter PAOX1 [27]) [27, 29, 44, 48]
<4> (expression of endothelial isoform in human embryonic kidney cells [51]; expression of endothelial isoform III in Spodoptera frugiperda cells via baculovirus infection [40]) [40, 51]

Engineering

C415A <2> (<2> contains no heme, no bound tetrahydrobiopterin, unable to oxidize NADPH and to synthezise nitric oxide, unaltered ability to reduce cytochrome c [45]) [45]
C415H <2> (<2> contains nearly no heme, no bound tetrahydrobiopterin, unable to oxidize NADPH and to synthezise nitric oxide, unaltered ability to reduce cytochrome c [45]) [45]
Additional information <2> (<2> construction of a heterodimer with one subunit being His-tagged [32]; <2> monomers encoding NOS oxygenase domain lacking the N-terminal 114 residues, His-tagged, expression in Escherichia coli [30]; <2> naturally occuring neuronal mutant with a 105-amino acid deletion in the heme-binding domain as a result of in-frame mutation by specific alternative splicing, contains heme, but shows no L-arginine and NADPH-dependent citrulline-forming activity in presence of Ca^{2+}-promoted calmodulin [28]) [28, 30, 32]

Application

pharmacology <3> (<3> NO synthase can be used to gain insight into the biological role of endogenous agmatine [44]) [44]

6 Stability

pH-Stability

7.4 <1> (<1> purified, $t_{1/2}$: 6 h, 4°C [18]; <1> purified, $t_{1/2}$: 3 h, 4°C [19,24]) [18, 19, 24]

Temperature stability

37 <2> (<2> tetrahydrobiopterin-free enzyme is not stable, therefore reaction is performed at 25°C [33]; <2> $t_{1/2}$: about 6 h [12]) [12, 33]

General stability information

<1>, ammonium sulfate precipitation results in remarkable loss of tetrahydrobiopterin [7]
<1>, bovine serum albumin stabilizes [2]
<1>, dithiothreitol stabilizes [18]
<1>, heat-, diethylether- and alkali-labile, slightly acid resistant cytosolic factor of MW above 100000 and of little charge at neutral pH stabilizes [24]
<1>, unstable during purification [19]
<2>, (6R)-tetrahydro-L-biopterin stabilizes during purification at 4°C [6]
<2>, Mg^{2+} does not stabilize [17]
<2>, bovine serum albumin stabilizes during refolding [49]
<2>, freezing inactivates, 50% v/v glycerol protects [17]
<3>, inducible isoform is unstable during purification in absence of L-arginine and tetrahydrobiopterin toward loss of the heme group and formation of low-spin species [48]
<4>, tetrahydrobiopterin stabilizes dimeric form [40]
<1, 2, 5>, glycerol stabilizes [2, 6, 17, 22]

Storage stability

<1>, 0°C, crude, $t_{1/2}$: 2 days, purified $t_{1/2}$: 2 h, with 1 mg/ml bovine serum albumin and 20% v/v glycerol as stabilizing agents, $t_{1/2}$: 7 days [2]
<1>, 4°C, crude, stable for at least 24 h [18, 19]
<1>, 4°C, in the presence of one or more cytosolic factors at least 24 h stable [24]
<1>, 4°C, purified, $t_{1/2}$: 3 h at pH 7.4 [19, 24]
<1>, 4°C, purified, $t_{1/2}$: 6 h at pH 7.4 [18]
<2>, -80°C, diluted preparation stable overnight [12]
<2>, -80°C, stable overnight in 50% v/v glycerol [17]
<2>, 4°C, 6 h stable [12]
<2>, 4°C, $t_{1/2}$: 2 h [17]
<5>, -70°C, with 20% v/v glycerol at least 4 weeks stable [22]
<5>, 4°C, unstable [22]

References

[1] Knowles, R.G.; Merrett, M.; Salter, M.; Moncada, S.: Differential induction of brain, lung and liver nitric oxide synthase by endotoxin in the rat. Biochem. J., **270**, 833-836 (1990)

[2] Bredt, D.S.; Snyder, S.H.: Isolation of nitric oxide synthetase, a calmodulin-requiring enzyme. Proc. Natl. Acad. Sci. USA, **87**, 682-685 (1990)

[3] Stuehr, D.J.; Ikeda-Saito, M.: Spectral characterization of brain and macrophage nitric oxide synthases. Cytochrome P-450-like hemeproteins that contain a flavin semiquinone radical. J. Biol. Chem., **267**, 20547-20550 (1992)

[4] Pufahl, R.A.; Nanjappan, P.G.; Woodard, R.W.; Marletta, M.A.: Mechanistic probes of N-hydroxylation of L-arginine by the inducible nitric oxide synthase from murine macrophages. Biochemistry, **31**, 6822-6828 (1992)

[5] Klatt, P.; Heinzel, B.; Mayer, B.; Ambach, E.; Werner-Felmayer, G.; Wachter, H.; Werner, E.R.: Stimulation of human nitric oxide synthase by tetrahydrobiopterin and selective binding of the cofactor. FEBS Lett., **305**, 160-162 (1992)

[6] Hevel, J.M.; Marletta, M.A.: Macrophage nitric oxide synthase: relationship between enzyme-bound tetrahydrobiopterin and synthase activity. Biochemistry, **31**, 7160-7165 (1992)

[7] Schmidt, H.H.H.W.; Smith, R.M.; Nakane, M.; Murad, F.: Ca^{2+}/calmodulin-dependent NO synthase type I: a biopteroflavoprotein with Ca^{2+}/calmodulin-independent diaphorase and reductase activities. Biochemistry, **31**, 3243-3249 (1992)

[8] Bush, P.A.; Gonzalez, N.E.; Griscavage, J.M.; Ignarro, L.J.: Nitric oxide synthase from cerebellum catalyzes the formation of equimolar quantities of nitric oxide and citrulline from L-arginine. Biochem. Biophys. Res. Commun., **185**, 960-966 (1992)

[9] White, K.A.; Marletta, M.A.: Nitric oxide synthase is a cytochrome P-450 type hemoprotein. Biochemistry, **31**, 6627-6631 (1992)

[10] Klatt, P.; Heinzel, B.; John, M.; Kastner, M.; Böhme, E.; Mayer, B.: Ca^{2+}/calmodulin-dependent cytochrome c reductase activity of brain nitric oxide synthase. J. Biol. Chem., **267**, 11374-11378 (1992)

[11] Xie, Q.W.; Hearn, J.C.; Calaycay, J.; Mumford, R.A.; Swiderek, K.M.; Lee, T.D.; Ding, A.; Troso, T.; Nathan, C.: Cloning and characterization of inducible nitric oxide synthase from mouse macrophages. Science, **256**, 225-228 (1992)

[12] Stuehr, D.J.; Cho, H.J.; Kwon, N.S.; Weise, M.F.; Nathan, C.F.: Purification and characterization of the cytokine-induced macrophage nitric oxide synthase: an FAD- and FMN-containing flavoprotein. Proc. Natl. Acad. Sci. USA, **88**, 7773-7777 (1991)

[13] Hiki, K.; Hattori, R.; Kawai, C.; Yui, Y.: Purification of insoluble nitric oxide synthase from rat cerebellum. J. Biochem., **111**, 556-558 (1992)

[14] Ohshima, H.; Oguchi, S.; Adachi, H.; Iida, S.; Suzuki, H.; Sugimura, T.; Esumi, H.: Purification of nitric oxide synthase from bovine brain: immunolo-

gical characterization and tissue distribution. Biochem. Biophys. Res. Commun., **183**, 238-244 (1992)
[15] Hecker, M.; Walsh, D.T.; Vane, J.R.: On the substrate specificity of nitric oxide synthase. FEBS Lett., **294**, 221-224 (1991)
[16] Schmidt, H.H.H.W.; Murad, F.: Purification and characterization of a human NO synthase. Biochem. Biophys. Res. Commun., **181**, 1372-1377 (1991)
[17] Hevel, J.M.; White, K.A.; Marletta, M.A.: Purification of the inducible murine macrophage nitric oxide synthase. Identification as a flavoprotein. J. Biol. Chem., **266**, 22789-22791 (1991)
[18] Yui, Y.; Hattori, R.; Kosuga, K.; Eizawa, H, Hiki, K.; Kawai, C.: Purification of nitric oxide synthase from rat macrophages. J. Biol. Chem., **266**, 12544-12547 (1991)
[19] Yui, Y.; Hattori, R.; Kosuga, K.; Eizawa, H, Hiki, K.; Ohkawa, S.; Ohnishi, K.; Terao, S.; Kawai, C.: Calmodulin-independent nitric oxide synthase from rat polymorphonuclear neutrophils. J. Biol. Chem., **266**, 3369-3371 (1991)
[20] Stuehr, D.J.; Fasehun, O.A.; Kwon, N.S.; Gross, S.S.; Gonzalez, J.A.; Levi, R.; Nathan, C.F.: Inhibition of macrophage and endothelial cell nitric oxide synthase by diphenyleneiodonium and its analogs. FASEB J., **5**, 98-103 (1991)
[21] Knowles, R.G.; Palacios, M.; Palmer, R.M.J.; Moncada, S.: Kinetic characteristics of nitric oxide synthase from rat brain. Biochem. J., **269**, 207-210 (1990)
[22] Mayer, B.; John, M.; Böhme, E.: Purification of a Ca^{2+}/calmodulin-dependent nitric oxide synthase from porcine cerebellum. Cofactor-role of tetrahydrobiopterin. FEBS Lett., **277**, 215-219 (1990)
[23] Murphy, M.: Kinetic characteristics of nitric oxide synthase from rat brain. Biochem. J., **271**, 563-564 (1990)
[24] Kosuga, K.; Yui, Y.; Hattori, R.; Eizawa, H, Hiki, K.; Kawai, C.: Stabilizing factor(s) of nitric oxide (NO) synthetase. Biochem. Biophys. Res. Commun., **172**, 705-708 (1990)
[25] Wolff, D.J.; Datto, G.A.: Identification and characterization of a calmodulin-dependent nitric oxide synthase from GH3 pituitary cells. Biochem. J., **285**, 201-206 (1992)
[26] Werner-Felmayer, G.; Werner, E.R.; Fuchs, D.; Hausen, A.; Mayer, B.; Reibnegger, G.; Weiss, G.; Wachter, H.: Calcium/calmodulin-dependent nitric oxide synthase activity in the human cervix carcinoma cell line ME-180. Biochem. J., **289**, 357-361 (1993)
[27] Leber, A.; Hemmens, B.; Klosch, B.; Goessler, W.; Raber, G.; Mayer, B.; Schmidt, K.: Characterization of recombinant human endothelial nitric-oxide synthase purified from the yeast Pichia pastoris. J. Biol. Chem., **274**, 37658-37664 (1999)
[28] Iwasaki, T.; Hori, H.; Hayashi, Y.; Nishino, T.; Tamura, K.; Oue, S.; Iizuka, T.; Ogura, T.; Esumi, H.: Characterization of mouse nNOS2, a natural variant of neuronal nitric-oxide synthase produced in the central nervous system by selective alternative splicing. J. Biol. Chem., **274**, 17559-17566 (1999)

[29] Xia, Y.; Tsai, A.L.; Berka, V.; Zweier, J.L.: Superoxide generation from endothelial nitric-oxide synthase. A Ca^{2+}/calmodulin-dependent and tetrahydrobiopterin regulatory process. J. Biol. Chem., **273**, 25804-25808 (1998)
[30] McMillan, K.; Adler, M.; Auld, D.S.; Baldwin, J.J.; Blasko, E.; Browne, L.J.; Chelsky, D.; Davey, D.; Dolle, R.E.; Eagen, K.A.; Erickson, S.; Feldman, R.I.; Glaser, C.B.; Mallari, C.; Morrissey, M.M.; Ohlmeyer, M.H.J.; Pan, G.; Parkinson, J.F.; Phillips, G.B.; Polokoff, M.A.; Sigal, N.H.; Vergona, R.; Whitlow, M.; Young, T.A. and Devlin, J.J: Allosteric inhibitors of inducible nitric oxide synthase dimerization discovered via combinatorial chemistry. Proc. Natl. Acad. Sci. USA, **97**, 1506-1511 (2000)
[31] Campos, K.L.; Giovanelli, J.; Kaufman, S.: Characteristics of the nitric oxide synthase-catalyzed conversion of arginine to N-hydroxyarginine, the first oxygenation step in the enzymic synthesis of nitric oxide. J. Biol. Chem., **270**, 1721-1728 (1995)
[32] Siddhanta, U.; Wu, C.; Abu-Soud, H.M.; Zhang, J.; Ghosh, D.K.; Stuehr, D.J.: Heme iron reduction and catalysis by a nitric oxide synthase heterodimer containing one reductase and two oxygenase domains. J. Biol. Chem., **271**, 7309-7312 (1996)
[33] Rusche, K.M.; Spiering, M.M.; Marletta, M.A.: Reactions catalyzed by tetrahydrobiopterin-free nitric oxide synthase. Biochemistry, **37**, 15503-15512 (1998)
[34] Baek, K.J.; Thiel, B.A.; Lucas, S.; Stuehr, D.J.: Macrophage nitric oxide synthase subunits. Purification, characterization, and role of prosthetic groups and substrate in regulating their association into a dimeric enzyme. J. Biol. Chem., **268**, 21120-21129 (1993)
[35] Cote, J.F.; Roberge, A.G.: Nitric oxide synthase in cat brain: cofactors - enzyme-substrate interaction. Free Radic. Biol. Med., **21**, 109-115 (1996)
[36] Ghosh, D.K.; Stuehr, D.J.: Macrophage NO synthase: characterization of isolated oxygenase and reductase domains reveals a head-to-head subunit interaction. Biochemistry, **34**, 801-807 (1995)
[37] Foerstermann, U.; Closs, E.I.; Pollock, J.S.; Nakane, M.; Schwarz, P.; Gath, I.; Kleinert, H.: Nitric oxide synthase isoenzymes. Characterization, purification, molecular cloning, and functions. Hypertension, **23**, 1121-1131 (1994)
[38] Nam, S.W.; Seo, D.W.; Sung, D.S.; Han, J.W.; Hong, S.Y.; Lee, H.W.: Nitric oxide synthase from bovine pancreas: purification and characterization. Arch. Pharmacol. Res., **21**, 128-134 (1998)
[39] Mittal, C.K.; Jadhav, A.L.: Calcium-dependent inhibition of constitutive nitric oxide synthase. Biochem. Biophys. Res. Commun., **203**, 8-15 (1994)
[40] Wever, R.M.F.; Van Dam, T.; Van Rijn, H.J.M.; De Groot, F.; Rabelink, T.J.: Tetrahydrobiopterin regulates superoxide and nitric oxide generation by recombinant endothelial nitric oxide synthase. Biochem. Biophys. Res. Commun., **237**, 340-344 (1997)
[41] Conrad, K.P.; Powers, R.W.; Davis, A.K.; Novak, J.: Citrulline is not the major product using the standard NOS activity assay on renal cortical homogenates. Am. J. Physiol., **282**, R303-R310 (2002)

[42] Garcia-Pascual, A.; Costa, G.; Labadia, A.; Persson, K.; Triguero, D.: Characterization of nitric oxide synthase activity in sheep urinary tract: functional implications. Br. J. Pharmacol., **118**, 905-914 (1996)
[43] Sheng, H.; Ignarro, L.J.: Biochemical and immunohistochemical characterization of nitric oxide synthase in the rat retina. Pharm. Res., **33**, 29-34 (1996)
[44] Demady, D.R.; Jianmongkol, S.; Vuletich, J.L.; Bender, A.T.; Osawa, Y.: Agmatine enhances the NADPH oxidase activity of neuronal NO synthase and leads to oxidative inactivation of the enzyme. Mol. Pharmacol., **59**, 24-29 (2001)
[45] Richards, M.K.; Clague, M.J.; Marletta, M.A.: Characterization of C415 mutants of neuronal nitric oxide synthase. Biochemistry, **35**, 7772-7780 (1996)
[46] Wolff, D.J.; Lubeskie, A.; Umansky, S.: The inhibition of the constitutive bovine endothelial nitric oxide synthase by imidazole and indazole agents. Arch. Biochem. Biophys., **314**, 360-366 (1994)
[47] Reif, D.W.; McCreedy, S.A.: N-Nitro-L-arginine and N-monomethyl-L-arginine exhibit a different pattern of inactivation toward the three nitric oxide synthases. Arch. Biochem. Biophys., **320**, 170-176 (1995)
[48] Gerber, N.C.; Nishida, C.R.; de Montellano, P.R.O.: Characterization of human liver inducible nitric oxide synthase expressed in Escherichia coli. Arch. Biochem. Biophys., **343**, 249-253 (1997)
[49] Abu-Soud, H.M.; Loftus, M.; Stuehr, D.J.: Subunit dissociation and unfolding of macrophage NO synthase: relationship between enzyme structure, prosthetic group binding, and catalytic function. Biochemistry, **34**, 11167-11175 (1995)
[50] Lacza, Z.; Puskar, M.; Figueroa, J.P.; Zhang, J.; Rajapakse, N.; Busija, D.W.: Mitochondrial nitric oxide synthase is constitutively active and is functionally upregulated in hypoxia. Free Radic. Biol. Med., **31**, 1609-1615 (2001)
[51] Presta, A.; Liu, J.; Sessa, W.C.; Stuehr, D.J.: Substrate binding and calmodulin binding to endothelial nitric oxide synthase coregulate its enzymic activity. Nitric Oxide, **1**, 74-87 (1997)

Anthraniloyl-CoA monooxygenase 1.14.13.40

1 Nomenclature

EC number
1.14.13.40

Systematic name
2-aminobenzoyl-CoA,NAD(P)H:oxygen oxidoreductase (de-aromatizing)

Recommended name
anthraniloyl-CoA monooxygenase

Synonyms
2-aminobenzoyl-CoA monooxygenase/reductase
reductase, anthraniloyl coenzyme A

CAS registry number
112692-57-6

2 Source Organism

<1> *Pseudomonas sp.* (strain KB740 [1, 2, 4, 7]; strain K172 and K 740 [8]) [1-4, 7, 8]
<2> *Azoarcus evansii* (strain KB740 [5]) [5, 6]

3 Reaction and Specificity

Catalyzed reaction
2-aminobenzoyl-CoA + 2 NAD(P)H + 2 H^+ + O_2 = 2-amino-5-oxocyclohex-1-enecarboxyl-CoA + H_2O + 2 $NAD(P)^+$ (<1> reaction mechanism [2])

Reaction type
oxidation
redox reaction
reduction

Natural substrates and products
S 2-aminobenzoyl-CoA + NAD(P)H + O_2 <1> (<1> pathway of aerobic breakdown of 2-aminobenzoate [1]) (Reversibility: ? <1> [1, 3]) [1, 3]
P 2-amino-5-oxocyclohex-1-enecarboxyl-CoA + H_2O + $NAD(P)^+$ <1> (<1> product formed under physiological NADH concentration [3]) [3]

Substrates and products

S 2-aminobenzoyl-CoA + NAD(P)H + O_2 <1, 2> (<2> reaction mechanism [6]; <1> i.e. anthraniloyl-CoA [7]) (Reversibility: ? <1, 2> [1-4, 6, 7, 8]) [1-4, 6, 7, 8]

P 2-amino-5-oxocyclohex-1-enecarboxyl-CoA + H_2O + $NAD(P)^+$ <1> (<1> at least 2 different reaction products depending on the concentration of NADH [2]; <1> formation of 3 different products, the non-aromatic product is unstable and releases CO_2 and NH_3, forming 1,4-cyclohexadione [3]; <1> 7 different products are formed, pattern is dependent on pH [8]) [2, 3, 8]

S 2-hydroxybenzoyl-CoA + NAD(P)H + O_2 <1> (<1> i.e. salicoyl-CoA, weak substrate, is probably monooxygenated but not hydrogenated [7]) (Reversibility: ? <1> [7]) [7]

P ?

S N-ethylmaleimide + NAD(P)H <1> (Reversibility: ? <1> [1, 4, 7]) [1, 4, 7]

P N-ethylsuccinimide + $NAD(P)^+$ [4]

S N-ethylmaleimide + NAD(P)H <1> (Reversibility: ? <1> [3]) [3]

P succinimide + $NAD(P)^+$ <1> [3]

S maleimide + NAD(P)H <1> (Reversibility: ? <1> [1]) [1]

P succinimide + $NAD(P)^+$

Inhibitors

2-aminobenzoyl-CoA <1> (<1> substrate inhibition above 0.15 mM [1]) [1]

4-hydroxymercuribenzoate <1> (<1> at least 90% inhibition at 0.2 mM [1]) [1]

5,5'-dithiobis(2-nitrobenzoate) <1> (<1> at least 90% inhibition at 0.2 mM [1]) [1]

$AgNO_3$ <1> (<1> at least 90% inhibition at 0.01 mM [1]) [1]

$HgSO_4$ <1> (<1> at least 90% inhibition at 0.1 mM [1]) [1]

Cofactors/prosthetic groups

ATP <1> (<1> requirement [8]) [8]

FAD <1> (<1> flavoprotein [1-3]; <1> approximately 2 mol noncovalently bound FAD per mol of enzyme [1]; <1> the three dimeric forms $\alpha\alpha$, $\alpha'\alpha'$ and $\alpha\alpha'$ differ in their mode of binding FAD [7]) [1-5, 7]

NADH <1> [1-3, 6, 7]

NADPH <1> (<1> enzyme is less active with NADPH than with NADH [1]) [1, 3]

Metals, ions

Mg^{2+} <1> (<1> requirement [8]) [8]

Turnover number (min^{-1})

4250 <1> (2-aminobenzoyl-CoA) [1]

Specific activity (U/mg)

0.015 <1> (<1> cell extract [8]) [8]

25 <1> [1]

Additional information <1> (<1> specific activity of isolated and reconstituted enzyme in different fractions after purification [7]) [7]

K_m-Value (mM)
0.005 <1> (O_2, <1> value below [1]) [1]
0.02 <1> (2-aminobenzoyl-CoA, <1> value below, cell extract [8]) [8]
0.025 <1> (2-aminobenzoyl-CoA, <1> value below [1]) [1]
0.026 <1> (NADH) [7]
0.042 <1> (NADH) [1]
0.25 <1> (N-ethylmaleimide) [1]
0.47 <1> (N-ethylmaleimide) [7]
0.5 <1> (NADPH) [1]

pH-Optimum
7 <1> [8]
8 <1> (<1> 2-aminobenzoyl-CoA + NADH [1]) [1]

4 Enzyme Structure

Molecular weight
85000 <1> (<1> SDS-PAGE [1]) [1]
87000 <2> (<2> calculated from deduced amino acid sequence [5]) [5]
170000 <1> (<1> gel filtration [1]) [1]

Subunits
dimer <1> (<1> 2 * 85000 SDS-PAGE, probably α_2 dimer, may exist in three dimeric forms: 1. α,α, 2. α',α', 3. α,α', where α' may be a subunit with a different conformation [1]; <1> α_2 homodimer [4]; <1> three dimeric forms: 1. α,α, 2. α',α', 3. α,α' [7]) [1, 4]

5 Isolation/Preparation/Mutation/Application

Purification
<1> (apparent homogeneity [2]) [1, 2, 7]
<2> [5]

Cloning
<1> (plasmid encodes 2 enzymes: 2-aminobenzoate CoA ligase and 2-aminobenzoyl CoA monooxygenase/reductase [4]) [4]
<2> (enzyme is a fusion protein of a monooxygenase and a reductase, plasmid also encodes enzymes of β-oxidation [5]) [5]

6 Stability

Temperature stability
0 <1> (<1> 50% loss of activity in 2 days when stored highly concentrated, in dilute solutions it loses 50% of activity in 2 min, NADH, FAD and 2-aminobenzoyl CoA enhance stability [7]) [7]

25 <1> (<1> 50% loss of activity in 2 days when stored highly concentrated [7]) [7]
40 <1> (<1> unstable above [1]) [1]

Storage stability
<1>, -196°C stable [7]
<1>, -196°C stable for at least 6 months [1]

References

[1] Buder, R.; Fuchs, G.: 2-Aminobenzoyl-CoA monooxygenase/reductase, a novel type of flavoenzyme. Purification and some properties of the enzyme. Eur. J. Biochem., **185**, 629-635 (1989)

[2] Buder, R.; Ziegler, K.; Fuchs, G.; Langkau, B.; Ghisla, S.: 2-Aminobenzoyl-CoA monooxygenase/reductase, a novel type of flavoenzyme. Studies on the stoichiometry and the course of the reaction. Eur. J. Biochem., **185**, 637-643 (1989)

[3] Langkau, B.; Ghisla, S.; Buder, R.; Ziegler, K.; Fuchs, G.: 2-Aminobenzoyl-CoA monooxygenase/reductase, a novel type of flavoenzyme. Identification of the reaction products. Eur. J. Biochem., **191**, 365-371 (1990)

[4] Altenschmidt, U.; Eckerskorn, C.; Fuchs, G.: Evidence that enzymes of a novel aerobic 2-amino-benzoate metabolism in denitrifying Pseudomonas are coded on a small plasmid. Eur. J. Biochem., **194**, 647-653 (1990)

[5] Schuhle, K.; Jahn, M.; Ghisla, S.; Fuchs, G.: Two similar gene clusters coding for enzymes of a new type of aerobic 2-aminobenzoate (anthranilate) metabolism in the bacterium Azoarcus evansii. J. Bacteriol., **183**, 5268-5278 (2001)

[6] Torres, R.A.; Bruice, T.C.: Theoretical investigation of the [1,2]-sigmatropic hydrogen migration in the mechanism of oxidation of 2-aminobenzoyl-CoA by 2-aminobenzoyl-CoA monooxygenase/reductase. Proc. Natl. Acad. Sci. USA, **96**, 14748-14752 (1999)

[7] Langkau, B.; Vock, P.; Massey, V.; Fuchs, G.; Ghisla, S.: 2-Aminobenzoyl-CoA monooxygenase/reductase. Evidence for two distinct loci catalyzing substrate monooxygenation and hydrogenation. Eur. J. Biochem., **230**, 676-685 (1995)

[8] Lochmeyer, C.; Koch, J.; Fuchs, G.: Anaerobic degradation of 2-aminobenzoic acid (anthranilic acid) via benzoyl-coenzyme A (CoA) and cyclohex-1-ene-carboxyl-CoA in a denitrifying bacterium. J. Bacteriol., **174**, 3621-3628 (1992)

Tyrosine N-monooxygenase 1.14.13.41

1 Nomenclature

EC number
1.14.13.41

Systematic name
L-tyrosine,NADPH:oxygen oxidoreductase (N-hydroxylating)

Recommended name
tyrosine N-monooxygenase

Synonyms
cytochrome P450Tyr
tyrosine N-monooxygenase
cytochrome P-450tyr
tyrosine N-hydroxylase

CAS registry number
159447-19-5

2 Source Organism

<1> *Sorghum bicolor* [1, 2, 3, 4, 5]
<2> *Sinapis alba* [6]

3 Reaction and Specificity

Catalyzed reaction
N,N-dihydroxytyrosine = (Z)-[*p*-hydroxyphenylacetaldehyde oxime] + CO_2 + H_2O (the second reaction is followed by spontaneous eliminative decarboxylation)
N-hydroxytyrosine + O_2 + NADPH + H^+ = N,N-dihydroxytyrosine + $NADP^+$ + H_2O (second reaction)
tyrosine + O_2 + NADPH + H^+ = N-hydroxytyrosine + $NADP^+$ + H_2O (first reaction)

Reaction type
oxidation
redox reaction
reduction

Natural substrates and products

S L-tyrosine + NADPH + O_2 <1> (<1> enzyme in biosynthesis of cyanogenic glucosides [1,3,4,5]) [1, 3, 4, 5]

P N-hydroxy-L-tyrosine + $NADP^+$ + H_2O

Substrates and products

S L-tyrosine + NADPH + O_2 <1> (<1> high substrate specificity [2]) (Reversibility: ? <1> [2, 3]) [1, 2, 3]

P N-hydroxy-L-tyrosine + $NADP^+$ + H_2O <1> [1, 2, 3]

S L-tyrosine + NADPH + O_2 <1> (<1> intermediate products: N-hydroxytyrosine, N,N-dihydroxytyrosine, (E)-*p*-hydroxyphenylacetaldoxime [5]) (Reversibility: ? <1> [4, 5]) [4, 5]

P *p*-hydroxyphenylacetaldoxime + $NADP^+$ + H_2O [4, 5]

Cofactors/prosthetic groups

NADPH <1> (<1> better cofactor than NADH [5]) [1, 5]

heme <1, 2> (<1> a heme-thiolate protein [1]; <2> heme binding domain [6]) [1, 6]

Activating compounds

glutathione <1> (<1> at 3 mM, activation rate differs between experiments [4]) [4]

Turnover number (min^{-1})

49.2 <1> (L-tyrosine, <1> in reconstitution experiments using Sorghum bicolor [4]) [4]

350 <1> (L-tyrosine, <1> in E. coli membranes [4]) [4]

Specific activity (U/mg)

0.0233 <1> [4]

49.2 <1> [4]

K_m-Value (mM)

0.013 <1> (NADPH) [5]

0.22 <1> (L-tyrosine, <1> in reconstitution experiments using Sorghum bicolor [4]) [4]

0.3 <1> (NADH) [5]

4 Enzyme Structure

Molecular weight

57000 <1> (<1> SDS-PAGE [2]) [2]

61700 <1> (<1> SDS-PAGE [4]) [4]

61760 <1> (<1> calculated from amino acid sequence [3]) [3]

61890 <1> (<1> calculated from DNA sequence [3]) [3]

Posttranslational modification

no modification <1> (<1> no posttranslational modifications at the N- and C-terminal ends except for the N-terminal methionine removal [3]) [3]

5 Isolation/Preparation/Mutation/Application

Source/tissue

seedling <1> [1]

Localization

membrane <1> [4]
microsome <1> [2]

Purification

<1> (combined use of Renex 690, CHAPS and RTX-100 is optimal for maximal recovery and avoidance of conversion into cytochrome P-420 [2]) [2]
<1> (homogeneity [4]) [4]

Cloning

<1> (full length clone [3]; various N-terminal modifications [4]) [3, 4]
<2> [6]

Engineering

Additional information <1> (<1> mutant 1: first codons of Escherichia coli mRNA are enriched for A's and T's, second codon is changed into GCT, first 8 codons of P_{450} sequence are replaced with the N-terminal sequence of bovine $P_{450}17\alpha$, mutant 2: deletion of 14 amino acids, mutant 3: deletion of 25 amino acids, mutant 4: deletion of 75 amino acids [4]) [4]

6 Stability

Temperature stability

7-9.5 <1> (<1> unstable below pH 7 and above pH 9.5 [2]) [2]

General stability information

<1>, fairly stable [5]

References

[1] Halkier, B.A.; Moller, B.L.: The biosynthesis of cyanogenic glucosides in higher plants. Identification of three hydroxylation steps in the biosynthesis of dhurrin in Sorghum bicolor (L.) Moench and the involvement of 1-ACI$^-$nitro-2-(*p*-hydroxyphenyl)ethane as an intermediate. J. Biol. Chem., **265**, 21114-21121 (1990)

[2] Libbesen, O.; Koch, B.; Halkier, B.A.; Moller, B.L.: Isolation of the heme-thiolate enzyme chytochrome P-450tyr, which catalyzes the committed step in the biosynthesis of the cyanogenic glucoside dhurrin in Sorghum bicolor (L.) Moench. Proc. Natl. Acad. Sci. USA, **91**, 9740-9744 (1994)

[3] Koch, B.M.; Sibbesen, O.; Halkier B.A.; Svendsen, I.; Möller B.L.: The primary sequence of cytochrome P_{450}tyr, the multifunctional N-hydroxylase catalyzing the conversion of L-tyrosine to *p*-hydroxyphenylacetaldehyde oxime in

the biosynthesis of the cyanogenic glucoside dhurrin in Sorghum bicolor (L.) Moench. Arch. Biochem. Biophys., **323**, 177-186 (1995)
[4] Halkier, B.A.; Nielsen, H.L.; Koch, B.; Moller B.L.: Purification and characterization of recombinant cytochrome P_{450}tyr expressed at high levels in Escherichia coli. Arch. Biochem. Biophys., **322**, 369-377 (1995)
[5] Kahn, R.A.; Fahrendorf, T.; Halkier, B.A.; Moller, B.L.: Substrate specificity of the cytochrome P_{450} enzymes CYP79A1 and CYP71E1 involved in the biosynthesis of the cyanogenic glucoside dhurrin in Sorghum bicolor (L.) Moench. Arch. Biochem. Biophys., **363**, 9-18 (1999)
[6] Bak, S.; Nielsen, H.L.; Halkier, B.A.: The presence of CYP79 homologues in glucosinolate-producing plants shows evolutionary conservation of the enzymes in the conversion of amino acid to aldoxime in the biosynthesis of cyanogenic clucosides and glucosinolates. Plant Mol. Biol., **38**, 725-734 (1998)

Hydroxyphenylacetonitrile 2-monooxygenase 1.14.13.42

1 Nomenclature

EC number
1.14.13.42

Systematic name
4-hydroxyphenylacetonitrile,NADPH:oxygen oxidoreductase (2-hydroxylating)

Recommended name
hydroxyphenylacetonitrile 2-monooxygenase

Synonyms
4-HPAN hydroxylase
4-hydroxyphenylacetonitrile hydroxylase
oxygenase, 4-hydroxyphenylacetonitrile mono-

CAS registry number
89287-41-2

2 Source Organism

<1> *Triglochin maritima* [1]
<2> *Sorghum bicolor* [2, 3]

3 Reaction and Specificity

Catalyzed reaction
4-hydroxyphenylacetonitrile + NADPH + H^+ + O_2 = 4-hydroxymandelonitrile + $NADP^+$ + H_2O

Reaction type
oxidation
redox reaction
reduction

Natural substrates and products
S 4-hydroxyphenylacetonitrile + NADPH + O_2 <1, 2> (<1,2>, enzyme is involved in biosynthesis of cyanogenic glucosides [1-3]) [1-3]
P 4-hydroxymandelonitrile + $NADP^+$ + H_2O

Substrates and products

S 4-hydroxyphenylacetonitrile + NADPH + O_2 <1, 2> (Reversibility: ? <1, 2> [1-3]) [1-3]

P 4-hydroxymandelonitrile + $NADP^+$ + H_2O <1, 2> [1-3]

Cofactors/prosthetic groups

NADPH <1, 2> [1, 2, 3]

cytochrome P_{450} <2> (<2>, dependent on [3]) [3]

K_m-Value (mM)

0.2-0.3 <2> (4-hydroxyphenylacetonitrile) [3]

5 Isolation/Preparation/Mutation/Application

Source/tissue

seedling <1, 2> (<2>, etiolated [3]) [1, 2, 3]

Localization

membrane <1> (<1>, bound [1]) [1]

microsome <1> [1]

Purification

<2> [3]

References

[1] Hösel, W.; Schiel, O.: Biosynthesis of cyanogenic glucosides: in vitro analysis of the glucosylation step. Arch. Biochem. Biophys., **229**, 177-186 (1984)

[2] Halkier, B.A.; Moller, B.L.: The biosynthesis of cyanogenic glucosides in higher plants. Identification of three hydroxylation steps in the biosynthesis of dhurrin in Sorghum bicolor (L.) Moench and the involvement of 1-ACI-nitro-2-(*p*-hydroxyphenyl)ethane as an intermediate. J. Biol. Chem., **265**, 21114-21121 (1990)

[3] Sibbesen, O.; Lykkesfeldt, J.; Koch, B.; Moeller, B.L.: Purification of the hydroxylating enzyme system involved in the biosynthesis of the cyanogenic glucoside dhurrin in Sorghum bicolor (L.) Moench. Cytochrome P-450 (Proc. First Int. Sci. Meeting of the Cassava Biotechnology Network, August 25-28), **7**, 324-326 (1992)

Questin monooxygenase 1.14.13.43

1 Nomenclature

EC number
1.14.13.43

Systematic name
questin,NADPH:oxygen oxidoreductase (hydroxylating, anthraquinone-ring-opening)

Recommended name
questin monooxygenase

Synonyms
questin oxygenase

CAS registry number
115232-45-6

2 Source Organism

<1> *Aspergillus terreus* [1]

3 Reaction and Specificity

Catalyzed reaction
questin + NADPH + H^+ + O_2 = sulochrin + $NADP^+$ + H_2O (The enzyme cleaves the anthraquinone ring of questin to form a benzophenone. Involved in the biosynthesis of the seco-anthraquinone (+)-geodin)

Reaction type
oxidation
redox reaction
reduction

Natural substrates and products
S questin + NADPH + O_2 <1> (<1> absolute requirement for oxygen and NADPH [1]) (Reversibility: ? <1> [1]) [1]
P sulochrin + $NADP^+$ + H_2O <1> [1]

Substrates and products
S questin + NADPH + O_2 <1> (<1> absolute requirement for oxygen and NADPH [1]) (Reversibility: ? <1> [1]) [1]
P sulochrin + $NADP^+$ + H_2O <1> [1]

Cofactors/prosthetic groups

NADPH <1> (<1> highly specific [1]) [1]

Specific activity (U/mg)

0.03 <1> (<1> crude extract [1]) [1]

pH-Optimum

7.5 <1> [1]

5 Isolation/Preparation/Mutation/Application

Source/tissue

mycelium <1> [1]

Localization

cytosol <1> [1]

Purification

<1> (partial, additional functional protein or proteins required for full activity [1]) [1]

6 Stability

Temperature stability

4 <1> (<1> complete loss of activity overnight [1]) [1]

General stability information

<1>, very unstable, polyols, non-ionic detergents and EDTA have stabilizing effects [1]

Storage stability

<1>, cold room, 0.01 M Tris-HCl, pH 7.5, 50% ethylene glycol, v/v, 1 mM EDTA, 0.05% Tween 80, w/v, 70-80% of its original activity after 2 days [1]

References

[1] Fuji, I.; Ebizuka, Y.; Sankawa, U.: A novel anthraquinone ring cleavage enzyme from Aspergillus terreus. J. Biochem., **103**, 878-883 (1988)

2-Hydroxybiphenyl 3-monooxygenase 1.14.13.44

1 Nomenclature

EC number
1.14.13.44

Systematic name
2-hydroxybiphenyl,NADH:oxygen oxidoreductase (3-hydroxylating)

Recommended name
2-hydroxybiphenyl 3-monooxygenase

Synonyms
HbpA
oxygenase, 2-hydroxybiphenyl 3-mono-

CAS registry number
118251-39-1

2 Source Organism

<1> *Pseudomonas azelaica* (HBP1 [1,2,3,4]) [1, 2, 3, 4]

3 Reaction and Specificity

Catalyzed reaction
2-hydroxybiphenyl + NADH + H^+ + O_2 = 2,3-dihydroxybiphenyl + NAD^+ + H_2O (<1>, ternary complex mechanism [3])

Reaction type
oxidation
redox reaction
reduction

Natural substrates and products
S 2-hydroxybiphenyl + $NADH_2$ + O_2 <1> (<1>, first enzyme of 2-hydroxybiphenyl degradation [4]) [4]
P 2,3-dihydroxybiphenyl + NAD^+ + H_2O

Substrates and products
S 2,2'-dihydroxybiphenyl + NADH + O_2 <1> (Reversibility: ? <1> [4]) [4]
P 2,2',3-trihydroxybiphenyl + NAD^+ + H_2O
S 2,2'-dihydroxybiphenyl + NADPH + O_2 <1> (Reversibility: ? <1> [4]) [4]

P 2,2',3-trihydroxybiphenyl + $NADP^+$ + H_2O
S 2,5-dihydroxybiphenyl + NADH + O_2 <1> (Reversibility: ? <1> [4]) [4]
P 2,3,5-trihydroxybiphenyl + NAD^+ + H_2O
S 2-ethylphenol + NADH + O_2 <1> (Reversibility: ? <1> [4]) [4]
P 1,2-dihydroxy-3-ethylbenzene + NAD^+ + H_2O
S 2-hydroxybiphenyl + NADH + O_2 <1> (<1>, ternary complex mechanism in which the aromatic substrate has strict control in both the reductive and oxidative half-reaction in a way that reactions leading to substrate hydroxylation are favored over those leading to the futile formation of hydrogen peroxide. NAD^+ release from the reduced enzyme-substrate complex is the slowest step in catalysis [3]) (Reversibility: ? <1> [3, 4]) [3, 4]
P 2,3-dihydroxybiphenyl + NAD^+ + H_2O <1> [3, 4]
S 2-hydroxybiphenyl + NADH + O_2 <1> (<1>, the activity of the mutant enzyme HbpAind is six times lower than that of the wild-type enzyme [2]) (Reversibility: ? <1> [1, 2, 3]) [1, 2, 3]
P 2,3-dihydroxybiphenyl + NAD^+ + H_2O
S 2-hydroxybiphenyl + NADPH + O_2 <1> (Reversibility: ? <1> [4]) [4]
P 2,3-dihydroxybiphenyl + $NADP^+$ + H_2O
S 2-methylphenol + NADH + O_2 <1> (Reversibility: ? <1> [4]) [4]
P 1,2-dihydroxy-3-methylbenzene + NAD^+ + H_2O
S 2-propylphenol + NADH + O_2 <1> (Reversibility: ? <1> [3, 4]) [3, 4]
P 1,2-dihydroxy-3-propylbenzene + NAD^+ + H_2O
S 2-sec-butylphenol + NADH + O_2 <1> (Reversibility: ? <1> [1, 3, 4]) [1, 3, 4]
P 2-sec-butylcatechol + NAD^+ + H_2O
S 2-sec-butylphenol + NADPH + O_2 <1> (Reversibility: ? <1> [4]) [4]
P 2-sec-butylcatechol + $NADP^+$ + H_2O
S 2-tert-butylphenol + NADH + O_2 <1> (Reversibility: ? <1> [1, 4]) [1, 4]
P 1,2-dihydroxy-3-tert-butylbenzene + NAD^+ + H_2O
S guaiacol + NADH + O_2 <1> (<1>, i.e. 2-methoxyphenol [1]) (Reversibility: ? <1> [1]) [1]
P 2,3-dihydroxy-methoxybenzene + NAD^+ + H_2O
S indole + NADH + O_2 <1> (Reversibility: ? <1> [2]) [2]
P ?
S Additional information <1> (<1>, Ile244 is located in the substrate binding pocket and is involved in accomodating the phenyl substituent of the phenol [1]; <1>, Asp222 is involved in substrate activation in HbpA [2]; <1>, the substrates partially uncouple oxygen activation from hydroxylation with resultant reduction of both atoms of oxygen to form hydrogen peroxide [4]) [1, 2, 4]
P ?

Inhibitors

2,3-dihydroxybiphenyl <1> (<1>, inhibits reaction with 2-hydroxybiphenyl [4]) [4]
$AgNO_3$ <1> (<1>, 0.01 mM, complete inhibition [4]) [4]

$CuSO_4$ <1> (<1>, 0.01 mM, complete inhibition [4]) [4]
$FeSO_4$ <1> (<1>, 0.08 mM, 30% inhibition [4]) [4]
$HgCl_2$ <1> (<1>, 0.01 mM, complete inhibition [4]) [4]
NaCl <1> (<1>, 10 mM, 36% inhibition. 100 mM, 89% inhibition [4]) [4]
p-hydroxymercuribenzoate <1> (<1>, partial protection in presence of 2-hydroxybiphenyl, reversed by excess of dithiothreitol [4]) [4]

Cofactors/prosthetic groups

FAD <1> (<1>, flavoenzyme [1,3]; <1>, one molecule of noncovalently bound FAD per subunit [4]) [1, 3, 4]
NADH <1> [1-4]
NADPH <1> (<1>, can replace NADH as electron donor, K_m-value for NADPH is much higher than for NADH [4]) [4]

Turnover number (min^{-1})

0.3 <1> (indole, <1>, wild-type enzyme [2]) [2]
5.4 <1, 1> (indole, <1>, mutant enzyme HbpAind [2]) [2]
57 <1> (guaiacol) [1]
84 <1> (2-hydroxybiphenyl, <1>, turnover rate refers to the enzyme monomer and not to the tetramer [3]) [3]
138 <1> (2-hydroxybiphenyl, <1>, mutant enzyme HbpAind [2]) [2]
540 <1> (2,2'-dihydroxybiphenyl, <1>, reaction with NADH and O_2 [4]) [4]
564 <1> (2,2'-dihydroxybiphenyl, <1>, reaction with NADPH and O_2 [4]) [4]
588 <1> (NADH, <1>, reaction with 2,2'-dihydroxybiphenyl and O_2 [4]) [4]
612 <1> (2-sec-butylphenol, <1>, reaction with NADPH and O_2 [4]) [4]
672 <1> (NADPH, <1>, reaction with 2,2'-dihydroxybiphenyl and O_2 [4]) [4]
714 <1> (2-hydroxybiphenyl, <1>, wild-type enzyme [1]) [1]
768 <1> (2-hydroxybiphenyl, <1>, reaction with NADPH and O_2 [4]) [4]
870 <1> (2-sec-butylphenol, <1>, wild-type enzyme [1]) [1]
894 <1> (2-hydroxybiphenyl, <1>, reaction with NADH and O_2 [4]) [4]
936 <1> (2-hydroxybiphenyl, <1>, wild-type enzyme [2]) [2]
948 <1> (2-sec-butylphenol, <1>, reaction with NADH and O_2 [4]) [4]
972 <1> (NADH, <1>, reaction with 2-hydroxybiphenyl and O_2 [4]) [4]
972 <1> (O_2, reaction with 2-hydroxybiphenyl and NADH [4]) [4]
1128 <1> (NADPH, <1>, reaction with 2-hydroxybiphenyl and O_2 [4]) [4]
Additional information <1> (<1>, turnover-numbers for mutant enzymes [1]) [1]

Specific activity (U/mg)

Additional information <1> [4]

K_m-Value (mM)

0.0019 <1> (2-hydroxybiphenyl) [3]
0.0028 <1> (2-hydroxybiphenyl, <1>, reaction with NADH and O_2 [4]) [4]
0.0031 <1> (2-hydroxybiphenyl, <1>, reaction with NADPH and O_2 [4]) [4]
0.0034 <1> (2,2'-dihydroxybiphenyl, <1>, reaction with NADPH and O_2 [4]) [4]
0.004 <1> (2,2'-dihydroxybiphenyl, <1>, reaction with NADH and O_2 [4]) [4]

0.0057 <1> (2-sec-butylphenol, <1>, reaction with NADH or NADPH and O_2 [4]) [4]
0.0097 <1> (NADH, <1>, reaction with 2-hydroxybiphenyl [3]) [3]
0.0216 <1> (NADH, <1>, reaction with 2,2'-dihydroxybiphenyl and O_2 [4]) [4]
0.0268 <1> (NADH, <1>, reaction with 2-hydroxybiphenyl and O_2 [4]) [4]
0.0292 <1> (O_2, reaction with 2-hydroxybiphenyl and NADH [4]) [4]
0.0943 <1> (NADPH, <1>, reaction with 2,2'-dihydroxybiphenyl and O_2 [4]) [4]
0.137 <1> (NADPH, <1>, reaction with 2-hydroxybiphenyl and O_2 [4]) [4]
Additional information <1> (<1>, K_m-values for wild-type and mutant enzymes [1]) [1]

K_i-Value (mM)
0.9 <1> (2,3-dihydroxybiphenyl) [4]

pH-Optimum
7.5 <1> [4]

pH-Range
7.2-7.8 <1> (<1>, more than 80% of maximal activity at pH 7.2 and pH 7.8, beyond pH 7.8 activity declines abruptly with increasing pH [4]) [4]

4 Enzyme Structure

Molecular weight
256000 <1> (<1>, gel filtration [4]) [4]

Subunits
tetramer <1> (<1>, 4 * 60000, SDS-PAGE [4]) [4]

5 Isolation/Preparation/Mutation/Application

Purification
<1> (recombinant enzyme [3]; native and recombinant enzyme [4]) [3, 4]

Engineering
I244V <1> (<1>, mutant enzyme has a 30% higher specific activity with 2-sec-butylphenol, guaiacol, and 2-hydroxybiphenyl. The K_m-value for guaiacol decreases with this mutant, but the K_m-value for 2-hydroxybiphenyl increase [1]) [1]
V368A/L417F <1> (<1>, double replacement improves the efficiency of substrate hydroxylation by reducing the uncoupled oxidation of NADH. With guaiacol as substrate, the V_{max} is increased and the K_m-value is decreased. With 2-tert-butylphenol as substrate the turnover number is increased more than 5fold [1]) [1]

Additional information <1> (<1>, direct enzyme evolution of EC 1.14.13.44 results in an enzyme variant HbpAind, that hydroxylates indole and indole derivatives such as hydroxyindoles and 5-bromoindole. The wild-type protein does not catalyze theses reactions. HbpAind contains amino acid substitutions D222V and V368A. The activity for indole hydroxylation is increased 18fold in this variant [2]) [2]

6 Stability

Storage stability

<1>, -20°C, pure enzyme at concentration of 3.8 mg/ml in 50 mM phosphate buffer, pH 7.5, stable for at least 6 months [4]

References

[1] Meyer, A.; Schmid, A.; Held, M.; Westphal, A.H.; Rothlisberger, M.; Kohler, H.P.E.; Van Berkel, W.J.H.; Witholt, B.: Changing the substrate reactivity of 2-hydroxybiphenyl 3-monooxygenase from Pseudomonas azelaica HBP1 by directed evolution. J. Biol. Chem., **277**, 5575-5582 (2002)

[2] Meyer, A.; Wursten, M.; Schmid, A.; Kohler, H.P.; Witholt, B.: Hydroxylation of indole by laboratory-evolved 2-hydroxybiphenyl 3-monooxygenase. J. Biol. Chem., **277**, 34161-34167 (2002)

[3] Suske, W.A.; Van Berkel, W.J.H.; Kohler, H.P.E.: Catalytic mechanism of 2-hydroxybiphenyl 3-monooxygenase, a flavoprotein from Pseudomonas azelaica HBP1. J. Biol. Chem., **274**, 33355-33365 (1999)

[4] Suske, W.A.; Held, M.; Schmid, A.; Fleischmann, T.; Wubbolts, M.G.; Kohler, H.P.E.: Purification and characterization of 2-hydroxybiphenyl 3-monooxygenase, a novel NADH-dependent, FAD-containing aromatic hydroxylase from Pseudomonas azelaica HBP1. J. Biol. Chem., **272**, 24257-24265 (1997)

CMP-N-Acetylneuraminate monooxygenase 1.14.13.45

1 Nomenclature

EC number
1.14.13.45

Systematic name
CMP-N-acetylneuraminate,NAD(P)H:oxygen oxidoreductase (hydroxylating)

Recommended name
CMP-N-acetylneuraminate monooxygenase

Synonyms
CMAH
CMP-N-acetylneuraminic acid
CMP-N-acetylneuraminic acid hydroxylase
CMP-Neu$_5$Ac hydroxylase
oxygenase, cytidine monophosphoacetylneuraminate mono-

CAS registry number
116036-67-0

2 Source Organism

<-1> no activity in *Homo sapiens* (the CMAH gene is inactivated shortly before the time when the brain expansion began in humankind's ancestry, 2.1-2.2 million years ago [11]; the human CMP-N-acetylneuraminic acid hydroxylase is inactive because of a partial deletion in the hydroxylase gene [12]) [11, 12]
<1> *Sus scrofa* [1, 6, 7, 17, 18, 21, 24]
<2> *Mus musculus* [2, 3, 4, 7, 9, 10, 12, 14, 15, 16, 22, 23]
<3> *Asterias rubens* [5, 8, 13, 19, 20]
<4> *Bos taurus* [7]
<5> *Rattus norvegicus* [23]

3 Reaction and Specificity

Catalyzed reaction
CMP-N-acetylneuraminate + NAD(P)H + H^+ + O_2 = CMP-N-glycoloylneuraminate + $NAD(P)^+$ + H_2O

Reaction type

oxidation
redox reaction
reduction

Natural substrates and products

S CMP-N-acetylneuraminate + NADH + O_2 <1, 2, 3, 4> (<2>, the enzyme is the key for regulation of the overall velocity of CMP-NeuAc hydroxylation and consequently for the expression of N-glycoloylneuraminic acid glycoconjugates [9]; <2>, binding of CMP-N-acetylneuraminate to CMP-N-acetylneuraminate hydroxylase changes conformation of the enzyme so as to construct a recognition site for cytochrome b_5, followed by the formation of a ternary complex through this domain. Then the transport of electrons from NAD(P)H to the enzyme through cytochrome b_5 takes place, CMP-N-acetylneuraminate is converted to CMP-N-glycoloylneuraminic acid and finally the ternary complex dissociates into its components to release CMP-N-glycoloylneuraminic acid [14]; <2>, a regulation of CMP-N-acetylneuraminate hydroxylation and thus the ratio of glycoconjugate-bound N-acetylneuraminate and N-acetylglycoloylneuraminate might occur by varying the amount of hydroxylase protein within the cell, possibly by controlling the expression of the hydroxylase gene [15]; <2>, key enzyme for the expression of N-glycoloylneuraminic acid [16]; <1>, the biosynthesis of the sialic acid N-glycolylneuraminic acid occurs by the action of cytidine monophosphate-N-acetylneuraminate hydroxylase. Incorporation of N-glycoloylneuraminic acid into glycoconjugates is generally controlled by the amount of hydroxylase protein expressed in a tissue [17]; <2>, the enzyme plays a decisive role in governing the relative amounts of N-acetylneuraminate and N-acetylglycolylneuraminate occuring in the glycoconjugates of a tissue [23]) (Reversibility: ? <1, 2, 3, 4> [1, 2, 7, 8, 9, 14, 15, 16, 17, 23]) [1, 2, 7, 8, 9, 14, 15, 16, 17, 23]

P CMP-N-glycoloylneuraminate + NAD^+ + H_2O

Substrates and products

S CMP-N-acetylneuraminate + 6,7-dimethyl-5,6,7,8-tetrahydrobiopterin + O_2 <1> (Reversibility: ? <1> [1]) [1]

P CMP-N-glycoloylneuraminate + ?

S CMP-N-acetylneuraminate + NADH + O_2 <1, 2, 3, 4> (<1>, NADPH and NADH are by far the most effective cofactors [1]) (Reversibility: ? <1, 2, 3, 4> [1, 2, 3, 7, 8, 9, 13, 15, 17, 19, 20, 21, 22]) [1, 2, 3, 7, 8, 9, 13, 15, 17, 19, 20, 21, 22]

P CMP-N-glycoloylneuraminate + NAD^+ + H_2O

S CMP-N-acetylneuraminate + NADPH + O_2 <1, 2> (<1>, NADPH and NADH are by far the most effective cofactors [1]) (Reversibility: ? <1, 2> [1, 2, 3]) [1, 2, 3]

P CMP-N-glycoloylneuraminate + $NADP^+$ + H_2O

S CMP-N-acetylneuraminate + ascorbic acid + O_2 <1> (<2>, ascorbate is ineffective [2]) (Reversibility: ? <1> [1]) [1]

P CMP-N-glycoloylneuraminate + dehydroascorbate + H_2O

S Additional information <2> (<2>, no activity towards free or α-glycosidically bound N-acetylneuraminic acid [4]) [4]

P ?

Inhibitors

1,10-phenanthroline <1, 2, 3> (<3>, 2 mM, 88% inhibition [19]; <2>, 5 mM, complete inhibition [22]) [13, 19, 20, 21, 22]
2,2'-dipyridyl <3> (<3>, 2 mM, 27% inhibition [19]) [19, 20]
CHAPS <3> (<3>, 15 mM, complete inhibition [13]) [13]
CMP-N-glycoloylneuraminate <2> [22]
Ca^{2+} <3> (<3>, 0.5 mM, 12% inhibition [13]) [13]
Co^{2+} <2> [2]
Cu^{2+} <2> [2]
EDTA <2> [2, 12]
Hg^{2+} <3> (<3>, slight inhibition [13]) [13]
KCN <1, 2, 3> (<3>, 2 mM, 80% inhibition [19]; <2>, 5 mM, 49% inhibition [22]) [13, 19, 21, 22]
Mg^{2+} <2> [2]
Mn^{2+} <2, 3> (<3>, 0.5 mM, slight inhibition [13]) [2, 13]
Na_2HPO_4 <2> (<2>, 2 mM, 86% inhibition [22]) [22]
$Na_4P_2O_7$ <2> (<2>, 2 mM, 54% loss of activity [22]) [22]
Ni^{2+} <2> [2]
Tiron <1, 2, 3> (<2>, 5 mM, 82% inhibition [22]) [13, 21, 22]
Zn^{2+} <2> [2]
Zwittergent 3-12 <3> (<3>, 5 mM, 40% inhibition [13]) [13]
anti-(rat cytochrome b_5) antiserum <1, 2> [7, 22]
azide <3> [13]
cardiolipin <2> (<2>, 3 mM, 15% inhibition [22]) [22]
cholic acid <3> (<3>, 10 mM, 89% inhibition [13]) [13]
decylglucopyranoside <3> (<3>, 5 mM, 22% inhibition [13]) [13]
ferrozine <1, 2, 3> (<3>, 2 mM, 83% inhibition [19]; <2>, 0.7 mM, 82% inhibition [22]) [13, 19, 20, 21, 22]
octylglucopyranoside <3> (<3>, 30 mM, 48% inhibition [13]) [13]
phosphatidic acid <2> (<2>, 3 mM, 20% inhibition [22]) [22]
phosphatidylinositol <2> (<2>, 3 mM, 68% inhibition [22]) [22]
Additional information <3> (<3>, not inhibited by increased ionic strength, no inhibition by 1 M NaCl [13]) [13]

Cofactors/prosthetic groups

NADH <1, 2, 3, 4> (<1>, NADPH and NADH are by far the most effective cofactors [1]; <2>, NADH is much more effective than NADPH [3]; <3>, most effective cofactor, optimal activity at 0.4 mM NADH, higher concentrations are slightly inhibitory [13]) [1, 2, 3, 7, 8, 9, 13, 15, 17, 19, 20, 21, 22]
NADPH <1, 2> (<1>, NADPH and NADH are by far the most effective cofactors [1]; <2>, NADH is much more effective than NADPH [3]) [1, 2, 3]
cytochrome b_5 <2> (<2>, electron carrier, essential for activity [22]) [22]

Activating compounds

Nonidet P-40 <2> (<2>, effective inhibitor [22]) [22]
SDS <2> (<2>, 1 mM, modest activation [22]) [22]
Triton X-100 <2> (<2>, effective inhibitor [22]) [22]
ascorbate <3> (<3>, activates [13]) [13]
decyl glucoside <2> (<2>, activation [22]) [22]
dithiothreitol <3> (<3>, activates [13]) [13]
glutathione <3> (<3>, activates [13]) [13]
octanoic acid <2> (<2>, 1 mM, modest activation [22]) [22]
octyl glucoside <2> (<2>, effective inhibitor [22]) [22]
Additional information <2, 3> (<3>, no activation by non-ionic detergents [13]; <2>, highest activity in 50 mM Hepes buffer, significant inhibition at increasing concentrations [22]) [13, 22]

Metals, ions

iron <1, 2, 3> (<1>, iron-sulfur protein of the Rieske type [6]; <2>, contains non-heme iron as an electron acceptor [9]; <3>, the enzyme contains a non-haem iron cofactor [13]) [6, 9, 13]

Specific activity (U/mg)

0.00093 <3> [20]
0.126 <2> [15]
0.816 <1> [21]
6.8 <2> [9]

K_m-Value (mM)

0.003 <1> (CMP-N-acetylneuraminate) [21]
0.005 <2> (CMP-N-acetylneuraminate, <2>, amphiphilic system [15]) [9, 15]
0.0072 <3> (CMP-N-acetylneuraminate) [19]
0.013 <2> (CMP-N-acetylneuraminate, <2>, soluble system [15]) [15]
0.018 <3> (CMP-N-acetylneuraminate) [13]
0.6-2.6 <2> (CMP-N-acetylneuraminate) [4]
Additional information <2> [2]

pH-Optimum

6-6.4 <3> [13]
6-6.6 <3> [19]
6.8-7.4 <2> [22]

pH-Range

5.6-6.8 <3> (<3>, about 60% of maximal activity at pH 5.6 and pH 6.8 [13]) [13]

Temperature optimum (°C)

22-27 <3> [19]
25-33 <3> [13]

Temperature range (°C)

15-43 <3> (<3>, 15°C: about 55% of maximal activity, 43°C: about 30% of maximal activity [13]) [13]

4 Enzyme Structure

Molecular weight

17000 <2> (<2>, gel filtration [22]) [22]
56000 <2> (<2>, gel filtration [7]) [7]
58000 <2> (<2>, gel filtration [9,15]) [9, 15]
60000 <1> (<1>, gel filtration [21]) [21]

Subunits

monomer <1, 2> (<2>, 1 * 64000, SDS-PAGE [9,15]; <1>, 1 * 65000, SDS-PAGE [21]) [9, 15, 21]

5 Isolation/Preparation/Mutation/Application

Source/tissue

alimentary canal <3> [13]
body wall <3> [13]
brain <-1> (<-1>, the CMAH gene is inactivated shortly before the time when the brain expansion began in humankind's ancestry, 2.1-2.2 million years ago [11]) [11]
gonad <3> [13, 19, 20]
heart <1> (<1>, weak activity [17]) [17]
kidney <1> (<1>, weak activity [17]) [17]
liver <1, 2, 5> (<1>, weak activity [17]) [2, 3, 7, 9, 10, 14, 15, 16, 17, 22, 23]
lung <1> [17]
lymph node <1> [17, 18]
lymphocyte <1> (<1>, from thymus, spleen, lymph node and peripheral blood. Highest activity in peripheral blood lymphocytes [18]) [18]
myeloma cell <2> [4]
small intestine <1> (<1>, no significant temporal alterations in the activity in foetal and newborn small intestine. Birth is followed by a 2-8fold decrease in activity, depending on the region of the small intestine. Increase in activity from duodenum to ileum [24]) [17, 24]
spleen <1> [17]
submandibular gland <1> [1, 6, 7, 17, 21]
thymus <1> [17]

Localization

cytosol <1, 2> (<2>, full-length enzyme with normal enzymatic activity [16]; <1>, in the vicinity of the nuclear membrane and the outer membrane of the mitochondria [18]) [3, 4, 6, 9, 10, 16, 18]
endoplasmic reticulum <2> (<2>, naturally occuring truncated protein lacking 46 amino acids in the middle of the normal full-length protein [16]) [16]
membrane <3> (<3>, bound to [8,19]) [8, 19]
microsome <2> [15]
soluble <3> [5]

Purification

<1> [21]
<2> [9, 10, 15, 17]
<3> [8, 20]

Cloning

<-1> (the human CMP-N-acetylneuraminic acid hydroxylase is inactive because of a partial deletion in the hydroxylase gene [12]) [12]
<1> [6]
<2> (expression in COS-1 cells [10,16]) [10, 12, 16]
<3> (expression in Escherichia coli [8]) [8]

Engineering

Additional information <2> (<2>, naturally occuring truncated protein, lacking 46 amino acids in the middle of the normal full-length protein, causes a change in intracellular distribution of the enzyme from cytosol to endoplasmic reticulum and a loss in activity [16]) [16]

6 Stability

Temperature stability

4 <3> (<3>, 48 h, 10% loss of activity [20]) [20]
25 <3> (<3>, 2 h, 50% loss of activity [20]) [20]

General stability information

<2>, enzyme is greatly stabilized by CMP-N-acetylneuraminate [9]
<3>, after each cycle of freezing and thawing, 25% loss of activity [20]

Storage stability

<2>, -80°C, 0.2 mM CMP-N-acetylneuraminate, stable for at least 6 months [9]
<3>, -80°C, very stable [20]
<3>, 4°C, 48 h, 10% loss of activity [20]

References

[1] Shaw, L.; Schauer, R.: The biosynthesis of N-glycoloylneuraminic acid occurs by hydroxylation of the CMP-glycoside of N-acetylneuraminic acid. Biol. Chem. Hoppe-Seyler, **369**, 477-486 (1988)
[2] Shaw, L.; Schauer, R.: Detection of CMP-N-acetylneuraminic acid hydroxylase activity in fractionated mouse liver. Biochem. J., **263**, 355-363 (1989)
[3] Kozutsumi, Y.; Kawano, T.; Yamakawa, T.; Suzuki, A.: Participation of cytochrome b_5 in CMP-N-acetylneuraminic acid hydroxylation in mouse liver cytosol. J. Biochem., **108**, 704-706 (1990)
[4] Muchmore, E.A.; Milewski, M.; Varki, A.; Diaz, S.: Biosynthesis of N-glycolyneuraminic acid. The primary site of hydroxylation of N-acetylneurami-

nic acid is the cytosolic sugar nucleotide pool. J. Biol. Chem., **264**, 20216-20223 (1989)

[5] Bergwerff, A.A.; Hulleman, S.H.D.; Kamerling, J.P.; Vliegenthart, J.F.G.; Shaw, L.; Reuter, G.; Schauer, R.: Nature and biosynthesis of sialic acids in the starfish Asterias rubens. Identification of sialo-oligomers and detection of S-adenosyl-L-methionine: N-acylneuraminate 8-O-methyltransferase and CMP-N-acetylneuraminate monooxygenase activities. Biochimie, **74**, 25-37 (1992)

[6] Schlenzka, W.; Shaw, L.; Kelm, S.; Schmidt, C.L.; Bill, E.; Trautwein, A.X.; Lottspeich, F.; Schauer, R.: CMP-N-acetylneuraminic acid hydroxylase: the first cytosolic Rieske iron-sulfur protein to be described in Eukarya. FEBS Lett., **385**, 197-200 (1996)

[7] Shaw, L.; Schneckenburger, P.; Schlenzka, W.; Carlsen, J.; Christiansen, K.; Juergensen, D.; Schauer, R.: CMP-N-acetylneuraminic acid hydroxylase from mouse liver and pig submandibular glands. Interaction with membrane-bound and soluble cytochrome b_5-dependent electron transport chains. Eur. J. Biochem., **219**, 1001-1011 (1994)

[8] Martensen, I.; Schauer, R.; Shaw, L.: Cloning and expression of a membrane-bound CMP-N-acetylneuraminic acid hydroxylase from the starfish Asterias rubens. Eur. J. Biochem., **268**, 5157-5166 (2001)

[9] Kawano, T.; Kozutsumi, Y.; Kawasaki, T.; Suzuki, A.: Biosynthesis of N-glycolylneuraminic acid-containing glycoconjugates. Purification and characterization of the key enzyme of the cytidine monophospho-N-acetylneuraminic acid hydroxylation system. J. Biol. Chem., **269**, 9024-9029 (1994)

[10] Kawano, T.; Koyama, S.; Takematsu, H.; Kozutsumi, Y.; Kawasaki, H.; Kawashima, S.; Kawasaki, T.; Suzuki, A.: Molecular cloning of cytidine monophospho-N-acetylneuraminic acid hydroxylase. Regulation of species- and tissue-specific expression of N-glycolylneuraminic acid. J. Biol. Chem., **270**, 16458-16463 (1995)

[11] Chou, H.H.; Hayakawa, T.; Diaz, S.; Krings, M.; Indriati, E.; Leakey, M.; Paabo, S.; Satta, Y.; Takahata, N.; Varki, A.: Inactivation of CMP-N-acetylneuraminic acid hydroxylase occurred prior to brain expansion during human evolution. Proc. Natl. Acad. Sci. USA, **99**, 11736-11741 (2002)

[12] Irie, A.; Suzuki, A.: CMP-N-acetylneuraminic acid hydroxylase is exclusively inactive in humans. Biochem. Biophys. Res. Commun., **248**, 330-333 (1998)

[13] Schlenzka, W.; Shaw, L.; Schauer, R.: Catalytic properties of the CMP-N-acetylneuraminic acid hydroxylase from the starfish Asterias rubens: comparison with the mammalian enzyme. Biochim. Biophys. Acta, **1161**, 131-138 (1993)

[14] Takematsu, H.; Kawano, T.; Koyama, S.; Kozutsumi, Y.; Suzuki, A.; Kawasaki, T.: Reaction mechanism underlying CMP-N-acetylneuraminic acid hydroxylation in mouse liver: formation of a ternary complex of cytochrome b_5, CMP-N-acetylneuraminic acid, and a hydroxylation enzyme. J. Biochem., **115**, 381-386 (1994)

[15] Schneckenburger, P.; Shaw, L.; Schauer, R.: Purification, characterization and reconstitution of CMP-N-acetylneuraminate hydroxylase from mouse liver. Glycoconjugate J., **11**, 194-203 (1994)
[16] Koyama, S.; Yamaji, T.; Takematsu, H.; Kawano, T.; Kozutsumi, Y.; Suzuki, A.; Kawasaki, T.: A naturally occurring 46-amino acid deletion of cytidine monophospho-N-acetylneuraminic acid hydroxylase leads to a change in the intracellular distribution of the protein. Glycoconjugate J., **13**, 353-358 (1996)
[17] Malykh, Y.N.; Shaw, L.; Schauer, R.: The role of CMP-N-acetylneuraminic acid hydroxylase in determining the level of N-glycolylneuraminic acid in porcine tissues. Glycoconjugate J., **15**, 885-893 (1998)
[18] Malykh, Y.N.; Krisch, B.; Shaw, L.; Warner, T.G.; Sinicrop, D.; Smith, R.; Chang, J.; Schauer, R.: Distribution and localization of CMP-N-acetylneuraminic acid hydroxylase and N-glycolylneuraminic acid-containing glycoconjugates in porcine lymph node and peripheral blood lymphocytes. Eur. J. Cell Biol., **80**, 48-58 (2001)
[19] Gollub, M.; Schauer, R.; Shaw, L.: Cytidine monophosphate-N-acetylneuraminate hydroxylase in the starfish Asterias rubens and other echinoderms. Comp. Biochem. Physiol. B, **120**, 605-615 (1998)
[20] Gollub, M.; Shaw, L.: Isolation and characterization of cytidine-5'-monophosphate-N-acetylneuraminate hydroxylase from the starfish Asterias rubens. Comp. Biochem. Physiol. B, **134**, 89-101 (2003)
[21] Schlenzka, W.; Shaw, L.; Schneckenburger, P.; Schauer, R.: Purification and characterization of CMP-N-acetylneuraminic acid hydroxylase from pig submandibular glands. Glycobiology, **4**, 675-683 (1994)
[22] Shaw, L.; Schneckenburger, P.; Carlsen, J.; Christiansen, K.; Schauer, R.: Mouse liver cytidine-5-monophosphate-N-acetylneuraminic acid hydroxylase. Catalytic function and regulation. Eur. J. Biochem., **206**, 269-277 (1992)
[23] Lepers, A.; Shaw, L.; Schneckenburger, P.; Cacan, R.; Verbert, A.; Schauer, R.: A study on the regulation of N-glycoloylneuraminic acid biosynthesis and utilization in rat and mouse liver. Eur. J. Biochem., **193**, 715-723 (1990)
[24] Malykh, Y.N.; King, T.P.; Logan, E.; Kelly, D.; Schauer, R.; Shaw, L.: Regulation of N-glycolylneuraminic acid biosynthesis in developing pig small intestine. Biochem. J., **22**, 2-29 (2002)

(-)-Menthol monooxygenase 1.14.13.46

1 Nomenclature

EC number
1.14.13.46

Systematic name
(-)-menthol,NADPH:oxygen oxidoreductase (8-hydroxylating)

Recommended name
(-)-menthol monooxygenase

Synonyms
l-menthol monooxygenase

CAS registry number
117590-75-7

2 Source Organism

<1> *Rattus sp.* [1]

3 Reaction and Specificity

Catalyzed reaction
(-)-menthol + NADPH + H^+ + O_2 = *p*-menthane-3,8-diol + $NADP^+$ + H_2O

Reaction type
oxidation
redox reaction
reduction

Natural substrates and products
S (-)-menthol + NADPH + O_2 <1> (Reversibility: ? <1> [1]) [1]
P *p*-menthane-3,8-diol + $NADP^+$ + H_2O <1> [1]

Substrates and products
S (-)-menthol + NADPH + O_2 <1> (Reversibility: ? <1> [1]) [1]
P *p*-menthane-3,8-diol + $NADP^+$ + H_2O <1> [1]

Cofactors/prosthetic groups
NADPH <1> (<1> cannot be replaced by NADH [1]) [1]

Specific activity (U/mg)
0.002 <1> (<1> microsomal fraction [1]) [1]

5 Isolation/Preparation/Mutation/Application

Source/tissue
liver <1> [1]

Localization
microsome <1> [1]

References

[1] Madhava Madyastha, K.; Srivatsan, V.: Studies on the metabolism of l-menthol in rats. Drug Metab. Dispos., **16**, 765-772 (1988)

(-)-Limonene 3-monooxygenase 1.14.13.47

1 Nomenclature

EC number
1.14.13.47

Systematic name
(-)-limonene,NADPH:oxygen oxidoreductase (3-hydroxylating)

Recommended name
(-)-limonene 3-monooxygenase

Synonyms
(-)-limonene 3-hydroxylase
oxygenase, (-)-limonene 3-mono-

CAS registry number
138066-92-9

2 Source Organism

<1> *Mentha piperita* (peppermint [1-3,5,6]) [1-3, 5-7]
<2> *Mentha x gracilis* (scotch spearmint, radiation-induced mutant 643-10-74 [2]) [2]
<3> *Mentha spicata* (spearmint [4]) [4]

3 Reaction and Specificity

Catalyzed reaction
(-)-limonene + NADPH + H^+ + O_2 = (-)-trans-isopiperitenol + $NADP^+$ + H_2O (<1> mixed-function oxygenase, ring-hydroxylation [1])

Reaction type
oxidation
redox reaction
reduction

Natural substrates and products
S (-)-limonene + NADPH + O_2 <1> (<1> one of the key reactions of oxygenated monoterpenes, biosynthesis of (-)-menthone [1]) (Reversibility: ? <1> [1]) [1]
P (-)-trans-isopiperitenol + $NADP^+$ + H_2O

Substrates and products

S (+)-limonene + NADPH + O_2 <1, 2> (<1> hydroxylation at 50%, [1]; <2> 53% the rate of (-)-limonene hydroxylation [2]) (Reversibility: ? <1, 2> [1, 2]) [1, 2, 7]

P (+)-trans-isopiperitenol + $NADP^+$ + H_2O <1, 2> [1, 2, 7]

S (+)-*p*-menth-1-ene + NADPH + O_2 <1, 2> (<1> i.e. (+)-8,9-dihydro-limonene, hydroxylation at 37% the rate of (-)-limonene hydroxylation [1]) (Reversibility: ? <1, 2> [1, 2]) [1, 2]

P (+)-trans-piperitol + $NADP^+$ + H_2O <1, 2> [1, 2]

S (-)-(4S)-limonene + NADPH + O_2 <1-3> (<1> highly specific, regiospecific and stereospecific reaction [1,6,7]; <3> limonene 6-hydroxylase mutant F363I [4]) (Reversibility: ? <1, 2> [1-4, 6]) [1-4, 6, 7]

P (-)-trans-(3S,4R)-isopiperitenol + $NADP^+$ + H_2O <1, 2> [1-4, 6, 7]

S (-)-*p*-menth-1-ene + NADPH + O_2 <1, 2> (<1> i.e. (-)-8,9-dihydrolimonene, hydroxylation at 37% the rate of (-)-limonene hydroxylation [1]) (Reversibility: ? <1, 2> [1, 2]) [1, 2]

P (-)-trans-piperitol + $NADP^+$ + H_2O <1, 2> [1, 2]

S Additional information <1> (<1> no substrates: isolimonenes, terpinolene, α- or β-phellandrene, α- or β-terpinene, bicyclic monoterpenes: pinene, sabinene, α-thujene, *p*-cymene, cis- or trans-*p*-menthane, highly specific, absolute requirement for a reduced pyridine nucleotide and molecular oxygen [1]) [1]

P ?

Inhibitors

5,11-dimethyl-6H-pyrido[4,3-b]carbazole <1> [1]
CO <1> (<1> CO:O_2 ratio of 9:1, photoreversible [1]) [1]
$NADP^+$ <1> (<1> 2 mM [1]) [1]
SKF 525A <1> (<1> i.e. 2-diethylaminoethyl-2,2-diphenylvalerate, moderate [1]) [1]
clotrimazole <1> (<1> i.e. 1-[chloro-α,α-diphenyl]imidazole, mixed-type, strong [1]) [1]
cytochrome c <1> (<1> strong [1]) [1]
metyrapone <1> (<1> i.e. 2-methyl-1,2-di-3-pyridyl-1-propanone, inhibition [1]) [1]
miconazole <1> (<1> i.e. 1-[2,4-dichloro-β-([2,4-di-chlorobenzyl]oxy)phenethyl]-imidazole, mixed-type, weak [1]) [1]
Additional information <1> (<1> no inhibition: ancymidol, imidazole, up to 5 mM [1]) [1]

Cofactors/prosthetic groups

NADH <1> (<1> 14% as effective as NADPH [1]) [1]
NADPH <1, 2> (<1> absolute requirement [1-3]) [1-3]
cytochrome P_{450} <1, 2> (<1> heme-thiolate protein, 0.2-0.9 nmol per mg protein [1]) [1-3]

Activating compounds

FAD <1> (<1> plus FMN, 0.005 mM each, activation [1]) [1]
FMN <1> (<1> plus FAD, 0.005 mM each, activation [1]) [1]

Specific activity (U/mg)

0.0003333 <1> [3]
0.000685 <1> [1]

K_m-Value (mM)

0.018 <1> ((-)-limonene) [1]

pH-Optimum

7.4 <1> [1, 3]

pH-Range

6.9-7.9 <1> (<1> about half-maximal activity at pH 6.9 and 7.9 [1]) [1]

Temperature optimum (°C)

32 <1> [1]

5 Isolation/Preparation/Mutation/Application

Source/tissue

leaf epidermis <1, 2> (<1> oil gland secretory cells [3]; <1> of both surfaces [1]) [1-3]

Localization

endoplasmic reticulum <1> [3]
microsome <1, 2> [1, 2]

Purification

<1> (<1> recombinant protein from Escherichia coli [6]) [6]
<1> (<1> sonication of secretory cells leads to most active crude extracts [3]) [3]

Cloning

<1> [5]
<1> (expression in Escherichia coli and Saccharomyces cerevisiae [6]) [6]

6 Stability

General stability information

<1>, DTT stabilizes during purification [3]
<1>, EDTA stabilizes during purification [3]
<1>, glycerol stabilizes during purification [3]

References

[1] Karp, F.; Mihaliak, C.A.; Harris, J.L.; Croteau, R.: Monoterpene biosynthesis: specificity of the hydroxylations of (-)-limonene by enzyme preparations from peppermint (Mentha piperita), spearmint (Mentha spicata), and perilla (Perilla frutescens) leaves. Arch. Biochem. Biophys., **276**, 219-226 (1990)
[2] Croteau, R.; Karp, F.; Wagschal, K.C.; Satterwhite, M.; Hyatt, D.C.; Skotland, C.B.: Biochemical characterization of a spearmint mutant that resembles peppermint in monoterpene content. Plant Physiol., **96**, 744-752 (1991)
[3] Gershenzon, J.; McCaskill, D.; Rajaonarivony, J.I.; Mihaliak, C.; Karp, F.; Croteau, R.: Isolation of secretory cells from plant glandular trichomes and their use in biosynthetic studies of monoterpenes and other gland products. Anal. Biochem., **200**, 130-138 (1992)
[4] Schalk, M.; Croteau, R.: A single amino acid substitution (F363I) converts the regiochemistry of the spearmint (-)-limonene hydroxylase from a C6- to a C3-hydroxylase. Proc. Natl. Acad. Sci. USA, **97**, 11948-11953 (2000)
[5] Lupien, S.; Karp, F.; Wildung, M.; Croteau, R.: Regiospecific cytochrome P_{450} limonene hydroxylases from mint (Mentha) species: cDNA isolation, characterization, and functional expression of (-)-4S-limonene-3-hydroxylase and (-)-4S-limonene-6-hydroxylase. Arch. Biochem. Biophys., **368**, 181-192 (1999)
[6] Haudenschild, C.; Schalk, M.; Karp, F.; Croteau, R.: Functional expression of regiospecific cytochrome P_{450} limonene hydroxylases from mint (Mentha spp.) in Escherichia coli and Saccharomyces cerevisiae. Arch. Biochem. Biophys., **379**, 127-136 (2000)
[7] Wust, M.; Little, D.B.; Schalk, M.; Croteau, R.: Hydroxylation of limonene enantiomers and analogs by recombinant (-)-limonene 3- and 6-hydroxylases from Mint (Mentha) species: Evidence for catalysis within sterically constrained active sites. Arch. Biochem. Biophys., **387**, 125-136 (2001)

(-)-Limonene 6-monooxygenase 1.14.13.48

1 Nomenclature

EC number
1.14.13.48

Systematic name
(-)-limonene,NADPH:oxygen oxidoreductase (6-hydroxylating)

Recommended name
(-)-limonene 6-monooxygenase

Synonyms
(-)-limonene 6-hydroxylase
oxygenase, (-)-limonene 6-mono-

CAS registry number
138066-93-0

2 Source Organism

<1> *Mentha spicata* (spearmint) [1, 3-10]
<2> *Mentha x gracilis* (scotch spearmint, wild type) [2]

3 Reaction and Specificity

Catalyzed reaction
(-)-limonene + NADPH + H^+ + O_2 = (-)-trans-carveol + $NADP^+$ + H_2O

Reaction type
oxidation
redox reaction
reduction

Natural substrates and products
S (-)-limonene + NADPH + O_2 <1> (<1> one of the key reactions of oxygenated monoterpenes, biosynthesis of (-)-carvone [1]) [1]
P (-)-trans-carveol + $NADP^+$ + H_2O

Substrates and products
S (+)-limonene + NADPH + O_2 <1, 2> (<1> hydroxylation at 25% the rate of (-)-limonene [1]; <2> radiation induced mutant 86% the rate of (-)-limonene [2]) (Reversibility: ? <1, 2> [1, 2]) [1, 2]

P (+)-cis-carveol + NAD^+ + H_2O <1, 2> [1, 2]
S (+)-*p*-menth-1-ene + NADPH + O_2 <1, 2> (<1> i.e. (+)-8,9-dihydrolimonene, hydroxylation at 30% the rate of (-)-limonene [1]) (Reversibility: ? <1, 2> [1, 2]) [1, 2]
P (+)-cis-carvotanacetol + $NADP^+$ + H_2O <2> [2]
S (-)-(4S)-limonene + NADPH + O_2 <1> (<1> mutant F363I [10]) (Reversibility: ? <1> [10]) [10]
P (-)-trans-isopiperitenol + $NADP^+$ + H_2O <1> (<1> mutant F363I [10]) [10]
S (-)-(4S)-limonene + NADPH + O_2 <1, 2> (<1,2> highly specific [1-3]) (Reversibility: ? <1, 2> [1-9]) [1-9]
P (-)-(4R,6S)-trans-carveol + $NADP^+$ + H_2O <1, 2> [1-9]
S (-)-*p*-menth-1-ene + NADPH + O_2 <1, 2> (<1> i.e. (-)-8,9-dihydrolimonene, hydroxylation at 74% the rate of (-)-limonene [1]) (Reversibility: ? <1, 2> [1, 2]) [1, 2]
P (-)-trans-carvotanacetol + $NADP^+$ + H_2O <2> [2]
S Additional information <1> (<1> no substrates: isolimonenes, terpinolene, α- or β-phellandrene, α- or β-terpinene, bicyclic monoterpenes: pinene, sabinene, α-thujene, *p*-cymene, cis- or trans-*p*-menthane, NADH can replace NADPH, but more slowly [1]; <1> (4S)-limonene is hydroxylated regiospecifically and stereospecifically at C_6 to give (-)-trans-carveol as the sole product, (4R)-limonene leads to multiple products indicating allylic rearrangement [5,9]; <1> overview [8]) [1, 5]
P ?

Inhibitors

5,11-dimethyl-6H-pyrido[4,3-b]-carbazole <1> [1]
CO <1> (<1> $CO:O_2$ ratio of 9:1, photoreversible [1]) [1]
$NADP^+$ <1> (<1> 2 mM [1]) [1]
SKF 525A <1> (<1> i.e. 2-diethylaminoethyl-2,2-diphenylvalerate, moderate [1]) [1]
clotrimazole <1, 2> (<1> i.e. 1-[chloro-α,α-diphenyl]imidazole, mixed-type, strong [1]) [1, 2]
cytochrome c <1> (<1> strong [1]) [1]
metyrapone <1> (<1> i.e. 2-methyl-1,2-di-3-pyridyl-1-propanone, moderate [1]) [1]
miconazole <1, 2> (<1> i.e. 1-[2,4-dichloro-β-([2,4-di-chlorobenzyl]oxy)-phenethyl]-imidazole, mixed-type, strong [1]) [1, 2]
Additional information <1> (<1> no inhibition: ancymidol, imidazole, up to 5 mM [1]) [1]

Cofactors/prosthetic groups

cytochrome P_{450} <1> (<1> average content of 0.0005 mmol/mg protein [1]) [1, 6]

Specific activity (U/mg)

0.00068 <1> [1]
0.0007 <1> [3]

K_m-Value (mM)
0.02 <1> ((-)-limonene) [1]

pH-Optimum
7.4 <1> [1, 3]

pH-Range
6.9-7.9 <1> (<1> about half-maximal activity at pH 6.9 and 7.9 [1]) [1]

Temperature optimum (°C)
30 <1> (<1> assay at [4]) [4]
32 <1> [1]

4 Enzyme Structure

Molecular weight
57000 <1> (<1> SDS-PAGE [6]) [6]
57000 <1> (<1> SDS-PAGE, recombinant protein expressed in Saccharomyces cerevisiae [7]) [7]
Additional information <1> (<1> overview [8]) [8]

5 Isolation/Preparation/Mutation/Application

Source/tissue
leaf epidermis <1, 2> (<1> oil glands on upper and lower surface [1,6]; <1,2> glandular trichomes [2,4]) [1-4]

Localization
endoplasmic reticulum <1> [3]
microsome <1, 2> [1, 2, 6]

Purification
<1> [6]
<1> (<1> overview [8]) [8]
<1> (<1> protein expressed in Escherichia coli [7]) [7]
<1> (<1> protein expressed in Saccharomyces cerevisiae [7]) [7]
<1> (<1> sonication of secretory cells leads to most active crude extracts [3]) [3]
<1, 2> (partial) [1-3]

Cloning
<1> [6]
<1> (expression in Escherichia coli and in Saccharomyces cerevisiae [7]) [7]
<1> (overexpression in Saccharomyces cerevisiae [5]) [5]
<1> (overview [8]) [8]

6 Stability

General stability information

<1>, DTT stabilizes during purification [3]
<1>, EDTA stabilizes during purification [3]
<1>, glycerol stabilizes during purification [3]

References

[1] Karp, F.; Mihaliak, C.A.; Harris, J.L.; Croteau, R.: Monoterpene biosynthesis: specificity of the hydroxylations of (-)-limonene by enzyme preparations from peppermint (Mentha piperita), spearmint (Mentha spicata), and perilla (Perilla frutescens) leaves. Arch. Biochem. Biophys., **276**, 219-226 (1990)

[2] Croteau, R.; Karp, F.; Wagschal, K.C.; Satterwhite, M.; Hyatt, D.C.; Skotland, C.B.: Biochemical characterization of a spearmint mutant that resembles peppermint in monoterpene content. Plant Physiol., **96**, 744-752 (1991)

[3] Gershenzon, J.; McCaskill, D.; Rajaonarivony, J.I.; Mihaliak, C.; Karp, F.; Croteau, R.: Isolation of secretory cells from plant glandular trichomes and their use in biosynthetic studies of monoterpenes and other gland products. Anal. Biochem., **200**, 130-138 (1992)

[4] Gershenzon, J.; Maffei, M.; Croteau, R.: Biochemical and histochemical localization of monoterpene biosynthesis in the glandular trichomes of spearmint (Mentha spicata). Plant Physiol., **89**, 1351-1357 (1989)

[5] Wuest, M.; Croteau, R.B.: Hydroxylation of specifically deuterated limonene enantiomers by cytochrome P_{450} limonene-6-hydroxylase reveals the mechanism of multiple product formation. Biochemistry, **41**, 1820-1827 (2002)

[6] Lupien, S.; Karp, F.; Wildung, M.; Croteau, R.: Regiospecific cytochrome P_{450} limonene hydroxylases from mint (Mentha) species: cDNA isolation, characterization, and functional expression of (-)-4S-limonene-3-hydroxylase and (-)-4S-limonene-6-hydroxylase. Arch. Biochem. Biophys., **368**, 181-192 (1999)

[7] Haudenschild, C.; Schalk, M.; Karp, F.; Croteau, R.: Functional Expression of regiospecific cytochrome P_{450} limonene hydroxylases from mint (Mentha spp.) in Escherichia coli and Saccharomyces cerevisiae. Arch. Biochem. Biophys., **379**, 127-136 (2000)

[8] Lupien, S.; Karp, F.; Ponnamperuma, K.; Wildung, M.; Croteau, R.: Cytochrome P_{450} limonene hydroxylases of Mentha species. Drug Metab. Drug Interact., **12**, 245-260 (1995)

[9] Wust, M.; Little, D.B.; Schalk, M.; Croteau, R.: Hydroxylation of limonene enantiomers and analogs by recombinant (-)-limonene 3- and 6-hydroxylases from Mint (Mentha) species: Evidence for catalysis within sterically constrained active sites. Arch. Biochem. Biophys., **387**, 125-136 (2001)

[10] Schalk, M.; Croteau, R.: A single amino acid substitution (F363I) converts the regiochemistry of the spearmint (-)-limonene hydroxylase from a C_6- to a C_3-hydroxylase. Proc. Natl. Acad. Sci. USA, **97**, 11948-11953 (2000)

(-)-Limonene 7-monooxygenase 1.14.13.49

1 Nomenclature

EC number
1.14.13.49

Systematic name
(-)-limonene,NADPH:oxygen oxidoreductase (7-hydroxylating)

Recommended name
(-)-limonene 7-monooxygenase

Synonyms
(-)-limonene hydroxylase
oxygenase, (-)-limonene mono-

CAS registry number
122653-75-2

2 Source Organism

<1> *Perilla frutescens* [1]

3 Reaction and Specificity

Catalyzed reaction
(-)-limonene + NADPH + H^+ + O_2 = (-)-perillyl alcohol + $NADP^+$ + H_2O (<1> side-chain hydroxylation, mixed-function oxygenase [1])

Reaction type
oxidation
redox reaction
reduction

Natural substrates and products
S (-)-limonene + NADPH + O_2 <1> (<1> one of the key reactions of oxygenated monoterpenes, perillyl aldehyde biosynthesis [1]) (Reversibility: ? <1> [1]) [1]
P perillyl alcohol + $NADP^+$ + H_2O

Substrates and products
S (+)-limonene + NADPH + O_2 <1> (<1> hydroxylation at the same rate as (-)-limonene [1]) (Reversibility: ? <1> [1]) [1]

P perillyl alcohol + $NADP^+$ + H_2O <1> [1]
S (-)-limonene + NADPH + O_2 <1> (<1> highly specific [1]) (Reversibility: ? <1> [1]) [1]
P perillyl alcohol + $NADP^+$ + H_2O <1> [1]
S Additional information <1> (<1> no substrates: isolimonenes, terpinolene, α- or β-phellandrene, α- or β-terpinene, bicyclic monoterpenes: pinene, sabinene, α-thujene, *p*-cymene, cis- or trans-*p*-menthane, *p*-menth-1-ene, i.e. 8,9-dihydrolimonene [1]) [1]
P ?

Inhibitors

5,11-dimethyl-6H-pyrido[4,3-b]carbazole <1> [1]
CO <1> (<1> CO:O_2 ratio of 9:1, photoreversible [1]) [1]
$NADP^+$ <1> (<1> 2 mM [1]) [1]
SKF 525A <1> (<1> i.e. 2-diethyl-aminoethyl-2,2-diphenylvalerate, moderate [1]) [1]
clotrimazole <1> (<1> i.e. 1-[chloro-α,α-diphenyl]imidazole, mixed-type, weak [1]) [1]
cytochrome c <1> (<1> strong [1]) [1]
metyrapone <1> (<1> i.e. 2-methyl-1,2-di-3-pyridyl-1-propanone, moderate [1]) [1]
miconazole <1> (<1> i.e. 1-[2,4-dichloro-β-([2,4-di-chlorobenzyl]oxy)phenethyl]-imidazole, mixed-type, weak [1]) [1]
Additional information <1> (<1> no inhibition: ancymidol, imidazole, up to 5 mM [1]) [1]

Cofactors/prosthetic groups

NADH <1> (<1> 5.3% as effective as NADPH [1]) [1]
NADPH <1> [1]
cytochrome P_{450} <1> (<1> heme-thiolate protein, 0.2-0.9 nmol per mg protein [1]) [1]

Activating compounds

FAD <1> (<1> plus FMN, 0.005 mM each, activation [1]) [1]
FMN <1> (<1> plus FAD, 0.005 mM each, activation [1]) [1]

Specific activity (U/mg)

0 <1> [1]

K_m-Value (mM)

0.021 <1> ((-)-limonene) [1]

pH-Optimum

7.8 <1> [1]

pH-Range

7.3-8.3 <1> (<1> about half-maximal activity at pH 7.3 and 8.3 [1]) [1]

Temperature optimum (°C)

32 <1> [1]

5 Isolation/Preparation/Mutation/Application

Source/tissue

leaf epidermis <1> (<1> oil glands on abaxial surface [1]) [1]

References

[1] Karp, F.; Mihaliak, C.A.; Harris, J.L.; Croteau, R.: Monoterpene biosynthesis: specificity of the hydroxylations of (-)-limonene by enzyme preparations from peppermint (Mentha piperita), spearmint (Mentha spicata), and perilla (Perilla frutescens) leaves. Arch. Biochem. Biophys., **276**, 219-226 (1990)

Pentachlorophenol monooxygenase 1.14.13.50

1 Nomenclature

EC number
1.14.13.50

Systematic name
pentachlorophenol,NADPH:oxygen oxidoreductase (hydroxylating, dechlorinating)

Recommended name
pentachlorophenol monooxygenase

Synonyms
PCP hydroxylase
PCP4MO <2> [9, 10]
dechlorinase, pentachlorophenol
oxygenase, pentachlorophenol 4-mono-
pentachlorophenol 4-mono-oxygenase
pentachlorophenol 4-monooxygenase
pentachlorophenol dechlorinase
pentachlorophenol dehalogenase
pentachlorophenol hydroxylase

CAS registry number
124148-88-5
136111-57-4

2 Source Organism

<1> *Arthrobacter sp.* (strain ATCC 33790 [1,2]) [1, 2]
<2> *Flavobacterium sp.* (strain ATCC 39723 [3-5,8]; identical with Sphingomonas chlorophenolica strain ATCC 39723 [6,9,10]) [3-6, 8-10]
<3> *Sphingomonas sp.* (UG30 [6]) [6]
<4> *Flavobacterium sp.* (strain ATCC 39723 [7]; inducible by PCP on DNA level [7]) [7]
<5> *Pseudomonas sp.* (strain IST 103, i.e. PCP 103 [11]) [11]

3 Reaction and Specificity

Catalyzed reaction

pentachlorophenol + NADPH + H^+ + O_2 = tetrachlorohydroquinone + $NADP^+$ + chloride (some tetrachlorophenols and trichlorophenols can act as substrates; <1> mechanism [2])

Reaction type

dehalogenation
oxidation
redox reaction
reduction

Natural substrates and products

S pentachlorophenol + NADPH + O_2 <1, 2, 4> (<2> wild-type and mutant strains [8]; <2> solitary oxygenolytic pathway for removal of the primary chlorine during PCP degradation, not a hydrolytic one [8]; <4> wild-type and transformed E. coli cells [7]) (Reversibility: ? <1, 2, 4> [2, 7, 8]) [2, 7, 8]
P tetrachlorohydroquinone + $NADP^+$ + H_2O <1, 2, 4> [2, 7, 8]
S Additional information <2> (<2> biodegradation of the herbicide bromoxynil, i.e. 3,5-dibromo-4-hydroxybenzonitrile [5]) [5]
P ?

Substrates and products

S 2,3,4,5-tetrachlorophenol + NADPH + O_2 <1> (Reversibility: ? <1> [1]) [1]
P 2,3,5-trichlorohydroquinone + Cl^- + NDP+ + H_2O
S 2,3,4-trichlorophenol + NADPH + O_2 <1, 2> (<2> low activity [4]) (Reversibility: ? <1, 2> [1, 4]) [1, 4]
P ?
S 2,3,5,6-tetrachlorophenol + NADPH + O_2 <2> (Reversibility: ? <2> [4]) [4]
P 2,3,5,6-tetrachloro-*p*-hydroquinone + $NADP^+$ <2> [4]
S 2,3,6-trichlorophenol + NADPH + O_2 <2> (Reversibility: ? <2> [4]) [4]
P 2,3,6-trichloro-*p*-hydroquinone + $NADP^+$ + H_2O <2> [4]
S 2,3-dichlorophenol + NADPH + O_2 <2> (<2> low activity [4]) (Reversibility: ? <2> [4]) [4]
P ?
S 2,4,5-trichlorophenol + NADPH + O_2 <1> (Reversibility: ? <1> [1]) [1]
P ?
S 2,4,6-tribromophenol + NADPH + O_2 <2> (<2> low activity [4]) (Reversibility: ? <2> [4]) [4]
P 2,6-dibromo-*p*-hydroquinone + Br^- + $NADP^+$ + H_2O <2> [4]
S 2,4,6-trichlorophenol + NADPH + O_2 <1, 2> (<2> low activity [4]) (Reversibility: ? <1, 2> [1, 4]) [1, 4]
P 2,6-dichloro-*p*-hydroquinone + Cl^- + $NADP^+$ + H_2O <2> [4]
S 2,6-dibromo-4-nitrophenol + NADPH + O_2 <2> (<2> low activity [4]) (Reversibility: ? <2> [4]) [4]

P 2,6-dibromo-*p*-hydroquinone + NO_2^- + $NADP^+$ + H_2O <2> [4]
S 2,6-dibromophenol + NADPH + O_2 <2> (Reversibility: ? <2> [4]) [4]
P 2,6-dibromo-*p*-hydroquinone + $NADP^+$ + H_2O <2> [4]
S 2,6-dichlorophenol + NADPH + O_2 <2> (Reversibility: ? <2> [4]) [4]
P 2,6-dichloro-*p*-hydroquinone + $NADP^+$ + H_2O <2> [4]
S 3,5-dibromo-4-hydroxybenzonitrile + NADPH + O_2 <2> (Reversibility: ? <2> [4]) [4]
P 2,6-dibromo-*p*-hydroquinone + CN^- + $NADP^+$ <2> [4]
S 3,5-diiodo-4-hydroxybenzonitrile + NADPH + O_2 <2> (Reversibility: ? <2> [4]) [4]
P ?
S 4-amino-2,6-dichlorophenol + NADPH + O_2 <2> (<2> low activity [4]) (Reversibility: ? <2> [4]) [4]
P 2,6-dichloro-*p*-hydroquinone + $NADP^+$ + NH_2OH <2> [4]
S 4-nitrocatechol + NADPH + O_2 <3> (Reversibility: ? <3> [6]) [6]
P ?
S *p*-nitrophenol + NADPH + O_2 <3> (<3> low activity [6]) (Reversibility: ? <3> [6]) [6]
P ?
S pentachlorophenol + NADPH + O_2 <1-5> (<1> oxygen and NADPH are essential [1]) (Reversibility: ? <1-5> [1-9, 11]) [1-9, 11]
P 2,3,5,6-tetrachlorohydroquinone + $NADP^+$ + chloride <2-4> [3-9]
S pentafluorophenol + NADPH + O_2 <2> (<2> low activity [4]) (Reversibility: ? <2> [4]) [4]
P ?
S triiodophenol + NADPH + O_2 <2> (Reversibility: ? <2> [3, 4]) [3, 4]
P diiodohydroquinone + iodide + $NADP^+$ + H_2O <2> [4]
S Additional information <1-4> (<2> broad substrate and catalytic reaction range [8]; <4> enzyme also removes amino, cyano and nitro groups from the *para*-position to the hydroxyl group of the phenol [7]; <2, 3> broad substrate specificity, hydroxylation in *para*-position [6]; <2> no activity with 3,4,5-trichlorophenol, 3,5-dichlorophenol [4]; <1> no activity with 3,4,5-trichlorophenol, 2,4-dichlorophenol, 3,4-dichlorophenol, 4-chlorophenol [1]; <2> overview: enzyme catalyzes a primary attack on a broad range of substituted phenols, hydroxylating the para position and removing halogen, nitro, amino and cyano groups to produce halide, nitrite, hydroxylamine and cyanide respectively [4]) [1, 4, 6-8]
P ?

Cofactors/prosthetic groups

FAD <2, 3> (<2> recombinant enzyme has no FAD, activity only in presence of added FAD [9]; <3> highly conserved FAD-binding site at the N-terminus [6]; <2> 1 FAD per enzyme molecule [3]) [3, 6, 8, 9]
NADH <2> (<2> can replace NADPH, less effective [3]) [3]
NADPH <1-4> (<2> requirement [8]; <1> essential [1]; <2> 2 mol of NADPH per mol of halogenated substrate, preferred electron donor [3]; <2> elimination of 1 mol of a halogen, nitro, or cyano group requires 2 mol of

NADPH, 1 mol of NADPH is required to remove 1 mol of an amino group or hydrogen [4]) [1-8]
Additional information <2> (<2> dithiothreitol can not be used as electron donor [3]) [3]

Activating compounds
EDTA <1> (<1> stimulates by 67% [1]) [1]

Metals, ions
Fe^{2+} <1> (<1> stimulates [1]) [1]
Additional information <2> (<2> no requirement for divalent cations [3]) [3]

Specific activity (U/mg)
Additional information <1-3> (<1> activity in crude extracts is undetectable or below the detection limit of 0.1 nU per mg with a chloride sensor [1]) [1, 3, 6]

K_m-Value (mM)
0.03 <2> (pentachlorophenol) [3]
0.05 <2> (pentachlorophenol, <2> recombinant enzyme [9]) [9]
0.082 <2> (NADPH) [3]
0.214 <2> (NADH) [3]

pH-Optimum
7.5 <1> [1]
7.5-8.5 <2> [3]

pH-Range
6-8.5 <2> (<2> pH 6.0: about 28% of activity maximum, pH 7.5-8.5: activity maximum [3]) [3]

Temperature optimum (°C)
25 <1> [1]
40 <2> [3]

Temperature range (°C)
40-45 <2> (<2> 40°C: activity maximum, 45°C: above 75% of activity maximum, 50°C: inactive [3]) [3]

4 Enzyme Structure

Molecular weight
30000 <5> (<5> gel filtration [11]) [11]
59930 <3> (<3> deduced from DNA sequence determination [6]) [6]
66000 <2> (<2> gel filtration [3]) [3]

Subunits
monomer <2> (<2> 1 * 63000, SDS-PAGE, under certain conditions dimer and multimer conformations are also observed [3]) [3]

5 Isolation/Preparation/Mutation/Application

Localization

cytoplasm <2> (<2> enzyme is synthesized in the cytoplasm, then integrated in the inner membranes and finally translocated to the periplasm [10]) [10]
membrane <2> (<2> enzyme is synthesized in the cytoplasm, then integrated in the inner membranes and finally translocated to the periplasm [10]) [10]
periplasm <2> (<2> enzyme is synthesized in the cytoplasm, then integrated in the inner membranes and finally translocated to the periplasm [10]) [10]
Additional information <1> (<1> probably not a soluble protein [1]) [1]

Purification

<1> (partially [1]) [1]
<2> (recombinant wild-type and mutant from Escherichia coli as glutathione-S-transferase fusion proteins [9]) [3, 9]
<5> [11]

Crystallization

<2> [3]

Cloning

<2> (pcpB gene, expression of wild-type and mutant in Escherichia coli as glutathione-S-transferase fusion protein [9]; pcpB gene, expression in Escherichia coli and Flavobacterium sp. ATCC 39723 of wild-type and isogenic mutant, DNA sequence analysis [8]) [6, 8, 9]
<3> (pcpB gene, expression in Escherichia coli BL21, DNA sequence analysis [6]) [6]
<4> (pcpB gene, constitutive overexpression in Escherichia coli, DNA sequence determination and analysis [7]) [7]
<5> (gene isolated from Pseudomonas sp. plasmid, recombinant expression in Escherichia coli XLBlue1α, DNA sequence analysis of the recombinant clones, comparison with other sequences [11]) [11]

Engineering

Additional information <2> (<2> introduction of tobacco etch virus protease cleavage site into the enzyme as a glutathione-S-transferase fusion protein [9]; <2> targeted mutagenesis via allelic exchange, construction of isogenic mutant without PCP 4-monooxygenase activity and 2 cointegration mutants with slow activity [8]) [8, 9]

Application

environmental protection <2> (<2> PCP-decontamination of soil and water, degradation of 3,5-dibromophenol derived in soil from the herbicide bromoxynil, i.e. 3,5-dibromo-4-hydroxybenzonitrile [5]; <2> development of biological methods for the decontamination of halophenol-polluted sites [4]) [4, 5]

6 Stability

Storage stability

<2>, -20°C, as ammonium sulfate precipitate or in solution at concentration greater than 1 mg/ml, 2 months [3]
<2>, 4°C, 50% loss of activity after 1 week [3]
<2>, 4°C, in cell extract or as 40-60% fraction of ammonium sulfate precipitation , total loss of activity after 3 days [3]

References

[1] Schenk, T.; Muller, R.; Mörsberger, F.; Otto, M.K.; Lingens, F.: Enzymatic dehalogenation of pentachlorophenol by extracts from Arthrobacter sp. strain ATCC 33790. J. Bacteriol., **171**, 5487-5491 (1989)
[2] Schenk, T.; Muller, R.; Lingens, F.: Mechanism of enzymatic dehalogenation of pentachlorophenol by Arthrobacter sp. strain ATCC 33790. J. Bacteriol., **172**, 7272-7274 (1990)
[3] Xun, L.; Orser, C.S.: Purification and properties of pentachlorophenol hydroxylase, a flavoprotein from Flavobacterium sp. strain ATCC 39723 [published erratum appears in J Bacteriol 1992 Aug;174(15):5176]. J. Bacteriol., **173**, 4447-4453 (1991)
[4] Xun, L.; Topp, E.; Orser, C.S.: Diverse substrate range of a Flavobacterium pentachlorophenol hydroxylase and reaction stoichiometries. J. Bacteriol., **174**, 2898-2902 (1992)
[5] Topp, E.; Xun, L.; Orser, C.S.: Biodegradation of the herbicide bromoxynil (3,5-dibromo-4-hydroxybenzonitrile) by purified pentachlorophenol hydroxylase and whole cells of Flavobacterium sp. strain ATCC 39723 is accompanied by cyanogenesis. Appl. Environ. Microbiol., **58**, 502-506 (1992)
[6] Leung, K.T.; Campbell, S.; Gan, Y.; White, D.C.; Lee, H.; Trevors, J.T.: The role of the Sphingomonas species UG30 pentachlorophenol-4-monooxygenase in *p*-nitrophenol degradation. FEMS Microbiol. Lett., **173**, 247-253 (1999)
[7] Orser, C.S.; Lange, C.C.; Xun, L.; Zahrt, T.C.; Schneider, B.J.: Cloning, sequence analysis, and expression of the Flavobacterium pentachlorophenol-4-monooxygenase gene in Escherichia coli. J. Bacteriol., **175**, 411-416 (1993)
[8] Lange, C.C.; Schneider, B.J.; Orser, C.S.: Verification of the role of PCP 4-monooxygenase in chlorine elimination from pentachlorophenol by Flavobacterium sp. strain ATCC 39723. Biochem. Biophys. Res. Commun., **219**, 146-149 (1996)
[9] Wang, H.; Tiirola, M.A.; Puhakka, J.A.; Kulomaa, M.S.: Production and characterization of the recombinant Sphingomonas chlorophenolica pentachlorophenol 4-monooxygenase. Biochem. Biophys. Res. Commun., **289**, 161-166 (2001)

[10] Wang, H.; Marjomaki, V.; Ovod, V.; Kulomaa, M.S.: Subcellular localization of pentachlorophenol 4-monooxygenase in Sphingobium chlorophenolicum ATCC 39723. Biochem. Biophys. Res. Commun., **299**, 703-709 (2002)
[11] Thakur, I.S.; Verma, P.; Upadhayaya, K.: Molecular cloning and characterization of pentachlorophenol-degrading monooxygenase genes of Pseudomonas sp. from the chemostat. Biochem. Biophys. Res. Commun., **290**, 770-774 (2002)

6-Oxocineole dehydrogenase 1.14.13.51

1 Nomenclature

EC number
1.14.13.51

Systematic name
6-oxocineole,NADPH:oxygen oxidoreductase

Recommended name
6-oxocineole dehydrogenase

Synonyms
6-oxocineole oxygenase

CAS registry number
122933-80-6

2 Source Organism

<1> *Rhodococcus* (strain C1 [1]) [1]

3 Reaction and Specificity

Catalyzed reaction
6-oxocineole + NADPH + H^+ + O_2 = 1,6,6-trimethyl-2,7-dioxabicyclo-[3.2.2]nonan-3-one + $NADP^+$ + H_2O

Reaction type
oxidation
redox reaction
reduction

Natural substrates and products
S 6-oxocineole + NADPH + O_2 <1> (<1> grown on 1,8-cineole or succinate as carbon source [1]) (Reversibility: ? <1> [1]) [1]
P 1,6,6-trimethyl-2,7-dioxabicyclo-[3.2.2]nonan-3-one + $NADP^+$ + H_2O <1> (<1> this product reacts spontaneously to 3-(1-hydroxy-1-methylethyl)-6-oxoheptanoic acid and subsequently lactonizes to 4,5-dihydro-5,5-dimethyl-4-(3'-oxobutyl)-furan-2(3H)-one [1]) [1]

Substrates and products

S 6-oxocineole + NADPH + O_2 <1> (Reversibility: ? <1> [1]) [1]

P 1,6,6-trimethyl-2,7-dioxabicyclo-[3.2.2]nonan-3-one + $NADP^+$ + H_2O <1> (<1> this product reacts spontaneously to 3-(1-hydroxy-1-methylethyl)-6-oxoheptanoic acid and subsequently lactonizes to 4,5-dihydro-5,5-dimethyl-4-(3'-oxobutyl)-furan-2(3H)-one [1]) [1]

S Additional information <1> (<1> no substrates: substrate analogues, i.e. (+)-camphor, (S)-(+)-carvone, cyclohexanene, 1,3-cyclohexanedione, dihydrocarvone, 2-, 3-, and 4-methylcyclohexanone, fenchone, menthone, pinan-3-one, pulegone, verbenone [1]) [1]

P ?

Cofactors/prosthetic groups

$NADP^+$ <1> [1]

NADPH <1> [1]

Specific activity (U/mg)

0.001 <1> (<1> succinate grown cells [1]) [1]

0.3-0.4 <1> (<1> 1,8-cineole-grown cells [1]) [1]

pH-Optimum

9 <1> [1]

5 Isolation/Preparation/Mutation/Application

Purification

<1> (partial [1]) [1]

6 Stability

General stability information

<1>, 5-10 vol% ethanol stabilizes the crude cell extract [1]

References

[1] Williams, D.R.; Trudgill, P.W.; Taylor, D.G.: Metabolism of 1,8-cineole by a Rhodococcus species: ring cleavage reactions. J. Gen. Microbiol., **135**, 1957-1967 (1989)

Isoflavone 3'-hydroxylase 1.14.13.52

1 Nomenclature

EC number
1.14.13.52

Systematic name
formononetin,NADPH:oxygen oxidoreductase (3'-hydroxylating)

Recommended name
isoflavone 3'-hydroxylase

Synonyms
isoflavone 3'-monooxygenase

CAS registry number
110183-50-1

2 Source Organism

<1> *Cicer arietinum* (strain ILC3279 resistant to Ascochyta rabiei and strain ILC susceptible to Ascochyta rabiei, elicitor-induced [2,3]) [1-3]

3 Reaction and Specificity

Catalyzed reaction
formononetin + NADPH + H^+ + O_2 = calycosin + NADP + H_2O

Reaction type
hydroxylation
oxidation
redox reaction
reduction

Natural substrates and products
S formononetin + NADPH + O_2 <1> (<1> involved in the biosynthesis of the pterocarpin phytoalexin maackiain, cytochrome P_{450}-dependent monooxygenase [1,3]) (Reversibility: ir [1, 3]) [1, 3]
P 3'-hydroxyformononetin + $NADP^+$ + H_2O

Substrates and products
S biochanin A + NADPH + O_2 <1> (Reversibility: ir [1-3]) [1-3]
P pratensein + $NADP^+$ + H_2O

S formononetin + NADPH + O_2 <1> (Reversibility: ir [1-3]) [1-3]
P 3'-hydroxyformononetin + $NADP^+$ + H_2O
S Additional information <1> (<1> monohydroxylation of 4'-methoxyisoflavones; no substrates are genistein and daidzein [1]) [1]
P ?

Inhibitors

BAS110 <1> [3]
CO <1> (<1> CO/O_2 9:1 lead to 88-96% inhibition [3]) [3]
cytochrome c <1> (<1> 0.01 mM reduces activity to 4%, 0.1 mM reduces activity to 1% [1]) [1]
juglone <1> (<1> 0.001 mM juglone reduces activity to 50% [3]) [3]
ketoconazol <1> [3]
tetcyclasis <1> [3]
triadimefone <1> [3]
Additional information <1> (<1> no inhibition by 5 mM KCN and 1 mM EDTA [1]) [1]

Cofactors/prosthetic groups

NADH <1> (<1> 8.5% of NADPH under same conditions [1]) [1]
NADPH <1> [1]
Additional information <1> (<1> no cofactors are FAD and FMN [1]) [1]

Activating compounds

NADH <1> (<1> synergistic effect with NADPH, 150% of NADPH activity [1]) [1]
Additional information <1> (<1> cells susceptible to Ascochyta rabiei can be elicitor-induced to 500% activity [1]) [1]

K_m-Value (mM)

11 <1> (formononetin) [3]
12.5 <1> (biochanin A) [3]

pH-Optimum

7.5 <1> (<1> assay conditions [1,2]) [1, 2]
8 <1> [3]

Temperature optimum (°C)

25 <1> (<1> assay conditions [2]) [2]
30 <1> [3]

5 Isolation/Preparation/Mutation/Application

Source/tissue

cell suspension culture <1> [1]

Localization

microsome <1> [1]

6 Stability

Storage stability

<1>, -20°C, 100 mM potassium phosphate, pH 7.5 containing 400 mM sucrose and 3.5 mM 2-mercaptoethanol, 24 h [1]

References

[1] Hinderer, W.; Flentje, U.; Barz, W.: Microsomal isoflavone 2'- and 3'-hydroxylases from chickpea (Cicer arietinum L.) cell suspensions induced for pterocarpan phytoalexin formation. FEBS Lett., **214**, 101-106 (1987)

[2] Daniel, S.; Tiemann, K.; Wittkampf, U.; Bless, W.; Hinderer, W.; Barz, W.: Elicitor-induced metabolic changes in cell cultures of chickpea (Cicer arietinum L.) cultivars resistant and susceptible to Ascochyta rabiei. Planta, **182**, 270-278 (1990)

[3] Clemens, S.; Hinderer, W; Wittkampf, U; Barz, W.: Characterization of cytochrome P_{450}-dependent isoflavone hydroxylases from chickpea. Phytochemistry, **32**, 653-657 (1993)

Isoflavone 2'-hydroxylase 1.14.13.53

1 Nomenclature

EC number
1.14.13.53

Systematic name
formononetin,NADPH:oxygen oxidoreductase (2'-hydroxylating)

Recommended name
isoflavone 2'-hydroxylase

Synonyms
isoflavone 2'-monooxygenase

CAS registry number
110183-49-8

2 Source Organism

<1> *Cicer arietinum* (strain ILC3279 resistant to Ascochyta rabiei and strain ILC susceptible to Ascochyta rabiei, elicitor-induced [2,6]) [1, 2, 6]
<2> *Glycyrrhiza echinata* (elicitor-induced [3]) [3]
<3> *Lotus japonicus* [4]
<4> *Glycine max* (Mervill cv. Harosoy 63, elicitor-induced [5]) [5]

3 Reaction and Specificity

Catalyzed reaction
formononetin + NADPH + H^+ + O_2 = 2'-hydroxyformononetin + $NADP^+$ + H_2O

Reaction type
hydroxylation
oxidation
redox reaction
reduction

Natural substrates and products
S formononetin + NADPH + O_2 <1, 2> (<1> involved in the biosynthesis of the pterocarpin phytoalexin medicarpin [1, 6]; <1, 2> cytochrome P_{450}-dependent monooxygenase [1, 3, 6]) (Reversibility: ir <1, 2> [1, 3]) [1, 3, 6]
P 2'-hydroxyformononetin + $NADP^+$ + H_2O

Substrates and products

- **S** biochanin A + NADPH + O_2 <1> (Reversibility: ir <1> [1, 2]) [1, 2]
- **P** 2'-hydroxybiochanin A + $NADP^+$ + H_2O <1> [1, 2]
- **S** daidzein + NADPH + O_2 <2> (Reversibility: ir <2> [3]) [3]
- **P** 2'-hydroxydaidzein + $NADP^+$ + H_2O <2> [3]
- **S** formononetin + NADPH + O_2 <1-3> (Reversibility: ir <1-3> [1-4]) [1-4]
- **P** 2'-hydroxyformononetin + $NADP^+$ + H_2O
- **S** genistein + NADPH + O_2 <2, 4> (Reversibility: ir <2> [3]; ? <4> [5]) [3, 5]
- **P** 2'-hydroxygenistein + $NADP^+$ + H_2O <2, 4> [3, 5]
- **S** Additional information <1> (<1> monohydroxylation of 4'-methoxyisoflavones, no substrates are genistein and daidzein [1,6]) [1, 6]
- **P** ?
- **S** Additional information <2> (<2> no substrates are 5,7,4'-trihydroxyflavone, 7,4'-dihydroxyflavone, trans-cinnamic acid and 4-coumaric acid [3]) [3]
- **P** ?

Inhibitors

BAS 110 <1> [6]
BAS 111 <1> [6]
CO <1> (<1> CO/O_2 9/1 leads to 88-96% inhibition [6]) [6]
cytochrome c <1> (<1> 0.01 mM cytochrome c reduces activity to 14%, 0.1 mM cytochrome c reduces activity to 4% [1]) [1]
juglone <1> (<1> 0.001 mM juglone reduces activity to 50% [6]) [6]
ketoconazol <1> [6]
tetcyclasis <1> [6]
triadimefone <1> [6]
Additional information <1> (<1> no inhibition by 5 mM KCN and 1 mM EDTA [1]) [1]

Cofactors/prosthetic groups

NADH <1> (<1> 13.5% of NADPH under same conditions [1]) [1]
NADPH <1> [1]
Additional information <1> (<1> no cofactors are FAD and FMN [1]) [1]

Activating compounds

NADH <1> (<1,4> synergistic effect with NADPH [1,5]; <1> 135% of NADPH activity [1]) [1, 5]
Additional information <1> (<1> cells susceptible to Ascochyta rabiei can be elicitor-induced to 500% activity [1]) [1]

Metals, ions

Mn^{2+} <1> (<1> cells treated with $MnCl_2$ exhibit 5fold activity after 12 hours [6]) [6]

K_m-Value (mM)

3.3 <1> (formononetin) [6]
13 <1> (biochanin A) [6]

pH-Optimum

7.4 <1> [6]
7.5 <1> (<1> assay conditions [1,2]) [1, 2]

Temperature optimum (°C)

25 <1> (<1> assay conditions [2]) [2]
30 <1> [6]

5 Isolation/Preparation/Mutation/Application

Source/tissue

cell suspension culture <1-4> [1-5]

Localization

microsome <1-4> [1-5]

Cloning

<2> (expressed in Saccharomyces cerevisae overexpressing CYP81E1 [3]) [3]
<3> (expressed in Saccharomyces cerevisae [4]) [4]

6 Stability

Storage stability

<1>, -20°C, 100 mM potassium phosphate, pH 7.5 containing 400 mM sucrose and 3.5 mM 2-mercaptoethanol, 24 h [1]

References

[1] Hinderer, W.; Flentje, U.; Barz, W.: Microsomal isoflavone 2'- and 3'-hydroxylases from chickpea (Cicer arietinum L.) cell suspensions induced for pterocarpan phytoalexin formation. FEBS Lett., **214**, 101-106 (1987)
[2] Daniel, S.; Tiemann, K.; Wittkampf, U.; Bless, W.; Hinderer, W.; Barz, W.: Elicitor-induced metabolic changes in cell cultures of chickpea (Cicer arietinum L.) cultivars resistant and susceptible to Ascochyta rabiei. Planta, **182**, 270-278 (1990)
[3] Akashi, T.; Aoki, T.; Ayabe, S.I.: CYP81E1, a cytochrome P_{450} cDNA of licorice (Glycyrrhiza echinata L.), encodes isoflavone 2'-hydroxylase. Biochem. Biophys. Res. Commun., **251**, 67-70 (1998)
[4] Shimada, N.; Akashi, T.; Aoki, T.; Ayabe, S.i.: Induction of isoflavonoid pathway in the model legume Lotus japonicus: molecular characterization of enzymes involved in phytoalexin biosynthesis. Plant Sci., **160**, 37-47 (2000)
[5] Kochs, G.; Grisebach, H.: Enzymic synthesis of isoflavones. Eur. J. Biochem., **155**, 311-318 (1986)
[6] Clemens, S.; Hinderer, W; Wittkampf, U; Barz, W.: Characterization of cytochrome P_{450}-dependent isoflavone hydroxylases from chickpea. Phytochemistry, **32**, 653-657 (1993)

Ketosteroid monooxygenase 1.14.13.54

1 Nomenclature

EC number
1.14.13.54

Systematic name
ketosteroid,NADPH:oxygen oxidoreductase (20-hydroxylating, ester-producing/20-hydroxylating, side-chain cleaving/17-hydroxylating, lactonizing)

Recommended name
ketosteroid monooxygenase

Synonyms
17α-hydroxyprogesterone, $NADPH_2$:oxygen oxidoreductase (20-hydroxylating, side-chain cleaving)
androgen hydroxylase
androstenedione, $NADPH_2$:oxygen oxidoreductase (17-hydroxylating, lactonizing)
hydroxylase, steroid
oxygenase, steroid mono-
progesterone, $NADPH_2$:oxygen oxidoreductase (20-hydroxylating, ester-producing)
steroid hormone hydroxylase
steroid hydroxylase
steroid monooxygenase
steroid-ketone monooxygenase

CAS registry number
9044-53-5

2 Source Organism

<1> *Cylindrocarpon radicicola* (ATTC 11011) [1, 2]

3 Reaction and Specificity

Catalyzed reaction
ketosteroid + NADPH + H^+ + O_2 = steroid ester/lactone + NADP + H_2O

Reaction type
oxidation
reduction

Substrates and products

S 1,4-pregnadiene-17α,21-diol-3,11,20-trione + NADPH + O_2 <1> [2]
P ?
S 11β-hydroxyprogesterone + NADPH + O_2 <1> [2]
P ?
S 17α-hydroxypregnenolone + NADPH + O_2 <1> [1]
P ?
S 17α-hydroxyprogesterone + NADPH + O_2 <1> (<1>, the 17-ketosteroid acts first with the enzyme, and then NADPH [1]) [1, 2]
P androstenedione + acetate + $NADP^+$ + H_2O <1> (<1>, 4-androstene-3,17-dione [1]) [1]
S 18-hydroxydeoxycorticosterone + NADPH + O_2 <1> [2]
P ?
S 21-deoxycortisol + NADPH + O_2 <1> [2]
P ?
S aldosterone + NADPH + O_2 <1> [2]
P ?
S androstenedione + NADPH + O_2 <1> [1, 2]
P 3-oxo-13,17-secoandrost-4-eno-17,13-α-lactone + $NADP^+$ + H_2O <1> (<1>, i.e. testololactone [1]) [1, 2]
S corticosterone + NADPH + O_2 <1> [2]
P ?
S cortisol + NADPH + O_2 <1> [2]
P ?
S cortisone + NADPH + O_2 <1> [2]
P ?
S deoxycorticosterone + NADPH + O_2 <1> [2]
P ?
S pregnenolone + NADPH + O_2 <1> [1]
P ?
S progesterone + NADPH + O_2 <1> [2]
P testosterone acetate + $NADP^+$ + H_2O <1> [2]

Inhibitors

5-pregnene-3β,20α-diol <1> (<1>, lactonization of androstenedione) [1]
pregnenolone <1> (<1>, and other C21-20-ketosteroids which are the substrates for oxygenative esterification of the enzyme, strongly inhibit lactonization of androstenedione) [1]
progesterone <1> (<1>, strong, competitive inhibitor of lactonization of 17-ketosteroids) [1]
testosterone <1> (<1>, lactonization of androstenedione) [1]
testosterone acetate <1> (<1>, lactonization of androstenedione) [1]

Cofactors/prosthetic groups

FAD <1> (<1>, enzyme contains 1.6 mol of FAD per mol of enzyme) [2]

Specific activity (U/mg)

Additional information <1> [1]

K_m-Value (mM)

0.0004 <1> (progesterone, <1>) [1, 2]
0.002 <1> (NADPH, with androstenedione and O_2 as cosubstrates, <1>) [1]
0.0043 <1> (NADPH, with progesterone and O_2 as cosubstrates, <1>) [1, 2]
0.04 <1> (androstenedione, <1>) [1]
0.07 <1> (O_2, with progesterone and NADPH as cosubstrates, <1>) [1, 2]

pH-Optimum

6.5 <1> (<1>, androstenedione monooxygenase reaction) [1]
7.8 <1> (<1>, progesterone monooxygenase reaction) [1, 2]

pH-Range

5.5-8 <1> (<1>, pH 5.5: about 20% of maximal activity, pH 8.0: about 40% of maximal activity, androstenedione monooxygenase activity) [1]
6-9 <1> (pH 6.0: about 35% of maximal activity, pH 9.0: about 65% of maximal activity, progesterone monooxygenase activity) [2]

Temperature optimum (°C)

45 <1> (<1>, progesterone monooxygenase activity) [1]

4 Enzyme Structure

Molecular weight

116000 <1> (<1>, gel filtration, gel electrophoresis) [2]

Subunits

dimer <1> (<1>, 2 * 56000, SDS-PAGE) [2]

5 Isolation/Preparation/Mutation/Application

Purification

<1> [2]

6 Stability

Temperature stability

4 <1> (<1>, half-life: 8 days) [1]

Storage stability

<1>, -80°C, Tris buffer, stable for more than 6 months, enzyme in crude extract [2]

References

[1] Itagaki, E.: Studies on steroid monooxygenase from Cylindrocarpon radicicola ATCC 11011. Oxygenative lactonization of androstenedione to testololactone. J. Biochem., **99**, 825-832 (1986)
[2] Itagaki, E.: Studies on steroid monooxygenase from Cylindrocarpon radicicola ATCC 11011. Purification and characterization. J. Biochem., **99**, 815-824 (1986)

Protopine 6-monooxygenase 1.14.13.55

1 Nomenclature

EC number
1.14.13.55

Systematic name
protopine,NADPH:oxygen oxidoreductase (6-hydroxylating)

Recommended name
protopine 6-monooxygenase

Synonyms
oxygenase, protropine 6-mono-
protopine-6-hydroxylase

CAS registry number
128561-60-4

2 Source Organism

<-2> no activity in *Berberis stolonifera* (elicited and non-elicited cells [1]) [1]
<-1> no activity in *Catharanthus roseus* (elicited and non-elicited cells [1]) [1]
<1> *Eschscholtzia californica* (strain AST, strain BB, strain ROT [1]) [1]
<2> *Eschscholtzia lobbi* [1]
<3> *Chelidonium majus* [1]
<4> *Fumaria parviflora* [1]
<5> *Dicentra cucularia* [1]

3 Reaction and Specificity

Catalyzed reaction
protopine + NADPH + H^+ + O_2 = 6-hydroxyprotopine + $NADP^+$ + H_2O

Reaction type
oxidation
reduction

Natural substrates and products
S Additional information <1-5> (<1-5>, the enzyme is specifically present in different plant species that produce benzo[c]phenanthridine alkaloids in culture [1]) [1]

Substrates and products

S protopine + NADPH + O_2 <1-5> [1]

P dihydrosanguinarine + $NADP^+$ <1-5> [1]

Inhibitors

CO <1> [1]
cytochrome c <1> [1]
ketoconazole <1> [1]
metyrapone <1> [1]
prochloraz <1> [1]

Cofactors/prosthetic groups

NADPH <1> (<1>, strictly dependent on [1]) [1]
cytochrome P_{450} <1> (<1>, the enzyme is linked to cytochrome P-450) [1]

K_m-Value (mM)

0.002 <1> (protopine, <1>) [1]

pH-Optimum

7.5 <1> [1]

pH-Range

6.3-8.5 <1> (<1>, pH 6.3: about 35% of maximal activity, pH 8.5: about 40% of maximal activity) [1]

Temperature optimum (°C)

30 <1> [1]

Temperature range (°C)

10-40 <1> (<1>, about 50% of maximal actvity at 10°C and 40°C) [1]

5 Isolation/Preparation/Mutation/Application

Source/tissue

cell suspension culture <1-5> (<1-5>, weak activity in non-elicited cells, higher activity in elicited cells [1]) [1]

Localization

microsome <1-5> (<1>, membrane-bound [1]) [1]

References

[1] Tanahashi, T.; Zenk, M.H.: Elicitor induction and characterization of microsomal protopine-6-hydroxylase, the central enzyme in benzophenanthridine alkaloid biosynthesis. Phytochemistry, **29**, 1113-1122 (1990)

Dihydrosanguinarine 10-monooxygenase 1.14.13.56

1 Nomenclature

EC number
1.14.13.56

Systematic name
dihydrosanguinarine,NADPH:oxygen oxidoreductase (10-hydroxylating)

Recommended name
dihydrosanguinarine 10-monooxygenase

Synonyms
dihydrosanguinarine 10-hydroxylase
oxygenase, dihydrosanguinarine 10-mono

CAS registry number
144388-41-0

2 Source Organism

<1> *Eschscholtzia californica* [1]

3 Reaction and Specificity

Catalyzed reaction
dihydrosanguinarine + NADPH + H^+ + O_2 = 10-hydroxydihydrosanguinarine + $NADP^+$ + H_2O

Reaction type
oxidation
reduction

Substrates and products
S dihydrosanguinarine + NADPH + O_2 <1> [1]
P 10-hydroxydihydrosanguinarine + $NADP^+$ <1> [1]

Inhibitors
CO <1> (<1>, strong inhibition in the dark, effect is partly reversible by illumination) [1]
ancymidol <1> [1]
cytochrome c <1> [1]
ketoconazole <1> [1]

Cofactors/prosthetic groups

NADPH <1> (<1>, required, no activity with NADH [1]) [1]
cytochrome P_{450} <1> (<1>, dependent on [1]) [1]

5 Isolation/Preparation/Mutation/Application

Source/tissue

cell suspension culture <1> [1]

Localization

microsome <1> (<1>, associated with) [1]

References

[1] De-Eknamkul, W.; Tanahashi, T.; Zenk, M.H.: Enzymic 10-hydroxylation and 10-O-methylation of dihydrosanguinarine in dihydrochelirubine formation by Eschscholtzia. Phytochemistry, **31**, 2713-2717 (1992)

Dihydrochelirubine 12-monooxygenase 1.14.13.57

1 Nomenclature

EC number
1.14.13.57

Systematic name
dihydrochelirubine,NADPH:oxygen oxidoreductase (12-hydroxylating)

Recommended name
dihydrochelirubine 12-monooxygenase

Synonyms
dihydrochelirubine 12-hydroxylase
oxygenase, dihydrochelirubine 12-mono

CAS registry number
158736-41-5

2 Source Organism

<1> *Thalictrum bulgaricum* [1]

3 Reaction and Specificity

Catalyzed reaction
dihydrochelirubine + NADPH + H^+ + O_2 = 12-hydroxydihydrochelirubine + $NADP^+$ + H_2O

Reaction type
oxidation
reduction

Natural substrates and products
S dihydrochelirubine + NADPH + O_2 <1> (<1>, involved the penultimate step in the biosynthesis of macarpine, the most highly oxidized benzophenanthridine alkaloid found in nature [1]) [1]

Substrates and products
S dihydrochelirubine + NADH + O_2 <1> (<1>, at 83% of the activity with NADPH [1]) [1]
P 12-hydroxydihydrochelirubine + NAD^+ <1> [1]

S dihydrochelirubine + NADPH + O_2 <1> [1]
P 12-hydroxydihydrochelirubine + $NADP^+$ <1> [1]

Inhibitors
2-methyl-1,2-di(3-pyridyl)-1-propanone <1> (<1>, i.e. methyrapone) [1]
5-hydroxy-1,4-naphthoquinone <1> (<1>, i.e. juglone [1]) [1]
CO <1> (<1>, inhibition in the dark, effect is partly reversible upon illumination [1]) [1]
cytochrome c <1> [1]
ketoconazole <1> [1]

Cofactors/prosthetic groups
NADH <1> (<1>, dependent on NADPH or NADH [1]) [1]
NADPH <1> (<1>, dependent on NADPH or NADH [1]) [1]
cytochrome P_{450} <1> (<1>, dependent on) [1]

pH-Optimum
8.5 <1> [1]

Temperature optimum (°C)
30 <1> [1]

5 Isolation/Preparation/Mutation/Application

Source/tissue
cell culture <1> (<1>, yeast-elicited cells [1]) [1]

Localization
microsome <1> (<1>, associated with [1]) [1]

References

[1] Kammerer, L.; De-Eknamkul, W.; Zenk, M.H.: Enzymic 12-hydroxylation and 12-O-methylation of dihydrochelirubine in dihydromacarpine formation by Thalictrum bulgaricum. Phytochemistry, **36**, 1409-1416 (1994)

Benzoyl-CoA 3-monooxygenase 1.14.13.58

1 Nomenclature

EC number
1.14.13.58

Systematic name
benzoyl-CoA, NADPH:oxygen oxidoreductase (3-hydroxylating)

Recommended name
benzoyl-CoA 3-monooxygenase

Synonyms
benzoyl-CoA 3-hydroxylase
oxygenase, benzoyl coenzyme A 3-mono

CAS registry number
151616-61-4

2 Source Organism

<1> *Pseudomonas sp.* (strain KB740 [1,2]) [1, 2]

3 Reaction and Specificity

Catalyzed reaction
benzoyl-CoA + NADPH + H^+ + O_2 = 3-hydroxybenzoyl-CoA + $NADP^+$ + H_2O

Reaction type
oxidation
reduction

Natural substrates and products
S benzoyl-CoA + NADPH + O_2 <1> (<1>, second enzyme in a variant of the aerobic benzoate degradation pathway [1,2]; <1>, enzyme is specifically induced during aerobic growth with benzoate [2]) [1, 2]

Substrates and products
S benzoyl-CoA + NADPH + O_2 <1> (<1>, no activity with NADH [1]) [1]
P 3-hydroxybenzoyl-CoA + NADP + H_2O <1> [1]

Inhibitors

EDTA <1> [1]

Cofactors/prosthetic groups

FAD <1> (<1>, reaction is strictly dependent on the presence of a flavin nucleotide, FMN or FAD as cofactor [1]; <1>, required [2]; <1>, K_m: 0.15 mM [2]) [1, 2]
FMN <1> (<1>, reaction is strictly dependent on the presence of a flavin nucleotide, FMN or FAD as cofactor) [1]
NADPH <1> (<1>, required [1,2]; <1>, no activity with NADH [1]) [1, 2]

Specific activity (U/mg)

0.025 <1> [2]
1.812 <1> [1]

K_m-Value (mM)

0.003 <1> (benzoyl-CoA, <1>) [2]
0.125 <1> (NADPH, <1>) [2]

pH-Optimum

8 <1> [1]

pH-Range

6.5-8.5 <1> (<1>, 50% of maximal activity at pH 6.5 and at pH 8.5) [1]

4 Enzyme Structure

Molecular weight

65000 <1> (<1>, gel filtration) [1]

Subunits

monomer <1> (<1>, 1 * 63000, SDS-PAGE) [1]

5 Isolation/Preparation/Mutation/Application

Source/tissue

culture condition:benzoate-grown cell <1> (<1>, grown aerobically) [1]

Purification

<1> (partial [2]) [1, 2]

6 Stability

Storage stability

<1>, -70°C, 50% loss of activity after 1 week [1]

References

[1] Niemetz, R.; Altenschmidt, U.; Brucker, S.; Fuchs, G.: Benzoyl-coenzyme-A 3-monooxygenase, a flavin-dependent hydroxylase. Purification, some properties and its role in aerobic benzoate oxidation via gentisate in a denitrifying bacterium. Eur. J. Biochem., **227**, 161-168 (1995)
[2] Altenschmidt, U.; Oswald, B.; Steiner, E.; Herrmann, H.; Fuchs, G.: New aerobic benzoate oxidation pathway via benzoyl-coenzyme A and 3-hydroxybenzoyl-coenzyme A in a denitrifying Pseudomonas sp.. J. Bacteriol., **175**, 4851-4858 (1993)

L-Lysine 6-monooxygenase (NADPH) 1.14.13.59

1 Nomenclature

EC number
1.14.13.59

Systematic name
L-Lysine,NADPH:oxygen oxidoreductase (6-hydroxylating)

Recommended name
L-Lysine 6-monooxygenase (NADPH)

Synonyms
EC 13.12.10 (formerly)
IucD
lysine N(6)-hydroxylase
lysine N^6-hydroxylase
lysine:N^6-hydroxylase
oxygenase, lysine N^6-mono-

CAS registry number
64295-82-5

2 Source Organism

<1> *Escherichia coli* (strain EN222 which contains the structural gene for the enzyme on a multicopy plasmid [1]; strain GR143 and EN222, and strain which produce recombinant enzymes [3]; recombinant enzyme form IucD398, with a deletion of 47 amino acids in the N-terminus [4]) [1-6]

3 Reaction and Specificity

Catalyzed reaction
L-lysine + NADPH + H^+ + O_2 = N^6-hydroxy-L-lysine + $NADP^+$ + H_2O

Reaction type
oxidation
reduction

Natural substrates and products
S L-Lys + NADPH + O_2 <1> (<1>, enzyme catalyzes the first step in aerobactin biosynthesis [3]) [3]

Substrates and products

S (S)-2-aminoethyl-L-Cys + NADPH + O_2 <1> (<1>, i.e. L-aminoethylcysteine [1]) [1, 4]
P ?
S 1,5-diaminopentane + NADPH + O_2 <1> [1]
P ?
S DL-4-selenalysine + NADPH + O_2 <1> [1]
P ?
S DL-homocysteine + NADPH + O_2 <1> [1]
P ?
S DL/DL-allo-δ-hydroxylysine + NADPH + O_2 <1> [1]
P ?
S L-Lys + NADH + O_2 <1> (<1>, with lower efficiency than NADPH, recombinant enzyme form IucD398, with a deletion of 47 amino acids in the N-terminus [4]) [4]
P ?
S L-Lys + NADPH + O_2 <1> (<1>, specific for NADPH [1]) [1-4]
P N^6-hydroxy-L-Lys + $NADP^+$ + H_2O <1> [1, 2]
S Additional information <1> (<1>, in absence of substrate, the enzyme has an NADPH oxidase activity which results in generation of H_2O_2 [2]) [2]
P ?

Inhibitors

Cl^- <1> (<1>, above 600 mM, enzyme exists in a reversible inactive conformation) [5]
FAD analogs <1> (<1>, complete loss of activity after prolonged incubation with 8-chloro-FAD, 8-fluoro-FAD, 8-mercapto-FAD or 8-methoxy-FAD) [2]
carbonylcyanide-*m*-chlorophenylhydrazone <1> [4]
carbonylcyanide-*p*-fluoromethoxyphenylhydrazone <1> [4]
cinnamylidene <1> [4]
p-chloromercuribenzoate <1> (<1>, 0.01 mM, 62% inhibition, reversed by dithiothreitol [1]) [1]

Cofactors/prosthetic groups

FAD <1> (<1>, contains 0.79 mol FAD per mol of subunit [1]; <1>, requires FAD [2,4-6]; <1>, binding is very weak [2]; <1>, K_m: 0.0051 mM [4]; <1>, K_m: 0.0007 mM [2]; <1>, K_m: 0.005 mM, parent enzyme protein rIucD and genetically engineered forms C51A rIucD, C51A/C158A rIucD and C158A rIucD [6]) [1-6]
NADH <1> (<1>, recombinant enzyme form IucD398, with a deletion of 47 amino acids in the N-terminus) [4]
NADPH <1> (<1>, required [1-5]) [1-5]
Additional information <1> (<1>, cofactor interactions [5]) [5]

Activating compounds

DL-2,3-Diaminopropionic acid <1> (<1>, stimulates, but remains unchanged [1]) [1]

DL-2,6-diaminopimelic acid <1> (<1>, stimulates, but remains unchanged [1]) [1]
L-Orn <1> (<1>, stimulates, but remains unchanged [1]) [1]
N^6-acetyl-L-Lys <1> (<1>, stimulates, but remains unchanged [1]) [1]

Specific activity (U/mg)
0.168 <1> [4]
0.189 <1> [1]
0.238 <1> [5]

K_m-Value (mM)
0.018 <1> (NADH, <1>) [1]
0.07 <1> (NADPH, parent enzyme protein rIucD and genetically engineered forms C51A rIucD, C51A/C158A rIucD and C158A rIucD, <1>) [6]
0.1 <1> (NADPH, recombinant enzyme form IucD398, with a deletion of 47 amino acids in the N-terminus, <1>) [4]
0.105 <1> (L-Lys, <1>) [1]

4 Enzyme Structure

Molecular weight
200000 <1> (<1>, gel filtration) [1]

Subunits
tetramer <1> (<1>, 4 * 50000, SDS-PAGE) [1]

5 Isolation/Preparation/Mutation/Application

Localization
cytoplasm <1> [4]
membrane <1> (<1>, enzyme from strain EN222 and strain GR143 [3]) [3]
soluble <1> (<1>, recombinant enzyme forms [3]) [1, 3]

Purification
<1> (recombinant enzyme form IucD439, in which the sequence encoding the IucD protein is fused in frame to the amino-terminal sequence of β-galactosidase) [3]

Engineering
C51A rlucD <1> (activity of the genetically engineered enzyme forms C51A rIucD, C51A/C158A eIucD is 1.5times that of the parent rIucD. The activity of C158A rIucD is similar to that of the parent enzyme form) [5]
C51A/C158A rlucD <1> (activity of the genetically engineered enzyme forms C51A rIucD, C51A/C158A eIucD is 1.5times that of the parent rIucD. The activity of C158A rIucD is similar to that of the parent enzyme form) [5]
Additional information <1> (construction of recombinant IucD proteins with modified amino termini by creating three in-frame gene fusions of IucD to

the amino-terminal amino acids of the cytoplasmic enzyme β-galactosidase. Two of these constructs result in the addition of the iucD coding region of a hydrophilic leader sequence of 13 and 30 amino acids. The other construct involves the deletion of the first 47 amino acids of the IucD amino terminus and the addition of 19 amino acids of the amino terminus of β-galactosidase. Cells expressing any of the three recombinant IucD forms produce soluble N^6-hydroxylysine) [3]

6 Stability

Storage stability

<1>, 4°C, medium of ionic strength 0.25 or higher, recombinant enzyme form IacD398, stable for 1 month [4]

References

[1] Plattner, H.J.; Pfefferle, P.; Romaguera, A.; Waschutza, S.; Dieckman, H.: Isolation and some properties of lysine N^6-hydroxylase from Escherichia coli strain EN222. Biol. Met., **2**, 1-5 (1989)

[2] Macheroux, P.; Plattner, H.J.; Romaguera, A.; Diekmann, H.: FAD and substrate analogs as probes for lysine N^6-hydroxylase from Escherichia coli EC 222. Eur. J. Biochem., **213**, 995-1002 (1993)

[3] Thariath, A.; Socha, D.; Valvano, M.A.; Viswanatha, T.: Construction and biochemical characterization of recombinant cytoplasmic forms of IucD protein (lysine:N^6-hydroxylase) encoded by the pColV-K30 aerobactin gene cluster. J. Bacteriol., **175**, 589-596 (1993)

[4] Thariath, A.M.; Fatum, K.L.; Valvano, M.A.; Viswanatha, T.: Physico-chemical characterization of a recombinant cytoplasmic form of lysine:N^6-hydroxylase. Biochim. Biophys. Acta, **1203**, 27-35 (1993)

[5] Marrone, L.; Beecroft, M.; Viswanatha, T.: Lysine:N^6-hydroxylase: cofactor interactions. Bioorg. Chem., **24**, 304-317 (1996)

[6] Marrone, L.; Viswanatha, T.: Effect of selective cysteine –] alanine replacements on the catalytic functions of lysine:N^6-hydroxylase. Biochim. Biophys. Acta, **1343**, 263-277 (1997)

27-Hydroxycholesterol 7α-monooxygenase 1.14.13.60

1 Nomenclature

EC number
1.14.13.60

Systematic name
27-hydroxycholesterol,NADPH:oxygen oxidoreductase (7α-hydroxylating)

Recommended name
27-hydroxycholesterol 7α-monooxygenase

Synonyms
27-hydroxycholesterol 7α-hydroxylase
27-hydroxycholesterol-7α-hydroxylase
27CHO 7α-OHase
hydroxycholesterol 7α-hydroxylase
oxygenase, 27-hydroxycholesterol 7α-mono
oxysterol 7α-hydroxylase

CAS registry number
149316-80-3

2 Source Organism

<1> *Mus musculus* [1, 5]
<2> *Rattus norvegicus* [2]
<3> *Homo sapiens* (gallstone patients [3]; infant with a severe neonatal cholestasis whose biological fluids are devoid of 7α-hydroxylated bile acids due to a mutation in the gene encoding 27-hydroxycholesterol 7α-monooxygenase [10]) [3, 8, 9, 10]
<4> *Mesocricetus auratus* [4, 6]
<5> *Sus scrofa* [7]

3 Reaction and Specificity

Catalyzed reaction
27-hydroxycholesterol + NADPH + H^+ + O_2 = 7α,27-dihydroxycholesterol + $NADP^+$ + H_2O

Reaction type
oxidation
reduction

Natural substrates and products

S 27-hydroxycholesterol + NADPH + O_2 <1, 3> (<1>, 27-hydroxycholesterol is a potent inhibitor of cholesterol synthesis, the 7α-hydroxylation leads to a complete loss of down-regulation of cholesterol synthesis [5]; <3>, the enzyme is important to fetal and neonatal bile acid synthesis. Deficiency in 27-hydroxycholesterol 7α-monooxygenase activity causes progressive neonatal intrahepatic cholestasis [10]) [5, 10]

S Additional information <1, 3> (<1>, first enzyme of one of two pathways which mediate the synthesis of essential 7α-hydroxylated bile acids [1]; <3>, the enzyme plays a role in regulating cholesterol synthesis in liver through modulating intracellular oxysterol composition by the 7α-hydroxylation [3]) [1, 3]

Substrates and products

S 25-hydroxycholesterol + NADPH + O_2 <1-3, 5> [1, 2, 7, 8]
P 7α,25-dihydroxycholesterol + $NADP^+$
S 27-hydroxycholesterol + NADPH + O_2 <1-5> [1-10]
P 5-cholestene-3β,7α,27 triol + $NADP^+$ <1, 3> [5, 9]

Inhibitors

N-bromosuccinimide <5> (<5>, 0.01 mM, 75-90% inhibition) [7]
α-naphthoflavone <5> (<5>, 0.2-0.5 mM, 15-30% inhibition) [7]
dehydroepiandrosterone <5> [7]
disulfiram <5> (<5>, poor [7]) [7]
interleukin-1β <2> (<2>, enhances activity) [2]
ketoconazole <5> (<5>, 0.1 mM, 70-80% inhibition) [7]
nafimidone <1> [1]
tumor necrosis factor α <2> (<2>, enhances activity) [2]

Cofactors/prosthetic groups

NADPH <1-5> (<1-5>, dependent on) [1-10]
cytochrome P_{450} <1, 5> (<1,5>, contains cytochrome P_{450} [5,7]) [5, 7]

K_m-Value (mM)

0.004 <5> (27-hydroxycholesterol, <5> [7]; 25-hydroxycholesterol, <5> [7]) [7]

4 Enzyme Structure

Subunits

? <5> (<5>, x * 51000-58000, SDS-PAGE) [7]

5 Isolation/Preparation/Mutation/Application

Source/tissue

Hep-G2 <4> [6]
fibroblast <3> (<3>, from gallstone patients [3]; <3>, cultured [8]) [3, 8]
liver <1, 4, 5> [1, 4, 6, 7, 9]
ovary <2> [2]

Localization

microsome <3-5> [3, 4, 6, 7, 9]

Purification

<5> [7]

Cloning

<1> (cDNA is originally isolated from hippocampus [1]) [1]
<3> (identification of a nonsensense mutation R388* in the fifth exon of the 27-hydroxycholesterol 7α-monooxygenase gene by cloning studies [10]) [10]

References

[1] Schwarz, M.; Lund, E.G.; Lathe, R.; Björkhem, I.; Russell, D.W.: Identification and characterization of a mouse oxysterol 7α-hydroxylase cDNA. J. Biol. Chem., **272**, 23995-24001 (1997)

[2] Payne, D.W.; Shackleton, C.; Toms, H.; Ben-Shlomo, I.; Kol, S.; DeMoura, M.; Strauss, J.F.; Adashi, E,Y.: A novel nonhepatic hydroxycholesterol 7α-hydroxylase that is markedly stimulated by interleukin-1β. Characterization in the immature rat ovary. J. Biol. Chem., **270**, 18888-18896 (1995)

[3] Shoda, J.; He, B.f.; Tanaka, N.; Matsuzaki, Y.; Osuga, T.: Novel sterol 7α-hydroxylase(s), microsomal 27-hydroxycholesterol 7α-hydroxylase, in cholesterol gallstone disease and its etiological significance. J. Gastroenterol., **29**, 241-243 (1994)

[4] Lee, C.; Martin, K.O.; Javitt, N.B.: Bile acid synthesis: 7α-hydroxylation of intermediates in the sterol 27-hydroxylase metabolic pathway. J. Lipid Res., **37**, 1356-1362 (1996)

[5] Martin, K.O.; Reiss, A.B.; Lathe, R.; Javitt, N.B.: 7α-Hydroxylation of 27-hydroxycholesterol: biologic role in the regulation of cholesterol synthesis. J. Lipid Res., **38**, 1053-1059 (1997)

[6] Martin, K.O.; Budai, K.; Javitt, N.B.: Cholesterol and 27-hydroxycholesterol 7α-hydroxylation: evidence for two different enzymes. J. Lipid Res., **34**, 581-588 (1993)

[7] Norlin, M.; Wikvall, K.: Biochemical characterization of the 7α-hydroxylase activities towards 27-hydroxycholesterol and dehydroepiandrosterone in pig liver microsomes. Biochim. Biophys. Acta, **1390**, 269-281 (1998)

[8] Zhang, J.; Larsson, O.; Sjövall, J.: 7α-Hydroxylation of 25-hydroxycholesterol and 27-hydroxycholesterol in human fibroblasts. Biochim. Biophys. Acta, **1256**, 353-359 (1995)

[9] Björkhem, I.; Nyberg, B.; Einarsson, K.: 7α-Hydroxylation of 27-hydroxycholesterol in human liver microsomes. Biochim. Biophys. Acta, **1128**, 73-76 (1992)

[10] Setchell, K.D.R.; Schwarz, M.; O'Connell, N.C.; Lund, E.G.; Davis, D.L.; Lathe, R.; Thompson, H.R.; Tyson, R.W.; Sokol, R.J.; Russell, D.W.: Identification of a new inborn error in bile acid synthesis: mutation of the oxysterol 7α-hydroxylase gene causes severe neonatal liver disease. J. Clin. Invest., **102**, 1690-1703 (1998)

2-Hydroxyquinoline 8-monooxygenase 1.14.13.61

1 Nomenclature

EC number
1.14.13.61

Systematic name
quinolin-2(1H)-one,NADH:oxygen oxidoreductase (8-oxygenating)

Recommended name
2-hydroxyquinoline 8-monooxygenase

Synonyms
2-oxo-1,2-dihydroquinoline 8-monooxygenase
2-oxo-1,2-dihydroquinoline 8-monooxygenase (Pseudomonas putida strain 86 gene oxoO subunit)
2-oxo-1,2-dihydroquinoline 8-monooxygenase (Pseudomonas putida strain 86 gene oxoR subunit)
GenBank Y12654-derived protein GI 2072729
GenBank Y12655-derived protein GI 2072732
oxygenase, 2-oxo-1,2-dihydroquinoline 8-mono-
oxygenase, 2-oxo-1,2-dihydroquinoline 8-mono- (Pseudomonas putida strain 86 gene oxoO subunit)
oxygenase, 2-oxo-1,2-dihydroquinoline 8-mono- (Pseudomonas putida strain 86 gene oxoR subunit)

CAS registry number
166799-89-9
191941-72-7 (GenBank Y12655-derived protein GI 2072732, 2-oxo-1,2-dihydroquinoline 8-monooxygenase (Pseudomonas putida strain 86 gene oxoR subunit))
191941-74-9 (GenBank Y12655-derived protein GI, 2-oxo-1,2-dihydroquinoline 8-monooxygenase (Pseudomonas putida strain 86 gene oxoO subunit))

2 Source Organism

<1> *Pseudomonas putida* (strain 86 [1-3]) [1-3]

3 Reaction and Specificity

Catalyzed reaction

quinolin-2-ol + NADH + H^+ + O_2 = quinolin-2,8-diol + NAD^+ + H_2O

Reaction type

oxidation
reduction

Natural substrates and products

S 2-oxo-1,2-dihydroquinoline+ NADH + O_2 <1> (<1>, second enzyme in the quinoline degradation pathway. The reductase component and the oxygenase component are inducible by the substrate 2-oxo-1,2-dihydroquinoline [1]) [1]

Substrates and products

S 2-oxo-1,2-dihydroquinoline + NADH + O_2 <1> (<1>, no catalytic activity under anaerobic conditions [1]) [1]
P 8-hydroxy-2-oxo-1,2-dihydroquinoline + NAD^+ + H_2O <1> [1]
S Additional information <1> (<1>, the reductase component shows NADH:acceptor reductase activity with: cytochrome c, ferricyanide, 2,6-dichlorophenol indophenol and 2-(4-iodophenyl)-3-(4-nitrophenyl)-5-phenyl-2H-tetrazolium chloride [1]) [1]
P ?

Inhibitors

1,10-phenanthroline <1> [1]
4,5-dihydroxy-1,3-benzene disulfonic acid <1> [1]
EDTA <1> [1]
iodoacetate <1> [1]
NEM <1> [1]
p-hydroxymercuribenzoate <1> [1]

Cofactors/prosthetic groups

FAD <1> (<1>, the reductase component contains 1 FAD [1]) [1]

Activating compounds

polyethylene glycol <1> (<1>, enhances activity) [1]

Metals, ions

Fe^{2+} <1> (<1>, enhances activity) [1]
Iron <1> (<1>, the reductase component contains 1 plant-type ferredoxin [2Fe-2S] cluster, the oxygenase component contains 6 Rieske-type [2Fe-2S]clusters and additional iron [1]; the oxygenase component is a Rieske [2Fe-2S]protein, the reductase component contains a [2Fe-2S]cluster. Study of the 2Fe-2S centres by EPR spectroscopy [3]) [1, 3]

Specific activity (U/mg)

0.97 <1> (<1>, oxygenase component) [1]
32.5 <1> (<1>, reductase component) [1]

pH-Optimum

7.5 <1> [1]

Temperature optimum (°C)

25-30 <1> [1]

4 Enzyme Structure

Molecular weight

39000 <1> (<1>, reductase component, gel filtration) [1]
330000 <1> (<1>, oxygenase component, gel filtration) [1]

Subunits

? <1> (<1>, the reductase component is a monomer, 1 * 37000, SDS-PAGE. The oxygenase component is a hexamer, 6 * 55000, SDS-PAGE [1]; <1>, x * 51200, calculation from nucleotide sequence, oxygenase component + x * 37000, calculation from nucleotide sequence, reductase component [2]) [1, 2]

5 Isolation/Preparation/Mutation/Application

Purification

<1> [1]

Cloning

<1> [2]

References

[1] Rosche, B.; Tshisuaka, B.; Fetzner, S.; Lingens, F.: 2-Oxo-1,2-dihydroquinoline 8-monooxygenase,a two-component enzyme system from Pseudomonas putida 86. J. Biol. Chem., **270**, 17836-17842 (1995)
[2] Rosche, B.; Tshisuaka, B.; Hauer, B.; Lingens, F.; Fetzner, S.: 2-Oxo-1,2-dihydroquinoline 8-monooxygenase: phylogenetic relationship to other multicomponent nonheme iron oxygenases. J. Bacteriol., **179**, 3549-3554 (1997)
[3] Rosche, B.; Fetzner, S.; Lingens, F.; Nitschke, W.; Riedel, A.: The 2Fe2S centres of the 2-oxo-1,2-dihydroquinoline 8-monooxygenase from Pseudomonas putida 86 studied by EPR spectroscopy. Biochim. Biophys. Acta, **1252**, 177-179 (1995)

4-Hydroxyquinoline 3-monooxygenase 1.14.13.62

1 Nomenclature

EC number
1.14.13.62

Systematic name
quinolin-4(1H)-one,NADH:oxygen oxidoreductase (3-oxygenating)

Recommended name
4-hydroxyquinoline 3-monooxygenase

Synonyms
1H-4-oxoquinoline monooxygenase
oxygenase, 4(1H)-oxoquinoline 3-mono-
quinolin-4(1H)-one 3-monooxygenase

CAS registry number
144378-37-0

2 Source Organism

<1> *Pseudomonas putida* (strain 33/1 [1,2]) [1, 2]

3 Reaction and Specificity

Catalyzed reaction
quinolin-4-ol + NADH + H^+ + O_2 = quinolin-3,4-diol + NAD^+ + H_2O

Reaction type
oxidation
reduction

Natural substrates and products
S Additional information <1> (<1>, inducible by 1H-4-oxoquinoline [1]) [1]

Substrates and products
S 1H-4-oxoquinoline + NADH + O_2 <1> [1, 2]
P 1H-3-hydroxy-4-oxoquinoline + NAD^+ + H_2O <1> [1, 2]
S 1H-4-oxoquinoline + NADPH + O_2 <1> (<1>, 50% of the activity with NADH [1]) [1]
P 1H-3-hydroxy-4-oxoquinoline + $NADP^+$ + H_2O <1> [1]

Inhibitors

4-chloromercuribenzoate <1> (<1>, partial inhibition) [1]
Cu^{2+} <1> [1]
H_2O_2 <1> [1]
Hg^{2+} <1> [1]
Zn^{2+} <1> [1]
iodoacetate <1> (<1>, partial inhibition) [1]
sodium dithionite <1> [1]

Cofactors/prosthetic groups

NADH <1> (<1>, required as cofactor [1,2]) [1, 2]
NADPH <1> (<1>, 50% of the activity compared to NADH [1]) [1]

Activating compounds

acetone <1> (<1>, 15%, 1.2fold activation. Strong decrease in activity above 15% [1]) [1]
ethanol <1> (<1>, 15%, 1.7fold activation. Strong decrease in activity above 15% [1]) [1]

Specific activity (U/mg)

6.8 <1> [1]

K_m-Value (mM)

0.025 <1> (1H-4-oxoquinoline, <1>) [1]
0.087 <1> (NADH, <1>) [1]

pH-Optimum

7.5-8 <1> [1]

Temperature optimum (°C)

35 <1> [1]

4 Enzyme Structure

Molecular weight

126000 <1> (<1>, gel filtration) [1]

Subunits

trimer <1> (<1>, 3 * 42000, SDS-PAGE [1]) [1]

5 Isolation/Preparation/Mutation/Application

Purification

<1> [1]

6 Stability

pH-Stability

8.5 <1> (<1>, strong inactivation above [1]) [1]

Organic solvent stability

acetone <1> (40% loss of activity after 2 h in 15% acetone) [1]
ethanol <1> (stable for 2 h in 15% ethanol, 50% loss of activity after 24 h) [1]
ethylene glycol <1> (completely stable after 96 h in 50% ethylene glycol) [1]
Additional information <1> (50% loss of enzyme activity after 96 h without addition of organic solvents) [1]

General stability information

<1>, 50% loss of enzyme activity after 96 h without additions [1]

References

[1] Block, D.W.; Lingens, F.: XIII. Purification and properties of 1H-4-oxoquinoline monooxygenase from Pseudomonas putida strain 33/1. Biol. Chem. Hoppe-Seyler, **373**, 249-254 (1992)
[2] Bott, G.; Schmidt, M.; Rommel, T.O.; Lingens, F.: Microbial metabolism of quinoline and related compounds. V. Degradation of 1H-4-oxoquinoline by Pseudomonas putida 33/1. Biol. Chem. Hoppe-Seyler, **371**, 999-1003 (1990)

3-Hydroxyphenylacetate 6-hydroxylase 1.14.13.63

1 Nomenclature

EC number
1.14.13.63

Systematic name
3-hydroxyphenylacetate,NAD(P)H:oxygen oxidoreductase (6-hydroxylating)

Recommended name
3-hydroxyphenylacetate 6-hydroxylase

Synonyms
3-hydroxyphenylacetate 6-monooxygenase
oxygenase, 3-hydroxyphenylacetate 6-mono
m-hydroxyphenylacetate 6-hydroxylase

CAS registry number
114705-01-0

2 Source Organism

<1> *Flavobacterium sp.* (strain JS-7 [1]) [1, 2]

3 Reaction and Specificity

Catalyzed reaction
3-hydroxyphenylacetate + NAD(P)H + H^+ + O_2 = 2,5-dihydroxyphenylacetate (homogentisate) + $NAD(P)^+$ + H_2O (homogentisate)

Reaction type
oxidation
reduction

Natural substrates and products
S 3-hydroxyphenylacetate + NAD(P)H + O_2 <1> (<1>, enzyme in the 3-hydroxyphenylacetate catabolism [2]) [2]

Substrates and products
S 3,4-dihydroxyphenylacetate + NADH + O_2 <1> [1]
P 2,4,5-trihydroxyphenylacetate + NAD^+ <1> [1]
S 3,4-dihydroxyphenylacetate + NADPH + O_2 <1> [1]
P 2,4,5-trihydroxyphenylacetate + $NADP^+$ <1> [1]

S 3-hydroxyphenylacetate + NADH + O_2 <1> (<1>, NADPH and NADH are oxidized at a comparable rate [1]) [1]
P 2,5-dihydroxyphenylacetate + NAD^+ <1> [1]
S 3-hydroxyphenylacetate + NADPH + O_2 <1> (<1>, NADPH and NADH are oxidized at a comparable rate [1]) [1]
P 2,5-dihydroxyphenylacetate + $NADP^+$ <1> [1]
S 4-hydroxyphenylacetate + NADH + O_2 <1> [1]
P 2,4-dihydroxyphenylacetate + NAD^+ <1> [1]
S Additional information <1> (<1>, in the absence of oxygen the free enzyme is slowly reduced by NADPH or NADH [1]) [1]
P ?

Inhibitors
2-chloromercuri-4-nitrophenol <1> [1]
5,5'-dithiobis(2-nitrobenzoate) <1> [1]
Cl^- <1> (<1>, competitive to the aromatic substrate) [1]
NEM <1> [1]
p-hydroxymercuribenzoate <1> [1]

Cofactors/prosthetic groups
FAD <1> (<1>, contains 1 mol non-covalently bound FAD per mol of subunit [1]) [1]

Turnover number (min^{-1})
936 <1> (3,4-dihydroxyphenylacetate, <1>) [1]
1344 <1> (3-hydroxyphenylacetate, <1>) [1]
2148 <1> (NADPH, <1>) [1]
2160 <1> (NADH, <1>) [1]

Specific activity (U/mg)
20.1 <1> [1]

K_m-Value (mM)
0.079 <1> (3-hydroxyphenylacetate, <1>) [1]
0.09 <1> (3,4-dihydroxyphenylacetate, <1>) [1]
0.14 <1> (NADPH, <1>) [1]
0.15 <1> (NADH, <1>) [1]

pH-Optimum
8.3-8.9 <1> [1]

pH-Range
7.3-8.9 <1> (<1>, 50% of maximal activity at pH 7.3 and 8.9) [1]

4 Enzyme Structure

Molecular weight
150000-200000 <1> (<1>, gel filtration) [1]

Subunits

dimer <1> (<1>, 2 * 63000, SDS-PAGE) [1]

5 Isolation/Preparation/Mutation/Application

Source/tissue

culture condition:phenylacetic acid-grown cell <1> [1]

Purification

<1> [1]

6 Stability

pH-Stability

6.5-7 <1> (<1>, 37°C, stable) [1]
9 <1> (<1>, strong inactivation above) [1]

General stability information

<1>, 3-hydroxyphenylacetate protects against thermal inactivation [1]

Storage stability

<1>, -70°C, stable for at least 1 month [1]

References

[1] van Berkel, W.J.H.; van den Tweel, W.J.J.: Purification and characterisation of 3-hydroxyphenylacetate 6-hydroxylase: a novel FAD-dependent monooxygenase from a Flavobacterium species. Eur. J. Biochem., **201**, 585-591 (1992)

[2] Van den Tweel, W.J.J.; Smits, J.P.; de Bont, J.A.M.: Catabolism of DL-α-phenylhydracrylic, phenylacetic and 3- and 4-hydroxyphenylacetic acid via homogentisic acid in a Flavobacterium sp.. Arch. Microbiol., **149**, 207-213 (1988)

4-Hydroxybenzoate 1-hydroxylase 1.14.13.64

1 Nomenclature

EC number
1.14.13.64

Systematic name
4-hydroxybenzoate,NAD(P)H:oxygen oxidoreductase (1-hydroxylating, decarboxylating)

Recommended name
4-hydroxybenzoate 1-hydroxylase

Synonyms
4-hydroxybenzoate 1-hydroxylase (decarboxylating)
4-hydroxybenzoate 1-monooxygenase
oxygenase, 4-hydroxybenzoate 1-mono

CAS registry number
134214-78-1

2 Source Organism

<1> *Candida parapsilosis* (strain CBS604 [1,2]) [1, 2]

3 Reaction and Specificity

Catalyzed reaction
4-hydroxybenzoate + NAD(P)H + H^+ + O_2 = hydroquinone + $NAD(P)^+$ + H_2O + CO_2

Reaction type
oxidation
reduction

Natural substrates and products
S 4-hydroxybenzoate + NAD(P)H + O_2 <1> (<1>, initial reaction of the catabolism of 4-hydroxybenzoate [1]; <1>, the enzyme is induced when the yeast is grown on either 4-hydroxybenzoate, 2,3-dihydroxybenzoate, or 3,4-dihydroxybenzoate as the sole carbon source [2]) [1, 2]

Substrates and products

S 2,3,5,6-tetrafluoro-4-hydroxybenzoate + NADH + O_2 <1> [2]
P ?
S 2,3,5,6-tetrafluoro-4-hydroxybenzoate + NADPH + O_2 <1> [2]
P ?
S 2,4-dihydroxybenzoate + NADH + O_2 <1> [2]
P 1,2,4-trihydroxybenzene + NAD^+ + CO_2 <1> [2]
S 2,4-dihydroxybenzoate + NADPH + O_2 <1> [2]
P 1,2,4-trihydroxybenzene + NAD^+ + CO_2 <1> [2]
S 2,5-difluoro-4-hydroxybenzoate + NADH + O_2 <1> [2]
P ?
S 2,5-difluoro-4-hydroxybenzoate + NADPH + O_2 <1> [2]
P ?
S 2,6-difluoro-4-hydroxybenzoate + NADH + O_2 <1> [2]
P ?
S 2,6-difluoro-4-hydroxybenzoate + NADPH + O_2 <1> [2]
P ?
S 2-chloro-4-hydroxbenzoate + NADH + O_2 <1> [2]
P ?
S 2-chloro-4-hydroxybenzoate + NADPH + O_2 <1> [2]
P ?
S 2-fluoro-4-hydroxybenzoate + NADH + O_2 <1> [2]
P 2-fluoro-1,4-dihydroxybenzene + NAD^+ + CO_2 <1> [2]
S 2-fluoro-4-hydroxybenzoate + NADPH + O_2 <1> [2]
P 2-fluoro-1,4-dihydroxybenzene + NAD^+ + CO_2 <1> [2]
S 3,4-dihydroxybenzoate + NADH + O_2 <1> [2]
P 1,2,4-trihydroxybenzene + NAD^+ + CO_2 <1> [2]
S 3,4-dihydroxybenzoate + NADPH + O_2 <1> [2]
P 1,2,4-trihydroxybenzene + $NADP^+$ + CO_2 <1> [2]
S 3,5-difluoro-4-hydroxybenzoate + NADH + O_2 <1> [2]
P ?
S 3,5-difluoro-4-hydroxybenzoate + NADPH + O_2 <1> [2]
P ?
S 3-amino-4-hydroxybenzoate + NADH + O_2 <1> [2]
P ?
S 3-amino-4-hydroxybenzoate + NADPH + O_2 <1> [2]
P ?
S 3-chloro-4-hydroxybenzoate + NADH + O_2 <1> [2]
P ?
S 3-chloro-4-hydroxybenzoate + NADPH + O_2 <1> [2]
P ?
S 3-fluoro-4-hydroxybenzoate + NADH + O_2 <1> [2]
P ?
S 3-fluoro-4-hydroxybenzoate + NADPH + O_2 <1> [2]
P ?
S 4-hydroxy-3-methoxybenzoate + NADH + O_2 <1> [2]
P ?

S 4-hydroxy-3-methoxybenzoate + NADPH + O_2 <1> [2]
P ?
S 4-hydroxy-3-nitrobenzoate + NADH + O_2 <1> [2]
P ?
S 4-hydroxy-3-nitrobenzoate + NADPH + O_2 <1> [2]
P ?
S 4-hydroxybenzoate + NADH + O_2 <1> (<1>, NADH is the preferred electron donor [2]) [1, 2]
P 1,4-dihydroxybenzene + NAD^+ + CO_2 <1> [1, 2]
S 4-hydroxybenzoate + NADPH + O_2 <1> (<1>, NADH is the preferred electron donor [2]) [1, 2]
P 1,4-dihydroxybenzene + $NADP^+$ + CO_2 <1> [1, 2]

Inhibitors
3,5-dichloro-4-hydroxybenzoate <1> [2]
4-hydroxy-3,5-dinitrobenzoate <1> [2]
4-hydroxyisophthalate <1> [2]

Cofactors/prosthetic groups
FAD <1> (<1>, required [1]; <1>, contains FAD as weakly bound cofactor [2]; <1>, K_m: 0.0015 mM [2]) [1, 2]
NADH <1> (<1>, NADH is preferred to NADPH) [1]
NADPH <1> (<1>, NADH is preferred to NADPH) [1]

Turnover number (min^{-1})
390 <1> (3,5-difluoro-4-hydroxybenzoate, <1>) [2]
450 <1> (2,3,5,6-tetrafluoro-4-hydroxybenzoate, <1>) [2]
498 <1> (3-amino-4-hydroxybenzoate, <1>) [2]
510 <1> (2,6-difluoro-4-hydroxybenzoate, <1>) [2]
576 <1> (2-chloro-4-hydroxybenzoate, <1>) [2]
600 <1> (4-hydroxybenzoate, <1>) [2]
708 <1> (4-hydroxy-3-methoxybenzoate, <1>) [2]
720 <1> (2,5-difluoro-4-hydroxybenzoate, <1>) [2]
840 <1> (2-fluoro-4-hydroxybenzoate, <1>) [2]
1008 <1> (3,4-dihydroxybenzoate, <1>) [2]
1020 <1> (3-chloro-4-hydroxybenzoate, <1>) [2]
1086 <1> (3-fluoro-4-hydroxybenzoate, <1>) [2]
1188 <1> (2,4-dihydroxybenzoate) [2]
1188 <1> (4-hydroxy-3-nitrobenzoate) [2]

Specific activity (U/mg)
12 <1> [2]

K_m-Value (mM)
0.004 <1> (2,6-difluoro-4-hydroxybenzoate, <1> [2]; 3-chloro-4-hydroxybenzoate, <1> [1]) [2]
0.005 <1> (2,5-difluoro-4-hydroxybenzoate, <1> [1]; 3,5-difluoro-4-hydroxybenzoate, <1> [1]) [2]
0.007 <1> (2,4-dihydroxybenzoate, <1>) [2]

0.008 <1> (3-fluoro-4-hydroxybenzoate, <1>) [2]
0.009 <1> (3,4-dihydroxybenzoate, <1>) [2]
0.01 <1> (4-hydroxybenzoate, <1>) [2]
0.013 <1> (3-amino-4-hydroxybenzoate, <1> [2]; 4-hydroxy-3-nitrobenzoate, <1> [2]) [2]
0.018 <1> (2-fluoro-4-hydroxybenzoate, <1>) [2]
0.019 <1> (NADH, <1>) [2]
0.025 <1> (2-chloro-4-hydroxybenzoate, <1>) [2]
0.03 <1> (2,3,5,6-tetrafluoro-4-hydroxybenzoate, <1>) [2]
0.125 <1> (4-hydroxy-3-methoxybenzoate, <1>) [2]
0.169 <1> (NADPH, <1>) [2]

pH-Optimum
8 <1> [2]

Temperature optimum (°C)
37 <1> [2]

4 Enzyme Structure

Molecular weight
49000 <1> (<1>, gel filtration) [2]

Subunits
monomer <1> (<1>, 1 * 52000, SDS-PAGE) [2]

5 Isolation/Preparation/Mutation/Application

Purification
<1> [1, 2]

References

[1] van Berkel, W.J.H.; Eppink, M.H.M.; Middelhoven, W.J.; Vervoort, J.; Rietjens, I.M.C.M.: Catabolism of 4-hydroxybenzoate in Candida parapsilosis proceeds through initial oxidative decarboxylation by a FAD-dependent 4-hydroxybenzoate 1-hydroxylase. FEMS Microbiol. Lett., **121**, 207-216 (1994)

[2] Eppink, M.H.M.; Boeren, S.A.; Vervoort, J.; van Berkel, W.J.H.: Purification and properties of 4-hydroxybenzoate 1-hydroxylase (decarboxylating), a novel flavin adenine dinucleotide-dependent monooxygenase from Candida parapsilosis CBS604. J. Bacteriol., **179**, 6680-6687 (1997)

2-Hydroxyquinoline 8-monooxygenase 1.14.13.65

1 Nomenclature

EC number
1.14.13.65

Systematic name
quinolin-2-ol,NADH:oxygen oxidoreductase (8-hydroxylating)

Recommended name
2-hydroxyquinoline 8-monooxygenase

Synonyms
2-oxo-1,2-dihydroquinoline 5,6-dioxygenase
oxygenase, 2(1H)-quinolinone 5,6-di

CAS registry number
172399-50-7 (not distinguished from EC 1.14.12.16)

2 Source Organism

<1> *Comamonas testosteroni* (strain 63) [1]

3 Reaction and Specificity

Catalyzed reaction
quinolin-2-ol + NADH + H^+ + O_2 = quinolin-2,8-diol + NAD^+

Reaction type
oxidation
reduction

Natural substrates and products
S (3-methyl-)2-oxo-1,2-dihydroquinoline + NADH + O_2 <1> (<1>, the second enzyme in the degradation of 3-methylquinoline) [1]

Substrates and products
S (3-methyl-)2-oxo-1,2-dihydroquinoline + NADH + O_2 <1> [1]
P 5,6-dihydro-5,6-dihydroxy-(3-methyl-)2-oxo-1,2-dihydroquinoline <1> [1]
S 2-oxo-1,2-dihydroquinoline + NADH + O_2 <1> [1]
P ?
S 6-hydroxy-2-oxo-1,2-dihydroquinoline <1> [1]

P ?
S 8-hydroxy-2-oxo-1,2-dihydroquinoline <1> [1]
P ?
S 8-hydroxyquinoline + NADH + O_2 <1> [1]
P ?

Inhibitors

1,10-phenanthroline <1> [1]
2,2'-dipyridyl <1> [1]
4-hydroxymercuribenzoate <1> [1]
$CaCl_2$ <1> (<1>, 0.2 M, 90% loss of activity) [1]
Cu^{2+} <1> [1]
EDTA <1> (<1>, inhibitory only after prolonged incubation time) [1]
KCl <1> (<1>, 0.2 M, 85% loss of activity) [1]
NaCl <1> (<1>, 0.2 M, 50% inhibition) [1]
Tiron <1> [1]
acriflavin <1> [1]
diethyldithiocarbaminate <1> (<1>, inhibitory only after prolonged incubation time) [1]
iodoacetate <1> [1]
quinacrine <1> [1]

Cofactors/prosthetic groups

NADH <1> (<1>, required, no activity with NADPH) [1]

Metals, ions

Fe^{2+} <1> (<1>, enhances activity) [1]

Specific activity (U/mg)

0.125 <1> [1]

pH-Optimum

7.3 <1> [1]

4 Enzyme Structure

Molecular weight

260000 <1> (<1>, gel filtration) [1]

5 Isolation/Preparation/Mutation/Application

Purification

<1> (partial [1]) [1]

6 Stability

Temperature stability

30 <1> (<1>, 5 min, 70% loss of activity) [1]
65 <1> (<1>, 5 min, complete loss of activity) [1]

General stability information

<1>, dithioerythritol, dithiothreitol or 2-oxo-1,2-dihydroxyquinoline at 0.1 mM stabilizes [1]

References

[1] Schach, S.; Tshisuaka, B.; Fetzner, S.; Lingens, F.: Quinoline 2-oxidoreductase and 2-oxo-1,2-dihydroquinoline 5,6-dioxygenase from Comamonas testosteroni 63. The first two enzymes in quinoline and 3-methylquinoline degradation. Eur. J. Biochem., **232**, 536-544 (1995)

2-Hydroxycyclohexanone 2-monooxygenase 1.14.13.66

1 Nomenclature

EC number
1.14.13.66

Systematic name
2-hydroxycyclohexan-1-one,NADPH:oxygen 2-oxidoreductase (1,2-lactonizing)

Recommended name
2-hydroxycyclohexanone 2-monooxygenase

Synonyms
2-hydroxycyclohexan-1-one monooxygenase
oxygenase, 2-hydroxycyclohexanone 2-mono-

CAS registry number
62628-31-3

2 Source Organism

<1> *Acinetobacter sp.* (TD63) [1]

3 Reaction and Specificity

Catalyzed reaction
2-hydroxycyclohexan-1-one + NADPH + H^+ + O_2 = 6-hydroxyhexan-6-olide + $NADP^+$ + H_2O

Reaction type
oxidation
reduction

Natural substrates and products
S 2-hydroxycyclohexan-1-one + NADPH + O_2 <1> (<1>, enzyme in the oxidation pathway of trans-cyclohexan-1,2-diol to adipate, enzyme activity is induced by trans-cyclohexan-1,2-diol [1]) [1]

Substrates and products
S 2-hydroxycyclohexan-1-one + NADPH + O_2 <1> [1]
P 1-oxa-2-oxo-7-hydroxycycloheptane + NADP + H_2O <1> (<1>, 1-oxa-2-oxo-7-hydroxycycloheptane spontaneously rearranges to yield 6-oxohexanoate [1]) [1]

S 2-methylcyclohexanone + NADPH + O_2 <1> [1]
P ?
S 3-methylcyclohexanone + NADPH + O_2 <1> [1]
P ?
S 4-methylcyclohexanone + NADPH + O_2 <1> [1]
P ?
S cyclohexanone + NADPH + O_2 <1> [1]
P ?

Cofactors/prosthetic groups

NADPH <1> [1]

K_m-Value (mM)

0.0053 <1> (NADPH, <1>) [1]
0.0062 <1> (2-hydroxycyclohexan-1-one, <1>) [1]

References

[1] Davey, J.F.; Trudgill, P.W.: The metabolism of trans-cyclohexan-1,2-diol by an Acinetobacter species. Eur. J. Biochem., **74**, 115-127 (1977)

Quinine 3-monooxygenase 1.14.13.67

1 Nomenclature

EC number
1.14.13.67

Systematic name
quinine, NADPH:oxygen oxidoreductase

Recommended name
quinine 3-monooxygenase

Synonyms
CYP2C19
CYP3A4
CYPIIIA4
NF-25
nifedipine oxidase
P450-PCN1
quinine 3-monooxygenase
cytochrome P450 isoform
quinine 3-hydroxylase

CAS registry number
205394-96-3

2 Source Organism

<1> *Homo sapiens* (human [1-4]) [1-4]
<2> *Mus musculus* (mouse [3]) [3]
<3> *Rattus norvegicus* (rat [3]) [3]
<4> *Canis familiaris* (dog [3]) [3]

3 Reaction and Specificity

Catalyzed reaction
quinine + NADPH + H^+ + O_2 = 3-hydroxyquinine + $NADP^+$ + H_2O

Reaction type
oxidation
redox reaction
reduction

Substrates and products

- **S** etoposide + NADPH + O_2 <1> (Reversibility: ? <1> [2, 4]) [2, 4]
- **P** 3'-demethyletoposide + $NADP^+$ + H_2O [2, 4]
- **S** quinine + NADPH + O_2 <1-4> (Reversibility: ? <1-4> [1-3]) [1-3]
- **P** 3-hydroxyquinine + $NADP^+$ + H_2O [1-3]
- **S** teniposide + NADPH + O_2 <1> (Reversibility: ? <1> [4]) [4]
- **P** teniposide catechol + $NADP^+$ + H_2O [4]

Inhibitors

S-mephenytoin <1, 2> (<1> maximum inhibition of 74% at 0.12 mM [1]; <2> inhibits more than 70% at 0.5 mM [3]) [1, 3]
α-naphthoflavone <2, 3> [3]
anti-CYP2C antibodies <1> (<1> inhibition of 20% [1]) [1]
anti-CYP3A4 antibodies <1-4> (<1> inhibition of 72% [1]; <2> inhibition of 96% [3]; <3> inhibition of 84% [3]; <4> inhibition of 92% [3]) [1, 3]
chloroquin <3, 4> [3]
diazepam <2, 3> [3]
doxycyclin <1, 3, 4> [3]
etoposide <1> (<1> maximum inhibition of quinine 3-hydroxylation of 60% at 0.1 mM [2]) [2]
ketoconazole <1-4> (<1> maximum inhibition of 90% at 0.0005 mM [1]; <1> maximum inhibition of 90% at 0.001 mM [2]; <2> maximum inhibition of 94% [3]; <3> maximum inhibition of 91% [3]; <4> maximum inhibition of 88% [3]; <1> maximum inhibition of 90% [3]) [1, 2, 3]
p-nitrophenol <1, 2> (<2> inhibits more than 75% at 10 mM [3]; <1> inhibition observed but not quantified [3]) [3]
primaquine <1, 3, 4> [3]
quinine <1> (<1> maximum inhibition of etoposide 3'-demethylation of 52% at 0.1 mM [2]) [2]
sulfaphenazole <2> (<2> inhibts more than 50% at 0.1 mM [3]) [3]
tetracyclin <1, 3, 4> [3]
troleandomycin <1> (<1> maximum inhibition of 70% at 0.08 mM [1]; <1> maximum inhibition of 80% at 0.1 mM [2]; <2> maximum inhibition of 85% [3]; <3> maximum inhibition of 66% [3]; <4> maximum inhibition of 93% [3]; <1> maximum inhibition of 70% [3]) [1, 2]

Activating compounds

α-naphthoflavone <1, 4> [3]
diazepam <1, 4> [3]

Turnover number (min^{-1})

0.196 <1> (quinine, <1> recombinant CYP2C19 [1]) [1]
7.49 <1> (quinine, <1> recombinant CYP3A4 [1]) [1]

K_m-Value (mM)

0.0227 <3> (quinine) [3]
0.0297 <2> (quinine) [3]
0.046 <1> (quinine, <1> recombinant CYP2C19 [1]) [1]
0.105 <1> (quinine) [3]

0.106 <1> (quinine, <1> between 0.08 mM and 0.145 mM in 10 different human livers [1]) [1]
0.114 <1> (quinine, <1> recombinant CYP3A4 [1]) [1]
0.138 <4> (quinine) [3]
Additional information <1> (etoposide, <1> between 0.0421 mM and 0.1151 mM [4]) [4]
Additional information <1> (tenoposide, <1> between 0.0197 mM and 0.0435 mM [4]) [4]

5 Isolation/Preparation/Mutation/Application

Source/tissue
liver <1-4> [1-4]

Localization
microsome <1-4> [1-4]

Cloning
<1> (expression of CYP3A4 in human B lymphoblastoid cell line AHH-1 [1]) [1]

References

[1] Zhao, X.J.; Yokoyama, H.; Chiba, K.; Wanwimolruk, S.; Ishizaki, T.: Identification of human cytochrome P450 isoforms involved in the hydroxylation of quinine by human liver microsomes and nine recombinant human cytochromes P450. J. Pharmacol. Exp. Ther., **279**, 1327-1334 (1996)
[2] Zhao, X.J.; Kawashiro, T.; Ishizaki, T.: Mutual inhibition between quinine and etoposide by human liver microsomes. Evidence for cytochrome P4503A4 involvement in their major metabolic pathways. Drug Metab. Dispos., **26**, 188-191 (1997)
[3] Zhao, X.J.; Ishizaki, T.: The in vitro hepatic metabolism of quinine in mice, rats and dogs: comparison with human liver microsomes. J. Pharmacol. Exp. Ther., **283**, 1168-1176 (1997)
[4] Relling, M.V.; Evans, R.; Dass, C.; Desiderio, D.M.; Nemec, J.: Human cytochrome P450 metabolism of teniposide and etoposide. J. Pharmacol. Exp. Ther., **261**, 491-496 (1992)

4-Hydroxyphenylacetaldehyde oxime monooxygenase 1.14.13.68

1 Nomenclature

EC number
1.14.13.68

Systematic name
4-hydroxyphenylacetaldehyde oxime, NADPH:oxygen oxidoreductase

Recommended name
4-hydroxyphenylacetaldehyde oxime monooxygenase

Synonyms
4-hydroxybenzeneacetaldehyde oxime monooxygenase
4-hydroxyphenylacetaldehyde oxime monooxygenase
CYP71E1
NADPH-cytochrome P_{450} reductase
cytochrome P450II-dependent monooxygenase

CAS registry number
213017-82-4

2 Source Organism

<1> *Sorghum bicolor* [1-3]

3 Reaction and Specificity

Catalyzed reaction
4-hydroxyphenylacetaldehyde oxime + NADPH + H^+ + O_2 = 4-hydroxymandelonitrile + $NADP^+$ + 2 H_2O

Reaction type
oxidation
redox reaction
reduction

Substrates and products
S 2-hydroxy(*p*-hydroxyphenyl)acetaldoxime <1> (<1> poor substrate [2]) (Reversibility: ? <1> [2]) [2]
P 4-hydroxymandelonitrile + H_2O

S 4-hydroxyphenylacetaldehyde oxime + NADPH + O_2 <1> (Reversibility: ? <1> [1-3]) [1-3]
P 4-hydroxymandelonitrile + $NADP^+$ + 2 H_2O
S 4-hydroxyphenylacetonitrile + NADPH + O_2 <1> (Reversibility: ? <1> [2]) [2]
P 4-hydroxymandelonitrile + $NADP^+$ + H_2O

Inhibitors
2-hydroxy(*p*-hydroxyphenyl)acetaldoxime <1> (<1> in higher concentrations [2]) [2]

Cofactors/prosthetic groups
NADH <1> (<1> less effective than NADPH [1]) [1]
NADPH <1> [1, 2]
O_2 <1> [1, 2]

Specific activity (U/mg)
0.003 <1> (<1> conversion of *p*-hydroxyphenylacetaldoxime to *p*-hydrobenzaldehyde [1]) [1]

K_m-Value (mM)
0.2 <1> (*p*-hydroxyphenylacetaldoxime) [2]
0.35 <1> (*p*-hydroxyphenylacetonitrile) [2]

pH-Optimum
7.8-8.4 <1> (<1> formation of *p*-hydroxybenzaldehyde from L-tyrosine in microsomal fractions [1]) [1]

5 Isolation/Preparation/Mutation/Application

Source/tissue
seedling <1> [1, 2]

Localization
microsome <1> [1, 2]

Purification
<1> (purification of heterologous expressed protein [3]) [3]

Cloning
<1> (in Escherichia coli strain JM109 [3]) [3]

References

[1] McFarlane, I.J.; Lees, E.M: Conn, E.E.: The in vitro biosynthesis of dhurrin, the cyanogenic glycoside of Sorghum bicolor. J. Biol. Chem., **250**, 4708-4713 (1975)

[2] Shimada, M.; Conn, E.E.: The enzymatic conversion of *p*-hydroxyphenylacetaldoxime to *p*-hydroxymandelonitrile. Arch. Biochem. Biophys., **180**, 199-207 (1977)

[3] Bak, S.; Kahn, R.A.; Nielsen, H.L.; Moeller, B.L.; Halkier, B.A.: Cloning of three A-type cytochromes P450, CYP71E1, CYP98, and CYP99 from Sorghum bicolor (L:) Moench by a PCR approach and identification by expression in Escherichia coli of CYP71E1 as a multifunctional cytochrome P450 in the biosynthesis of the cyanogenic glucoside dhurrin. Plant Mol. Biol., **36**, 393-405 (1998)

Alkene monooxygenase 1.14.13.69

1 Nomenclature

EC number
1.14.13.69

Systematic name
alkene,NADH:oxygen oxidoreductase

Recommended name
alkene monooxygenase

Synonyms
AMO
alkene epoxygenase
propene monooxygenase

CAS registry number
63439-50-9

2 Source Organism

<1> *Mycobacterium sp.* (E3 [1, 7]; E20 [1]; multi-component enzyme [7]) [1, 7]
<2> *Rhodococcus corallinus* (i.e. Nocardia corallina [5]; B-276 [2, 4, 5]; multi-component enzyme [2]; enzyme is encoded by the linear plasmid pNC30 [5]) [2, 4, 5]
<3> *Xanthobacter sp.* (strain Py2 [3, 5, 6, 8, 9, 10]) [3, 5, 6, 8, 9, 10]
<4> *Rhodococcus ruber* (P-IV-B-11 and P-V-B-171 [5]) [5]
<5> *Mycobacterium aurum* (L1 [7]; multi-component enzyme [7]) [7]

3 Reaction and Specificity

Catalyzed reaction
propene + NADH + H^+ + O_2 = 1,2-epoxypropane + NAD^+ + H_2O

Reaction type
epoxidation
oxidation
redox reaction
reduction

Natural substrates and products

S propene + NADH + H^+ + O_2 <2, 3> (<3>, induction by propylene and propylene oxide and a variety of aliphatic and chlorinated alkenes and epoxides [3]; <2>, constitutive enzyme [4]; <3>, the inducible enzyme is central to the bacterial metabolism of aliphatic alkenes. Enzyme is expressed during growth of Xanthobacter on aliphatic alkenes or epoxides and repressed during growth on other carbon sources [9]) (Reversibility: ? <2, 3> [3, 4, 9]) [3, 4, 9]

P 1,2-epoxypropane + NAD^+ + H_2O

Substrates and products

S chloroethene + NADH + H^+ + O_2 <3> (Reversibility: ? <3> [3]) [3]

P 2-chlorooxirane + NAD^+ + H_2O

S propene + NADH + H^+ + O_2 <1-5> (Reversibility: ? <1-5> [1-10]) [1-10]

P 1,2-epoxypropane + NAD^+ + H_2O <1-5> [1-10]

S trans-1,2-dichloroethylene + NADH + H^+ + O_2 <3> (Reversibility: ? <3> [3]) [3]

P 2,3-dichlorooxirane + NAD^+ + H_2O

S trichloroethylene + NADH + H^+ + O_2 <2, 3> (Reversibility: ? <2, 3> [3, 5]) [3, 5]

P 2,2,3-trichlorooxirane + NAD^+ + H_2O

Inhibitors

1,10-phenanthroline <1> (<1>, 2 mM, 40% inhibition. Activity is completely restored by 2 mM Fe^{2+} [7]) [7]

1,2-epoxypropane <1> [1]

8-hydroxyquinoline <1> [7]

KCN <1> [7]

acetylene <1> [7]

propyne <2, 3> (<2>, weak [4]; <3>, mechanism-based inactivator of the 21200 Da protein [9]) [4, 9]

Additional information <2> (<2>, no inhibition by ethyne [4]) [4]

Cofactors/prosthetic groups

FAD <2, 3> (<2>, the reductase component contains two prosthetic groups, an FAD centre and a [2Fe-2S] cluster. The FAD moiety is reduced by bound NADH in a two-electron reaction. The electrons are then transported to the [2Fe-2S] centre one at a time, which reduces the di-iron centre of the epoxydase. Reduction of the di-iron centre is required for oxygen binding and substrate oxidation [2]; <3>, the 35500 Da NADH reductase component contains 1 mol of FAD [9]) [2, 9]

Metals, ions

[2Fe-2S]cluster <2> (<2>, the reductase component contains two prosthetic groups, an FAD centre and a [2Fe-2S] cluster. The FAD moiety is reduced by bound NADH in a two-electron reaction. The electrons are then transported to the [2Fe-2S] centre one at a time, which reduces the di-iron centre of the epoxydase. Reduction of the di-iron centre is required for oxygen binding and substrate oxidation [2]) [2] iron <2, 3> (<2>, the reductase component contains two prosthetic groups, an FAD centre and a [2Fe-2S] cluster. The FAD moiety is reduced by bound NADH in a two-electron reaction. The electrons are then transported to the [2Fe-2S] centre one at a time, which reduces the di-iron centre of the epoxydase. Reduction of the di-iron centre is required for oxygen binding and substrate oxidation [2]; <2>, enzyme contains a bridged dinuclear iron centre [4]; <3>, α-helical structure that surrounds the binuclear iron binding site [6]; <3>, enzyme contains 4 components: 1. a monomeric 35500 Da NADH reductase containing 1 mol of FAD and a probable 2Fe-2S center, 2. a 13300 Da ferredoxin containing a Rieske-type 2Fe-2S cluster, 3. a 11000 Da monomeric protein that contains no detectable cofactors, 4. a 212000 Da $\alpha_2\beta_2\gamma_2$ multimeric protein containing 4 atoms of nonheme iron. The physiological electron acceptor for the reductase is the Rieske-type ferredoxin, which is proposed to be an intermediate electron carrier between the reductase and terminal catalytic component of the system [9]) [2, 4, 6, 9]

Specific activity (U/mg)

0.085 <2> [4]
Additional information <3> [9]

K_m-Value (mM)

0.016 <2> (propene) [5]
0.187 <2> (trichlorethene) [5]
0.235 <2> (propene) [4]

4 Enzyme Structure

Subunits

Additional information <1, 2, 3, 5> (<1,2,3,5>, multi-component enzyme [2, 7, 8]; <3>, enzyme contains 4 components: 1. a monomeric 35500 Da NADH reductase containing 1 mol of FAD and a probable 2Fe-2S center, mass spectrometry 2. a dimeric ferredoxin consisting of two 13300 Da subunits, each containing a Rieske-type 2Fe-2S cluster, SDS-PAGE 3. a 11000 Da monomeric protein that contains no detectable cofactors, mass spectrometry 4. a 212000 Da $\alpha_2\beta_2\gamma_2$ multimeric protein containing 4 atoms of nonheme iron [9]; <3>, x * 58037, oxygenase α-subunit, + x * 9740, γ-subunit, + x * 13359, ferredoxin, + x * 11193, coupling or effector protein, + 38188, oxygenase β-subunit, + x * 34171, reductase subunit, calculation from nucleotide sequence [8]) [2, 7, 8, 9]

5 Isolation/Preparation/Mutation/Application

Purification

<2> [4]
<3> [9]

Cloning

<3> (expression in Xanthobacter autotrophicus [10]) [10]

References

[1] Habets-Cruetzen, A.Q.H.; De Bont, J.A.M.: Inactivation of alkene oxidation by epoxides in alkene- and alkane-grown bacteria. Appl. Microbiol. Biotechnol., **22**, 428-433 (1985)

[2] Gallagher, S.C.; Cammack, R.; Dalton, H.: Electron transfer reactions in the alkene mono-oxygenase complex from Nocardia corallina B-276. Biochem. J., **339**, 79-85 (1999)

[3] Ensign, S.A.: Aliphatic and chlorinated alkenes and epoxides as inducers of alkene monooxygenase and epoxidase activities in Xanthobacter strain Py2. Appl. Environ. Microbiol., **62**, 61-66 (1996)

[4] Gallagher, S.C.; Cammack, R.; Dalton, H.: Alkene monooxygenase from Nocardia corallina B-276 is a member of the class of dinuclear iron proteins capable of stereospecific epoxygenation reactions. Eur. J. Biochem., **247**, 635-641 (1997)

[5] Saeki, H.; Akira, M.; Furuhashi, K.; Averhoff, B.; Gottschalk, G.: Degradation of trichloroethene by a linear-plasmid-encoded alkene monooxygenase in Rhodococcus corallinus (Nocardia corallina) B-276. Microbiology, **145**, 1721-1730. (1999)

[6] Zhou, N.Y.; Jenkins, A.; Chan Kwo Chion, C.K.N.; Leak, D.J.: The alkene monooxygenase from Xanthobacter Py2 is a binuclear non-heme iron protein closely related to toluene 4-monooxygenase. FEBS Lett., **430**, 181-185 (1998)

[7] Hartmans, S.; Weber, F.J.; Somhorst, D.P.M.; De Bont, J.A.M.: Alkene monooxygenase from Mycobacterium: a multicomponent enzyme. J. Gen. Microbiol., **137**, 2555-2560 (1991)

[8] Zhou, N.Y.; Jenkins, A.; Chan Kwo Chion, C.K.; Leak, D.J.: The alkene monooxygenase from Xanthobacter strain Py2 is closely related to aromatic monooxygenases and catalyzes aromatic monohydroxylation of benzene, toluene, and phenol. Appl. Environ. Microbiol., **65**, 1589-1595. (1999)

[9] Small, F.J.; Ensign, S.A.: Alkene monooxygenase from Xanthobacter strain Py2. Purification and characterization of a four-component system central to the bacterial metabolism of aliphatic alkenes. J. Biol. Chem., **272**, 24913-24920 (1997)

[10] Zhou, N.Y.; Chan Kwo Chion, C.K.N.; Leak, D.J.: Cloning and expression of the genes encoding propene monooxygenase from Xanthobacter Py2. Appl. Microbiol. Biotechnol., **44**, 582-588 (1996)

Sterol 14-demethylase 1.14.13.70

1 Nomenclature

EC number

1.14.13.70

Systematic name

sterol,NADPH:oxygen oxidoreductase (14-methyl cleaving)

Recommended name

sterol 14-demethylase

Synonyms

14-demethylase
14DM
14α-demethylase
14α-methylsterol 14α-demethylase
14α-sterol demethylase
CYP51
CYPL1
LDM
obtusifoliol 14-α demethylase
P 450 lanosterol C-14 demethylase
P-450 lanosterol demethylase
P-45014DM
P-45014DM-containing monooxygenase system
P-450OBT 14DM
P450(14DM)
P450-14DM
P450-L1A1
P45014DM
cytochrome CYP51
cytochrome P 450 CYP51
cytochrome P-450 lanosterol 14α-demethylase
cytochrome P-450-dependent 14α-sterol demethylase
cytochrome P-450-dependent obtusifoliol 14α-demethylase
cytochrome P-450/14DM
cytochrome P-45014DM
cytochrome P450 14DM
cytochrome P450 51
cytochrome P450 CYP51
cytochrome-P450 14α-demethylase
demethylase, methylsterol 14α-

eburicol 14 α-demethylase
eburicol 14α-demethylase
lanosterol 14 α-demethylase
lanosterol 14-demethylase
lanosterol 14α-demethylase
lanosterol 14α-methyldemethylase
lanosterol C-14 demethylase
lanosterol demethylase
methylsterol 14α-demethylase (P450 CYP51)
obtusifoliol 14-demethylase
obtusifoliol 14α-demethylase
obtusifoliol-metabolizing 14α-demethylase
obtusufoliol 14-demethylase
rLDM <3> (<3> rat lanosterol 14α-demethylase [10]) [10]
sterol 14-demethylase
sterol 14-demethylase P450
sterol 14α-demethylase
sterol 14α-demethylase (CYP51)
sterol C_{14} demethylase

CAS registry number

138674-19-8 (deleted registry number)
341989-59-1 (deleted registry number)
60063-87-8
90463-45-9 (deleted registry number)

2 Source Organism

<-1> no activity in *Escherichia coli* (no endogenous sterol 14α-demethylase [1]) [1]
<1> *Sorghum bicolor* (L., Moench [1,7]) [1, 7]
<2> *Saccharomyces cerevisiae* [1, 2, 3, 4, 5, 8, 9, 11]
<3> *Rattus norvegicus* (male Sprague-Dawley rats [2]; Sprague-Dawley rat [6]; male rat [11]) [1, 2, 5, 6, 10, 11]
<4> *Sinapis alba* (L. [1]) [1]
<5> *Manihot esculenta* (Crantz [1]) [1]
<6> *Homo sapiens* (human [1,11]) [1, 11]
<7> *Candida albicans* [1, 5, 9, 11]
<8> *Candida tropicalis* [1, 5, 9, 11]
<9> *Candida glabrata* [1]
<10> *Penicillium italicum* (filamentous fungi [1]; filamentous fungal plant pathogen [9]) [1, 9, 11]
<11> *Schizosaccharomyces pombe* [1, 11]
<12> *Ustilago maydis* [1, 11]
<13> *Zea mays* (maize [1,12]; LG11 [12]) [1, 12]
<14> *Arabidopsis thaliana* [1]
<15> *plants* (higher plants [3, 5, 7, 11]) [3, 5, 7, 11]

3 Reaction and Specificity

Catalyzed reaction

obtusifoliol + 3 O_2 + 3 NADPH + 3 H^+ = 4α-methyl-5α-ergosta-8,14,24(28)-trien-3β-ol + formate + 3 $NADP^+$ + 3 H_2O (The heme-thiolate enzyme, P-450, catalyses successive hydroxylations of the 14α-methyl group and C-15, followed by elimination as formate leaving the 14(15) double bond. This enzyme acts on a range of steroids with a 14α-methyl group; <2, 13> mechanism [4, 8, 12]; <2> stoichiometry [8]; <3> regulation [10])

Reaction type

demethylation (14α-demethylation)
hydroxylation
monooxygenation
oxidation
oxygenation
redox reaction
reduction

Natural substrates and products

S 24,25-dihydrolanosterol + O_2 + NADPH <2, 3> (4,4,14α-trimethyl-5α-cholesta-8-en-3β-ol; 8-lanosta-3β-ol; 8-lanosten-3β-ol; DHL; <2> not the natural substrate [2]; <13,15> not a substrate [1, 3, 12]) (Reversibility: ? <2, 3> [1, 2, 4]) [1, 2, 4]

P 4,4-dimethyl-5α-cholesta-8,14-dien-3β-ol + formate + $NADP^+$ + H_2O

S eburicol + O_2 + NADPH <10> (<13> not a substrate [1]) (Reversibility: ? <10> [1, 9]) [1, 9]

P ? + formate + $NADP^+$ + H_2O

S lanosterol + O_2 + NADPH <2, 3, 6> (4,4,14α-trimethyl-5α-cholesta-8,24-dien-3β-ol; lanosta-8,24-dien-3β-ol; <2,3> natural substrate [2, 3, 8]; <2> ergosterol synthesis in yeast involves oxidative removal of the 14α-methyl group, C-32, of lanosterol [8]; <2> P-45014DM catalyzes all three oxygenation steps from lanosterol to dimethylcholestatrienol [8]; <1, 13> not a substrate [1, 7, 12]) (Reversibility: ? <2, 3, 6> [1, 2, 3, 4, 6, 8, 10, 11]) [1, 2, 3, 4, 6, 8, 10, 11]

P 4,4-dimethyl-5α-cholesta-8,14,24-trien-3β-ol + formate + $NADP^+$ + H_2O

S obtusifoliol + O_2 + NADPH <1, 13, 15> (4α,14α-dimethyl-5α-ergosta-8,24(28)-dien-3β-ol; 4α,14α-dimethyl-24-methylene-5α-cholesta-8-en-3β-ol) (Reversibility: ? <1, 13, 15> [1, 3, 7, 12]) [1, 3, 7, 12]

P 4α-methyl-5α-ergosta-8,14,24(28)-trien-3β-ol + formate + $NADP^+$ + H_2O

S Additional information <1, 2, 3, 6, 7, 8, 10, 11, 12, 13, 15> (<1, 2, 3, 6, 7, 8, 10, 11, 12, 13, 15> enzyme of sterol biosynthetic pathway [1, 2, 3, 5, 6, 11, 12]; <2, 15> 14α-demethylation is a key step of sterol biosynthesis in eukaryotes [3]; <2> enzyme of ergosterol biosynthesis [2]; <3, 6> enzyme of cholesterol biosynthesis [2, 10, 11]; <1, 13, 15> enzyme of plant sterol, phytosterol, biosynthesis [1, 3, 12]; <1> catalyzes an essential step in sterol biosynthesis as evidenced by the absence of a 14α-methyl group in all known functional sterols, removal of the 14α-methyl group is essential [1]; <1> biosynthetic enzyme with very narrow substrate specificity [1,7]; <2,3> removal of 14α-methyl group, C32, from 14α-methylated precursor sterols is an essential step of sterol biosynthesis in eukaryotes [2]; <3> lanosterol 14-demethylation is situated at the root of sterol-biosynthetic branch of mevalonic acid pathway [6]; <3> brain microsomes, existence of sterol biosynthetic pathway in brain, cholesterol is synthesized de novo in brain [6]; <1> key enzyme in plant sterol, phytosterol, biosynthesis [7]; <1> enzyme is a multifunctional cytochrome P_{450}, which as the same active site catalyze demethylation in three consecutive NADPH- and O_2-dependent hydroxylation reactions, resulting in the elimination of the methyl group as formic acid and the introduction of a double bond at the Δ14 position [1, 7]; <3> enzyme for regulation of cholesterol biosynthesis [10]; <3,6> housekeeping enzyme essential for viability of mammals, essential step in cholesterol biosynthesis [11]; <2,3,6-8,10-12> enzyme of sterol biosynthesis, sterol 14-demethylation occurs in all organism exhibiting de novo sterol biosynthesis [11]; <3,6> there is a possibility that P45014DM participates not only in sterol biogenesis but also in production of biosignal substance regulating meiosis of mammalian oocytes [11]) [1, 2, 3, 5, 6, 7, 10, 11, 12]

P ?

Substrates and products

S 14α-methyl-24,28-dihydrofecosterol + O_2 + NADPH <13> (<13> 50% of activity compared to obtusifoliol [12]) (Reversibility: ? <13> [12]) [12]

P ? + formate + $NADP^+$ + H_2O

S 24,25-dihydro-31-norlanosterol + O_2 + NADPH <13> (<13> good substrate, 67% of activity compared to obtusifoliol [12]) (Reversibility: ? <13> [12]) [12]

P ? + formate + $NADP^+$ + H_2O

S 24,25-dihydrolanosterol + O_2 + NADPH <2, 3, 6, 7, 8, 9> (4,4,14α-trimethyl-5α-cholesta-8-en-3β-ol; 8-lanosta-3β-ol; 8-lanosten-3β-ol; DHL; <13, 15> not a substrate [1, 3, 12]) (Reversibility: ? <2, 3, 6, 7, 8, 9> [1, 2, 3, 4]) [1, 2, 3, 4]

P 4,4-dimethyl-5α-cholesta-8,14-dien-3β-ol + formate + $NADP^+$ + H_2O

S 24,28-dihydroobtusifoliol + O_2 + NADPH <2, 13, 15> (4α,14α-dimethyl-5α-ergosta-8-en-3β-ol; DHO; <2> very poor substrate, about 10% of obtusifoliol demethylation, activity disappears in the presence of same concentration of lanosterol, 24-methylene-24,25-dihydrolanosterol, obtusifoliol or 24,25-dihydrolanosterol [3]; <13, 15> good substrate, 75% of activity to obtusifoliol [3, 12]) (Reversibility: ? <2, 13, 15> [3, 12]) [3, 12]

P 4α-methyl-5α-ergosta-8,14-dien-3β-ol + formate + $NADP^+$ + H_2O

S 24-methylene-24,25-dihydrolanosterol + O_2 + NADPH <2, 3> (4,4,14α-trimethylergosta-8,24(28)-dien-3β-ol; 24-methylenelanost-8-en-3β-ol, 24-methylene-DHL; <2> activity for 24-methylene-DHL is considerably higher, 4fold, than that for 24,25-dihydrolanosterol, DHL [2]; <2> about 60% activity to that of lanosterol [2]; <3> poorest substrate, catalyzes

14α-demethylation of 24-methylene-DHL, but activity is considerably lower than that for lanosterol and for 24,25-dihydrolanosterol, DHL [2]; <2> good substrate [3]) (Reversibility: ? <2, 3> [2, 3]) [2, 3]

P 4,4-dimethyl-ergosta-8,14,24(28)-trien-3β-ol + formate + $NADP^+$ + H_2O

S 7-lanosten-3β-ol + O_2 + NADPH <2> (<2> very low activity [4]) (Reversibility: ? <2> [4]) [4]

P 4,4-dimethylcholesta-7,14-dien-3β-ol + formate + $NADP^+$ + H_2O (<2> 32-nor-14,15-unsaturated metabolite [4])

S 7-lanostene-3β,32-diol + O_2 + NADPH <2> (<2> very low activity [4]) (Reversibility: ? <2> [4]) [4]

P 4,4-dimethylcholesta-7,14-dien-3β-ol + formate + $NADP^+$ + H_2O (<2> 32-nor-14,15-unsaturated metabolite [4])

S 8-lanostene-3β,32-diol + O_2 + NADPH <2> (Reversibility: ? <2> [4]) [4]

P 4,4-dimethylcholesta-8,14-dien-3β-ol + formate + $NADP^+$ + H_2O (<2> 32-nor-14-unsaturated metabolite [4])

S eburicol + O_2 + NADPH <10> (<13> not a substrate [1,12]) (Reversibility: ? <10> [1, 9]) [1, 9]

P ? + formate + $NADP^+$ + H_2ONAD^+ NO_2NAD^+ NO_2

S lanosterol + O_2 + NADPH <2, 3, 6, 7, 8, 9> (4,4,14α-trimethyl-5α-cholesta-8,24-dien-3β-ol; lanosta-8,24-dien-3β-ol; <2, 3> best substrate [2, 3, 4, 6]; <1, 13, 15> not a substrate [1, 3, 7, 12]) (Reversibility: ? <2, 3, 6, 7, 8, 9> [1, 2, 3, 5, 4, 6, 8, 9, 10, 11]) [1, 2, 3, 4, 5, 6, 8, 9, 10, 11]

P 4,4-dimethyl-5α-cholesta-8,14,24-trien-3β-ol + formate + $NADP^+$ + H_2O

S obtusifoliol + O_2 + NADPH <1, 2, 13, 15> (4α,14α-dimethyl-5α-ergosta-8,24(28)-dien-3β-ol; 4α,14α-dimethyl-24-methylene-5α-cholesta-8-en-3β-ol; <1, 2, 13, 15> catalyzes 14α-demethylation of obtusifoliol [1, 3, 7, 12]; <13> best substrate [12]) (Reversibility: ? <1, 2, 13, 15> [1, 3, 7, 12]) [1, 3, 7, 12]

P 4α-methyl-5α-ergosta-8,14,24(28)-trien-3β-ol + formate + $NADP^+$ + H_2O

S Additional information <1, 2, 3, 13, 15> (<1, 2, 3, 13, 15> substrate specificity [1, 2, 3, 7, 12]; <2, 15> substrate recognition [3]; <1, 13> plant sterol 14α-demethylase have high substrate specificity [1, 7, 12]; <2, 13, 15> narrow substrate selectivity [3, 4, 12]; <1, 13> biosynthetic enzyme with very narrow substrate specificity [1, 12]; <13> enzyme with high degree of substrate and product specificity [12]; <1> enzyme is a multifunctional cytochrome P_{450}, which as the same active site catalyze demethylation in three consecutive NADPH- and O_2-dependent hydroxylation reactions, resulting in the elimination of the methyl group as formic acid and the introduction of a double bond at the Δ14 position [1, 7]; <1> substrate binding spectra [1, 7]; <15> can not catalyze demethylation of sterols having 4β-methyl group, favorably interacts with sterols having saturated side chain [3]; <2> 3-hydroxy group, the 8-lanostene conformation of sterol ring and the side-chain terminal, C25, C26, C27, are the essential structures of substrates for interacting with the yeast enzyme [3]; <2> 4β-methyl group, C31, does not affect the activity of yeast P-45014DM, although removal reduces affinity for enzyme in some extent [3]; <2, 15> yeast enzyme poorly metabolizes sterols having saturated side chain, plant enzyme shows considerable activity for such sterols [3]; <2> 8-double bond of lanosterol plays an important critical role in enzyme-substrate interaction of cytochrome P-45014DM [4]; <2> enzyme recognizes 8-lanostene structure and favourably interacts with 8-lanostene derivatives, can act also with substrates having 7-lanostene structure, utilizes them with lower efficiency than 8-lanostene derivatives [4]; <2> reaction reqires molecular oxygen, does not occur anaerobically [8]; <1, 15> substrate for 14α-demethylation reaction in plants is different from that in animals and fungi [7]; <13> P-450OBT 14DM has probably a specific apolar binding site for the side chain. Δ^8-double bond is absolute required for substrate demethylation and the 3-hydroxy group plays a critical role in enzyme-substrate interaction [12]; <2> cycloartenol: not or very poor substrate [3]; <2> no activity with 6-lanostene-3β,32-diol and lanostane-3β,32-diol [4]; <1> no activity with lanosterol, campesterol, sitosterol, or stigmasterol [7]; <13> no activity with 31-norlanosterol, cycloeucalenol, 4α,14α-dimethyl-5α-ergost-9(11)-en-3β-ol, 4α,14α-dimethyl-5α-ergost-7-en-3β-ol, 8(9),24(25)-tetrahydro-31-norlanosterol, 24-methylenelanosterol, 24,28-dihydro-4β-methyl-30-norobtusifoliol, 24,25-dihydrolanosterol, lanosterol, obtusifoliyl-3β-methoxy, obtusifoliyl-3β-acetoxy, obtusifoliyl-3β-amino [12]) [1, 2, 3, 4, 7, 8, 12]

P ?

Inhibitors

24,25-dihydrolanosterol <2> (<2> 8.3% inhibition of 24-methylene-24,25-dihydrolanosterol demethylation, no inhibition of lanosterol demethylation [2]; <2> 16.4% inhibition of obtusifoliol 14α-demethylation [3]) [2, 3]

24-methylene-24,25-dihydrolanosterol <2> (<2> 21.6% inhibition of lanosterol demethylation, 55.3% inhibition of 24,25-dihydrolanosterol demethylation [2]; <2> 47.4% inhibition of obtusifoliol 14α-demethylation [3]) [2, 3]

CO <1, 2, 3> (<3> high partial pressure of CO, ratio CO/O_2 95/5, 48% inhibition [6]; <2> ratio CO/O_2 90/10, 51.1% inhibition, ration CO/O_2 95/5, 100% inhibition [8]) [6, 7, 8]

SKF-525A <2> (<2> potent inhibitor, 1.0 mM: 100% inhibition [8]) [8]

anti-P-45014DM antibodies <2> (<2> complete inhibition [8]) [8]

azalanstat <3> (<3> specific inhibitor, IC_{50} less than 0.000002 mM [10]) [10]

azole antifungal agents <1, 2, 3> (<1, 2, 3> specific inhibitors [1, 2]) [1, 2]

fungicides <10> [9]

ketoconazole <3> (<3> potent inhibitor, 0.015 mM: complete inhibition [6]) [6]

lanosterol <2> (<2> 63% inhibition of 24-methylene-24,25-dihydrolanosterol demethylation, 74.9% inhibition of 24,25-dihydrolanosterol demethylation [2]; <2> 53.1% inhibition of obtusifoliol 14α-demethylation [3]) [2, 3]

menadione <2> (<2> 0.125 mM: 62.1% inhibition [8]) [8]

metyrapone <2> (<2> strong inhibition, 0.1 mM: 57.3% inhibition [8]) [8]

nitrogen <2> (<2> nitrogen atmosphere [8]) [8]

obtusifoliol <2> (<2> 24.4% inhibition of 24,25-dihydrolanosterol, DHL, demethylation, no inhibition of lanosterol and 24-methylene-24,25-dihydrolanosterol demethylation [3]) [3]
Additional information <2> (<2> no inhibition by 24,28-dihydroobtusifoliol [3]) [3]

Cofactors/prosthetic groups
NADPH <1-15> (<1-15> required [1-8]; <13> totally dependent upon NADPH, NADH extremely poor reductant [12]) [1-8, 11, 12]
cytochrome P_{450} <1-15> [1-12]
heme <1-15> (<1> hemoprotein [1]; <1-15> heme-thiolate enzyme [1-12]) [1-12]

Activating compounds
7-lanostene-3β,32-diol <2> (<2> activation [4]) [4]
7-lanostene-3β-ol <2> (<2> activation [4]) [4]
8-lanostene-3β,32-diol <2> (<2> activation [4]) [4]
8-lanostene-3β-ol <2> (<2> activation [4]) [4]
cycloartenol <2> (<2> activation of enzymatic reaction [3]) [3]
cytosolic carrier <3> (<3> required for maximum activity [6]) [6]
lanosterol <2> (<2> induction and activation [8]) [8]

Turnover number (min^{-1})
Additional information <3> [11]

Specific activity (U/mg)
0.0000084 <3> (<3> brain microsomes, substrate: lanosterol [6]) [6]
Additional information <2, 3> [2, 3, 4, 5]

K_m-Value (mM)
0.0001 <2> (8-lanostene-3β,32-diol) [4]
0.003 <2> (7-lanostene-3β,32-diol) [4]
0.005 <2> (lanosterol) [3]
0.0067 <2> (lanosterol) [2]
0.0077 <2> (24-methylene-24,25-dihydrolanosterol) [3]
0.0087 <2> (24-methylene-24,25-dihydrolanosterol) [2]
0.012 <2> (obtusifoliol) [3]
0.017 <2> (24,25-dihydrolanosterol) [3]
0.02 <2> (24,25-dihydrolanosterol, <2> 8-lanosten-3β-ol [4]) [4]
0.116 <13> (24,25-dihydro-31-norlanosterol) [12]
0.16 <13> (obtusifoliol) [12]
Additional information <2, 13> (<2, 13> kinetic parameters [3, 12]; <2> kinetics [4, 8]; <2> enzyme shows higher affinity for 8-lanostene conformation, such as lanosterol and 24,25-dihydrolanosterol, than for 7-lanostene one [4]; <2> affinity and activity for 7-lanosten-3β-ol is very low, no exact K_m [4]) [3, 4, 8, 12]

pH-Optimum
7.5 <2, 3> (<2,3> assay at [4, 6, 8]) [4, 6, 8]
7.5-8.5 <13> [12]
7.9 <1> (<1> assay at [1]) [1]

pH-Range
6.5-9.5 <13> [12]

Temperature optimum (°C)
30 <1, 2, 13> (<1, 2, 13> assay at [1, 2, 3, 4, 8, 12]; <2, 13> aerobic conditions [2,4,8,12]) [1, 2, 3, 4, 8, 12]
37 <3> (<3> assay at [2,6,11]; <3> aerobic conditions [6]) [2, 6, 11]

4 Enzyme Structure

Molecular weight
53000 <1> (<1> SDS-PAGE and amino acid sequence analysis [7]) [7]
55050 <3> (<3> amino acid analysis [10]) [10]
57300 <10> (<10> amino acid analysis [9]) [9]
Additional information <3, 6> (<3> protein consisting of 503 amino acids, amino acid sequencing [11]; <6> enzyme consisting of 509 amino acids, amino acid sequencing [11]) [11]

Subunits
monomer <1> (<1>, 1 * 53000, SDS-PAGE [7]) [7]

5 Isolation/Preparation/Mutation/Application

Source/tissue
brain <3> [6]
liver <3, 6> [2, 5, 6, 10, 11]
seedling <1, 13> (<1> etiolated sorghum seedlings [7]) [1, 7, 12]

Localization
cytoplasm <3> [2]
endoplasmic reticulum <13> [12]
endoplasmic reticulum <13> (<13> light membrane fraction of endoplasmic reticulum, not in plasma membrane [12]) [12]
membrane <1, 3, 13> (<1,13> membrane-bound [1,12]) [1, 6, 7, 12]
microsome <1, 2, 3, 13, 15> (<13> microsomal-bound [12]) [1, 2, 6, 7, 8, 11, 12]
Additional information <13> (<13> subcellular localization, distribution [12]; <13> not in plasma membrane [12]) [12]

Purification
<1> (recombinant sorghum CYP51 [1]; <1> purification and reconstitution [7]) [1, 7]
<2> [2, 3, 4, 8]
<3> [2, 5, 10, 11]

Cloning

<1> (cDNA cloning of sorghum CYP51and functional expression in Escherichia coli JM109 in high levels [1]) [1]
<3> (pRT-9 clone, pRT-9 protein is P45014DM [5,11]; pRT-9 cDNA [5,11]; cDNA, cloning and functional expression [10,11]; cDNA expression in COS7 cells and cloning of pRT-9 cDNA [11]) [5, 10, 11]
<6> (cDNA cloning [11]) [11]
<10> (cDNA, introduced into Aspergillus niger by transformation [9]) [9]

Application

agriculture <1, 2, 13> (<1, 2, 13> target enzyme for azole antifungal agents. These specific inhibitors are of great importance as plant growth regulators, fungicides and herbicides in the agricultural and medical fields [1,2,7,12]; <1> all known functional sterols lack a 14α-methyl group, and therefore the 14α-demethylation reaction has received much attention from the pharmaceutical and agriculture-chemical industry as a possible means to specifically control and inhibit sterol biosynthesis in mammals, fungi, and plant [7]; <13> target of important agrochemicals such as fungicides, plant growth regulators and herbicides [12]) [1, 2, 7, 12]
medicine <1, 2, 3, 13> (<1, 2, 13> target enzyme for azole antifungal agents. These specific inhibitors are of great importance as plant growth regulators, fungicides and herbicides in the agricultural and medical fields [1,2,7,12]; <3> target for cholesterol-lowering drugs [10]) [1, 2, 7, 10, 12]
pharmacology <1, 2, 3, 13> (<1, 2, 13> target enzyme for azole antifungal agents. These specific inhibitors are of great importance as plant growth regulators, fungicides and herbicides in the agricultural and medical fields [1,2,7,12]; <1> target enzyme for the design of phyla-specific sterol 14α-demethylase inhibitors [7]; <1> all known functional sterols lack a 14α-methyl group, and therefore the 14α-demethylation reaction has received much attention from the pharmaceutical and agriculture-chemical industry as a possible means to specifically control and inhibit sterol biosynthesis in mammals, fungi, and plant [7]; <3> target for cholesterol-lowering drugs [10]) [1, 2, 7, 10, 12]

6 Stability

General stability information

<13>, plant demethylase is remarkably stable [12]

Storage stability

<1>, -80°C, frozen in liquid nitrogen [7]

References

[1] Bak, S.; Kahn, R.A.; Olsen, C.E.; Halkier, B.A.: Cloning and expression in Escherichia coli of the obtusifoliol 14α-demethylase of Sorghum bicolor

(L.) Moench, a cytochrome P_{450} orthologous to the sterol 14α-demethylases (CYP51) from fungi and mammals. Plant J., **11**, 191-201 (1997)
[2] Aoyama, Y.; Yoshida, Y.: Different substrate specificities of lanosterol 14α-demethylase (P-45014DM) of Saccharomyces cerevisiae and rat liver for 24-methylene-24,25-dihydrolanosterol and 24,25-dihydrolanosterol. Biochem. Biophys. Res. Commun., **178**, 1064-1071 (1991)
[3] Aoyama, Y.; Yoshida, Y.: The 4β-methyl group of substrate does not affect the activity of lanosterol 14α-demethylase (P-45014DM) of yeast: Difference between the substrate recognition by yeast and plant sterol 14α-demethylases. Biochem. Biophys. Res. Commun., **183**, 1266-1272 (1992)
[4] Aoyama, Y.; Yoshida, Y.; Sonoda, Y.; Sato, Y.: Role of the 8-double bond of lanosterol in the enzyme-substrate interaction of cytochrome P-45014DM (lanosterol 14α-demethylase). Biochim. Biophys. Acta, **1001**, 196-200 (1989)
[5] Aoyama, Y.; Funae, Y.; Noshiro, M.; Horiuchi, T.; Yoshida, Y.: Occurrence of a P_{450} showing high homology to yeast lanosterol 14-demethylase (P45014DM) in the rat liver. Biochem. Biophys. Res. Commun., **201**, 1320-1326 (1994)
[6] Aoyama, Y.; Horiuchi, T.; Yoshida, Y.: Lanosterol 14-demethylase activity expressed in rat brain microsomes. J. Biochem., **120**, 982-986 (1996)
[7] Kahn, R.A.; Bak, S.; Olsen, C.E.; Svendsen, I.; Moller, B.L.: Isolation and reconstitution of the heme-thiolate protein obtusifoliol 14α-demethylase from Sorghum bicolor (L.) Moench. J. Biol. Chem., **271**, 32944-32950 (1996)
[8] Aoyama, Y.; Yoshida, Y.; Sato, R.: Yeast cytochrome P-450 catalyzing lanosterol 14α-demethylation. II. Lanosterol metabolism by purified P-45014DM and by intact microsomes. J. Biol. Chem., **259**, 1661-1666 (1984)
[9] Van Nistelrooy, J.G.M.; Van den Brink, J.M.; Van Kan, J.A.L.; Van Gorcom, R.F.M.; de Waard, M.A.: Isolation and molecular characterisation of the gene encoding eburicol 14α-demethylase (CYP51) from Penicillium italicum. Mol. Gen. Genet., **250**, 725-733 (1996)
[10] Sloane, D.L.; So, O.Y.; Leung, R; Scarafia, L.E.; Saldou, N.; Jarnagin, K.; Swinney, D.C.: Cloning and functional expression of the cDNA encoding rat lanosterol 14α-demethylase. Gene, **161**, 243-248 (1995)
[11] Aoyama, Y.; Noshiro, M.; Gotoh, O.; Imaoka, S.; Funae, Y.; Kurosawa, N.; Horiuchi, T.; Yoshida, Y.: Sterol 14-demethylase P_{450} (P45014DM*) is one of the most ancient and conserved P_{450} species. J. Biochem., **119**, 926-933 (1996)
[12] Taton, M.; Rahier, A.: Properties and structural requirements for substrate specificity of cytochrome P-450-dependent obtusifoliol 14α-demethylase from maize (Zea mays) seedlings. Biochem. J., **277**, 483-492 (1991)

N-Methylcoclaurine 3'-monooxygenase 1.14.13.71

1 Nomenclature

EC number
1.14.13.71

Systematic name
(S)-N-methylcoclaurine, NADPH:oxygen oxidoreductase (3'-hydroxylating)

Recommended name
N-methylcoclaurine 3'-monooxygenase

Synonyms
cytochrome P_{450} 80B1
cytochrome P_{450} 80B2
N-methylcoclaurine 3'-hydroxylase

CAS registry number
202420-37-9

2 Source Organism

<1> *Eschscholtzia californica* (california poppy [1]) [1]

3 Reaction and Specificity

Catalyzed reaction
(S)-N-methylcoclaurine + NADPH + H^+ + O_2 = (S)-3'-hydroxy-N-methylcoclaurine + $NADP^+$ + H_2O

Reaction type
hydroxylation

Natural substrates and products
S (S)-N-methylcoclaurine + NADPH + H^+ + O_2 <1> (Reversibility: ? <1> [1]) [1]
P (S)-3'-hydroxy-N-methylcoclaurine + $NADP^+$ + H_2O <1> [1]

Substrates and products
S (S)-N-methylcoclaurine + NADPH + H^+ + O_2 <1> (<1> absolute requirement for an N-methylgroup, a 4'-hydroxyl moiety and S-configuration at carbon atom 1 [1]) (Reversibility: ? <1> [1]) [1]
P (S)-3'-hydroxy-N-methylcoclaurine + $NADP^+$ + H_2O <1> [1]

Cofactors/prosthetic groups
NADPH <1> [1]

K_m-Value (mM)
0.015 <1> ((S)-N-methylcoclaurine) [1]

pH-Optimum
7.5 <1> (<1> Tris-HCl buffer [1]) [1]

Temperature optimum (°C)
35 <1> [1]

5 Isolation/Preparation/Mutation/Application

Cloning
<1> (expressed in Saccharomyces cerevisiae and Spodoptera frugiperda [1]) [1]

References

[1] Pauli, H.H.; Kutchan, T.M.: Molecular cloning and functional heterologous expression of two alleles encoding (S)-N-methylcoclaurine 3'-hydroxylase (CYP80B1), a new methyl jasmonate-inducible cytochrome P-450-dependent mono-oxygenase of benzylisoquinoline alkaloid biosynthesis. Plant J., **13**, 793-801 (1998)

Methylsterol monooxygenase 1.14.13.72

1 Nomenclature

EC number
1.14.13.72

Systematic name
4,4-dimethyl-5α-cholest-7-en-3β-ol,hydrogen-donor:oxygen oxidoreductase (hydroxylating)

Recommended name
methylsterol monooxygenase

Synonyms
4,4-dimethylsterol-4α-methyl oxidase <2> [3]
4-methylsterol oxidase
4α-methylsterole-4α-methyl oxidase <2> [3]
EC 1.14.99.16
methylsterol hydroxylase
methylsterol monooxygenase

CAS registry number
37256-80-7

2 Source Organism

<1> *Rattus norvegicus* (Sprague-Dawley [1]; male, adult [1,2]) [1, 2]
<2> *Zea mays* (variety LG11 [3,4]; embryos [4]) [3, 4]

3 Reaction and Specificity

Catalyzed reaction
3β-hydroxy-4β-methyl-5α-cholest-7-ene-4α-carbaldehyde + NAD(P)H + H^+ + O_2 = 3β-hydroxy-4β-methyl-5α-cholest-7-ene-4α-carboxylate + $NAD(P)^+$ + H_2O (requires cytochrome b_5. Also acts on 4α-methyl-5α-cholest-7-en-3β-ol. the sterol can be based on cycloartenol as well as lanosterol. Formerly EC 1.14.99.16; <1> aerobic hydroxylation [1]; <2> initial oxidative step is rate-limiting step. Suggested that demethylation is not catalyzed by a cytochrome P-450 dependent monooxigenase. possible participation of cytochrome b_5 as intermediate electron carrier [4])

4,4-dimethyl-5α-cholest-7-en-3β-ol + NAD(P)H + H^+ + O_2 = 4β-hydroxymethyl-4α-methyl-5α-cholest-7-en-3β-ol + $NAD(P)^+$ + H_2O
4β-hydroxymethyl-4α-methyl-5α-cholest-7-en-3β-ol + NAD(P)H + H^+ + O_2 = 3β-hydroxy-4β-methyl-5α-cholest-7-ene-4α-carbaldehyde + $NAD(P)^+$ + 2 H_2O

Reaction type

oxidation
redox reaction
reduction

Substrates and products

S 24-ethylidenelophenol + NAD(P)H + O_2 <2> (<2> best substrate, 4α-carbomethoxy-5α-stigmasta-7,24(241)-dien-3β-yl-acteate and avenastenone as intermediates [4]) (Reversibility: ? <2> [4]) [4]
P avenasterol + $NAD(P)^+$ + H_2O <2> [4]
S 24-methylenecycloartanol + NAD(P)H + O_2 <2> (<2> best substrate, 4β,14α-dimethyl-4α-acetoxymethyl-5α-ergosta-9β,19-cyclo-24(241)-en-3β-yl-acetate, 4β-,14α-dimethyl-4α-carbomethoxy-5α-ergosta-9β,19-cyclo-24(241)-en-3β-yl-acetate and cycloeucalenone as intermediates [4]) (Reversibility: ? <2> [4]) [4]
P cycloeucalenol + $NAD(P)^+$ + H_2O <2> [4]
S 4,4-dimethyl-5α-cholest-7-en-3β-ol + NAD(P)H + O_2 <1> (Reversibility: ? <1> [1, 2]) [1, 2]
P 4α-methyl-5α-cholest-7-en-3β-ol + $NAD(P)^+$ + CO_2 <1> (<1> products of two step hydroxylase and demethylase activity, labile intermediate 4α-hydroxymethyl-4β-methyl-5α-cholest-7-en-3β-ol and 4β-hydroxymethyl-4α-methyl-5α-cholest-7-en-3β-ol [2]) [2]
S Additional information <2> (<2> not catalyzed by P-450 but cytochrome b_5 is suggested to be electron donor, NADPH and 24-methylenecycloartanol or 24-ethylidenelophenol [3]; <2> 25 different sterol derivatives were tested as substrate. cyclolaudenol, 24xi-methylcycloartenol, 24(25)-dihydrocycloartenol, 24-methylenelophenol, lophenol, cycloeucalenol, 24xi(241)-dihydrocycloeucalenol determined as further possible substrates. Strict structural requirements for the 4α-methyl oxidation of 4,4-dimethylsterols and 4α-methylsterols [4]) [3, 4]
P ?

Inhibitors

IgG <2> (<2> polyclonal antibody raised against recombinant cytochrome b_5, strongly inhibited of enzyme activity by low amounts of IgG, this results from IgG specific to cytochrome b_5 [3]) [3]
cyanide <1, 2> [1, 3]
cyanide <2> (<2> strong inhibition [4]) [4]
cytochrome c <1, 2> [1, 3, 4]
metal chelating agent <2> [4]
o-phenanthroline <2> [4]

stigmastan-3β,5α,6α-triol <2> (<2> metabolism of 24-ethylidenelophenol effectively inhibited. Metabolism of 24-methylenecycloartenol unaffected [4]) [4]
trypsin <1> (<1> 30 min treatment, 18% of original concentration of cytochrome b_5 left [1]) [1]
Additional information <2> (<2> insensitive to azide ions and CO [4]) [4]

Cofactors/prosthetic groups
NADH <1, 2> (<1> 10 to 15% more active with NADH than with NADPH [1]) [1, 2, 4]
NADPH <1, 2> [1-4]

Metals, ions
Fe^{2+} <2> (<2> presumably, or other metal ion suggested to be enzyme-boud in hydroxylating system [4]) [4]

Specific activity (U/mg)
0.000201 <1> (<1> recovered after Lubrol treatment, 1% of original cytochrome b_5 concentration in microsomes, 4-fold that observed for microsomes, maybe cytochrome b_5 does not participate in hydroxylation [1]) [1]
0.00356 <1> [1]
Additional information <1> (<1> enzyme does not involve P-450 [1]) [1]

K_m-Value (mM)
0.09 <2> (cycloeucalenol, <2> +/-0.01 [4]) [4]
0.15 <2> (24-methylcycloartenol, <2> +/-0.01 [4]) [4]
0.25 <2> (cyclolaudenol, <2> +/-0.01 [4]) [4]
0.34 <2> (24-methylenelophenol, <2> +/-0.02 [4]) [4]
0.48 <2> (24-ethylidenelophenol, <2> +/-0.02 [4]) [4]
0.5 <2> (24-cycloeucalenol, <2> +/-0.02 [4]) [4]

pH-Optimum
7.5-8.5 <2> [4]

Temperature optimum (°C)
30 <2> (<2> assay at [4]) [4]

5 Isolation/Preparation/Mutation/Application

Source/tissue
liver <1> [1, 2]

Localization
microsome <1, 2> (<2> membrane bound [4]) [1-4]

Purification
<1> (gel filtration, ammonium sulfate fractionation, affinity chromatography [1]) [1]

6 Stability

General stability information

<2>, remarkable stability during 4 to 6 h incubation periods [4]

Storage stability

<1>, microsomes frozen, buffers contain 1 mM glutathione, no loss of activity [1]

References

[1] Gaylor, J.L.; Mason, H.S.: Investigation of the component reactions of oxidative sterol demethylation. Evidence against participation of cytochrome P-450. J. Biol. Chem., **243**, 4966-4972 (1968)
[2] Miller, W.L.; Kalafer, M.E.; Gaylor, J.L.; Delwiche, C.V.: Investigation of the component reactions of oxidative sterol demethylation. Study of the aerobic and anaerobic processes. Biochemistry, **6**, 2673-2678 (1967)
[3] Rahier, A.; Smith, M.; Taton, M.: The role of cytochrome b_5 in 4α-methyl-oxidation and C5(6) desaturation of plant sterol precursors. Biochem. Biophys. Res. Commun., **236**, 434-437 (1997)
[4] Pascal, S.; Taton, M.; Rahier A.: Plant sterol viosynthesis. Identification and characterization of two destinct microsomal oxidative enzymatic systems involved in sterol C4-demethylation. J. Biol. Chem., **268**, 11639-11654 (1993)

Tabersonine 16-hydroxylase 1.14.13.73

1 Nomenclature

EC number
1.14.13.73

Systematic name
tabersonine,NADPH:oxygen oxidoreductase (16-hydroxylating)

Recommended name
tabersonine 16-hydroxylase

Synonyms
T16H
oxygenase, tabersonine 16-mono-
tabersonine 16-monooxygenase

CAS registry number
250378-34-8

2 Source Organism

<1> *Catharanthus roseus* (line CP3a [1]) [1, 2]

3 Reaction and Specificity

Catalyzed reaction
tabersonine + NADPH + H^+ + O_2 = 16-hydroxytabersonine + $NADP^+$ + H_2O

Reaction type
oxidation
redox reaction
reduction

Natural substrates and products
S tabersonine + NADPH + O_2 <1> (<1> indole alkaloid synthesis [1]; <1> first step in vindoline boxynthesis [2]) (Reversibility: ? <1> [1, 2]) [1, 2]
P 16-hydroxytabersonine + NAD^+ + H_2O

Substrates and products
S tabersonine + NADPH + O_2 <1> (<1> NADH is less active as substrate [2]) (Reversibility: ? <1> [1, 2]) [1, 2]
P 16-hydroxytabersonine + NAD^+ + H_2O <1> [2]

Inhibitors

CO <1> (<1> concentration dependent competitive inhibition, 77% inhibition at 90% [2]) [2]
clotrimazole <1> (<1> 50% inhibition at 0.05 mM [2]) [2]
cytochrome c <1> (<1> 50% inhibition at 0.001 mM [2]) [2]
miconazole <1> (<1> 50% inhibition at 0.3 mM [2]) [2]
tabersonine <1> (<1> inhibition above 0.03 mM [2]) [2]
tricliphane <1> (<1> 50% inhibition at 0.5 mM [2]) [2]

Cofactors/prosthetic groups

cytochrome P_{450} <1> (<> K []) [1]

K_m-Value (mM)

0.011 <1> (tabersonine) [2]
0.014 <1> (NADPH) [2]

pH-Optimum

7.5 <1> (<1> phosphate buffer [2]) [2]
8 <1> (<1> Tris buffer [2]) [2]

Temperature optimum (°C)

30 <1> (<1> assay at [2]) [2]

4 Enzyme Structure

Molecular weight

129000 <1> (<1> SDS-PAGE [1]) [1]

5 Isolation/Preparation/Mutation/Application

Source/tissue

bud <1> [2]
leaf <1> [2]
root <1> [2]
seedling <1> [2]

Localization

endoplasmic reticulum <1> [2]

Cloning

<1> [1]

References

[1] Schroder, G.; Unterbusch, E.; Kaltenbach, M.; Schmidt, J.; Strack, D.; De Luca, V.; Schroder, J.: Light-induced cytochrome P_{450}-dependent enzyme in

indole alkaloid biosynthesis: tabersonine 16-hydroxylase. FEBS Lett., **458**, 97-102 (1999)
[2] St-Pierre, B.; De Luca, V.: A cytochrome P-450 monooxygenase catalyzes the first step in the conversion of tabersonine to vindoline in Catharanthus roseus. Plant Physiol., **109**, 131-139 (1995)

7-Deoxyloganin 7-hydroxylase 1.14.13.74

1 Nomenclature

EC number
1.14.13.74

Systematic name
7-deoxyloganin,NADPH:oxygen oxidoreductase (7α-hydroxylating)

Recommended name
7-deoxyloganin 7-hydroxylase

CAS registry number
335305-40-3

2 Source Organism

<1> *Catharanthus roseus* [1]
<2> *Lonicera japonica* [2]

3 Reaction and Specificity

Catalyzed reaction
7-deoxyloganin + NADPH + H^+ + O_2 = loganin + $NADP^+$ + H_2O

Reaction type
oxidation
redox reaction
reduction

Natural substrates and products
S 7-deoxyloganin + NADPH + H^+ + O_2 <2> (<2> enzyme is involved in loganin biosynthesis [2]) (Reversibility: ? <2> [2]) [2]
P loganin + $NADP^+$ + H_2O <2> [2]

Substrates and products
S 7-deoxyloganic acid + NADPH + H^+ + O_2 <2> (<2> 50% of activity with 7-deoxyloganin [2]) (Reversibility: ? <2> [2]) [2]
P loganic acid + $NADP^+$ + H_2O <2> [2]
S 7-deoxyloganin + NADPH + H^+ + O_2 <1, 2> (<2> NADPH is essential for activity, 45% activity with NADH, substitution of O_2 with N_2 reduces activity by 50% [2]) (Reversibility: ? <1, 2> [1]) [1, 2]
P loganin + $NADP^+$ + H_2O <1, 2> [1, 2]

Inhibitors

CO <2> (<2> 60% inhibition in the dark, partially reversible by light [2]) [2]
ancymidol <2> (<2> 0.01 mM, 58.4% inhibition, 0.1 mM, 77.0% inhibition [2]) [2]
cytochrome c <2> (<2> 0.01 mM, 44.9% inhibition, 0.1 mM, 59.6% inhibition [2]) [2]
ketoconazole <2> (<2> 0.01 mM, 93% inhibition, 0.1 mM, complete inhibition [2]) [2]
metyrapone <2> (<2> 0.01 mM, 73.6% inhibition, 0.1 mM, 92.4% inhibition [2]) [2]
miconazole <2> (<2> 0.01 mM, 26.2% inhibition, 0.1 mM, 78.5% inhibition [2]) [2]

Cofactors/prosthetic groups

NADH <2> (<2> 45% of activity with NADPH [2]) [2]
NADPH <2> (<2> essential for activity [2]) [2]

Specific activity (U/mg)

9.9e-005 <2> (<2> activity in cell extracts, cofactor NADPH [2]) [2]

K_m-Value (mM)

0.018 <2> (NADPH) [2]
0.17 <2> (7-deoxyloganin) [2]

pH-Optimum

7.5 <2> (<2> 50% of maximal activity at pH 7.1 and pH 9.4 [2]) [2]

5 Isolation/Preparation/Mutation/Application

Source/tissue

cell culture <1, 2> [1, 2]

Localization

microsome <1, 2> [1, 2]

References

[1] Irmler, S.; Schroder, G.; St-Pierre, B.; Crouch, N.P.; Hotze, M.; Schmidt, J.; Strack, D.; Matern, U.; Schroder, J.: Indole alkaloid biosynthesis in Catharanthus roseus: new enzyme activities and identification of cytochrome P450 CYP72A1 as secologanin synthase. Plant J., **24**, 797-804 (2000)
[2] Katano, N.; Yamamoto, H.; Iio, R.; Inoue, K.: 7-Deoxyloganin 7-hydroxylase in Lonicera japonica cell cultures. Phytochemistry, **58**, 53-58 (2001)

Vinorine hydroxylase 1.14.13.75

1 Nomenclature

EC number
1.14.13.75

Systematic name
vinorine,NADPH:oxygen oxidoreductase (21α-hydroxylating)

Recommended name
vinorine hydroxylase

CAS registry number
162875-03-8

2 Source Organism

<1> *Rauwolfia serpentina* [1]

3 Reaction and Specificity

Catalyzed reaction
vinorine + NADPH + H^+ + O_2 = vomilenine + $NADP^+$ + H_2O

Reaction type
oxidation
redox reaction
reduction

Natural substrates and products
S vinorine + NADPH + H^+ + O_2 <1> (<1> enzyme is involved in the biosynthesis of the alkaloid ajmaline [1]) (Reversibility: ? <1> [1]) [1]
P vomilenine + $NADP^+$ + H_2O <1> [1]

Substrates and products
S vinorine + NADPH + H^+ + O_2 <1> (Reversibility: ? <1> [1]) [1]
P vomilenine + $NADP^+$ + H_2O <1> [1]

Inhibitors
CO <1> [1]
cytochrome c <1> [1]
ketoconazole <1> [1]

Cofactors/prosthetic groups
NADPH <1> (<1> strictly dependent on [1]) [1]

K_m-Value (mM)
3 <1> (NADPH) [1]
26 <1> (vinorine) [1]

pH-Optimum
8.3 <1> [1]

Temperature optimum (°C)
40 <1> [1]

6 Stability

Storage stability
<1>, -28°C, 20% sucrose, no loss of activity [1]

References

[1] Falkenhagen, H.; Stoeckigt, J.: Enzymic biosynthesis of vomilenine, a key intermediate of the ajmaline pathway, catalyzed by a novel cytochrome P_{450}-dependent enzyme from plant cell cultures of Rauwolfia serpentina. Z. Naturforsch. C, **50**, 45-53 (1995)

Taxane 10β-hydroxylase 1.14.13.76

1 Nomenclature

EC number
1.14.13.76

Systematic name
taxa-4(20),11-dien-5α-yl acetate,NADPH:oxygen oxidoreductase (10β-hydroxylating)

Recommended name
taxane 10β-hydroxylase

Synonyms
oxygenase, taxane 10β-mono-
taxa-4(20),11-dien-5α-yl acetate 10β-hydroxylase
taxadien-5α-yl-acetate 10β-hydroxylase
taxane 10β-hydroxylase
taxane 10β-monooxygenase

CAS registry number
337514-75-7

2 Source Organism

<1> *Taxus sp.* [1]

3 Reaction and Specificity

Catalyzed reaction
taxa-4(20),11-dien-5α-yl acetate + NADPH + H^+ + O_2 = 10β-hydroxytaxa-4(20),11-dien-5α-yl acetate + $NADP^+$ + H_2O

Reaction type
oxidation
redox reaction
reduction

Natural substrates and products
S Additional information <1> (<1> catalyzes early step in taxol biosynthesis [1]) [1]
P ?

Substrates and products

S taxa-4(20),11-dien-5α-yl acetate + NADPH + H^+ + O_2 <1> (Reversibility: ? <1> [1]) [1]

P 10β-hydroxytaxa-4(20),11-dien-5α-yl acetate + NADP + H_2O

Cofactors/prosthetic groups

cytochrome P_{450} <1> [1]

K_m-Value (mM)

Additional information <1> (<1> in vivo and in vitro assay in yeast for recombinant enzyme [1]) [1]

5 Isolation/Preparation/Mutation/Application

Source/tissue

cell culture <1> [1]

Cloning

<1> [1]

References

[1] Schoendorf, A.; Rithner, C.D.; Williams, R.M.; Croteau, R.B.: Molecular cloning of a cytochrome P_{450} taxane 10β-hydroxylase cDNA from Taxus and functional expression in yeast. Proc. Natl. Acad. Sci. USA, **98**, 1501-1506 (2001)

Taxane 13α-hydroxylase 1.14.13.77

1 Nomenclature

EC number
1.14.13.77

Systematic name
taxa-4(20),11-dien-5α-ol,NADPH:oxygen oxidoreductase (13α-hydroxylating)

Recommended name
taxane 13α-hydroxylase

CAS registry number
399030-58-1

2 Source Organism

<1> *Taxus cuspidata* (yew [1]) [1, 2]

3 Reaction and Specificity

Catalyzed reaction
taxa-4(20),11-dien-5α-ol + NADPH + H^+ + O_2 = taxa-4(20),11-dien-5α,13α-diol + $NADP^+$ + H_2O

Reaction type
oxidation
redox reaction
reduction

Substrates and products
- **S** taxa-4(20),11-dien-5α-ol + NADPH + H^+ + O_2 <1> (Reversibility: ? <1> [1]) [1, 2]
- **P** taxa-4(20),11-dien-5α,13α-diol + $NADP^+$ + H_2O <1> [1, 2]
- **S** taxa-4(20),11-dien-5α-yl-acetate + NADPH + H^+ + O_2 <1> (<1> 6% of activity with taxa-4(20),11-dien-5α-ol [2]) (Reversibility: ? <1> [2]) [2]
- **P** taxa-4(20),11-dien-5α,13α-diol-5-acetate + $NADP^+$ + H_2O <1> [2]

K_m-Value (mM)
0.014 <1> (taxa-4(20),11-dien-5α-ol) [1, 2]
0.024 <1> (taxa-4(20),11-dien-5α-yl-acetate) [1, 2]

pH-Optimum

7.2 <1> [1]
7.5 <1> [2]

pH-Range

6.2-8.2 <1> (<1> half-maximal activity at pH 6.2 and pH 8.2 [1]) [1]

5 Isolation/Preparation/Mutation/Application

Source/tissue

cell culture <1> [1, 2]

Localization

microsome <1> [1]

Cloning

<1> (heterologous expression in Sf9 insect cells [2]) [2]

References

[1] Wheeler, A.L.; Long, R.M.; Ketchum, R.E.B.; Rithner, C.D.; Williams, R.M.; Croteau, R.: Taxol biosynthesis: differential transformations of taxadien-5α-ol and its acetate ester by cytochrome P_{450} hydroxylases from Taxus suspension cells. Arch. Biochem. Biophys., **390**, 265-278 (2001)
[2] Jennewein, S.; Rithner, C.D.; Williams, R.M.; Croteau, R.B.: Taxol biosynthesis: taxane 13α-hydroxylase is a cytochrome P_{450}-dependent monooxygenase. Proc. Natl. Acad. Sci. USA, **98**, 13595-13600 (2001)

ent-Kaurene oxidase 1.14.13.78

1 Nomenclature

EC number
1.14.13.78

Systematic name
ent-kaur-16-ene,NADPH:oxygen oxidoreductase (hydroxylating)

Recommended name
ent-kaurene oxidase

CAS registry number
149565-67-3

2 Source Organism

<1> *Marah macrocarpus* (wild cucumber [1]) [1]
<2> *Gibberella fujikuroi* (strain SG78 [2]) [2, 3, 6]
<3> *Cucurbita maxima* [2, 3]
<4> *Malus sp.* [2]
<5> *Arabidopsis thaliana* [4]
<6> *Arabidopsis thaliana* [5]
<7> *Gibberella fujikuroi* [6]

3 Reaction and Specificity

Catalyzed reaction
ent-kaur-16-ene + NADPH + H^+ + O_2 = ent-kaur-16-en-19-ol + $NADP^+$ + H_2O
ent-kaur-16-ene-19-al + NADPH + H^+ + O_2 = ent-kaur-16-en-19-oate + $NADP^+$ + H_2O
ent-kaur-16-ene-19-ol + NADPH + H^+ + O_2 = ent-kaur-16-en-19-al + $NADP^+$ + H_2O

Reaction type
oxidation
redox reaction
reduction

Natural substrates and products

S ent-kaur-16-ene + NADPH + H^+ + O_2 <7> (<7> involved in gibberillin biosynthesis pathway [6]) (Reversibility: ? <7> [6]) [6]

P ent-kaur-16-en-19-ol + $NADP^+$ + H_2O <7> [6]

S ent-kaur-16-ene-19-ol + NADPH + H^+ + O_2 <7> (<7> involved in gibberillin biosynthesis pathway [6]) (Reversibility: ? <7> [6]) [6]

P ent-kaur-16-en-19-al + $NADP^+$ + H_2O <7> [6]

S ent-kaur-16-ene-al + NADPH + H^+ + O_2 <7> (<7> involved in gibberillin biosynthesis pathway [6]) (Reversibility: ? <7> [6]) [6]

P ent-kaur-16-en-19-oate + $NADP^+$ + H_2O <7> [6]

Substrates and products

S ent-kaur-16-ene + NADPH + H^+ + O_2 <2-7> (Reversibility: ? <2-7> [2-6]) [2-6]

P ent-kaur-16-en-19-ol + $NADP^+$ + H_2O <2-7> [2-6]

S ent-kaur-16-ene-19-ol + NADPH + H^+ + O_2 <1-7> (Reversibility: ? <1-7> [1-6]) [1-6]

P ent-kaur-16-en-19-al + $NADP^+$ + H_2O <1-7> [1-6]

S ent-kaur-16-ene-al + NADPH + H^+ + O_2 <2-7> (Reversibility: ? <2-7> [2-6]) [2-6]

P ent-kaur-16-en-19-oate + $NADP^+$ + H_2O <2-7> [2-6]

Inhibitors

(2R,3R)-paclobutrazol <2> (<2> 0.0045 mM, 50% inhibition [2]) [2]
(2S,3S)-paclobutrazol <2> (<2> 0.00025 mM, 50% inhibition [2]) [2]
CO <2> (<2> 90% inhibition [2]) [2]

Cofactors/prosthetic groups

FAD <2> (<2> stimulates activity [2]) [2]
NADPH <1, 2> [1, 2]

Specific activity (U/mg)

Additional information <2> (<2> 2.19 Bq kaurenoids/min/mg, activity in cell extracts [2]) [2]

K_m-Value (mM)

0.0063 <2> (ent-kaur-16-ene) [2]

K_i-Value (mM)

0.000007 <2> ((2S,3S)-paclobutrazol) [2]
0.00002 <4> ((2S,3S)-paclobutrazol) [2]
0.00002 <3> (paclobutrazol racemate) [2]
0.00006 <4> ((2R,3R)-paclobutrazol) [2]
0.0015 <2> ((2R,3R)-paclobutrazol) [2]

4 Enzyme Structure

Molecular weight

50000-55000 <2> (<2> gel filtration [3]) [3]

5 Isolation/Preparation/Mutation/Application

Source/tissue
embryo <4> [2]
endosperm <3> [2]
mycelium <7> (<7> transcript detection [6]) [6]
seed <1> [1]

Localization
chloroplast <6> (<6> enzyme is targeted to the outer chloroplast envelope [5]) [5]
microsome <2, 3> (<2> most likely membrane-associated [2]) [2, 3]

Purification
<2> (25% polyethylene 8000, anion exchange, gel filtration [3]) [3]

Cloning
<5> (expression of cDNA in Saccharomyces cerevisiae [4]) [4]
<7> (cloning of cDNA [6]) [6]

6 Stability

General stability information
<2>, 20% glycerol reduces the degradation to P-420 [2]

References

[1] Sherwin, P.F.; Coates, R.M.: Stereospecificity of the oxidation of ent-kauren-19-ol to ent-kaurenal by a microsomal enzyme preparation from Marah macrocarpus. J. Chem. Soc., **1982**, 1013-1014 (1982)
[2] Ashman, P.J.; Mackenzie, A.; Bramley, P.M.: Characterization of ent-kaurene oxidase activity from Gibberella fujikuroi. Biochim. Biophys. Acta, **1036**, 151-157 (1990)
[3] Archer, C.; Ashman, P.J.; Hedden, P.; Bowyer, J.R.; Bramley, P.M.: Purification of ent-kaurene oxidase from Gibberella fujikuroi and Cucurbita maxima. Biochem. Soc. Trans., **20**, 218S (1992)
[4] Helliwell, C.A.; Poole, A.; Peacock, W.J.; Dennis, E.S.: Arabidopsis ent-kaurene oxidase catalyzes three steps of gibberellin biosynthesis. Plant Physiol., **119**, 507-510 (1999)
[5] Helliwell, C.A.; Sullivan, J.A.; Mould, R.M.; Gray, J.C.; Peacock, W.J.; Dennis, E.S.: A plastid envelope location of Arabidopsis ent-kaurene oxidase links the plastid and endoplasmic reticulum steps of the gibberellin biosynthesis pathway. Plant J., **28**, 201-208 (2001)
[6] Tudzynski, B.; Hedden, P.; Carrera, E.; Gaskin, P.: The P_{450}-4 gene of Gibberella fujikuroi encodes ent-kaurene oxidase in the gibberellin biosynthesis pathway. Appl. Environ. Microbiol., **67**, 3514-3522 (2001)

ent-Kaurenoic acid oxidase 1.14.13.79

1 Nomenclature

EC number
1.14.13.79

Systematic name
ent-kaur-16-en-19-oate,NADPH:oxygen oxidoreductase (hydroxylating)

Recommended name
ent-kaurenoic acid oxidase

Synonyms
KAO
ent-kaurenoate C-10 oxidase
ent-keurenoate oxidase
ent-keurenoic acid oxidase
gibberelin 12 synthase
synthase, gibberellin A12

CAS registry number
337507-95-6

2 Source Organism

<1> *Hordeum vulgare* (dwarf mutant of barley that accumulates ent-keurenoic acid in developing grains [1]) [1]
<2> *Arabidopsis thaliana* [1]
<3> *Zea mays* [1]
<4> *Pisum sativum* [2]

3 Reaction and Specificity

Catalyzed reaction
ent-kaur-16-en-19-oate + NADPH + H^+ + O_2 = ent-7α-hydroxykaur-16-en-19-oate + $NADP^+$ + H_2O (first step)
ent-7α-hydroxykaur-16-en-19-oate + NADPH + H^+ + O_2 = gibberellin A12 aldehyde + $NADP^+$ + 2 H_2O (second step, The second step includes a ring-B contraction giving the gibbane skeleton)
gibberellin A12 aldehyde + NADPH + O_2 = gibberellin A12 + $NADP^+$ + H_2O (third step)

Reaction type

oxidation
redox reaction
reduction

Natural substrates and products

S ent-7α-hydroxykaur-16-en-19-oate + NADPH + O_2 <1, 2, 3, 4> (<1, 2, 3>, the enzyme catalyzes the three steps of the gibberellin biosynthetic pathway from ent-kaurenoic acid to gibberellin A12 [1, 2]) (Reversibility: ? <1, 2, 3, 4> [1, 2]) [1, 2]
P gibberellin A12 aldehyde + $NADP^+$ + 2 H_2O
S ent-kaur-16-en-19-oate + NADPH + H^+ + O_2 <1, 2, 3, 4> (<1, 2, 3>, the enzyme catalyzes the three steps of the gibberellin biosynthetic pathway from ent-kaurenoic acid to gibberellin A12 [1, 2]) (Reversibility: ? <1, 2, 3, 4> [1, 2]) [1, 2]
P ent-7α-hydroxykaur-16-en-19-oate + $NADP^+$ + H_2O
S gibberellin A12 aldehyde + NADPH + O_2 <1, 2, 3, 4> (<1, 2, 3, 4>, the enzyme catalyzes the three steps of the gibberellin biosynthetic pathway from ent-kaurenoic acid to gibberellin A12 [1, 2]) (Reversibility: ? <1, 2, 3, 4> [1, 2]) [1, 2]
P gibberellin A12 + $NADP^+$ + H_2O

Substrates and products

S ent-7α-hydroxykaur-16-en-19-oate + NADPH + O_2 <1, 2, 3, 4> (<1, 2, 3, 4>, the enzyme catalyzes the three steps of the gibberellin biosynthetic pathway from ent-kaurenoic acid to gibberellin A12 [1, 2]) (Reversibility: ? <1, 2, 3, 4> [1, 2]) [1, 2]
P gibberellin A12 aldehyde + $NADP^+$ + 2 H_2O
S ent-kaur-16-en-19-oate + NADPH + H^+ + O_2 <1, 2, 3, 4> (<1, 2, 3, 4>, the enzyme catalyzes the three steps of the gibberellin biosynthetic pathway from ent-kaurenoic acid to gibberellin A12 [1, 2]) (Reversibility: ? <1, 2, 3, 4> [1, 2]) [1, 2]
P ent-7α-hydroxykaur-16-en-19-oate + $NADP^+$ + H_2O
S gibberellin A12 aldehyde + NADPH + O_2 <1, 2, 3, 4> (<1, 2, 3, 4>, the enzyme catalyzes the three steps of the gibberellin biosynthetic pathway from ent-kaurenoic acid to gibberellin A12 [1, 2]) (Reversibility: ? <1, 2, 3, 4> [1, 2]) [1, 2]
P gibberellin A12 + $NADP^+$ + H_2O

Cofactors/prosthetic groups

NADPH <1> [1]
cytochrome P_{450} <1, 3, 4> (<1, 3, 4>, cytochrome P_{450} protein [1, 2]) [1, 2]

5 Isolation/Preparation/Mutation/Application

Source/tissue

apical bud <4> (<4>, expression of the gene PsKAO1 [2]) [2]
leaf <4> (<4>, expression of the gene PsKAO1 [2]) [2]

pod <4> (<4>, expression of the gene PsKAO1 [2]) [2]
root <4> (<4>, expression of the gene PsKAO1 [2]) [2]
seed <4> (<4>, developing, expression of the gene PsKAO1, the gene PsKAO1 is expressed only in seed [2]) [2]

Cloning

<1> (expression in yeast [1]) [1]
<3> (expression in yeast [1]) [1]
<4> (both genes PsKAO1 and PsKAO$_2$ encode the ent-kaurenoic acid oxidase, expression in Saccharomyces cerevisiae [2]) [2]

References

[1] Helliwell, C.A.; Chandler, P.M.; Poole, A.; Dennis, E.S.; Peacock, W.J.: The CYP88A cytochrome P_{450}, ent-kaurenoic acid oxidase, catalyzes three steps of the gibberellin biosynthesis pathway. Proc. Natl. Acad. Sci. USA, **98**, 2065-2070 (2001)
[2] Davidson, S.E.; Elliott, R.C.; Helliwell, C.A.; Poole, A.T.; Reid, J.B.: The pea gene NA encodes ent-kaurenoic acid oxidase. Plant Physiol., **131**, 335-344 (2003)

(R)-Limonene 6-monooxygenase 1.14.13.80

1 Nomenclature

EC number
1.14.13.80

Systematic name
(R)-limonene,NADPH:oxygen oxidoreductase (6-hydroxylating)

Recommended name
(R)-limonene 6-monooxygenase

Synonyms
(+)-limonene 6-monooxygenase
(+)-limonene-6-hydroxylase

CAS registry number
221461-49-0

2 Source Organism

<1> *Carum carvi* [1, 2]

3 Reaction and Specificity

Catalyzed reaction
(+)-(R)-limonene + NADPH + H^+ + O_2 = (+)-trans-carveol + $NADP^+$ + H_2O (The reaction is stereospecific with over 95% yield of (+)-trans-carveol from (R)-limonene. (S)-Limonene, the substrate for EC 1.14.13.48, (S)-limonene 6-monooxygenase, is not a substrate. Forms part of the carvone biosynthesis pathway in Carum carvi (caraway) seeds)

Reaction type
oxidation
redox reaction
reduction

Natural substrates and products
S (+)-(R)-limonene + NADPH + H^+ + O_2 <1> (Reversibility: ? <1> [1, 2]) [1, 2]
P (+)-trans-carveol + $NADP^+$ + H_2O

Substrates and products

S (+)-(R)-limonene + NADPH + H^+ + O_2 <1> (Reversibility: ? <1> [1, 2]) [1, 2]

P (+)-trans-carveol + $NADP^+$ + H_2O

References

[1] Bouwmeester, H.J.; Gershenzon, J.; Konings, M.C.J.M.; Croteau, R.: Biosynthesis of the monoterpenes limonene and carvone in the fruit of caraway. I. Demonstration of enzyme activities and their changes with development. Plant Physiol., **117**, 901-912 (1998)

[2] Bouwmeester, H.J.; Konings, M.C.J.M.; Gershenzon, J.; Karp, F.; Croteau, R.: Cytochrome P-450 dependent (+)-limonene-6-hydroxylation in fruits of caraway (Carum Carvi). Phytochemistry, **50**, 243-248 (1999)

Magnesium-protoporphyrin IX monomethyl ester (oxidative) cyclase 1.14.13.81

1 Nomenclature

EC number
1.14.13.81

Systematic name
magnesium-protoporphyrin-IX 13-monomethyl ester,NADPH:oxygen oxidoreductase (hydroxylating)

Recommended name
magnesium-protoporphyrin IX monomethyl ester (oxidative) cyclase

Synonyms
Mg-protoporphyrin IX monomethyl ester (oxidative) cyclase

CAS registry number
92353-62-3

2 Source Organism

<1> *Chlamydomonas reinhardtii* [1]
<2> *Synechocystis sp. PCC 6803* [1]

3 Reaction and Specificity

Catalyzed reaction
131-hydroxy-magnesium-protoporphyrin IX 13-monomethyl ester + NADPH + H^+ + O_2 = 131-oxo-magnesium-protoporphyrin IX 13-monomethyl ester + $NADP^+$ + 2 H_2O (2)
131-oxo-magnesium-protoporphyrin IX 13-monomethyl ester + NADPH + H^+ + O_2 = divinylprotochlorophyllide + $NADP^+$ + 2 H_2O (3; Requires Fe(II) for activity. The cyclase activity in Chlamydomonas reinhardtii is associated exclusively with the membranes, whereas that from cucumber cotyledons requires both membrane and soluble fractions for activity)
magnesium-protoporphyrin IX 13-monomethyl ester + NADPH + H^+ + O_2 = 131-hydroxy-magnesium-protoporphyrin IX 13-monomethyl ester + $NADP^+$ + H_2O (1)

Natural substrates and products
S 131-hydroxy-magnesium-protoporphyrin IX 13-monomethyl ester + NADPH + H^+ + O_2 <1, 2> (Reversibility: ? <1, 2> [1]) [1]

P 131-oxo-magnesium-protoporphyrin IX 13-monomethyl ester + $NADP^+$ + H_2O
S 131-oxo-magnesium-protoporphyrin IX 13-monomethyl ester + NADPH + H^+ + O_2 <1, 2> (Reversibility: ? <1, 2> [1]) [1]
P divinylprotochlorophyllide + $NADP^+$ + H_2O
S magnesium-protoporphyrin IX 13-monomethyl ester <1, 2> (Reversibility: ? <1, 2> [1]) [1]
P 131-hydroxy-magnesium-protoporphyrin IX 13-monomethyl ester + $NADP^+$ + H_2O

Substrates and products
S 131-hydroxy-magnesium-protoporphyrin IX 13-monomethyl ester + NADPH + H^+ + O_2 <1, 2> (Reversibility: ? <1, 2> [1]) [1]
P 131-oxo-magnesium-protoporphyrin IX 13-monomethyl ester + $NADP^+$ + H_2O
S 131-oxo-magnesium-protoporphyrin IX 13-monomethyl ester + NADPH + H^+ + O_2 <1, 2> (Reversibility: ? <1, 2> [1]) [1]
P divinylprotochlorophyllide + $NADP^+$ + H_2O
S magnesium-protoporphyrin IX 13-monomethyl ester <1, 2> (Reversibility: ? <1, 2> [1]) [1]
P 131-hydroxy-magnesium-protoporphyrin IX 13-monomethyl ester + $NADP^+$ + H_2O

References

[1] Bollivar, D.W.; Beale, S.I.: The chlorophyll biosynthetic enzyme Mg-protoporphyrin IX monomethyl ester (oxidative) cyclase (characterization and partial purification from Chlamydomonas reinhardtii and Synechocystis sp. PCC 6803). Plant Physiol., **112**, 105-114 (1996)

Unspecific monooxygenase 1.14.14.1

1 Nomenclature

EC number

1.14.14.1

Systematic name

substrate,reduced-flavoprotein:oxygen oxidoreductase (RH-hydroxylating or -epoxidizing)

Recommended name

unspecific monooxygenase

Synonyms

3AH15
6 β-hydroxylase
6-β-testosterone hydroxylase
aldehyde oxygenase
arachidonic acid epoxygenase
brain aromatase
CYP102 <3> [4, 5]
CYP1A1
CYP1A2
CYP1A3
CYP2A3
CYP4502F4
CYP6B1V1/CYP6B1V2/ CYP6B1V3
CYP6B3V1/CYP6B3V2
CYP6B4V1/CYP6B4V2
CYP6B5V1
CYPIA1
CYPIA2
CYPIA4
CYPIA5
CYPIB1
CYPIIA1
CYPIIA10
CYPIIA11
CYPIIA12
CYPIIA13
CYPIIA2
CYPIIA3
CYPIIA4

CYPIIA5
CYPIIA6
CYPIIA7
CYPIIA8
CYPIIA9
CYPIIB1
CYPIIB10
CYPIIB11
CYPIIB12
CYPIIB19
CYPIIB2
CYPIIB20
CYPIIB3
CYPIIB4
CYPIIB5
CYPIIB6
CYPIIB9
CYPIIC1
CYPIIC10
CYPIIC11
CYPIIC12
CYPIIC13
CYPIIC14
CYPIIC15
CYPIIC16
CYPIIC17
CYPIIC18
CYPIIC19
CYPIIC2
CYPIIC20
CYPIIC21
CYPIIC22
CYPIIC23
CYPIIC24
CYPIIC25
CYPIIC26
CYPIIC27
CYPIIC28
CYPIIC29
CYPIIC3
CYPIIC30
CYPIIC31
CYPIIC37
CYPIIC38
CYPIIC39
CYPIIC4
CYPIIC40

CYPIIC41
CYPIIC42
CYPIIC5
CYPIIC6
CYPIIC7
CYPIIC8
CYPIIC9
CYPIID1
CYPIID10
CYPIID11
CYPIID14
CYPIID15
CYPIID16
CYPIID17
CYPIID18
CYPIID19
CYPIID2
CYPIID3
CYPIID4
CYPIID5
CYPIID6
CYPIID9
CYPIIE1
CYPIIF1
CYPIIF3
CYPIIF4
CYPIIG1
CYPIIH1
CYPIIH2
CYPIIIA1
CYPIIIA10
CYPIIIA11
CYPIIIA12
CYPIIIA13
CYPIIIA14
CYPIIIA15
CYPIIIA16
CYPIIIA17
CYPIIIA18
CYPIIIA19
CYPIIIA2
CYPIIIA21
CYPIIIA24
CYPIIIA25
CYPIIIA27
CYPIIIA28
CYPIIIA29

CYPIIIA3
CYPIIIA30
CYPIIIA31
CYPIIIA5
CYPIIIA6
CYPIIIA7
CYPIIIA8
CYPIIIA9
CYPIIJ1
CYPIIJ2
CYPIIJ3
CYPIIJ5
CYPIIJ6
CYPIIK1
CYPIIK3
CYPIIK4
CYPIIL1
CYPIIM1
CYPIVA4
CYPIVA8
CYPIVB1
CYPIVC1
CYPIVF1
CYPIVF11
CYPIVF12
CYPIVF4
CYPIVF5
CYPIVF6
CYPIVF8
CYPVIA1
CYPVIB1
CYPVIB2
CYPVIB4
CYPVIB5
CYPVIB6
CYPVIB7
CYPXIX
CYPXIXA1
CYPXIXA2
CYPXIXA3
clone PF26
clone PF3/46
coumarin 7-hydroxylase
cytochrome P450-D2
DAH1
DAH2
debrisoquine 4-hydroxylase

EC 1.14.1.1 (formerly)
EC 1.14.14.2 (formerly)
EC 1.14.99.8 (formerly)
EC 1.99.1.1 (formerly)
estrogen synthetase
HLp
hepatic cytochrome P-450MC1
IIA3
isozyme 3A
LMC1
laurate ω-1 hydroxylase
lauric acid ω-6-hydroxylase
mephenytoin 4-hydroxylase
OLF2
olfactive
ovarian aromatase
P(3)450
P-448
P-450 PHPAH1
P-450(M-1)
P-450-MK2
P-450AROM
P-450IB
P-450IIIAM1
P-450MC
P-450MP
P-450UT
P1-88
P24
P450 17-α
P450 2D-29/2D-35
P450 CM3A-10
P450 DUT2
P450 FA
P450 FI
P450 HSM1
P450 HSM2
P450 HSM3
P450 HSM4
P450 IIB1
P450 IIC2
P450 LM4
P450 LM6
P450 LMC2
P450 MD
P450 MP-12/MP-20
P450 P49

P450 PB1
P450 PB4
P450 PBC1
P450 PBC2
P450 PBC3
P450 PBC4
P450 PCHP3
P450 PCHP7
P450 TCDDAA
P450 TCDDAHH
P450 form 3B
P450 form HP1
P450 type B2
P450 types B0 and B1
P450(I)
P450-11A
P450-15-COH
P450-15-α
P450-16-α
P450-254C
P450-3C
P450-6B/29C
P450-A3
P450-AFB
P450-ALC
P450-CMF1A
P450-CMF1B
P450-CMF2
P450-CMF3
P450-DB1
P450-DB2
P450-DB3
P450-DB4
P450-DB5
P450-HFLA
P450-HP
P450-IIA10
P450-IIA11
P450-IIA3.1
P450-IIA3.2
P450-IIA4
P450-KP1
P450-LM2
P450-MC1
P450-MC4
P450-MK1
P450-MKJ1

P450-MKMP13
P450-MKNF2
P450-NMB
P450-OLF1
P450-OLF3
P450-P1
P450-P2/P450-P3
P450-P3
P450-PB1 and P450-PB2
P450-PCN1
P450-PCN2
P450-PCN3
P450-PM4
P450-PP1
P450-PROS2
P4501A1
P450CB
P450CMEF
P450E
P450EF
P450F
P450H
P450I
P450IIC5
P450MT2
P450RAP
P450RLM6
P52
PB15
PHP2
PHP3
PTF1
PTF2
progesterone 21-hydroxylase
prostaglandin ω-hydroxylase
S-mephenytoin 4-hydroxylase
steroid hormones 7-α-hydroxylase
testosterone 15-α-hydroxylase
testosterone 16-α hydroxylase
testosterone 6-β-hydroxylase
testosterone 7-α-hydroxylase
aromatase
aryl hydrocarbon hydroxylase
aryl-4-monooxygenase
cytochrome P-450 BM3 <3> (<3> enzyme contains a P-450 heme domain and an NADPH-cytochrome P-450 reductase flavoprotein domain in a single polypeptide chain [4]) [4, 5]

flavocytochrome P450BM-3 <3> [5]
flavoprotein monooxygenase
flavoprotein-linked monooxygenase
microsomal P-450
microsomal monooxygenase
oxygenase, flavoprotein-linked mono-
xenobiotic monooxygenase

CAS registry number
62213-32-5

2 Source Organism

<1> *Mus musculus* (mouse [1,2]) [1, 2]
<2> *Rattus norvegicus* [3]
<3> *Bacillus megaterium* [4, 5]

3 Reaction and Specificity

Catalyzed reaction
RH + reduced flavoprotein + O_2 = ROH + oxidized flavoprotein + H_2O (a group of heme-thiolate proteins (P-450), acting on a wide range of substrates including many xenobiotics, steroids, fatty acids, vitamins and prostaglandins; reactions catalysed include hydroxylation, epoxidation, N-oxidation, sulfooxidation, N-, S- and O-dealkylations, desulfation, deamination, and reduction of azo, nitro and N-oxide groups; together with EC 1.6.2.4 NADPH-ferrihemoprotein reductase, it forms a system in which two reducing equivalents are supplied by NADPH2; formerly EC 1.14.1.1, EC 1.14.99.8 and EC 1.99.1.1; some of the reactions attributed to EC 1.14.15.3 alkane 1-monooxygenase belong here)

Reaction type
N-dealkylation
N-oxidation
O-dealkylation
S-dealkylation
deamination
desulfation
epoxidation
hydroxylation
oxidation
redox reaction
reduction
reduction of azo, nitro, N-oxide groups
sulfoxidation

Substrates and products

S aryl hydrocarbons + reduced flavoprotein + O_2 <1, 2> (<1> e.g. benzo[a]pyrene, ethoxyresuforin, biphenyl, *p*-nitroanisole, acetanilide, 2-acetylaminofluorene, 2-ethoxycoumarin, estradiol-17β, testosterone [1, 2]; <2> prostaglandins [3]) [1-3]

P ?

S Additional information <3> (<3> enzyme catalyses hydroxylation in the ω-1, ω-2 and ω-3 positions and/or epoxidation of medium- and long-chain fatty acids [4]; self-sufficient fatty acid monooxygenase [5]) [4, 5]

P ?

Turnover number (min^{-1})

Additional information <3> (<3> K_{cat}-values of wild-type and mutant enzyme for different fatty acids and alkyl trimethylammonium compounds [4]) [4]

K_m-Value (mM)

Additional information <3> (<3> K_m-values of wild-type and mutant enzyme for different fatty acids and alkyl trimethylammonium compounds [4]) [4]

5 Isolation/Preparation/Mutation/Application

Source/tissue

liver <1, 2> [1-3]

Localization

microsome <1, 2> (<1> 16 different cytochrome P-450 have been isolated from mouse liver of which each contains numerous different forms of P-450 [1]) [1-3]

Purification

<1> (partial [1]) [1]

Crystallization

<3> (crystal structure of the complex between the heme- and FMN-binding domains of the enzyme, crystals are grown at room temperature by liquid-liquid free interface diffusion in a capillary, the flavodoxin-like flavin domain is positioned at the proximal face of the heme domain [5]) [5]

Cloning

<3> (<3> expression in Escherichia coli [4]) [4]

Engineering

R47E <3> (<3> the mutant enzyme retains significant hydroxylase activity towards saturated fatty acids and shows much increased activity towards C_{12}-C_{16} alkyl trimethylammonium compounds [4]) [4]

References

[1] Lang, M.A.; Nebert, D.W.: Structural gene products of the Ah locus. Evidence for many unique P-450-mediated monooxygenase activities reconstituted from 3-methylcholanthrene-treated C57BL/6N mouse liver microsomes. J. Biol. Chem., **256**, 12058-12067 (1981)
[2] Lang, M.A.; Gielen, J.E.; Nebert, D.W.: Genetic evidence for many unique liver microsomal P-450-mediated monooxygenase activities in heterogeneic stock mice. J. Biol. Chem., **256**, 12068-12075 (1981)
[3] Theoharides, A.D.; Kupfer, D.: Evidence for different hepatic microsomal monooxygenases catalyzing ω- and (ω-1)-hydroxylations of prostaglandins E1 and E2. Effects of inducers of monooxygenase on the kinetic constants of prostaglandin hydroxylation. J. Biol. Chem., **256**, 2168-2175 (1981)
[4] Oliver, C.F.; Modi, S.; Primrose, W.U.; Lian, L.Y.; Roberts, G.C.K.: Engineering the substrate specificity of Bacillus megaterium cytochrome P-450 BM3: hydroxylation of alkyl trimethylammonium compounds. Biochem. J., **327**, 537-544 (1997)
[5] Sevrioukova, I.F.; Li, H.; Zhang, H.; Peterson, J.A.; Poulos, T.L.: Structure of a cytochrome P_{450}-redox partner electron-transfer complex. Proc. Natl. Acad. Sci. USA, **96**, 1863-1868 (1999)

Benzopyrene 3-monooxygenase 1.14.14.2

1 Nomenclature

EC number
1.14.14.2 (deleted, included in EC 1.14.14.1)

Recommended name
benzopyrene 3-monooxygenase

Alkanal monooxygenase (FMN-linked) 1.14.14.3

1 Nomenclature

EC number
1.14.14.3

Systematic name
alkanal,reduced-FMN:oxygen oxidoreductase (1-hydroxylating, luminescing)

Recommended name
alkanal monooxygenase (FMN-linked)

Synonyms
Vibrio fischeri luciferase
aldehyde monooxygenase
bacterial luciferase
luciferase

CAS registry number
9014-00-0

2 Source Organism

<1> *Vibrio harveyi* (expressed in Escherichia coli [8]; formerly Beneckea harveyi [9, 19, 41]) [4-14, 16-19, 21, 23-36, 38, 40-44, 46-48, 56, 58-61, 63, 66, 67, 69]
<2> *Vibrio cholerae* [1]
<3> *Xenorhabdus luminescens* [2, 4, 57]
<4> *Vibrio sp.* (<4> symbiotic bacterium from Kryptophanaron alfredi, flashlight fish [3, 22]) [3, 22]
<5> *Vibrio fischeri* (formerly Photobacterium fischeri [9, 19, 41]) [9, 19, 39, 41, 44, 46, 55]
<6> *Photobacterium phosphoreum* [15, 19, 41, 45, 49, 50, 52]
<7> *Photobacterium leiognathi* [20, 44]
<8> *Photobacterium sp.* [51, 53, 54]
<9> *bacteria* (not classified) [37]
<10> *Photinus pyralis* (firefly) [62, 64]
<11> *Luciola mingrelica* (firefly) [65]
<12> *Pyrocystis lunula* [68]

3 Reaction and Specificity

Catalyzed reaction

RCHO + reduced FMNH + O_2 = RCOOH + FMN + H_2O + hv

Reaction type

oxidation
redox reaction
reduction

Natural substrates and products

S decanal + FMNH + O_2 <1-9> (Reversibility: ir <1-9> [1-60]) [1-60]
P decanoic acid + FMN + H_2O + light
S luciferin + O_2 + ATP <10-12> (Reversibility: ir <10-12> [62, 64, 65, 68]) [62, 64, 65, 68]
P oxyluciferin + AMP + CO_2 + light

Substrates and products

S FMNH + O_2 <1, 5, 7> (Reversibility: ? <1, 5, 7> [44]) [44]
P FMN + H_2O_2
S decanal + FMNH + O_2 <1-9> (Reversibility: ir <1-9> [1-61, 63, 66, 67]) [1-61, 63, 66, 67]
P decanoic acid + FMN + H_2O + light
S dodecanal + FMNH + O_2 <1-9> (Reversibility: ir <1-9> [1-61, 63, 66, 67]) [1-61, 63, 66, 67]
P dodecanoic acid + FMN + H_2O + light
S hexachlorethane + 2 e- <1> (Reversibility: ? <1> [61]) [61]
P tetrachlorethylene + 2 Cl^-
S luciferin + O_2 + ATP <10-12> (Reversibility: ir <10-12> [62, 64, 65, 68]) [62, 64, 65, 68]
P oxyluciferin + AMP + CO_2 + light
S myristic aldehyde + FMNH + O_2 <3> (Reversibility: ? <3> [57]) [57]
P myristic acid + FMN + H_2O + light
S octanal + FMNH + O_2 <1-9> (Reversibility: ir <1-9> [1-61, 63, 66, 67]) [1-61, 63, 66, 67]
P octanoic acid + FMN + H_2O + light
S pentachlorethane + 2 e- <1> (Reversibility: ? <1> [61]) [61]
P trichlorethylene + 2 Cl^-
S Additional information <1, 5, 6> (<1, 5, 6> aldehydes of chain-length 8 or more required [41]) [41]
P ?

Inhibitors

2,3-dichloro-(6-phenylphenoxy)ethylamine <1, 5, 8> [41, 54]
2,4-dinitrofluorobenzene <1> (<1> Sanger's reagent [35]) [35]
2-bromodecanal <1> (<1> protection by dithiothreitol or mercaptoethanol [28]l) [28]
2-diethylaminoethyl-2,2-diphenylvalerate <1, 5, 8> [41, 54]

8-anilino-1-naphthalenesulfonate <1, 5, 7> (<1> inhibitor binding site separate from FMN-binding site by 30 A [47]) [41, 44, 47]
CN^- <1> [5]
$FADH_2$ <6> [50]
N,N-diethyl-2,4-dichloro-(6-phenylphenoxy)ethylamine <8> [54]
N-ethylmaleimide <8> (<8> protection by substrates [51]) [51]
acetone <1> [6]
aliphatic alcohols <1> [6]
aliphatic alkanes <1> [6]
amino group reagents <1, 5, 6> [41]
benzylalcohol <1> [6]
butanoic acid <10> (<10> IC_{50}: 13.6 mM [64]) [64]
butanone <1> [6]
chloroform <1> [6]
decanoic acid <10> (<10> IC_{50}: 0.0132 mM [64]) [64]
diethylether <1, 10> [6, 62]
dodecanoic acid <10> (<10> IC_{50}: 0.0012 mM [64]) [64]
enflurane <1> [6]
ethoxyformic anhydride <1, 5, 7> [44]
fluroxene <1> [6]
halothane <1, 10> [6, 62]
hexadecanoic acid <10> (<10> IC_{50}: 0.00067 mM [64]) [64]
hexanoic acid <10> (<10> IC_{50}: 3.4 mM [64]) [64]
imidazole reagents <1, 5, 6> [41]
iodoacetamide <8> [51]
isoflurane <1> [6]
lumichrome <6> [50]
lumiflavin <6> [50]
methoxyflurane <1, 10> [6, 62]
n-butanol <10> [62]
n-decanal <1> (<1> reversible substrate inhibition, depending on phosphate concentration [30]) [30, 60]
n-heptanol <10> [62]
n-hexanol <10> [62]
nonal <1> [60]
octadecanoic <10> (<10> IC_{50}: 0.00063 mM [64]) [64]
octanoic acid <10> (<10> IC_{50}: 2.9 mM [64]) [64]
paraldehyde <1> [6]
proteases <1> (<1> trypsin, chymotrypsin [36]) [36]
reduced riboflavin <6> [50]
sulfhydryl reagents <1, 5, 6> [41]
tetradecanoic acid <10> (<10> IC_{50}: 0.00068 mM [64]) [64]
tridecanoic acid <6> [49]
undecanal <1> [60]
uracil <6> [50]
urethane <1> [6]

Cofactors/prosthetic groups

1-deaza-$FMNH_2$ <1> (<1> can replace $FMNH_2$ [11]) [11]
2',3'-diacetyl-$FMNH_2$ <1> (<1> as substitute for $FMNH_2$ [33]) [33]
2-thio-$FMNH_2$ <1> (<1> as substitute for $FMNH_2$ [33]) [33]
3'-carboxymethyl-$FMNH_2$ <1> (<1> as substitute for $FMNH_2$ [33]) [33]
$FMNH_2$ <1, 5-7> (<1,5,7> specific for, low activity with other flavins or flavin analogs [44]; <6> 8-substituted FMN-analogs [45]) [11, 38, 44, 45]
iso-$FMNH_2$ <1> [33]

Activating compounds

2-fluoroethylamine <1> (<1> αH44A mutant, works as catalytic base [59]) [59]
ammonia <1> (<1> αH44A mutant, works as catalytic base [59]) [59]
cyanomethylamine <1> (<1> αH44A mutant, works as catalytic base [59]) [59]
ethanolamine <1> (<1> αH44A mutant, works as catalytic base [59]) [59]
ethylamine <1> (<1> αH44A mutant, works as catalytic base [59]) [59]
imidazole <1> (<1> αH44A mutant, works as catalytic base [59]) [59]
methylamine <1> (<1> αH44A mutant, works as catalytic base [59]) [59]
ω-carboxypentylflavin <1> (<1> as substitute for $FMNH_2$ [33]) [33]
propylamine <1> (<1> αH44A mutant, works as catalytic base [59]) [59]

Turnover number (min^{-1})

0.0009 <3> (FMNH, <3> myristic aldehyde as second substrate [57]) [57]
0.0015 <3> (FMNH, <3> dodecanoic aldehyde as second substrate [57]) [57]
0.0019 <3> (FMNH, <3> decanoic aldehyde as second substrate [57]) [57]
Additional information <1, 3, 5, 7> (<3> tetradecanal + FMNH + O_2 (possibly) turnover rate determined by measuring the half-time for decay of luminescence between 70% and 35% of the maximum intensity, values depending on chain-length of aldehyde [4]; <1, 5, 7> slow turnover rate [44]) [4, 44]

K_m-Value (mM)

0.0001 <1> (O_2) [38]
0.00015 <3> (FMNH, <3> α subunit [57]) [57]
0.0003 <1> (decanal, <1> wild type, 50 mM phosphate [66]) [66]
0.0004-0.0008 <1, 6> (FMNH) [38, 50]
0.00058 <3> (FMNH, <3> β subunit [57]) [57]
0.0008 <1> (FMNH, <1> wild type, 10 mM phosphate [66]) [66]
0.001-0.01 <1> (aldehydes) [38]
0.0012-0.009 <1> (n-decanal, <1> depending on buffer system [30]) [30]
0.0018 <1> (FMNH, <1> wild type, 300 mM phosphate [66]) [66]
0.0021 <1> (decanal, <1> αR107S, 50 mM phosphate [66]) [66]
0.0022 <1> (decanal, <1> αR107E, 50 mM phosphate [66]) [66]
0.0031 <1> (decanal, <1> αR107A, 50 mM phosphate [66]) [66]
0.0038 <1> (FMNH, <1> αR107S, 10 mM phosphate [66]) [66]
0.0049 <1> (FMNH, <1> αR107S, 300 mM phosphate [66]) [66]
0.0081 <1> (FMNH, <1> αR107A, 10 mM phosphate [66]) [66]
0.0095 <1> (FMNH, <1> αR107A, 300 mM phosphate [66]) [66]
0.0114 <1> (FMNH, <1> αR107E, 300 mM phosphate [66]) [66]
0.0234 <1> (FMNH, <1> αR107E, 10 mM phosphate [66]) [66]

pH-Optimum

5.5-6 <12> [68]
6.5 <8> (<8> optimum for binding of $FMNH_2$ [51]) [51]
6.5-7 <8> (<8> optimum for reaction velocity [51]) [51]
8.1 <8> (<8> optimum for quantum yield [51]) [51]

Temperature optimum (°C)

Additional information <1> (<1> temperature dependence of thermodynamic parameters [48]) [48]

Temperature range (°C)

6-25 <10> [62]

4 Enzyme Structure

Molecular weight

77000-78000 <1, 3> (<3> gel filtration [2,4]; <1> nucleotide sequence data [24,35]) [2, 4, 24, 35]
84000 <6> (<6> renatured enzyme, osmometry [52]) [52]

Subunits

heterodimer <1, 3> (<3> $\alpha\beta$, 1 * 41389 + 1 * 37112, nucleotide sequence [2]; <1> $\alpha\beta$, 1 * 40108 + 1 * 36349 nucleotide sequence [24, 25]) [2, 7, 24, 25, 56, 58-60]
monomer <1> (<1> 1 * 78000, produced by gene fusion of luxA and luxB genes [8]) [8]

5 Isolation/Preparation/Mutation/Application

Localization

cytoplasm <1> [12]
membrane <1, 5> [9]
soluble <1, 5, 7> [20, 38, 44]

Purification

<1> (affinity methods [19]; preparation of subunits [42]; preparation of enzyme with modified subunits [43]) [19, 38, 41-43, 58, 63, 66, 67]
<5> (affinity methods [19]) [19]
<7> (HPLC) [20]
<8> [53]
<11> [65]

Renaturation

<1, 5, 6> [17, 18, 52, 55]

Crystallization

<1> [27]
<1> (<1> structure is determined in absence of substrate at low-salt concentrations [69]) [69]

Cloning

<1> (expression in Pseudomonas putida [61]) [61]
<1> (expression of fused luxA and luxB genes in Saccharomyces cerevisiae, Bacillus subtilis, plant cells, plasmid expression vector and in Escherichia coli [10, 13, 16, 21, 26, 29]) [4, 7, 10, 13, 14, 16, 21, 26, 29, 32]
<1> (expression of luxA gene in Escherichia coli [56]) [56]
<1> (expression of seperated luxA and luxB gene in Escherichia coli JM109 [58, 63]) [58, 63]
<1> (overexpression of mutant in XL1 blue MRF' cell line [67]) [67]
<2> (expression of fused luxA and luxB genes in Escherichia coli [1]) [1]
<3> (expression in Escherichia coli [57]) [57]
<3> (expression of fused luxA and luxB genes in Escherichia coli [2]) [2]
<6> (expression of fused luxA and luxB genes and also luxF gene in Escherichia coli and Nicotiana plumbaginifolia [15]) [15]
<11> (overexpression in Escherichia coli [65]) [65]

Engineering

$\alpha\Delta$262-290β <1> (<1> four times higher affinity for FMN than wild type [67]) [67]
αH44A <1> (<1> decreased bioluminescence [59]) [59]
αR107A <1> (<1> lower affinity for FMNH [66]) [66]
αR107E <1> (<1> lower affinity for FMNH [66]) [66]
αR107S <1> (<1> lower affinity for FMNH [66]) [66]
more coexpression of luciferase and cytochrome P-4 <1> (<1> reductive dehalogenation [61]) [61]

6 Stability

pH-Stability

6-8.5 <5> [41]
6-9.5 <1> [41]

Temperature stability

25 <10> (<10> stable below [62]) [62]
35 <1, 5, 6> (<1,5,6> stable below [41]) [41]
37 <1> (<1> comparison of stability of wild-type enzyme and gene-fusion monomeric enzyme [8]) [8]
45 <1, 3> (<3> half-life over 3 h [4]; <1> half-life 5 min [4]) [4]

General stability information

<1>, repeated freezing/thawing causes inactivation of immobilized enzyme [31]
<1, 5, 6>, citrate stabilizes against inactivation by proteases, heat, urea [41]
<1, 5, 6>, diphosphate stabilizes against inactivation by proteases, heat, urea [41]
<1, 5, 6>, inactivation by lyophilization [41]
<1, 5, 6>, labile to proteases [41]

<1, 5, 6>, no inactivation by repeated freezing/thawing [41]
<1, 5-7>, phosphate stabilizes against inactivation by proteases, heat, urea [41, 44]
<1, 5-7>, sulfate stabilizes against inactivation by proteases, heat, urea [41, 44]

Storage stability
<1>, -20°C, phosphate buffer [40]
<1>, -80°C, 0.5 mM dithiothreitol [28]
<1>, 0-4°C, immobilized enzyme, 0.1 mM dithiothreitol, 20% loss of activity in 3 days [31]
<1, 8>, -20°C, 50 mM potassium phosphate buffer, pH 7.0, protein concentration 1 mg/ml , -20°C, 0.1 M phosphate buffer, pH 7, 0.1 mM dithiothreitol, 1 mM EDTA [6, 53]

References

[1] Palmer, L.M.; Colwell, R.R.: Detection of luciferase gene sequence in nonluminescent Vibrio cholerae by colony hybridization and polymerase chain reaction. Appl. Environ. Microbiol., **57**, 1286-1293 (1991)
[2] Xi, L.; Cho, K.W.; Tu, S.C.: Cloning and nucleotide sequences of lux genes and characterization of luciferase of Xenorhabdus luminescens from a human wound. J. Bacteriol., **173**, 1399-1405 (1991)
[3] Haygood, M.G.: Relationship of the luminous bacterial symbiont of the Caribbean flashlight fish, Kryptophanaron alfredi (family Anomalopidae) to other luminous bacteria based on bacterial luciferase (luxA) genes. Arch. Microbiol., **154**, 496-503 (1990)
[4] Szittner, R.; Meighen, E.: Nucleotide sequence, expression, and properties of luciferase coded by lux genes from a terrestrial bacterium. J. Biol. Chem., **265**, 16581-16587 (1990)
[5] Makemson, J.C.: A cyanide-aldehyde complex inhibits bacterial luciferase. J. Bacteriol., **172**, 4725-4727 (1990)
[6] Curry, S.; Lieb, W.R.; Franks, N.P.: Effects of general anesthetics on the bacterial luciferase enzyme from Vibrio harveyi: an anesthetic target site with differential sensitivity. Biochemistry, **29**, 4641-4652 (1990)
[7] Xi, L.; Cho, K.W.; Herndon, M.E.; Tu, S.C.: Elicitation of an oxidase activity in bacterial luciferase by site-directed mutation of a noncatalytic residue. J. Biol. Chem., **265**, 4200-4203 (1990)
[8] Escher, A.; O'Kane, D.J.; Lee, J.; Szalay, A.A.: Bacterial luciferase $\alpha \beta$ fusion protein is fully active as a monomer and highly sensitive in vivo to elevated temperature. Proc. Natl. Acad. Sci. USA, **86**, 6528-6532 (1989)
[9] Angell, P.; Langley, D.; Chamberlain, A.H.L.: Localization of luciferase in luminous marine bacteria by gold immunocytochemical labelling. FEMS Microbiol. Lett., **65**, 177-182 (1989)

[10] Kirchner, G.; Roberts, J.L.; Gustafson, G.D.; Ingolia, T.D.: Active bacterial luciferase from a fused gene: expression of a Vibrio harveyi luxAB translational fusion in bacteria, yeast and plant cells. Gene, **81**, 349-354 (1989)
[11] Kurfurst, M.; Macheroux, P.; Ghisla, S.; Hastings, J. W.: Bioluminescence emission of bacterial luciferase with 1-deaza-FMN. Evidence for the noninvolvement of N(1)-protonated flavin species as emitters. Eur. J. Biochem., **181**, 453-457 (1989)
[12] Colepicolo, P.; Nicolas, M.T.; Bassot, J.M.; Hastings, J.W.: Expression and localization of bacterial luciferase determined by imunogold labeling. Arch. Microbiol., **152**, 72-76 (1989)
[13] Peabody, D.S.; Andrews, C.L.; Escudero, K.W.; Devine, J.H.; Baldwin, T.O.; Bear, D.G.: A plasmid vector and quantitative techniques for the study of transcription termination in Escherichia coli using bacterial luciferase. Gene, **75**, 289-296 (1989)
[14] Miyamoto, C.M.; Boylan, M.; Graham, A.F.; Meighen, E.A.: Organization of the lux structural genes of Vibrio harveyi. Expression under the T7 bacteriophage promoter, mRNA analysis, and nucleotide sequence of the luxD gene. J. Biol. Chem., **263**, 13393-13399 (1988)
[15] Soly, R.R.; Mancini, J.A.; Ferri, S.R.; Boylan, M.; Meighen, E.A.: A new lux gene in bioluminescent bacteria codes for a protein homologous to the bacterial luciferase subunits. Biochem. Biophys. Res. Commun., **155**, 351-358 (1988)
[16] Karp, M.: Expression of bacterial luciferase genes from Vibrio harveyi in Bacillus subtilis and in Escherichia coli. Biochim. Biophys. Acta, **1007**, 84-90 (1989)
[17] Sugihara, J.; Baldwin, T.O.: Effects of 3 end deletions from the Vibrio harveyi luxB gene on luciferase subunit folding and enzyme assembly: generation of temperature-sensitive polypeptide folding mutants. Biochemistry, **27**, 2872-2880 (1988)
[18] Waddle, J.J.; Johnston, T.C.; Baldwin, T.O.: Polypeptide folding and dimerization in bacterial luciferase occur by a concerted mechanism in vivo. Biochemistry, **26**, 4917-4921 (1987)
[19] Baldwin, T.O.; Holzman, T.F.; Holzman, R.B.; Riddle, V.A.: Purification of bacterial luciferase by affinity methods. Methods Enzymol., **133**, 98-108 (1986)
[20] O'Kane, D.J.; Ahmed, M.; Matheson, I.B.C.; Lee, J.: Purification of bacterial luciferase by high-performance liquid chromatography. Methods Enzymol., **133**, 109-128 (1986)
[21] Koncz, C.; Olsson, O.; Langridge, W.H.R.; Schell, J.; Szalay, A.A.: Expression and assembly of functional bacteria luciferase in plants. Proc. Natl. Acad. Sci. USA, **84**, 131-135 (1987)
[22] Haygood, MG.; Cohn, D.H.: Luciferase genes cloned from the unculturable luminous bacteroid symbiont of the Caribbean flashlight fish, Kryptophanaron alfredi. Gene, **45**, 203-209 (1986)
[23] Vervoort, J.; Muller, F.; Lee, J.; van den Berg, W.A. M.; Moonen, C.T.W.: Identificartions of the true carbon-13 nuclear magnetic resonance spec-

trum of the stable intermediate II in bacterial luciferase. Biochemistry, **25**, 8062-8067 (1986)
[24] Johnston, T.C.; Thompson, R.B.; Baldwin, T.O.: Nucleotide sequence of the luxB gene of Vibrio harveyi and the complete amino acid sequence of the β subunit of bacterial luciferase. J. Biol. Chem., **261**, 4805-4811 (1986)
[25] Cohn, D.H.; Milcham, A.J.; Simon, M.I.; Nealson, K.H.; Rausch, S.K.; Bonam, D.; Baldwin, T.O.: Nucleotide sequence of the luxA gene of Vibrio harveyi and the complete amino acid sequence of the α subunit of bacterial luciferase. J. Biol. Chem., **260**, 6139-6146 (1985)
[26] Gupta, S.C.; O'Brien, D.; Hastings, J.W.: Expression of the cloned subunits of bacterial luciferase from separate replicons. Biochem. Biophys. Res. Commun., **127**, 1007-1011 (1985)
[27] Swanson, R.; Weaver, L.H.; Remington, S.J.; Matthews, B.W.; Baldwin, T.O.: Crystals of luciferase from Vibrio harveyi. A preliminary characterization. J. Biol. Chem., **260**, 1287-1289 (1985)
[28] Fried, A.; Tu, S.C.: Affinity labeling of the aldehyde site of bacterial luciferase. J. Biol. Chem., **259**, 10754-10759 (1984)
[29] Baldwin, T.O.; Berends, T.; Bunch, T.A.; Holzman, T. F.; Rausch, S.K.; Shamansky, L.; Treat, M.L.; Ziegler, M. M.: Cloning of the luciferase structural genes from Vibrio harveyi and expression of bioluminescence in Escherichia coli. Biochemistry, **23**, 3663-3667 (1984)
[30] Holzman, T.F.; Baldwin, T.O.: Reversible Inhibition of the bacterial luciferase catalyzed bioluminescence reaction by aldehyde substrate: Kinetic mechanism and ligand effects. Biochemistry, **22**, 2838-2846 (1983)
[31] Wienhausen, G.K.; Kricka, L.J.; Hinkley, J.E.; DeLuca, M.: Properties of bacterial luciferase/NADH:FMN oxidoreductase and firefly luciferase immobilized onto Sepharose. Appl. Biochem. Biotechnol., **7**, 463-473 (1982)
[32] Cohn, D.H.; Ogden, R.C.; Abelson, J.N.; Baldwin, T.O.; Nealson, K.H.; Simon, M.I.; Mileham, A.J.: Cloning of the Vibrio harveyi luciferase genes: use of a synthetic oligonucleotide probe. Proc. Natl. Acad. Sci. USA, **80**, 120-123 (1983)
[33] Tu, S.C.: Isolation and properties of bacterial luciferase intermediates containing different oxygenated flavins. J. Biol. Chem., **257**, 3719-3725 (1982)
[34] Kurfurst, M.; Ghisla, S.; Presswood, R.; Hastings, J. W.: Structure and catalytic inactivity of the bacterial luciferase neutral flavin radical. Eur. J. Biochem., **123**, 355-361 (1982)
[35] Welches, W.R.; Baldwin, T.O.: Active center studies on bacterial luciferase: modification of the enzyme with 2,4-dinitrofluorobenzene. Biochemistry, **20**, 512-517 (1981)
[36] Holzman, T.F.; Riley, P.L.; Baldwin, T.O.: Inactivation of luciferase from the luminous marine bacterium Beneckea harveyi by proteases: evidence for a protease labile region and properties of the protein following inactivation. Arch. Biochem. Biophys., **205**, 554-563 (1980)
[37] Kosower, E.M.: A proposed mechanism for light emission by bacterial luciferase involving dissociative electron transfer. Biochem. Biophys. Res. Commun., **92**, 356-364 (1980)

[38] Hastings, J.W.; Presswood, R.P.: Bacterial luciferase: $FMNH_2$-aldehyde oxidase. Methods Enzymol., **53**, 558-570 (1978)
[39] Viswanathan, T.S.; Campling, M.R.; Cushley, R.J.: Interactions of long-chain aldehydes with luciferase. A carbon-13 nuclear magnetic resonance study. Biochemistry, **18**, 2504-2508 (1979)
[40] Baumstark, A.L.; Cline, T.W.; Hastings, J.W.: Reversible steps in the reaction of aldehydes with bacterial luciferase intermediates. Arch. Biochem. Biophys., **193**, 449-455 (1979)
[41] Hastings, J.W.; Baldwin, T.O.; Nicoli, M.Z.: Bacterial luciferase: Assay, purification, and properties. Methods Enzymol., **57**, 135-152 (1978)
[42] Tu, S.C.: Preparation of the subunits of bacterial luciferase. Methods Enzymol., **57**, 171-174 (1978)
[43] Meighen. E.A.: Preparation of luciferases containing chemically modified subunits. Methods Enzymol., **57**, 174-181 (1978)
[44] Hastings, J.W.: Bacterial bioluminescence light emission in the mixed function oxidation of reduced flavin and fatty aldehyde. CRC Crit. Rev. Biochem., **5**, 163-184 (1978)
[45] Watanabe, T.; Matsui, K.; Kasai, S.; Nakamura, T.: Studies on luciferase from Photobacterium phosphoreum. XI. Interaction of 8-substituted $FMNH_2$ with luciferase. J. Biochem., **84**, 1441-1446 (1978)
[46] Becvar, J.E.; Tu, S.C.; Hastings, J.W.: Activity and stability of the luciferase-flavin intermediate. Biochemistry, **17**, 1807-1812 (1978)
[47] Tu, S.C.; Wu, C.W.; Hastings, J.W.: Structural studies on bacterial luciferase using energy transfer and emission anisotropy. Biochemistry, **17**, 987-993 (1978)
[48] Mangold, A.; Langerman, N.: The enthalpy of oxidation of flavin mononucleotide. Temperature dependence of in vitro bacterial luciferase bioluminescence. Arch. Biochem. Biophys., **169**, 126-133 (1975)
[49] Yoshida, K.; Nakamura, T.: Studies on luciferase from Photobacterium phosphoreum. VII. Interaction with carboxylic acid. J. Biochem., **76**, 985-990 (1974)
[50] Watanabe, T.; Tomita, G.; Nakamura, T.: Studies on luciferase from Photobacterium phosphoreum. VI. Stoichiometry and mode of binding of $FMNH_2$ and O_2 to stripped luciferase. J. Biochem., **75**, 1249-1255 (1974)
[51] Nicoli, M.Z.; Meighen, E.A.; Hastings, J.W.: Bacterial luciferase. Chemistry of the reactive sulfhydryl. J. Biol. Chem., **249**, 2385-2392 (1974)
[52] Yoshida, K.; Nakamura, T.: Studies on luciferase from Photobacterium phosphoreum. IV. Preparation and properties of stripped luciferase. J. Biochem., **74**, 915-922 (1973)
[53] Gunsalus-Miguel, A.; Meighen, E.A.; Nicoli, M.Z.; Nealson, K.H.: Purification and properties of bacterial luciferases. J. Biol. Chem., **247**, 398-404 (1972)
[54] Nealson, K.H.; Hastings, J.W.: The inhibition of bacterial luciferase by mixed function oxidase inhibitors. J. Biol. Chem., **247**, 888-894 (1972)
[55] Friedland, J.; Hastings, J.W.: The reversibility of the denaturation of bacterial luciferase. Biochemistry, **6**, 2893-2900 (1967)

[56] Noland, B.W.; Lawrence, J.D.; Baldwin, T.O.: Folding, stability, and physical properties of the α subunit of bacterial luciferase. Biochemistry, **38**, 16136-16145 (1999)
[57] Li, Z.; Meighen, E.A.: The turnover of bacterial luciferase is limited by a slow decomposition of the ternary enzyme-product complex of luciferase, FMN, and fatty acid. J. Biol. Chem., **269**, 6640-6644 (1994)
[58] Choi, H.; Tang, C.K.; Tu, S.C.: Catalytically active forms of the individual subunits of Vibrio harveyi luciferase and their kinetic properties. J. Biol. Chem., **270**, 16813-16819 (1995)
[59] Huang, S.; Tu, S.C.: Identification and Characterization of a catalytic base in bacterial luciferase by chemical rescue of a dark mutant. Biochemistry, **36**, 14609-14615 (1997)
[60] Francisco, W.A.; Abu-Soud, H.M.; Baldwin, T.O.; Raushel, F.M.: Interaction of bacterial luciferase with aldehyde substrates and inhibitors. J. Biol. Chem., **268**, 24734-24741 (1993)
[61] Shanker, R.; Atkins, W.M.: Luciferase-dependent, cytochrome P-450-catalyzed dehalogenation in genetically engeneered Pseudomonas. Biotechnol. Prog., **12**, 474-479 (1996)
[62] Dickinson, R.; Franks, N.P.; Lieb, W.R.: Thermodynamics of anesthetic / protein interactions. Biophys. J., **64**, 1264-1271 (1993)
[63] Sinclair, J.F.; Waddle, J.J.; Waddill, E.F.; Baldwin, T.O.: Purified native subunits of bacterial luciferases are active in the bioluminescence reaction but fail to assemble into the $\alpha\beta$ structure. Biochemistry, **32**, 5036-5044 (1993)
[64] Matsuki, H.; Suzuki, A.; Kamaya, H.; Ueda, I.: Specific and non-specific binding of long-chain fatty acids to firefly luciferase: cutoff at octanate. Biochim. Biophys. Acta, **1426**, 143-150 (1999)
[65] Devine, J.H.; Kutuzova, G.D.; Green, V.A; Ugarova, N.N.; Baldwin, T.: Luciferase from the east european firefly Luciola mingrelica: cloning and nucleotide sequence of the cDNA, overexpression in Escherichia coli and purification of the enzyme. Biochim. Biophys. Acta, **1173**, 121-132 (1993)
[66] Moore, C.; Lei, B.; Tu, S.C.: Relationship between the conserved α subunit arginine 107 and effects of phosphate on the activity and stability of Vibrio harveyi luciferase. Arch. Biochem. Biophys., **370**, 45-50 (1999)
[67] Sparks, J.M.; Baldwin, T.O.: Functional implications of the unstructered loop in the $(\beta/\alpha)_8$ barrel structure of the bacterial luciferase α subunit. Biochemistry, **40**, 15436-15443 (2001)
[68] Morishita, H.; Ohashi, S.; Oku, T.; Nakajima, y.; Kojima, S.; Ryufuku, M.; Ohmija, Y.: Cloning and characterization of an active fragment of luciferase from luminescent marine alga, Pyrocystis lunula. Photochem. Photobiol., **75**, 311-315 (2002)
[69] Fisher, A.J.; Thompson, T.P.; Baldwin, T.O.; Rayment, I.: The 1.5-A resolution crystal structure of bacterial luciferase in low salt conditions. J. Biol. Chem., **271**, 21956-29168 (1996)

Choline monooxygenase 1.14.14.4

1 Nomenclature

EC number
1.14.14.4 (deleted, identical to EC 1.14.15.7)

Recommended name
choline monooxygenase

Alkanesulfonate monooxygenase 1.14.14.5

1 Nomenclature

EC number
1.14.14.5

Systematic name
alkanesulfonate, reduced-FMN:oxygen oxidoreductase

Recommended name
alkanesulfonate monooxygenase

Synonyms
SsuD
alkanesulfonate α-hydroxylase
oxygenase, alkanesulfonate 1-mono-
sulfate starvation-induced protein 6

CAS registry number
54596-24-6

2 Source Organism

<1> *Escherichia coli* (EC1250, chromosomal gene cluster ssuEADCB [1, 2]) [1, 2, 3, 4]

3 Reaction and Specificity

Catalyzed reaction
an alkanesulfonate (R-CH_2-SO_3H) + $FMNH_2$ + O_2 = an aldehyde (R-CHO) + FMN + sulfite + H_2O (<1> absolutely dependent on oxygen [1]; <1> catalysis by SsuD probably results in an unstable 1-hydroxysulfonate that spontaneously decomposes to the corresponding aldehyde and sulfite [3])

Reaction type
desulfonation

Natural substrates and products
S R-CH_2-SO_3H + $FMNH_2$ + O_2 <1> (Reversibility: ? <1> [1]) [1]
P R-CHO + FMN + sulfite + H_2O <1> [1]

Substrates and products

S 1,3-dioxo-2-isoindolineethanesulfonic acid + $FMNH_2$ + O_2 <1> (Reversibility: ? <1> [1, 2]) [1, 2]
P (1,3-dioxo-1,3-dihydro-2H-isoindol-2-yl)acetaldehyd + FMN + sulfite + H_2O
S 2-(4-pyridyl)ethanesulfonic acid + $FMNH_2$ + O_2 <1> (Reversibility: ? <1> [1, 2]) [1, 2]
P pyridin-4-ylacetaldehyde
S 4-phenyl-1-butanesulfonic acid + $FMNH_2$ + O_2 <1> (Reversibility: ? <1> [1, 2]) [1, 2]
P 4-phenylbutanol + FMN + sulfite + H_2O
S MOPS + $FMNH_2$ + O_2 <1> (Reversibility: ? <1> [1, 2]) [1, 2]
P ?
S N-phenyltaurine + $FMNH_2$ + O_2 <1> (Reversibility: ? <1> [1, 2]) [1, 2]
P anilinoacetaldehyde + FMN + sulfite + H_2O
S PIPES + $FMNH_2$ + O_2 <1> (Reversibility: ? <1> [1, 2]) [1, 2]
P ?
S butanesulfonic acid + $FMNH_2$ + O_2 <1> (Reversibility: ? <1> [1, 2]) [1, 2]
P butanal + FMN + sulfite + H_2O
S decanesulfonic acid + $FMNH_2$ + O_2 <1> (Reversibility: ? <1> [1, 2]) [1, 2]
P decanal + FMN + sulfite + H_2O
S hexanesulfonic acid + $FMNH_2$ + O_2 <1> (Reversibility: ? <1> [1, 2]) [1, 2]
P hexanal + FMN + sulfite + H_2O
S octanesulfonic acid + $FMNH_2$ + O_2 <1> (Reversibility: ? <1> [1, 2]) [1, 2]
P octanal + FMN + sulfite + H_2O
S pentanesulfonic acid + $FMNH_2$ + O_2 <1> (Reversibility: ? <1> [1, 2]) [1, 2]
P pentaldehyde + FMN + sulfite + H_2O <1> [1]
S Additional information <1> (<1> no substrates are taurine, methanesulfonic acid, benzenesulfonic acid, L-cysteic acid, ethanedisulfonic acid, toluene-4-sulfonic acid, *p*-sulfobenzoic acid, benzenesulfonic acid, 4-hydroxybenzenesulfonic acid, SsuD is able to desulfonate C-2 to C-10 unsubstituted alkanesulfonates, substituted ethanesulfonic acids and HEPES, the catalytic efficiency increases with increasing chain length up to decanesulfonic acid [1]; <1> further substrates: sulfoacetate, ethanesulfate, propanesulfonate, 2-hydroxyethanesulfonic acid, 3-aminopropanesulfate, no substrate: taurine [2]) [1, 2]
P ?

Cofactors/prosthetic groups

$FMNH_2$ <1> [1, 2, 3]

Specific activity (U/mg)

2.5 <1> [1]

K_m-Value (mM)

0.035 <1> (decanesulfonic acid) [1]
0.044 <1> (octanesulfonic acid) [1]

0.095 <1> (hexanesulfonic acid) [1]
0.11 <1> (4-phenyl-1-butanesulfonic acid) [1]
0.114 <1> (1,3-dioxo-2-isoindolineethanesulfonic acid) [1]
0.139 <1> (2-(4-pyridyl)ethanesulfonic acid) [1]
0.189 <1> (pentanesulfonic acid) [1]
0.237 <1> (N-phenyltaurine) [1]
0.617 <1> (MOPS) [1]
0.87 <1> (butanesulfonic acid) [1]
1.11 <1> (PIPES) [1]

pH-Optimum
9.1 <1> (<1> in 10 mM Tris-HCl [1]) [1]

Temperature optimum (°C)
30 <1> (<1> enzyme assay [1]) [1]

4 Enzyme Structure

Molecular weight
181000 <1> (<1> gel filtration [1]) [1]

Subunits
homotetramer <1> (<1> 4 * 41200, SDS-PAGE [1]) [1]

5 Isolation/Preparation/Mutation/Application

Purification
<1> [1]

Crystallization
<1> (X-ray characterization, tetramer 96 A x 90 A x 66 A, comprises two homodimers, monomer 60A x 50 A x 40 A, TIM-barrel protein [4]) [4]

Cloning
<1> (expressed in Escherichia coli [1]) [1]

6 Stability

Storage stability
<1>, -20°C, 15% glycerol, the activity increases slightly during the first 2 to 3 weeks of storage [1]

References

[1] Eichhorn, E.; van der Ploeg, J.R.; Leisinger, T.: Characterization of a two-component alkanesulfonate monooxygenase from Escherichia coli. J. Biol. Chem., **274**, 26639-26646 (1999)
[2] Eichhorn, E.; van der Ploeg, J.R.; Leisinger, T.: Deletion analysis of the Escherichia coli taurine and alkanesulfonate transport systems. J. Bacteriol., **182**, 2687-2795 (2000)
[3] Van der Ploeg, J.R.; Eichhorn, E.; Leisinger, T.: Sulfonate-sulfur metabolism and its regulation in Escherichia coli. Arch. Microbiol., **176**, 1-8 (2001)
[4] Eichhorn, E.; Davey, C.A.; Sargent, D.F.; Leisinger, T.: Monooxygenase SsuD. J. Mol. Biol., **324**, 457-468 (2002)

Printing: Krips bv, Meppel
Binding: Stürtz, Würzburg